Copyright © 2014 National Fire Protection Association®. All Rights Reserved

NFPA® 99

Health Care Facilities Code

2015 Edition

This edition of NFPA 99, *Health Care Facilities Code*, was prepared by the Technical Committees on Electrical Systems, Fundamentals, Health Care Emergency Management and Security, Hyperbaric and Hypobaric Facilities, Mechanical Systems, Medical Equipment, and Piping Systems, released by the Correlating Committee on Health Care Facilities, and acted on by NFPA at its June Association Technical Meeting held June 9–12, 2014, in Las Vegas, NV. It was issued by the Standards Council on August 14, 2014, with an effective date of September 3, 2014, and supersedes all previous editions.

Tentative Interim Amendments (TIAs) to 10.2.3.6(5) and 11.5.1.1 were issued on August 14, 2014. For further information on tentative interim amendments, see Section 5 of the Regulations Governing the Development of NFPA Standards, available at http://www.nfpa.org/regs

This edition of NFPA 99 was approved as an American National Standard on September 3, 2014.

Origin and Development of NFPA 99

The idea for this document grew as the number of documents under the original NFPA Committee on Hospitals grew. By the end of 1980, there existed 12 documents on a variety of subjects, 11 directly addressing fire-related problems in and about health care facilities. These documents covered health care emergency preparedness, inhalation anesthetics, respiratory therapy, laboratories in health-related institutions, hyperbaric facilities, hypobaric facilities, inhalation anesthetics in ambulatory care facilities, home use of respiratory therapy, medical–surgical vacuum systems in hospitals, essential electrical systems for health care facilities, safe use of electricity in patient care areas of health care facilities, and safe use of high-frequency electricity in health care facilities.

A history on the documents that covered these topics can be found in the "Origin and Development of NFPA 99" in the 1984 edition of NFPA 99.

What was then the Health Care Facilities Correlating Committee reviewed the matter beginning in late 1979 and concluded that combining all the documents under its jurisdiction would be beneficial to those who used those documents, for the following reasons:

(1) The referenced documents were being revised independently of each other. Combining all the individual documents into one document would place all of them on the same revision cycle.
(2) It would place in one unit many documents that referenced each other.
(3) It would be an easier and more complete reference for the various users of the document (e.g., hospital engineers, medical personnel, designers and architects, and the various types of enforcing authorities).

To learn if this proposal was desired or desirable to users of the individual documents, the Committee issued a request for public comments in the spring of 1981, asking whether purchasers of the individual documents utilized more than one document in the course of their activities and whether combining these individual documents would be beneficial. Seventy-five percent of responses supported such a proposal, with 90 percent of health care facilities and organizations supportive of it. Based on this support, the Correlating Committee proceeded with plans to combine all the documents under its jurisdiction into one document.

In January, 1982, a compilation of the latest edition of each of the 12 individual documents under the jurisdiction of the Correlating Committee was published. It was designated NFPA 99, *Health Care Facilities Code*. The Correlating Committee also entered the document into the revision cycle reporting to the 1983 Fall Meeting for the purpose of formally adopting the document.

NFPA and National Fire Protection Association are registered trademarks of the National Fire Protection Association, Quincy, Massachusetts 02169.

For the 1984 edition of NFPA 99, in addition to technical changes, administrative and organizational changes were made.

For the 1987 edition of NFPA 99, the third and final step in the process of combining the previous individual documents took place — that of integrating the content of these individual documents into a cohesive document. In addition, there were again technical changes made. The 1987 edition also saw the incorporation of NFPA 56F, *Standard on Nonflammable Medical Piped Gas Systems*, into NFPA 99.

For the 1990 edition of NFPA 99, some structural changes were made and some modifiers were added to make it easier to determine where requirements are applicable. Technical changes made included the following: correlation with NFPA 101®, *Life Safety Code*®; changes for compressed medical air systems on the use of gas-powered medical devices operating at a gauge pressure of 200 psi, and piped gas systems in general; changes in leakage current limits for patient care electrical appliances; clarification that patient care areas and wet locations are mutually exclusive; and further guidance on the effects of a disaster on staff.

For the 1993 edition of NFPA 99 there were further efforts to make the document more user-friendly (e.g., placing all "recommended" guidance either in notes or in the appendix). Significant technical changes included the following: adding requirements and recommendations to further prevent or minimize fires in operating rooms; making major changes to requirements in Chapter 4 for installing, testing, inspecting, verifying, and maintaining nonflammable medical piped gas systems; adding new sections on dental compressed air and dental vacuum requirements in Chapter 4; changing leakage current limits of patient care–related electrical appliances to correlate more closely with an international document on the subject; revising laboratory requirements to correlate more closely with NFPA 45, *Standard for Laboratories Using Chemicals*; changing essential electrical system requirements in ambulatory health care clinics and medical/dental offices; and extensively revising hyperbaric chamber requirements (Chapter 19).

For the 1996 edition of NFPA 99, further changes to make the document more user-friendly were made. These included restructuring Chapters 3 and 4 so that all requirements for a Type 1, 2, or 3 essential electrical system, or a Level 1, 2, 3, or 4 piped gas or vacuum system, were contained in one section.

Other technical changes included the following:

(1) Moving requirements on flammable anesthetizing locations and the use of flammable inhalation anesthetics to a new Appendix 2
(2) Upgrading the subject of emergency preparedness from guidance to a new chapter containing requirements
(3) Adding a new chapter (Chapter 18) on home health care
(4) Revising Section 1-1 to reflect the intent that NFPA 99 applies only to facilities treating human beings
(5) In Chapter 3, revising load testing requirements for emergency generators to reference NFPA 110, *Standard for Emergency and Standby Power Systems*, and revising emergency lighting criteria for operating rooms
(6) In Chapter 4, revising requirements for medical compressed air systems, dental compressed air systems, waste anesthetic gas disposal systems, and dental piped gas/vacuum systems; adding a new section on "headwall units" ("manufactured assemblies"); and clarifying and moving requirements for transfilling containers of liquid oxygen to Chapter 8
(7) In Chapter 8, adding requirements for storage rooms containing cylinders and containers totaling less than 3000 ft^3
(8) In Chapters 12 to 17, revising criteria for gas and vacuum systems
(9) In Chapter 19, in addition to many technical changes, adding criteria for mobile hyperbaric facilities

For the 1999 edition, significant technical and structural changes included the following:

(1) Chapters 13, 14, and 15 (on ambulatory health care centers, clinics, and medical/dental offices, respectively) were replaced completely by new Chapter 13 covering health care facilities other than hospitals, nursing homes, and limited care facilities as defined in Chapter 2.
(2) Requirements for Level 2 gas and vacuum systems were developed (Section 4.4 in Chapter 4).
(3) Subsections 12.3.4, 16.3.4, and 7.3.4 were revised to correlate with the two significant changes in (1) and (2).
(4) In Chapter 3, load testing requirements for emergency power supplies of the essential electrical system were changed through reference, and the testing interval ("monthly") was reworded to be more responsive to needs of health care facilities.
(5) Clarification of transfer switches and branches of the emergency system was made.
(6) Clarification on the use of emergency power supplies other than for emergency power was made in 3.4.1.1.5.
(7) Paragraph 4.3.1.2, Distribution Requirements for Level 1 Gas Systems, was completely revised and restructured.
(8) Chapter 4 was made more user-friendly by reducing the number of internal cross-references between Sections 4.3 and 4.5.
(9) The order of installation and testing requirements for piped gas and vacuum systems was revised.
(10) Emphasis on emergency preparedness was made in Chapter 11 and its appendix material.
(11) Chapter 19, "Hyperbaric Facilities," was extensively revised in the areas of electrical wiring, air quality, ventilation lighting, equipment, communication, and safety management.
(12) A new chapter (Chapter 20) on freestanding birthing centers was added.

The 2002 edition included format and technical revisions. The *Manual of Style for NFPA Technical Committee Documents*, April 2000 edition, was applied to this document, resulting in changes to its structure and format. Introductory material in Chapter 1 was formatted for consistency among all NFPA documents. Referenced publications that apply to the document were relocated from the last chapter to Chapter 2, resulting in the renumbering of chapters. Informational references remained in the last annex. Appendices were designated as annexes. Definitions in Chapter 3 were reviewed for consistency with definitions in other NFPA documents, were systematically aligned, and were individually numbered. Paragraph structuring was revised with the intent of one mandatory requirement per section, subsection, or paragraph. Information that often accompanied many of the requirements was moved to Annex A, Explanatory Material. Exceptions were deleted or rephrased in mandatory text, unless the exception represented an allowance or required alternate procedure to a general rule when limited specified conditions exist. The reformatted appearance and structure provided continuity among NFPA documents, clarity of mandatory text, and greater ease in locating specific mandatory text.

The document scope and individual chapter scopes defining the intent of each chapter and document as a whole were located in Chapter 1.

The occupancy Chapters 13–21 stated what is required, while Chapters 4–12 prescribed how those requirements are achieved. Each chapter began with a section explaining applicability. Information concerning the nature of hazards was moved to Annex B. Annexes A and C retained explanatory information, and Annexes 1 and 2 became Annexes D and E. Informational references were in Annex F.

The changes in Chapter 4, Electrical Systems, addressed electrical wiring, transfer switches, inspection, and application.

Chapter 5 on Piping Systems was realigned so that Level 1 requirements were found in Section 5.1, and concurrently Level 2 in Section 5.2 and Level 3 in Section 5.3. Level 4 associated with laboratories was deleted, with requirements realigned in Chapter 11 on laboratories. Definitions were developed for vacuum systems and Levels 1, 2, and 3 gas systems in Chapter 3. Revisions were made to compressed gas cylinder identification and restraint; valve venting; ventilation of storage rooms; alarms; connection of the electrical supply for central supply systems with the essential electrical system; allowance of a three-way full port ball valve to isolate one branch or component; provisions for a monitored and audible low-content alarm on the surge gas while brazing; the allowance of medical air systems for application with human respiration; and deletion of 20-year-old appendix information.

Gas Delivery, Chapter 8, included a new section on the storage of compressed gas cylinders in patient care areas.

Chapter 11, Laboratories, clarified the structural protection of exits, and intent of portable fire extinguishers. Revisions were made concerning flammable and combustible liquids handling requirements.

An increased focus on the total process of maintaining services during a disaster, mitigating damage from a disaster, and recovery from a disaster was reflected in Chapter 12, Emergency Management. Annexed security program information was expanded.

Chapter 20, Hyperbaric Facilities, contained revised emergency depressurization requirements, safety director responsibilities, and emergency procedure performance.

The changes made to the 2005 edition were mainly for clarity and were editorial in nature. In Chapter 3, the definitions for medical gas, patient medical gas, and medical support gas were modified to differentiate between the different types of gases.

In Chapter 4, the requirements for switches and receptacles in anesthetizing locations were moved to Chapter 13, Hospital Requirements. The extracted material from NFPA 110, *Standard for Emergency and Standby Power Systems*, was updated.

In Chapter 5, the requirements for construction materials for filters, dryers, regulators, vacuum pumps, and aftercoolers were changed to allow the manufacturers to choose the materials.

A centralized computer was allowed to be used in lieu of one of the master alarms. Cylinders were allowed to be fitted with a means to slow the initial opening pressure. The requirement to individually secure the cylinders was changed to no longer require the cylinders to be secured individually. Two new methods for making joints were added to the requirements. Stainless steel tubing was added as an approved material for vacuum systems. The requirement to braze a joint within 1 hour after cleaning was changed to 8 hours. Vacuum joints were required to be leak tested, and operational pressure testing was permitted to be conducted with the source gas.

Chapters 6, 7, 8, 9, 10, and 11 underwent minor changes for clarity or for editorial reasons.

Chapter 12 was revised to update the techniques used in emergency management in health care facilities.

In Chapters 13, 14, 15, 16, 17, 18, and 19 editorial corrections were made.

Chapter 20 was revised to include requirements for heating and ventilation changes in the chamber. Additional restrictions to the types of materials that are allowed in the chamber were added.

The 2012 edition went through a major overhaul. The premise of an occupancy-based document was modified to become a risk-based document. NFPA 99 was changed to a "code" instead of a "standard" to reflect how the document is used and adopted. This change was made to reflect how health care is delivered. The risk to the patient does not change for a given procedure. If the procedure is performed in a doctor's office versus a hospital, the risk remains the same. Therefore, NFPA 99 eliminated the occupancy chapters and transitioned to a risk-based approach. New Chapter 4 outlined the parameters for this approach. The Code now reflected the risk to the patient in defined categories of risk.

Chapter 5, Gas and Vacuum Systems, went through some editorial changes in the 2012 edition as well as adding new material on the testing and maintenance of gas and vacuum systems. In addition, the administrative details for care, maintenance, and handling of cylinders moved to chapters under the responsibility of the new Technical Committee on Medical Equipment. Several new chapters were added for the 2012 edition on Information Technology and Communications Systems; Plumbing; Heating, Ventilation, and Air Conditioning; Security Management; and Features of Fire Protection. Many of these systems were not previously addressed by NFPA 99. These are important systems and protection features in health care and needed to be addressed. The Technical Committees on Gas Delivery Equipment and the Technical Committee on Electrical Equipment were combined into a single Technical Committee on Medical Equipment. The hyperbaric chapter had relatively minor changes for clarity.

The 2015 edition of NFPA 99 builds upon the major change that the 2012 edition presented. The way that risk categories are defined has been revised to be more inclusive, and the categories can now be applied to equipment and activities, rather than being applicable only to chapters that deal with systems. The requirements for Category 3 medical gas and vacuum systems, while originally aimed specifically for dental applications, have been expanded to include the possibility that other gases might fall under Category 3, based on the facility's risk assessment.

The Technical Committee on Electrical Systems has continued the task of correlating requirements with *NFPA 70®*, *National Electrical Code®*, and Chapter 6, and they have removed the requirements for Level 3 essential electrical systems (EES), determining that if there is not a need for a Level 1 or 2 EES, then the requirements in *NFPA 70* that apply to all buildings will provide the necessary level of safety. Each of the technical committees has made a concerted effort to specifically identify how each of its chapters is to apply to existing buildings or installations and to list the sections that apply.

Correlating Committee on Health Care Facilities (HEA-AAC)

Michael A. Crowley, Chair
The RJA Group, Inc., TX [SE]

Chad E. Beebe, ASHE - AHA, WA [U]
Constance Bobik, B&E Fire Safety Equipment Inc., FL [IM]
Wayne L. Brannan, Medical University of South Carolina, SC [U]
  Rep.
Gordon D. Burrill, Teegor Consulting Inc., Canada [U]
  Rep. Canadian Healthcare Engineering Society
David A. Dagenais, Wentworth-Douglass Hospital, NH [U]
  Rep. NFPA Health Care Section
Tolga Durak, Radford University, VA [SE]
Keith Ferrari, Praxair, Inc., NC [M]
Daniel P. Finnegan, Siemens Industry, Inc., NJ [M]
Robert M. Gagnon, Gagnon Engineering, MD [SE]

Michael D. Hancock, Shermco Industries, TX [IM]
Michael S. Jensen, U.S. Department of Health & Human Services, AZ [E]
Frank L. Keisler, Jr., CNA Insurance Company, GA [I]
William E. Koffel, Koffel Associates, Inc., MD [SE]
Denise L. Pappas, Valcom, Inc., VA [M]
  Rep. National Electrical Manufacturers Association
Eric R. Rosenbaum, Hughes Associates, Inc., MD [U]
  Rep. American Health Care Association
Grayson Sack, Cashins and Associates, Inc., MA [SE]
Ronald A. Schroeder, ASCO/Power Switching & Controls, NJ [M]
Joseph H. Versteeg, Versteeg Associates, CT [E]
  Rep. DNV Health Care Inc.
Mayer D. Zimmerman, Randallstown, MD [SE]

Alternates

H. Shane Ashby, West Tennessee Healthcare, TN [U]
  (Alt. to W. L. Brannan)
Tracy J. Donohue, Siemens Industry, Inc., IL [M]
  (Alt. to D. P. Finnegan)
Sharon S. Gilyeat, Koffel Associates, Inc., MD [SE]
  (Alt. to W. E. Koffel)

Donald D. King, Kaiser Permanente, CA [U]
  (Alt. to C. E. Beebe)
Mark Pustejovsky, Shermco Industries, TX [IM]
  (Alt. to M. D. Hancock)
Rodger Reiswig, Tyco/SimplexGrinnell, FL [M]
  (Alt. to D. L. Pappas)

Nonvoting

David P. Klein, U.S. Department of Veterans Affairs, DC [U]
  Rep. TC on Fundamentals
James K. Lathrop, Koffel Associates, Inc., CT [SE]
  Rep. TC on Piping Systems
Roger W. Lautz, Affiliated Engineers, Inc., WI [SE]
  Rep. TC on Mechanical Systems
Alan Lipschultz, HealthCare Technology Consulting LLC, DE [M]
  Rep. TC on Medical Equipment

Susan B. McLaughlin, MSL Healthcare Consulting, Inc., IL [U]
  Rep. TC on Health Care Emergency Management & Security
Robert B. Sheffield, International ATMO, Inc., TX [U]
  Rep. TC on Hyperbaric and Hypobaric Facilities
Walter N. Vernon, IV, Mazzetti, CA [SE]
  Rep. TC on Electrical Systems
Marvin J. Fischer, Monroe Township, NJ [SE]
  (Member Emeritus)

Jonathan Hart, NFPA Staff Liaison

This list represents the membership at the time the Committee was balloted on the final text of this edition. Since that time, changes in the membership may have occurred. A key to classifications is found at the back of the document.

NOTE: Membership on a committee shall not in and of itself constitute an endorsement of the Association or any document developed by the committee on which the member serves.

Committee Scope: This Committee shall have primary responsibility for documents that contain criteria for safeguarding patients and health care personnel in the delivery of health care services within health care facilities: a) from fire, explosion, electrical, and related hazards resulting either from the use of anesthetic agents, medical gas equipment, electrical apparatus, and high frequency electricity, or from internal or external incidents that disrupt normal patient care; b) from fire and explosion hazards; c) in connection with the use of hyperbaric and hypobaric facilities for medical purposes; d) through performance, maintenance and testing criteria for electrical systems, both normal and essential; and e) through performance, maintenance and testing, and installation criteria: (1) for vacuum systems for medical or surgical purposes, and (2) for medical gas systems; and f) through performance, maintenance and testing of plumbing, heating, cooling, and ventilating in health care facilities.

Technical Committee on Electrical Systems (HEA-ELS)

Walter N. Vernon, IV, *Chair*
Mazzetti, CA [SE]

Paul Acre, Arkansas Department Of Health, AR [E]
Steven J. Barker, University of Arizona, AZ [C]
    Rep. American Society of Anesthesiologists
Nancy W. Chilton, Schneider Electric, NC [M]
Dan Chisholm, Sr., MGI Systems, Inc., FL [IM]
James H. Costley, Jr., Newcomb & Boyd, GA [SE]
    Rep. NFPA Health Care Section
Jason D'Antona, Thompson Consultants, Inc., MA [SE]
David A. Dagenais, Wentworth-Douglass Hospital, NH [U]
Herbert H. Daugherty, Electric Generating Systems Association, FL [M]
Daniel T. DeHanes, Ascom Wireless Solutions, FL [M]
    Rep. National Electrical Manufacturers Association
Tony Easty, University Health Network, Canada [U]
Jan Ehrenwerth, Yale University, CT [C]
Chris M. Finen, Eaton Electrical Corporation, TN [M]
William T. Fiske, Intertek Testing Services, NY [RT]
    Rep. National Electrical Code Correlating Committee
Don W. Jhonson, Interior Electric, Inc., FL [IM]
    Rep. National Electrical Contractors Association

Burton R. Klein, Burton Klein Associates, MA [SE]
Gary J. Krupa, U.S. Department of Veterans Affairs, NE [U]
Stephen M. Lipster, The Electrical Trades Center, OH [L]
    Rep. International Brotherhood of Electrical Workers
Dale Martinson, Via Christi Health System, KS [U]
James E. Meade, U.S. Army Corps of Engineers, MD [U]
Joseph P. Murnane, Jr., UL LLC, NY [RT]
Hugh O. Nash, Jr., Nash-Consult, TN [SE]
Thomas J. Parrish, Telgian Corporation, MI [SE]
John Peterson, Utility Service Corporation, AL [IM]
    Rep. InterNational Electrical Testing Association
Vincent M. Rea, TLC Engineering for Architecture, FL [SE]
Michael L. Savage, Sr., Middle Department Inspection Agency, Inc., MD [E]
Ronald M. Smidt, Carolinas HealthCare System, NC [U]
    Rep. American Society for Healthcare Engineering
Leonard W. White, Stanford White Associates Consulting Engineers, Inc., NC [SE]
Robert Wolff, IES Engineers-Dewberry, NC [SE]

Alternates

Gary A. Beckstrand, Utah Electrical JATC, UT [L]
    (Alt. to S. M. Lipster)
Chad E. Beebe, ASHE - AHA, WA [U]
    (Alt. to R. M. Smidt)
H. David Chandler, Newcomb & Boyd, GA [SE]
    (Alt. to J. H. Costley, Jr.)
Dan Chisholm, Jr., MGI Systems, Inc., FL [IM]
    (Alt. to D. Chisholm, Sr.)
Michael S. Goodheart, Ascom (US), FL [M]
    (Alt. to D. T. DeHanes)

Chad Kennedy, Schneider Electric, SC [M]
    (Alt. to N. W. Chilton)
Robert G. Loeb, University of Arizona, AZ [C]
    (Alt. to S. J. Barker)
Gary A. Spivey, US Army Corps of Engineers, VA [U]
    (Alt. to J. E. Meade)
Herbert V. Whittall, Electrical Generating Systems Association, FL [M]
    (Alt. to H. H. Daugherty)

Jonathan Hart, NFPA Staff Liaison

This list represents the membership at the time the Committee was balloted on the final text of this edition. Since that time, changes in the membership may have occurred. A key to classifications is found at the back of the document.

NOTE: Membership on a committee shall not in and of itself constitute an endorsement of the Association or any document developed by the committee on which the member serves.

Committee Scope: This Committee shall have primary responsibility for documents or portions of documents covering the minimum requirements for performance, testing, maintenance, operations, and failure management of electrical systems, low voltage systems, wireless technologies, informatics, and telemedicine to safeguard patients, staff, and visitors within health care facilities based on established risk categories.

Technical Committee on Fundamentals (HEA-FUN)

David P. Klein, *Chair*
U.S. Department of Veterans Affairs, DC [U]
Rep. U.S. Department of Veterans Affairs

Bruce L. Abell, U.S. Army Corps of Engineers, VA [U]
Chad E. Beebe, ASHE - AHA, WA [U]
  Rep. American Society for Healthcare Engineering
Jeff N. Besel, HGA Architects and Engineers, MN [M]
  Rep. Automatic Fire Alarm Association, Inc.
Gordon D. Burrill, Teegor Consulting Inc., Canada [U]
  Rep. Canadian Healthcare Engineering Society
Michael A. Crowley, The RJA Group, Inc., TX [SE]
Richard E. Cutts, NORAD/NORTHCOM (USAF), GA [U]
August F. DiManno, Jr., Program Risk Management, Inc., NY [I]
Sean S. Donohue, [SE]
Carl J. Ferlitch, Jr., Chubb Group of Insurance Companies, PA [I]
Michael S. Jensen, U.S. Department of Health & Human Services, AZ [E]
  Rep. U.S. Dept. of Health & Human Services/IHS
Frank L. Keisler, Jr., CNA Insurance Company, GA [I]
Henry Kowalenko, Illinois Department of Public Health, IL [E]
James K. Lathrop, Koffel Associates, Inc., CT [SE]
Stephen M. Lipster, The Electrical Trades Center, OH [L]
  Rep. International Brotherhood of Electrical Workers
Bret M. Martin, [U]
Jack McNamara, Bosch Security Systems, NY [M]
  Rep. National Electrical Manufacturers Association
Thomas G. McNamara, North Shore LIJ Health System-Southside Hospital, NY [U]
James S. Peterkin, Heery International, PA [SE]
Grayson Sack, [SE]
Frank L. Van Overmeiren, FP&C Consultants, Inc., IN [SE]
Robert F. Willey III, North Carolina Dept. of Public Instruction, NC [E]
John L. Williams, [E]
Jennifer Zaworski, Aon Fire Protection Engineering, D [I]

Alternates

Beth A. Alexander, FP&C Consultants, Inc., IN [SE]
  (Alt. to F. L. Van Overmeiren)
Gary A. Beckstrand, Utah Electrical JATC, UT [L]
  (Alt. to S. M. Lipster)
Daniel P. Finnegan, Siemens Industry, Inc., NJ [M]
  (Alt. to J. N. Besel)
Sharon S. Gilyeat, Koffel Associates, Inc., MD [SE]
  (Alt. to J. K. Lathrop)
Peter Leszczak, U.S. Department of Veterans Affairs, CT [U]
  (Alt. to D. P. Klein)
Michael Mansfield, CNA Insurance Company, FL [I]
  (Alt. to F. L. Keisler, Jr.)
Terrence Pintar, The RJA Group, Inc., IL [SE]
  (Alt. to M. A. Crowley)
Kevin A. Scarlett, [E]
  (Alt. to J. L. Williams)

Jonathan Hart, NFPA Staff Liaison

This list represents the membership at the time the Committee was balloted on the final text of this edition. Since that time, changes in the membership may have occurred. A key to classifications is found at the back of the document.

NOTE: Membership on a committee shall not in and of itself constitute an endorsement of the Association or any document developed by the committee on which the member serves.

Committee Scope: This Committee shall have primary responsibility for documents or portions of documents on the scope, application, and intended use of documents under the Health Care Facilities Project, including reference standards, performance, the protection from fire and explosion hazards, protection of special hazards, establishing criteria for levels of health care services based on risk, as well as definitions not assigned to other committees in the Health Care Facilities Project.

Technical Committee on Health Care Emergency Management and Security (HEA-HES)

Susan B. McLaughlin, Chair
MSL Healthcare Consulting, Inc., IL [U]
Rep. American Society for Healthcare Engineering

Robert M. Becker, Incident Management Solutions, Inc., NY [SE]
Pete Brewster, U.S. Department of Veterans Affairs, V [U]
David A. Dagenais, Wentworth-Douglass Hospital, NH [U]
Steve Ennis, Virginia Hospital & Healthcare Association, VA [U]
Jon M. Evenson, The RJA Group, Inc., IL [SE]
Scott R. Fernhaber, Johnson Controls, Inc., WI [M]
Sharon S. Gilyeat, Koffel Associates, Inc., MD [SE]
Frank L. Keisler, Jr., CNA Insurance Company, GA [I]
William C. McPeck, State of Maine Employee Health & Safety, ME [E]

James L. Paturas, Yale New Haven Health System, CT [U]
Jack Poole, Poole Fire Protection, Inc., KS [SE]
Ronald C. Reynolds, Virginia State Fire Marshal's Office, VA [E]
Tom Mayer Scheidel, Centers for Medicare and Medicaid Services, TX [E]
Hulbert P. L. Silver, Central Newfoundland Regional Health Centre, Canada [U]
James P. Simpson, National Joint Apprentice & Training Committee, MN [L]
  Rep. International Brotherhood of Electrical Workers
Michael D. Widdekind, Zurich Services Corporation, MD [I]

Alternates

Chad E. Beebe, ASHE - AHA, WA [U]
  (Alt. to S. B. McLaughlin)
Jennifer L. Frecker, Koffel Associates, Inc., MD [SE]
  (Alt. to S. S. Gilyeat)
Nicholas Gabriele, Russell Phillips & Associates, CT [SE]
  (Alt. for Russell Phillips & Associates)

Zachary Goldfarb, Incident Management Solutions, Inc., NY [SE]
  (Alt. to R. M. Becker)
Richard A. Mahnke, The RJA Group, Inc., IL [SE]
  (Alt. to J. M. Evenson)

Nonvoting

Reginald D. Jackson, U.S. Department of Labor, DC [E]

Jonathan Hart, NFPA Staff Liaison

This list represents the membership at the time the Committee was balloted on the final text of this edition. Since that time, changes in the membership may have occurred. A key to classifications is found at the back of the document.

NOTE: Membership on a committee shall not in and of itself constitute an endorsement of the Association or any document developed by the committee on which the member serves.

Committee Scope: This Committee shall have primary responsibility for documents or portions of documents covering the framework for emergency management and security of health care facilities proportionate to the risk of the patient and health care staff. This Committee shall have primary responsibility for the elements of planning over a continuum from minor incidences to catastrophic events, including: management controls, mitigation practices, incident response, continuity of services, recovery, stored capacity, staff training, and program evaluation based on established risk categories.

Technical Committee on Hyperbaric and Hypobaric Facilities (HEA-HYP)

Robert B. Sheffield, *Chair*
International ATMO, Inc., TX [U]

Michael W. Allen, Life Support Technologies Group Inc., PA [U]
Peter Atkinson, Royal Brisbane and Womens Hospital, Australia [C]
    Rep. Hyperbaric Technicians & Nurses Association Inc.
Richard C. Barry, Diversified Clinical Services & National Healing, FL [SE]
Chad E. Beebe, ASHE - AHA, WA [U]
    Rep. American Society for Healthcare Engineering
James Bell, Intermountain Healthcare, UT [U]
W. Robert Bryant, Perry Baromedical Corporation, TX [M]
Mario Caruso, Comprehensive Healthcare Solutions, Inc., FL [SE]
Keith Ferrari, Praxair, Inc., NC [M]

Angela M. Fuqua, Chubb Group Insurance Companies, TX [I]
William C. Gearhart, University of Maryland Medical Systems, MD [U]
W. T. Gurnée, OxyHeal Health Group, CA [M]
John R. Kitchens, Los Angeles City Fire Department, CA [E]
Barry E. Newton, Wendell Hull & Associates, Inc., NM [SE]
Stephen D. Reimers, Reimers Systems, Inc., VA [M]
Rachael Sheets, The Linde Group, NJ [IM]
John M. Skinner, Medical Equipment Technology, Inc., GA [IM]
Deepak Talati, Sechrist Industries, Inc., CA [M]
Wilbur T. Workman, Undersea & Hyperbaric Medical Society, TX [U]

Alternates

Justin Callard, Hyperbaric Technicians & Nurses Association Inc., Australia [C]
    (Alt. to P. Atkinson)
Mark Chipps, Life Support Technologies Group Inc., NY [U]
    (Alt. to M. W. Allen)

Kevin I. Posey, International ATMO, Inc., TX [U]
    (Alt. to R. B. Sheffield)
Michael P. Powers, OxyHeal Health Group, CT [M]
    (Alt. to W. T. Gurnée)

Jonathan Hart, NFPA Staff Liaison

This list represents the membership at the time the Committee was balloted on the final text of this edition. Since that time, changes in the membership may have occurred. A key to classifications is found at the back of the document.

NOTE: Membership on a committee shall not in and of itself constitute an endorsement of the Association or any document developed by the committee on which the member serves.

Committee Scope: This Committee shall have primary responsibility for documents or portions of documents covering the construction, installation, testing, performance, and maintenance of hyperbaric and hypobaric facilities for safeguarding staff and occupants of chambers.

Technical Committee on Mechanical Systems (HEA-MEC)

Roger W. Lautz, Chair
Affiliated Engineers, Inc., WI [SE]

Hugo Aguilar, International Association of Plumbing & Mechanical Officials, CA [E]
Kimberly A. Barker, Siemens Building Technologies, Inc., IL [M]
Chad E. Beebe, ASHE - AHA, WA [U]
Christopher Bernecker, H. T. Lyons, Inc., PA [IM]
Gordon D. Burrill, Teegor Consulting Inc., Canada [U]
  Rep. Canadian Healthcare Engineering Society
Raj Daswani, Arup, CA [SE]
Robert J. Dubiel, Luther Midelfort Mayo Health, WI [U]

Keith Ferrari, Praxair, Inc., NC [M]
Ronald E. Galloway, Moses Cone Health System, NC [U]
Phil Gioia, Mazzetti & Associates Inc., CO [SE]
Jonathan Hartsell, Rodgers, NC [SE]
Gustavo Olano, Phoenix Controls Corporation, MA [M]
Charles Seyffer, Camfil Farr, NY [M]
Michael P. Sheerin, TLC Engineering for Architecture, FL [SE]
Allan D. Volz, OSF HealthCare System, IL [U]

Alternates

Kenneth Crooks, Phoenix Controls Corporation, MA [M]
  (Alt. to G. Olano)

Jonathan Hart, NFPA Staff Liaison

This list represents the membership at the time the Committee was balloted on the final text of this edition. Since that time, changes in the membership may have occurred. A key to classifications is found at the back of the document.

NOTE: Membership on a committee shall not in and of itself constitute an endorsement of the Association or any document developed by the committee on which the member serves.

Committee Scope: This committee shall have primary responsibility for documents or portions of documents covering the performance, operations, testing, and maintenance, for air quality, temperature, humidity, critical space pressure relationships, water and waste water, and their associated systems based on established risk categories.

Technical Committee on Medical Equipment (HEA-MED)

Alan Lipschultz, Chair
HealthCare Technology Consulting LLC, DE [M]
Rep. Association for the Advancement of Medical Instrumentation

Michael E. Brousseau, Intertek Testing Services, MA [RT]
Charles Connor, [U]
Keith Ferrari, Praxair, Inc., NC [M]
  Rep. Compressed Gas Association
William C. Fettes, Airgas, Inc., KS [IM]
Gerald R. Goodman, Texas Woman's University, TX [SE]
Donald D. King, Kaiser Permanente, CA [U]
  Rep. American Society for Healthcare Engineering
Harvey Kostinsky, ECRI Institute, PA [RT]
Joseph P. Murnane, Jr., UL LLC, NY [RT]

Ronald C. Reynolds, Virginia State Fire Marshal's Office, VA [E]
Ezra R. Safdie, U.S. Department of Veterans Affairs, CA [U]
Lawrence S. Sandler, Bonita Springs, FL [SE]
Kevin A. Scarlett, [E]
Hulbert P. L. Silver, Central Newfoundland Regional Health Centre, Canada [U]
Robert M. Sutter, B&R Compliance Associates, PA [SE]

Alternates

Gary L. Bean, Air Products & Chemicals, Inc., GA [M]
  (Alt. to K. Ferrari)
Chad E. Beebe, ASHE - AHA, WA [U]
  (Alt. to D. D. King)

Barry E. Brown, Airgas, Inc., GA [IM]
  (Alt. to W. C. Fettes)
Kenneth Gerard Funk, [SE]
  (Alt. to R. M. Sutter)

Nonvoting

Saul Aronow, Auburndale, MA [SE]
  (Member Emeritus)

Jonathan Hart, NFPA Staff Liaison

This list represents the membership at the time the Committee was balloted on the final text of this edition. Since that time, changes in the membership may have occurred. A key to classifications is found at the back of the document.

NOTE: Membership on a committee shall not in and of itself constitute an endorsement of the Association or any document developed by the committee on which the member serves.

Committee Scope: This committee shall have primary responsibility for documents or portions of documents covering the maintenance, performance, and testing of electrical medical equipment and portable patient-related gas equipment for the purpose of safeguarding patients and health care personnel within patient care areas of health care facilities from the hazards of fire, explosion, electricity, nonionizing radiation, heat, and electrical interference based on established risk categories.

Technical Committee on Piping Systems (HEA-PIP)

James K. Lathrop, *Chair*
Koffel Associates, Inc., CT [SE]

Mark W. Allen, Beacon Medaes, SC [M]
Grant A. Anderson, Bard, Rao & Athanas Consulting Engineers, LLC, MA [SE]
Steven J. Barker, University of Arizona, AZ [C]
 Rep. American Society of Anesthesiologists
Chad E. Beebe, ASHE - AHA, WA [U]
 Rep. American Society for Healthcare Engineering
David L. Brittain, ProVac, OH [M]
Dana A. Colombo, PIPE/National ITC Corporation, LA [L]
Keith Ferrari, Praxair, Inc., NC [M]
 Rep. Compressed Gas Association
William C. Fettes, Airgas, Inc., KS [IM]
Michael Frankel, Utility Systems Consultants, FL [SE]
 Rep. American Society of Plumbing Engineers
Ed Golla, TRI/Air Testing, TX [RT]
John C. Gregory, HDR Architecture Inc., AZ [SE]
Daniel Patrick Kelly, The Cleveland Clinic, OH [U]
Anthony Lowe, Allied Hospital Systems, MD [IM]
James L. Lucas, Tri-Tech Medical Inc., OH [M]
Jeffery F. McBride, Red Lion Medgas Consultants, Inc., DE [SE]
Donald R. McIlroy, Providence Health System, OR [U]
Spiro Megremis, [U]
David B. Mohile, Medical Engineering Services, Inc., VA [SE]
Thomas J. Mraulak, Plumbing Industry Training Center, MI [L]
 Rep. American Society of Sanitary Engineering
Ronald J. Schwipps, Hill-Rom, Inc., IN [M]
E. Daniel Shoemaker, Accutron Inc., AZ [M]
Matt Sigler, International Association of Plumbing & Mechanical Officials, CA [E]
Ronald M. Smidt, Carolinas HealthCare System, NC [U]
 Rep. NFPA Health Care Section
Russell C. Thomason, U.S. Army Corps of Engineers, VA [U]
Allan D. Volz, OSF HealthCare System, IL [U]
J. Richard Wagner, J. Richard Wagner, PE, LLC, MD [IM]
 Rep. Mechanical Contractors Association of America, Inc.
Jonathan C. Willard, Certified Medical Gas Services, NH [SE]

Alternates

Gary L. Bean, Air Products & Chemicals, Inc., GA [M]
 (Alt. to K. Ferrari)
Barry E. Brown, Airgas, Inc., GA [IM]
 (Alt. to W. C. Fettes)
Laurence T. Coleman, Pipe Fitters Local Union 597-Chicago, IL [L]
 (Alt. to T. J. Mraulak)
Gary Currence, Allied Hospital Systems, MD [IM]
 (Alt. to A. Lowe)
Marc Dodson, [SE]
 (Alt. to J. C. Gregory)
P. L. Fan, Santa Fe, NM [U]
 (Alt. to S. Megremis)
Mark T. Franklin, Sherman Engineering Company, PA [M]
 (Alt. to M. W. Allen)
Neil Gagné, [SE]
 (Alt. to D. B. Mohile)
Robert G. Loeb, University of Arizona, AZ [C]
 (Alt. to S. J. Barker)
Edward J. Lyczko, [U]
 (Alt. to D. P. Kelly)
Michael T. Massey, PIPE/National ITC Corporation, CA [L]
 (Alt. to D. A. Colombo)

Jonathan Hart, NFPA Staff Liaison

This list represents the membership at the time the Committee was balloted on the final text of this edition. Since that time, changes in the membership may have occurred. A key to classifications is found at the back of the document.

NOTE: Membership on a committee shall not in and of itself constitute an endorsement of the Association or any document developed by the committee on which the member serves.

Committee Scope: This Committee shall have primary responsibility for documents or portions of documents covering the performance, maintenance, installation, and testing of medical and dental related gas piping systems and medical and dental related vacuum piping systems based on established risk categories.

Contents

Chapter 1	Administration	99– 14
1.1	Scope	99– 14
1.2	Purpose	99– 14
1.3	Application	99– 14
1.4	Equivalency	99– 15
1.5	Units	99– 15
1.6	Code Adoption Requirements	99– 15

Chapter 2	Referenced Publications	99– 15
2.1	General	99– 15
2.2	NFPA Publications	99– 16
2.3	Other Publications	99– 16
2.4	References for Extracts in Mandatory Sections	99– 17

Chapter 3	Definitions	99– 18
3.1	General	99– 18
3.2	NFPA Official Definitions	99– 18
3.3	General Definitions	99– 18
3.4	BICSI Definitions	99– 25

Chapter 4	Fundamentals	99– 25
4.1	Risk Categories	99– 25
4.2	Risk Assessment	99– 26
4.3	Application	99– 26
4.4	Materials	99– 26

Chapter 5	Gas and Vacuum Systems	99– 26
5.1	Category 1 Piped Gas and Vacuum Systems	99– 26
5.2	Category 2 Piped Gas and Vacuum Systems	99– 59
5.3	Category 3 Piped Gas and Vacuum Systems	99– 60

Chapter 6	Electrical Systems	99– 67
6.1	Applicability	99– 67
6.2	Nature of Hazards	99– 68
6.3	Electrical System	99– 68
6.4	Essential Electrical System Requirements — Type 1	99– 72
6.5	Essential Electrical System Requirements — Type 2	99– 80

Chapter 7	Information Technology and Communications Systems	99– 81
7.1	Applicability	99– 81
7.2	Reserved	99– 81
7.3	Category 1 Systems	99– 81
7.4	Category 2 Systems	99– 84
7.5	Category 3 Systems	99– 85

Chapter 8	Plumbing	99– 85
8.1	Applicability	99– 85
8.2	System Category Criteria	99– 85
8.3	General Requirements	99– 85
8.4	Category 1. (Reserved)	99– 86
8.5	Category 2. (Reserved)	99– 86
8.6	Category 3. (Reserved)	99– 86

Chapter 9	Heating, Ventilation, and Air Conditioning (HVAC)	99– 86
9.1	Applicability	99– 86
9.2	System Category Criteria	99– 86
9.3	General	99– 86
9.4	Category 1. (Reserved)	99– 87
9.5	Category 2. (Reserved)	99– 87
9.6	Category 3. (Reserved)	99– 87

Chapter 10	Electrical Equipment	99– 88
10.1	Applicability	99– 88
10.2	Performance Criteria and Testing for Patient Care–Related Electrical Appliances and Equipment	99– 88
10.3	Testing Requirements — Fixed and Portable	99– 89
10.4	Nonpatient Electrical Appliances and Equipment	99– 90
10.5	Administration	99– 90

Chapter 11	Gas Equipment	99– 93
11.1	Applicability	99– 93
11.2	Cylinder and Container Source	99– 93
11.3	Cylinder and Container Storage Requirements	99– 94
11.4	Performance Criteria and Testing	99– 94
11.5	Administration	99– 95
11.6	Operation and Management of Cylinders	99– 96
11.7	Liquid Oxygen Equipment	99– 98

Chapter 12	Emergency Management	99– 98
12.1	Applicability	99– 98
12.2	Responsibilities	99– 99
12.3	Emergency Management Categories	99– 99
12.4	General	99– 99
12.5	Emergency Management Category 1 and Emergency Management Category 2 Requirements	99– 99

Chapter 13	Security Management	99–103
13.1	Applicability	99–103
13.2	Security Management Plan	99–103
13.3	Security Vulnerability Assessment (SVA)	99–103
13.4	Responsible Person	99–103

13.5	Security-Sensitive Areas	**99–103**
13.6	Access and Egress Security Measures	**99–104**
13.7	Media Control	**99–104**
13.8	Crowd Control	**99–104**
13.9	Security Equipment	**99–104**
13.10	Employment Practices	**99–104**
13.11	Security Operations	**99–104**
13.12	Program Evaluation	**99–104**

Chapter 14 Hyperbaric Facilities **99–105**
14.1	Scope	**99–105**
14.2	Construction and Equipment	**99–105**
14.3	Administration and Maintenance	**99–114**

Chapter 15 Features of Fire Protection **99–117**
15.1	Applicability	**99–117**
15.2	Construction and Compartmentation	**99–117**
15.3	Special Hazard Protection for Flammable Liquids and Gases	**99–117**
15.4	Laboratories	**99–118**
15.5	Utilities	**99–118**
15.6	Waste Chutes, Incinerators, and Linen Chutes	**99–118**
15.7	Fire Detection, Alarm, and Communications Systems	**99–118**
15.8	Automatic Sprinklers and Other Extinguishing Equipment	**99–119**
15.9	Manual Extinguishing Equipment	**99–120**
15.10	Compact Storage	**99–120**
15.11	Compact Mobile Storage	**99–120**
15.12	Maintenance and Testing	**99–120**
15.13	Fire Loss Prevention in Operating Rooms	**99–120**

Annex A	**Explanatory Material**	**99–121**
Annex B	**Additional Explanatory Notes**	**99–171**
Annex C	**Sample Ordinance Adopting NFPA 99**	**99–188**
Annex D	**Informational References**	**99–189**
Index		**99–193**

NFPA 99

Health Care Facilities Code

2015 Edition

IMPORTANT NOTE: This NFPA document is made available for use subject to important notices and legal disclaimers. These notices and disclaimers appear in all publications containing this document and may be found under the heading "Important Notices and Disclaimers Concerning NFPA Standards." They can also be obtained on request from NFPA or viewed at www.nfpa.org/disclaimers.

NOTICE: An asterisk (*) following the number or letter designating a paragraph indicates that explanatory material on the paragraph can be found in Annex A.

A reference in brackets [ ] following a section or paragraph indicates material that has been extracted from another NFPA document. As an aid to the user, the complete title and edition of the source documents for extracts in mandatory sections of the document are given in Chapter 2 and those for extracts in informational sections are given in Annex D. Extracted text may be edited for consistency and style and may include the revision of internal paragraph references and other references as appropriate. Requests for interpretations or revisions of extracted text shall be sent to the technical committee responsible for the source document.

Information on referenced publications can be found in Chapter 2 and Annex D.

Chapter 1 Administration

1.1 Scope.

1.1.1 The scope of this code is to establish minimum criteria as follows in 1.1.2 through 1.1.13.

1.1.2 Fundamentals. Chapter 4 establishes criteria for levels of health care services or systems based on risk to the patients, staff, or visitors in health care facilities.

1.1.3 Gas and Vacuum Systems.

1.1.3.1 Chapter 5 covers the performance, maintenance, installation, and testing of the following:

(1) Nonflammable medical gas systems with operating pressures below a gauge pressure of 2068 kPa (300 psi)
(2) Vacuum systems in health care facilities
(3) Waste anesthetic gas disposal (WAGD) systems, also referred to as scavenging
(4) Manufactured assemblies that are intended for connection to the medical gas, vacuum, or WAGD systems (also referred to as scavenging)

1.1.3.2 Requirements for portable compressed gas systems are covered in Chapter 11.

1.1.4 Electrical Systems.

1.1.4.1 Chapter 6 covers the performance, maintenance, and testing of electrical systems (both normal and essential) in health care facilities.

1.1.4.2 The following areas are not addressed in this code, but are addressed in other NFPA documents:

(1) Specific requirements for wiring and installation of equipment are covered in *NFPA 70, National Electrical Code.*
(2) Requirements for illumination and identification of means of egress in health care facilities are covered in NFPA *101, Life Safety Code.*
(3) Requirements for installation, testing, and maintenance of fire protection signaling systems are covered in *NFPA 72, National Fire Alarm and Signaling Code.*
(4) Requirements for installation of fire pumps are covered in NFPA 20, *Standard for the Installation of Stationary Pumps for Fire Protection,* except that the alternate source of power are permitted to be the essential electrical system.
(5) Requirements for installation of stationary engines and gas turbines are covered in NFPA 37, *Standard for the Installation and Use of Stationary Combustion Engines and Gas Turbines.*

1.1.5 Information Technology and Communications Systems. Chapter 7 covers the performance, maintenance, and testing of information technology and communications systems in health care facilities.

1.1.6 Plumbing. Chapter 8 covers the performance, maintenance, and testing of plumbing systems in health care facilities.

1.1.7 HVAC Systems. Chapter 9 covers the performance, maintenance, and testing of heating, cooling, and ventilating in health care facilities.

1.1.8 Electrical Equipment. Chapter 10 covers the performance, maintenance, and testing of electrical equipment in health care facilities.

1.1.9 Gas Equipment. Chapter 11 covers the performance, maintenance, and testing of gas equipment in health care facilities.

1.1.10* Emergency Management. Chapter 12 establishes criteria for emergency management in the development of a program for effective disaster preparedness, response, mitigation, and recovery in health care facilities.

1.1.11 Security Management. Chapter 13 establishes criteria for security management, including management controls, mitigation practices, staff training, and program evaluation in health care facilities.

1.1.12* Hyperbaric Facilities. Chapter 14 covers the recognition of, and protection against, hazards of an electrical, explosive, or implosive nature, as well as fire hazards associated with hyperbaric chambers and associated facilities that are used, or intended to be used, for medical applications and experimental procedures at gauge pressures from 0 kPa to 690 kPa (0 psi to 100 psi).

1.1.13 Features of Fire Protection. Chapter 15 covers the performance, maintenance, and testing of fire protection equipment in health care facilities.

1.2 Purpose. The purpose of this code is to provide minimum requirements for the installation, inspection, testing, maintenance, performance, and safe practices for facilities, material, equipment, and appliances, including other hazards associated with the primary hazards.

1.3 Application.

1.3.1 This code shall apply to all health care facilities other than home care and veterinary care.

1.3.1.1 This document is intended for use by those persons involved in the design, construction, inspection, and operation of health care facilities and in the design, manufacture, and testing of appliances and equipment used in patient care rooms of health care facilities.

1.3.2 Construction and equipment requirements shall be applied only to new construction and new equipment, except as modified in individual chapters.

1.3.2.1 Only the altered, renovated, or modernized portion of an existing system or individual component shall be required to meet the installation and equipment requirements stated in this code.

1.3.2.2 If the alteration, renovation, or modernization adversely impacts the existing performance requirements of a system or component, additional upgrading shall be required.

1.3.2.3 An existing system that is not in strict compliance with the provisions of this code shall be permitted to be continued in use, unless the authority having jurisdiction has determined that such use constitutes a distinct hazard to life.

1.3.3 Policies.

1.3.3.1 The health care organization shall ensure that policies are established and maintained that permit the attending physician to satisfy the emergency needs of any patient that supersede the requirements of this code.

1.3.3.2 Each such special use shall be clearly documented and reviewed to attempt to have future similar needs met within the requirements of this code.

1.3.4 Patient Care Spaces.

1.3.4.1 The governing body of the facility or its designee shall establish the following areas in accordance with the type of patient care anticipated *(see definition of patient care spaces in Chapter 3)*:

(1) Category 1 spaces
(2) Category 2 spaces
(3) Category 3 spaces
(4) Category 4 spaces

1.3.4.2 Anesthesia. It shall be the responsibility of the governing body of the health care organization to designate anesthetizing locations.

1.3.4.3 Wet Procedure Locations. It shall be the responsibility of the governing body of the health care organization to designate wet procedure locations.

1.4 Equivalency.

1.4.1 Nothing in this code is intended to prevent the use of systems, methods, or devices of equivalent or superior quality, strength, fire resistance, effectiveness, durability, and safety to those prescribed by this code. Technical documentation shall be submitted to the authority having jurisdiction to demonstrate equivalency. The system, method, or device shall be approved for the intended purpose by the authority having jurisdiction.

1.4.2 Alternative systems, methods, or devices approved as equivalent by the authority having jurisdiction shall be recognized as being in compliance with this code.

1.4.3 The authority having jurisdiction shall be permitted to grant exceptions to this code.

1.5* Units. Primary units will be trade units, and secondary units will be the conversion.

1.6 Code Adoption Requirements.

1.6.1 The effective date of application of any provision of this document is not determined by the National Fire Protection Association. All questions related to applicability shall be directed to the authority having jurisdiction.

1.6.2 Enforcement. This code shall be administered and enforced by the authority having jurisdiction. *(See Annex C for a sample wording for enabling legislation.)*

Chapter 2 Referenced Publications

2.1* General. The documents referenced in this chapter, or portions of such documents, are referenced within this code and shall be considered part of the requirements of this code, and the following shall also apply:

(1) Documents referenced in this chapter, or portion of such documents, shall only be applicable to the extent called for within other chapters of this code.
(2) Where the requirements of a referenced code or standard differ from the requirements of this code, the requirements of this code shall govern.
(3) Existing buildings or installations that do not comply with the provisions of the codes or standards referenced in this chapter shall be permitted to be continued in service, provided that the lack of conformity with these documents does not present a serious hazard to the occupants as determined by the authority having jurisdiction. [*101*: 2.1]

2.2 NFPA Publications. National Fire Protection Association, 1 Batterymarch Park, Quincy, MA 02169-7471.

NFPA 10, *Standard for Portable Fire Extinguishers*, 2013 edition.

NFPA 13, *Standard for the Installation of Sprinkler Systems*, 2013 edition.

NFPA 14, *Standard for the Installation of Standpipe and Hose Systems*, 2013 edition.

NFPA 20, *Standard for the Installation of Stationary Pumps for Fire Protection*, 2013 edition.

NFPA 25, *Standard for the Inspection, Testing, and Maintenance of Water-Based Fire Protection Systems*, 2014 edition.

NFPA 30, *Flammable and Combustible Liquids Code*, 2015 edition.

NFPA 31, *Standard for the Installation of Oil-Burning Equipment*, 2011 edition.

NFPA 37, *Standard for the Installation and Use of Stationary Combustion Engines and Gas Turbines*, 2015 edition.

NFPA 45, *Standard on Fire Protection for Laboratories Using Chemicals*, 2011 edition.

NFPA 54, *National Fuel Gas Code*, 2015 edition.

NFPA 55, *Compressed Gases and Cryogenic Fluids Code*, 2013 edition.

NFPA 58, *Liquefied Petroleum Gas Code*, 2014 edition.

NFPA 70®, *National Electrical Code*®, 2014 edition.

NFPA 72®, *National Fire Alarm and Signaling Code*, 2013 edition.

NFPA 82, *Standard on Incinerators and Waste and Linen Handling Systems and Equipment*, 2014 edition.

NFPA 90A, *Standard for the Installation of Air-Conditioning and Ventilating Systems*, 2015 edition.

NFPA 91, *Standard for Exhaust Systems for Air Conveying of Vapors, Gases, Mists, and Noncombustible Particulate Solids*, 2010 edition.

NFPA 96, *Standard for Ventilation Control and Fire Protection of Commercial Cooking Operations*, 2014 edition.

NFPA *101*®, *Life Safety Code*®, 2015 edition.

NFPA 110, *Standard for Emergency and Standby Power Systems*, 2013 edition.

NFPA 111, *Standard on Stored Electrical Energy Emergency and Standby Power Systems*, 2013 edition.

NFPA 211, *Standard for Chimneys, Fireplaces, Vents, and Solid Fuel–Burning Appliances*, 2013 edition.

NFPA 259, *Standard Test Method for Potential Heat of Building Materials*, 2013 edition.

NFPA 260, *Standard Methods of Tests and Classification System for Cigarette Ignition Resistance of Components of Upholstered Furniture*, 2013 edition.

NFPA 261, *Standard Method of Test for Determining Resistance of Mock-Up Upholstered Furniture Material Assemblies to Ignition by Smoldering Cigarettes*, 2013 edition.

NFPA 286, *Standard Methods of Fire Tests for Evaluating Contribution of Wall and Ceiling Interior Finish to Room Fire Growth*, 2011 edition.

NFPA 701, *Standard Methods of Fire Tests for Flame Propagation of Textiles and Films*, 2010 edition.

NFPA 750, *Standard on Water Mist Fire Protection Systems*, 2015 edition.

NFPA 853, *Standard for the Installation of Stationary Fuel Cell Power Systems*, 2010 edition.

NFPA 1600®, *Standard on Disaster/Emergency Management and Business Continuity Programs*, 2013 edition.

NFPA 5000®, *Building Construction and Safety Code®*, 2015 edition.

2.3 Other Publications.

2.3.1 ANSI Publications. American National Standards Institute, Inc., 22 West 43rd Street, 4th Floor, New York, NY 10036.

ANSI B57.1, *Compressed Gas Cylinder Valve Outlet and Inlet Connections*, 1965.

ANSI Z136.3, *Safe Use of Optical Fiber Communication Systems Utilizing Laser Diode and LED Sources*, 2011.

ANSI/UL 723, *Standard for Test for Surface Burning Characteristics of Building Materials*, 2010.

ANSI/AAMI ES 60601-1, *Medical Electrical Equipment*, 2012.

ANSI/UL 1069, *Safety Standard for Hospital Signaling and Nurse Call Equipment*, 2012.

2.3.2 ASHRAE Publications. ASHRAE, 1791 Tullie Circle, NE, Atlanta, GA 30329-2305.

ASHRAE 90.1, *Energy Standard for Buildings Except Low-Rise Residential Buildings*, 2010.

ASHRAE 170, *Ventilation of Health Care Facilities*, 2013.

2.3.3 ASME Publications. American Society of Mechanical Engineers, Two Park Avenue, New York, NY 10016-5990.

ASME A.17.1, *Safety Code for Elevators and Escalators*, 2010.

ASME A.17.3, *Safety Code for Existing Elevators and Escalators*, 2011.

ASME B1.20.1, *Pipe Threads, General Purpose, Inch*, 2006.

ASME B16.22, *Wrought Copper and Copper Alloy Solder-Joint Pressure Fittings*, 2010.

ASME B16.26, *Cast Copper Alloy Fittings for Flared Copper Tubes*, 2011.

ANSI/ASME B16.50, *Wrought Copper and Copper Alloy Braze-Joint Pressure Fittings*, 2008.

ASME B31.3, *Pressure Process Piping*, 2010.

ASME B40.100, *Pressure Gauges and Gauge Attachments*, 2011.

ASME *Boiler and Pressure Vessel Code*, Sections VIII and IX, 2010.

ANSI/ASME PVHO-1, *Safety Standard for Pressure Vessels for Human Occupancy*, 2012.

2.3.4 ASSE Publications. American Society of Sanitary Engineering, 901 Canterbury Road, Suite A, Westlake, OH 44145-1480.

ASSE 6010, *Professional Qualification Standard for Medical Gas Systems Installers*, 2012.

ASSE 6015, *Professional Qualification Standard for Bulk Medical Gas Systems Installer*, 2012.

ASSE 6030, *Professional Qualification Standard for Medical Gas Systems Verifiers*, 2012.

ASSE 6040, *Professional Qualification Standard for Medical Gas Maintenance Personnel*, 2012.

2.3.5 ASTM Publications. ASTM International, 100 Barr Harbor Drive, P.O. Box C700, West Conshohocken, PA 19428-2959.

ASTM A 269, *Standard Specification for Seamless and Welded Austenitic Stainless Steel Tubing for General Service*, 2010.

ASTM A 312, *Standard Specification for Seamless, Welded, and Heavily Cold Worked Austenitic Stainless Steel Pipes*, 2013a.

ASTM B 32, *Standard Specification for Solder Metal*, 2008.

ASTM B 88, *Standard Specification for Seamless Copper Water Tube*, 2009.

ASTM B 280, *Standard Specification for Seamless Copper Tubing for Air Conditioning and Refrigeration Field Service*, 2008.

ASTM B 819, *Standard Specification for Seamless Copper Tube for Medical Gas Systems*, 2000 (2011).

ASTM B 828, *Standard Practice for Making Capillary Joints by Soldering of Copper and Copper Alloy Tube and Fittings*, 2002 (2010).

ASTM D 5, *Standard Test Method for Penetration of Bituminous Materials*, 2006 e1.

ASTM D 1785, *Standard Specification for Poly(Vinyl Chloride) (PVC) Plastic Pipe, Schedules 40, 80, and 120*, 2012.

ASTM D 2466, *Standard Specification for Poly(Vinyl Chloride) (PVC) Plastic Pipe Fittings, Schedule 40*, 2006.

ASTM D 2467, *Standard Specification for Poly(Vinyl Chloride) (PVC) Plastic Pipe Fittings, Schedule 80*, 2006.

ASTM D 2672, *Standard Specification for Joints for IPS PVC Pipe Using Solvent Cement*, 1996a (2009).

ASTM D 2846, *Standard Specification for Chlorinated Poly(Vinyl Chloride) (CPVC) Plastic Hot- and Cold-Water Distribution Systems*, 2009b e1.

ASTM D 2863, *Standard Test Method for Measuring the Minimum Oxygen Concentration to Support Candle-Like Combustion of Plastics (Oxygen Index)*, 2012.

ASTM D 4359, *Standard Test Method for Determining Whether a Material Is a Liquid or a Solid*, 2012.

ASTM E 84, *Standard Test Method for Surface Burning Characteristics of Building Materials*, 2012c.

ASTM E 136, *Standard Test Method for Behavior of Materials in a Vertical Tube Furnace at 750°C*, 2012.

ASTM E 1352, *Standard Test Method for Cigarette Ignition Resistance of Mock-Up Upholstered Furniture Assemblies*, 2008.

ASTM E 1353, *Standard Test Methods for Cigarette Ignition Resistance of Components of Upholstered Furniture*, 2008.

ASTM E 1537, *Standard Test Method for Fire Testing of Upholstered Furniture*, 2013.

ASTM E 1590, *Standard Test Method for the Fire Testing of Mattresses*, 2012.

ASTM E 2652, *Standard Test Method for Behavior of Materials in a Tube Furnace with a Cone-shaped Airflow Stabilizer, at 750°C*, 2012.

2.3.6 AWS Publications. American Welding Society, 550 NW LeJeune Road, Miami, FL 33126.

ANSI/AWS A5.8, *Specification for Filler Metals for Brazing and Braze Welding*, 2011.

AWS B2.2, *Standard for Brazing Procedure and Performance Qualification*, 2010.

2.3.7 BICSI Publications. BICSI, 8610 Hidden River Parkway, Tampa, FL 33637-1000.

The BICSI Information Transport Systems (ITS) Dictionary, 3rd edition.

2.3.8 CDA Publications. Copper Development Association Inc., 260 Madison Avenue, New York, NY 10016.

Copper Tube Handbook, 2010.

2.3.9 CGA Publications. Compressed Gas Association, 14501 George Carter Way, Suite 103, Chantilly, VA 20151-2923.

CGA C-4, *Standard Method of Marking Portable Compressed Gas Containers to Identify the Material Contained*, 1954.

CGA C-7, *Guide to the Preparation of Precautionary Labeling and Marking of Compressed Gas Containers*, 2011.

CGA G-4, *Oxygen*, 2008.

CGA G-4.1, *Cleaning Equipment for Oxygen Service*, 2009.

CGA G-6.1, *Standard for Insulated Carbon Dioxide Systems at Consumer Sites*, 2005.

CGA G-6.5, *Standard for Small, Stationary, Insulated Carbon Dioxide Supply Systems*, 2007.

CGA G-8.1, *Standard for Nitrous Oxide Systems at Consumer Sites*, 2007.

CGA M-1, *Guide for Medical Gas Installations at Consumer Sites*, 2007.

CGA O2-DIR, *Directory of Cleaning Agents for Oxygen Service*, Edition 4.

CGA P-2.5, *Transfilling of High Pressure Gaseous Oxygen to Be Used for Respiration*, 2011.

CGA P-2.6, *Transfilling of Liquid Oxygen to Be Used for Respiration*, 2011.

CGA P-18, *Standard for Bulk Inert Gas Systems at Consumer Sites*, 2006.

CGA V-1, *Compressed Gas Association Standard for Compressed Gas Cylinder Valve Outlet and Inlet Connections* (ANSI B57.1), 2005.

CGA V-5, *Diameter-Index Safety System (Noninterchangeable Low Pressure Connections for Medical Gas Applications)*, 2008.

2.3.10 CSA Publications. Canadian Standards Association, 5060 Spectrum Way, Mississauga, ON, L4W 5N6, Canada.

CSA C22.2 No. 0.3, *Test Methods for Electrical Wires and Cables*, 2009.

2.3.11 FGI Publications. Facility Guidelines Institute, 1919 McKinney Avenue, Dallas, TX 75201.

Guidelines for Design and Construction of Hospitals and Outpatient Facilities, 2014.

2.3.12 IEC Publications. International Electrotechnical Commission, 3, rue de Varembé, P.O. Box 131, CH-1211 Geneva 20, Switzerland.

IEC 60601-1, *Medical Electrical Equipment—Part 1: General Requirements for Basic Safety and Essential Performance*, 2007.

2.3.13 ISA Publications. Instrumentation, Systems, and Automation Society (ISA), 67 Alexander Drive, Research Triangle Park, NC 27709.

ANSI/ISA S-7.0.01, *Quality Standard for Instrument Air*, 1996.

2.3.14 MSS Publications. Manufacturer's Standardization Society of the Valve and Fittings Industry, Inc., 127 Park Street NE, Vienna, VA 22180.

SP-58, *Pipe Hangers and Supports — Materials, Design, Manufacture, Selection, Application and Installation*, 2009.

2.3.15 TC Publications. Transport Canada, 330 Sparks Street, Ottawa, ON, K1A/ON5, Canada.

Transportation of Dangerous Goods Regulations.

2.3.16 TIA Publications. Telecommunications Industry Association, 2500 Wilson Boulevard, Suite 300, Arlington, VA 22201.

TIA/EIA 568-B, *Commercial Building Telecommunications Cabling Standard*, 2012.

TIA/EIA 606-A, *Administration Standard for Commercial Telecommunications Infrastructure*, 2009.

2.3.17 UL Publications. Underwriters Laboratories Inc., 333 Pfingsten Road, Northbrook, IL 60062-2096.

UL 723, *Standard for Test for Surface Burning Characteristics of Building Materials*, 2008, Revised 2010.

UL 1685, *Standard for Vertical-Tray Fire-Propagation and Smoke-Release Test for Electrical and Optical-Fiber Cables*, 2007, Revised 2010.

2.3.18 U.S. Government Publications. Document Automation and Production Service (DAPS), Building 4D, 700 Robbins Avenue, Philadelphia, PA 19111-5094, www.dodssp.daps.mil.

21 USC 9, United States Food, Drug, and Cosmetic Act.

U.S. Government Commercial Standard 223-59, *Casters, Wheels, and Glides for Hospital Equipment*.

16 CFR 1632, *Standard for the Flammability of Mattresses and Mattress Pads (FF 4-72)*, 2000.

16 CFR Part 1633, *Standard for the Flammability (Open Flame) of Mattress Sets*, 2000.

2.3.19 Other Publications.

Merriam-Webster's Collegiate Dictionary, 11th edition, Merriam-Webster, Inc., Springfield, MA, 2003.

California Technical Bulletin 117, *Requirements, Test Procedure and Apparatus for Testing the Flame Retardance of Resilient Filling Materials Used in Upholstered Furniture*, 2000.

California Technical Bulletin 129, *Flammability Test Procedure for Mattresses for Use in Public Buildings*, 1992.

California Technical Bulletin 133, *Flammability Test Procedure for Seating Furniture for Use in Public Occupancies*, State of California, Department of Consumer Affairs, 3485 Orange Grove Avenue, North Highlands, CA 95660-5595.

2.4 References for Extracts in Mandatory Sections.

NFPA 13, *Standard for the Installation of Sprinkler Systems*, 2013 edition.

NFPA 30, *Flammable and Combustible Liquids Code*, 2015 edition.

NFPA 55, *Compressed Gases and Cryogenic Fluids Code*, 2013 edition.
NFPA 70®, *National Electrical Code®*, 2014 edition.
NFPA 99B, *Standard for Hypobaric Facilities*, 2015 edition.
NFPA 101®, *Life Safety Code®*, 2015 edition.
NFPA 110, *Standard for Emergency and Standby Power Systems*, 2013 edition.
NFPA 1670, *Standard on Operations and Training for Technical Search and Rescue Incidents*, 2014 edition.
NFPA 5000®, *Building Construction and Safety Code®*, 2015 edition.

Chapter 3   Definitions

3.1 General. The definitions contained in this chapter shall apply to the terms used in this code. Where terms are not defined in this chapter or within another chapter, they shall be defined using their ordinarily accepted meanings within the context in which they are used. *Merriam-Webster's Collegiate Dictionary*, 11th edition, shall be the source for the ordinarily accepted meaning.

3.2 NFPA Official Definitions.

3.2.1* Approved. Acceptable to the authority having jurisdiction.

3.2.2* Authority Having Jurisdiction (AHJ). An organization, office, or individual responsible for enforcing the requirements of a code or standard, or for approving equipment, materials, an installation, or a procedure.

3.2.3* Code. A standard that is an extensive compilation of provisions covering broad subject matter or that is suitable for adoption into law independently of other codes and standards.

3.2.4 Guide. A document that is advisory or informative in nature and that contains only nonmandatory provisions. A guide may contain mandatory statements such as when a guide can be used, but the document as a whole is not suitable for adoption into law.

3.2.5 Labeled. Equipment or materials to which has been attached a label, symbol, or other identifying mark of an organization that is acceptable to the authority having jurisdiction and concerned with product evaluation, that maintains periodic inspection of production of labeled equipment or materials, and by whose labeling the manufacturer indicates compliance with appropriate standards or performance in a specified manner.

3.2.6* Listed. Equipment, materials, or services included in a list published by an organization that is acceptable to the authority having jurisdiction and concerned with evaluation of products or services, that maintains periodic inspection of production of listed equipment or materials or periodic evaluation of services, and whose listing states that either the equipment, material, or service meets appropriate designated standards or has been tested and found suitable for a specified purpose.

3.2.7 Shall. Indicates a mandatory requirement.

3.2.8 Should. Indicates a recommendation or that which is advised but not required.

3.2.9 Standard. A document, the main text of which contains only mandatory provisions using the word "shall" to indicate requirements and which is in a form generally suitable for mandatory reference by another standard or code or for adoption into law. Nonmandatory provisions are not to be considered a part of the requirements of a standard and shall be located in an appendix, annex, footnote, informational note, or other means as permitted in the *Manual of Style for NFPA Technical Committee Documents*.

3.3 General Definitions.

3.3.1 Adiabatic Heating. The heating of a gas caused by its compression. (HYP)

3.3.2 Aerosol. An intimate mixture of a liquid or a solid in a gas; the liquid or solid, called the dispersed phase, is uniformly distributed in a finely divided state throughout the gas, which is the continuous phase or dispersing medium. (MED)

3.3.3 Alarm System.

3.3.3.1 *Area Alarm System.* A warning system within an area of use that provides continuous visible and audible surveillance of Category 1 and Category 2 medical gas and vacuum systems. (PIP)

3.3.3.2 *Category 3 Alarm System.* A warning system within an area of use that provides continuous visible and audible surveillance of Category 3 medical gas systems. (PIP)

3.3.3.3 *Local Alarm System.* A warning system that provides continuous visible and audible surveillance of medical gas and vacuum system source equipment at the equipment site. (PIP)

3.3.3.4 *Master Alarm System.* A warning system that monitors the operation and condition of the source of supply, the reserve source (if any), and the pressure in the main lines of each medical gas and vacuum piping system. (PIP)

3.3.4 Alternate Power Source. One or more generator sets, or battery systems where permitted, intended to provide power during the interruption of the normal electrical service; or the public utility electrical service intended to provide power during interruption of service normally provided by the generating facilities on the premises. (ELS)

3.3.5 Ambulatory Health Care Occupancy. An occupancy used to provide services or treatment simultaneously to four or more patients that provides, on an outpatient basis, one or more of the following: (1) treatment for patients that renders the patients incapable of taking action for self-preservation under emergency conditions without the assistance of others; (2) anesthesia that renders the patients incapable of taking action for self-preservation under emergency conditions without the assistance of others; (3) emergency or urgent care for patients who, due to the nature of their injury or illness, are incapable of taking action for self-preservation under emergency conditions without the assistance of others. [*101*, 2015] (FUN)

3.3.6 Ampacity. The maximum current, in amperes, that a conductor can carry continuously under the conditions of use without exceeding its temperature rating. [**70**, 2014] (ELS)

3.3.7 Anesthetic. As used in this code, applies to any inhalational agent used to produce sedation, analgesia, or general anesthesia. (MED)

3.3.8 Appliance. Utilization equipment, generally other than industrial, normally built in standardized sizes or types, that is installed or connected as a unit to perform one or more functions. (MED)

3.3.9 Applicable Code. The building code, fire code, or NFPA *101, Life Safety Code,* adopted by the jurisdiction, or NFPA *101* where no such code has been adopted by the jurisdiction.

3.3.10* Applicator. A means of applying high-frequency energy to a patient other than by an electrically conductive connection. (MED)

3.3.11 Area of Administration. Any point within a room within 4.3 m (15 ft) of oxygen equipment or an enclosure containing or intended to contain an oxygen-enriched atmosphere. (MED)

3.3.12* Atmosphere. The pressure exerted by, and gaseous composition of, an environment. (HYP)

3.3.12.1 *Atmosphere Absolute (ATA).* The pressure of the earth's atmosphere, 760.0 mmHg, 101.325 kPa, or 14.7 psia. Two ATA = two atmospheres. *(See also 3.3.12, Atmosphere.)* (HYP)

3.3.12.2* *Atmosphere of Increased Burning Rate.* Any atmosphere containing a percentage of oxygen or oxygen and nitrous oxide greater than the quotient of 23.45 divided by the square root of the total pressure in atmospheres. (HYP)

3.3.12.3 *Chamber Atmosphere.* The environment inside a chamber. (HYP)

3.3.13 Automatic. Providing a function without the necessity of human intervention. (ELS)

3.3.14 Bathrooms. An area including a basin with one or more of the following: a toilet, a urinal, a tub, a shower, a bidet, or similar plumbing fixtures. [**70**, 2014] (FUN)

3.3.15 Battery-Powered Lighting Units. Individual unit equipment for backup illumination consisting of a rechargeable battery, battery-charging means, provisions for one or more lamps mounted on the equipment, or with terminals for remote lamps, or both, and relaying device arranged to energize the lamps automatically upon failure of the supply to the unit equipment. [**70**, 2014] (ELS)

3.3.16 Bends. Decompression sickness; caisson worker's disease. (HYP)

3.3.17 Branch Circuit. The circuit conductors between the final overcurrent device protecting the circuit and the outlet(s). [**70**, 2014] (ELS)

3.3.18 Branch Line. See 3.3.133, Piping.

3.3.19 Bulk System. An assembly of equipment for supplying compressed gas (consisting of, but not limited to, storage containers, pressure regulators, pressure relief devices, vaporizers, manifolds, and interconnecting piping) that terminates where the gas, at service pressure, first enters the main line. The storage containers are either stationary or movable and include unconnected reserves on hand at the site, and the source gas is stored as a compressed gas or cryogenic fluid.

3.3.19.1 *Bulk Inert Gas System.* A bulk system with a storage capacity of more than 566 m^3 [20,000 ft^3 (scf)] of inert gas.

3.3.19.2 *Bulk Nitrous Oxide System.* A bulk system with a storage capacity of more than 1452 kg (3200 lb) [approximately 793 m^3 (28,000 ft^3) at normal temperature and pressure] of nitrous oxide (PIP).

3.3.19.3* *Bulk Oxygen System.* A bulk system with a storage capacity of more than 566 m^3 (20,000 ft^3) at normal temperature and pressure of oxygen.

3.3.19.4 *Micro-Bulk Cryogenic System.* An assembly of equipment including a container that is permanently installed through anchoring to a foundation, pressure regulators, pressure relief devices, vaporizers, manifolds, and interconnecting piping that is designed to be filled at the health care facility with a cryogenic gas, that has a storage capacity of less than or equal to 566 m^3 [20,000 ft^3 (scf)] of oxygen, including unconnected reserves on hand at the site, and that terminates at the source valve.

3.3.20 Category 3 Drive Gas System. An assembly of component parts including, but not limited to, the source, pressure and operating controls, filters and purification equipment, valves, alarm warning systems, alarm wiring, gauges, and a network of piping and suitable outlets that produces and distributes compressed air from cylinders, compressed air from compressors, or nitrogen from cylinders less than 1100 kPa gauge (less than 160 psi gauge) to power devices (hand pieces, syringes, cleaning devices, delivery system chairs, and so forth) as a power source. The system includes the compressor intakes and ends with the service outlet where the user connects their clinical equipment. (PIP)

3.3.21 Category 3 Vacuum System. A Category 3 vacuum distribution system that can be either a wet system designed to remove liquids, air–gas, or solids from the treated area; or a dry system designed to trap liquid and solids before the service inlet and to accommodate air–gas only through the service inlet. (PIP)

3.3.22 Combustible. Capable of undergoing combustion. (MED)

3.3.23* Combustible Liquid. Any liquid that was a closed-cup flash point at or above 37.8°C (100°F). Combustible liquids are classified as follows: (a) Class II liquid. Any liquid that has a flash point at or above 37.8°C (100°F) and below 60°C (140°F); (b) Class IIIA liquid. Any liquid that has a flash point at or above 60°C (140°F) and below 93°C (200°F); (c) Class IIIB liquid. Any liquid that has a flash point at or above 93°C (200°F).

3.3.24* Combustion. A chemical process of oxidation that occurs at a rate fast enough to produce heat and usually light in the form of either a glow or flame. [**5000**, 2015] (HYP)

3.3.25 Compact Storage. Storage on solid shelves not exceeding 0.9 m (36 in.) in total depth, arranged as part of a compact storage module, with no more than 0.76 m (30 in.) between shelves vertically and with no internal vertical flue spaces other than those between individual shelving sections. [**13**, 2013] (FUN)

3.3.26 Container. A low-pressure, vacuum-insulated vessel containing gases in liquid form. (MED)

3.3.26.1 *Liquid Oxygen Ambulatory Container.* A container used for liquid oxygen not exceeding 1.5 L (0.396 gal) specifically designed for use as a medical device as defined by 21 USC Chapter 9, the United States Food, Drug and Cosmetic Act, that is intended for portable therapeutic use and to be filled from its companion base unit, which is a liquid oxygen home care container. (MED)

3.3.26.2 *Liquid Oxygen Base Reservoir Container.* A container used for liquid oxygen not exceeding 60 L (15.8 gal) specifically designed for use as a medical device as defined by 21 USC Chapter 9, the United States Food, Drug and

Cosmetic Act, that is intended to deliver gaseous oxygen for therapeutic use, transfilling, or both. (MED)

3.3.26.3 *Liquid Oxygen Portable Container.* A container used for liquid oxygen not exceeding 1.5 L (0.396 gal) specifically designed for use as a medical device as defined by 21 USC Chapter 9, the United States Food, Drug and Cosmetic Act, that is intended for portable therapeutic use and to be filled from its companion base unit, which is a liquid oxygen base reservoir container. (MED)

3.3.27 Critical Branch. A system of feeders and branch circuits supplying power for task illumination, fixed equipment, select receptacles, and select power circuits serving areas and functions related to patient care that are automatically connected to alternate power sources by one or more transfer switches during interruption of the normal power source. (ELS)

3.3.28 Critical Care Area. A room or space in which failure of equipment or a system is likely to cause major injury or death to patients or caregivers (Category 1). *(See 3.3.127.)*

3.3.29 Critical Equipment. That equipment essential to the safety of the occupants of the facility. (HYP)

3.3.30 Cylinder. A supply tank containing high-pressure gases or gas mixtures at pressures that can be in excess of 13.8 kPa gauge (2000 psi gauge). (MED)

3.3.31 Decompression Sickness. A syndrome due to evolved gas in the tissues resulting from a reduction in ambient pressure. (HYP)

3.3.32* Defend in Place. The operational response to an emergency in a building, in which the initial action does not involve evacuation of the building occupants. (FUN)

3.3.33 Demand Check. A paired set of fittings that permit gas flow when correctly mated but interrupt flow when separated. (PIP)

3.3.34 Detonation. An exothermic reaction wherein the reaction propagates through the unreacted material at a rate exceeding the velocity of sound, hence the explosive noise. (MED)

3.3.35* Direct Electrical Pathway to the Heart. An externalized conductive pathway, insulated except at its ends, one end of which is in direct contact with heart muscle while the other is outside the body and is accessible for inadvertent or intentional contact with grounded objects or energized, ground-referenced sources. (MED)

3.3.36* Disaster. Within the context of this code, any unusual occurrence or unforeseen situation that seriously overtaxes or threatens to seriously overtax the routine capabilities of a health care facility. (HES)

3.3.37 D.I.S.S. Connector. A system of noninterchangeable medical gas and vacuum connectors complying with CGA V-5, *Diameter-Index Safety System (Noninterchangeable Low Pressure Connections for Medical Gas Applications)*. (PIP)

3.3.38* Double-Insulated Appliances. Appliances where the primary means of protection against electrical shock is not grounding. The primary means is by the use of combinations of insulation and separation spacings in accordance with an approved standard. (MED)

3.3.39 Electrical Life Support Equipment. Electrically powered equipment whose continuous operation is necessary to maintain a patient's life. (ELS)

3.3.40 Electrode. An electrically conductive connection to a patient. (MED)

3.3.41 Emergency Management. The act of developing procedures and plans to create effective preparedness, mitigation, response, and recovery during a disaster affecting a health care facility. (HES)

3.3.42 Emergency Oxygen Supply Connection. An assembly of equipment that permits a gas supplier to make a temporary connection to supply oxygen to a building that has had its normal source of oxygen interrupted. (PIP)

3.3.43 Equipment Branch. A system of feeders and branch circuits arranged for delayed, automatic, or manual connection to the alternate power source and that serves primarily 3-phase power equipment. (ELS)

3.3.44 Equipment Grounding Bus. A grounding terminal bus in the feeder circuit of the branch circuit distribution panel that serves a particular area. (MED)

3.3.45* Essential Electrical System. A system comprised of alternate sources of power and all connected distribution systems and ancillary equipment, designed to ensure continuity of electrical power to designated areas and functions of a health care facility during disruption of normal power sources, and also to minimize disruption within the internal wiring system. (ELS)

3.3.46 Evacuation — Waste Gas. See 3.3.169, Waste Anesthetic Gas Disposal.

3.3.47 Exposed Conductive Surfaces. Those surfaces that are capable of carrying electric current and that are unprotected, uninsulated, unenclosed, or unguarded, permitting personal contact. (ELS)

3.3.48* Facility Fire Plan. A plan developed by the health care facility to describe the actions to be taken during a fire emergency.

3.3.49* Failure. An incident that increases the hazard to personnel or patients or that affects the safe functioning of electric appliances or devices. (MED)

3.3.50 Fault Current. A current in an accidental connection between an energized and a grounded or other conductive element resulting from a failure of insulation, spacing, or containment of conductors. (ELS)

3.3.51 Feeder. All circuit conductors between the service equipment, the source of a separately derived system, or other power supply source and the final branch-circuit overcurrent device. (ELS)

3.3.52* Flammable. A combustible that is capable of easily being ignited and rapidly consumed by fire.

3.3.53 Flammable Gas. Any substance that exists in the gaseous state at normal atmospheric temperature and pressure and is capable of being ignited and burned when mixed with proper proportion of air, oxygen, or other oxidizers. (HYP)

3.3.54 Flammable Liquid. A liquid that has a closed-cup flash point that is below 37.8°C (100°F) and a maximum vapor pressure of 2068 mmHg (40 psi absolute) at 37.8°C (100°F).

3.3.55* Flash Point. The minimum temperature at which a liquid or a solid emits vapor sufficient to form an ignitable mixture with air near the surface of the liquid or the solid. (FUN)

3.3.56 Flow-Control Valve. A valve, usually a needle valve, that precisely controls flow of gas. (MED)

3.3.57* Flowmeter. A device for measuring volumetric flow rates of gases and liquids. (MED)

3.3.58* Frequency. The number of oscillations, per unit time, of a particular current or voltage waveform. The unit of frequency is the hertz. (MED)

3.3.59* Fume Hood. An enclosure designed to draw air inward by means of mechanical ventilation.

3.3.60 Gas-Powered System. A Level 3 gas distribution system comprised of component parts including but not limited to cylinders, manifolds, air compressor, motor, receivers, controls, filters, dryers, valves, and piping that delivers compressed air or nitrogen at pressures less than 1100 kPa (less than 160 psi) gauge to power devices (e.g., hand pieces, syringes, cleaning devices) as a power source. (PIP)

3.3.61* General Anesthesia and Levels of Sedation/Analgesia.

3.3.61.1 *General Anesthesia.* A drug-induced loss of consciousness during which patients are not arousable, even by painful stimulation. The ability to independently maintain ventilatory function is often impaired. Patients often require assistance in maintaining a patent airway, and positive pressure ventilation may be required because of depressed spontaneous ventilation or drug-induced depression of neuromuscular function. Cardiovascular function may be impaired. (MED)

3.3.61.2 *Deep Sedation/Analgesia.* A drug-induced depression of consciousness during which patients cannot be easily aroused but respond purposefully following repeated or painful stimulation. The ability to independently maintain ventilatory function may be impaired. Patients may require assistance in maintaining a patent airway, and spontaneous ventilation may be inadequate. Cardiovascular function is usually maintained. (MED)

3.3.61.3 *Moderate Sedation/Analgesia (Conscious Sedation).* A drug-induced depression of consciousness during which patients respond purposefully to verbal commands, either alone or accompanied by light tactile stimulation. No interventions are required to maintain a patient airway, and spontaneous ventilation is adequate. Cardiovascular function is usually maintained. (MED)

3.3.61.4 *Minimal Sedation (Anxiolysis).* A drug-induced state during which patients respond normally to verbal commands. Although cognitive function and coordination may be impaired, ventilatory and cardiovascular functions are unaffected. (MED)

3.3.62 Governing Body. The person or persons who have the overall legal responsibility for the operation of a health care facility. (FUN)

3.3.63* Ground-Fault Circuit Interrupter (GFCI). A device intended for the protection of personnel that functions to de-energize a circuit or portion thereof within an established period of time when a current to ground exceeds the values established for a Class A device. [**70**, 2014] (ELS)

3.3.64 Grounding. See 3.3.65, Grounding System.

3.3.65* Grounding System. A system of conductors that provides a low-impedance return path for leakage and fault currents. (ELS)

3.3.66 Hazard Current. For a given set of connections in an isolated power system, the total current that would flow through a low impedance if it were connected between either isolated conductor and ground. (ELS)

3.3.66.1 *Fault Hazard Current.* The hazard current of a given isolated power system with all devices connected except the line isolation monitor. (ELS)

3.3.66.2 *Monitor Hazard Current.* The hazard current of the line isolation monitor alone. (ELS)

3.3.66.3 *Total Hazard Current.* The hazard current of a given isolated system with all devices, including the line isolation monitor, connected. (ELS)

3.3.67* Health Care Facilities. Buildings, portions of buildings, or mobile enclosures in which human medical, dental, psychiatric, nursing, obstetrical, or surgical care is provided. (FUN)

3.3.68 Home Care. Medical services (equipment) provided in residential occupancies. (FUN)

3.3.69 Hospital. A building or portion thereof used on a 24-hour basis for the medical, psychiatric, obstetrical, or surgical care of four or more inpatients. [**101**, 2015] (FUN)

3.3.70 Humidifier. A device used for adding water vapor to inspired gas. (MED)

3.3.71 Hyperbaric Facility. Building, structure, or space used to house hyperbaric chambers and auxiliary service equipment for medical applications and procedures at pressures above normal atmospheric pressure. (HYP)

3.3.72 Hyperbaric Stand-Alone Oxygen System. The oxygen system is entirely separate from the hospital's Category 1 Oxygen System or is a freestanding hyperbaric facility. (HYP)

3.3.73 Hypobaric Facility. Building, structure, or space used to house hypobaric chambers and auxiliary service equipment for medical applications and procedures at pressures below atmospheric pressure. (HYP)

3.3.74 Hypoxia. A state of inadequate oxygenation of the blood and tissue sufficient to cause impairment of function. [**99B**, 2015] (HYP)

3.3.75 Immediate Restoration of Service. Automatic restoration of operation with an interruption of not more than 10 seconds. (ELS)

3.3.76* Impedance. Impedance is the ratio of the voltage drop across a circuit element to the current flowing through the same circuit element. The unit of impedance is the ohm. (MED)

3.3.77 Incident Command System (ICS). The combination of facilities, equipment, personnel, procedures, and communications operating within a common organizational structure that has responsibility for the management of assigned resources to effectively accomplish stated objectives pertaining to an incident or training exercise. [**1670**, 2014] (HES)

3.3.78* Instrument Air. A medical support gas that falls under the general requirements for medical gases. Medical air and instrument air are distinct systems for mutually exclusive applications. (PIP)

3.3.79 Intermittent Positive-Pressure Breathing (IPPB). Ventilation of the lungs by application of intermittent positive pressure to the airway. (MED)

3.3.80* Intrinsically Safe. As applied to equipment and wiring, equipment and wiring that are incapable of releasing sufficient electrical energy under normal or abnormal conditions to cause ignition of a specific hazardous atmospheric mixture. (HYP)

3.3.81 Invasive Procedure. Any procedure that penetrates the protective surfaces of a patient's body (i.e., skin, mucous membrane, cornea) and that is performed with an aseptic field (procedural site). [Not included in this category are placement of peripheral intravenous needles or catheters used to administer fluids and/or medications, gastrointestinal endoscopies (i.e., sigmoidoscopies), insertion of urethral catheters, and other similar procedures.] (ELS)

3.3.82 Isolated Patient Lead. A patient lead whose impedance to ground or to a power line is sufficiently high that connecting the lead to ground, or to either conductor of the power line, results in current flow below a hazardous limit in the lead. (MED)

3.3.83* Isolated Power System. A system comprising an isolation transformer or its equivalent, a line isolation monitor, and its ungrounded circuit conductors. (ELS)

3.3.84 Isolation Transformer. A transformer of the multiple-winding type, with the primary and secondary windings physically separated, that inductively couples its ungrounded secondary winding to the grounded feeder system that energizes its primary winding. (ELS)

3.3.85* Laboratory. A building, space, room, or group of rooms intended to serve activities involving procedures for investigation, diagnosis, or treatment in which flammable, combustible, or oxidizing materials are to be used.

3.3.86 Leak Detectant. For purposes of this standard, a reagent, a solution, or an electronic or mechanical device suitable for the detection or visualization of escaping gas. (PIP)

3.3.87 Life Safety Branch. A system of feeders and branch circuits supplying power for lighting, receptacles, and equipment essential for life safety that is automatically connected to alternate power sources by one or more transfer switches during interruption of the normal power source. (ELS)

3.3.88* Limited-Combustible (Material). See 4.4.2.

3.3.89 Line Isolation Monitor. A test instrument designed to continually check the balanced and unbalanced impedance from each line of an isolated circuit to ground and equipped with a built-in test circuit to exercise the alarm without adding to the leakage current hazard. (ELS)

3.3.90* Liquid. Any material that (1) has a fluidity greater than that of 300 penetration asphalt when tested in accordance with ASTM D 5, *Standard Test Method for Penetration of Bituminous Materials*, or (2) is a viscous substance for which a specific melting point cannot be determined but that is determined to be a liquid in accordance with ASTM D 4359, *Standard Test for Determining Whether a Material is a Liquid or a Solid*. [30, 2015] (LAB)

3.3.91* Local Signal. A visible indication of the operating status of equipment. (PIP)

3.3.92 mA. Milliampere.

3.3.93 Manifold. A device for connecting the outlets of one or more gas cylinders to the central piping system for that specific gas. (PIP)

3.3.94* Manufactured Assembly. A factory-assembled product designed for aesthetics or convenience that contains medical gas or vacuum outlets, piping, or other devices related to medical gases. (PIP)

3.3.95 Mask. A device that fits over the mouth and nose (oronasal) or nose (nasal) used to administer gases to a patient. (MED)

3.3.96* Medical Air. For purposes of this code, medical air is air supplied from cylinders, bulk containers, or medical air compressors or reconstituted from oxygen USP and oil-free, dry nitrogen NF. (PIP)

> **3.3.96.1 *Proportioning System for Medical Air USP.*** A central supply that produces medical air (USP) reconstituted from oxygen USP and nitrogen NF by means of a mixer or blender. (PIP)

3.3.97 Medical Air Compressor. A compressor that is designed to exclude oil from the air stream and compression chamber and that does not under normal operating conditions or any single fault add any toxic or flammable contaminants to the compressed air. (PIP)

3.3.98* Medical/Dental Office. A building or part thereof in which the following occur: (1) examinations and minor treatments/procedures are performed under the continuous supervision of a medical/dental professional; (2) only sedation or local anesthesia is involved and treatment or procedures do not render the patient incapable of self-preservation under emergency conditions; and (3) overnight stays for patients or 24-hour operation are not provided. (FUN)

3.3.99 Medical Gas. A patient medical gas or medical support gas. *(See also 3.3.131, Patient Medical Gas and 3.3.101, Medical Support Gas.)* (PIP)

3.3.100 Medical Gas System. An assembly of equipment and piping for the distribution of nonflammable medical gases such as oxygen, nitrous oxide, compressed air, carbon dioxide, and helium. (PIP)

3.3.101 Medical Support Gas. Nitrogen or instrument air used for any medical support purpose (e.g., to remove excess moisture from instruments before further processing, or to operate medical–surgical tools, air-driven booms, pendants, or similar applications) and, if appropriate to the procedures, used in laboratories and are not respired as part of any treatment. Medical support gas falls under the general requirements for medical gases. (PIP)

3.3.102 Medical–Surgical Vacuum. A method used to provide a source of drainage, aspiration, and suction in order to remove body fluids from patients. (PIP)

3.3.103 Medical–Surgical Vacuum System. An assembly of central vacuum–producing equipment and a network of piping for patient suction in medical, medical–surgical, and waste anesthetic gas disposal (WAGD) applications. (PIP)

3.3.104 Multiple Treatment Facility. A diagnostic or treatment complex under a single management comprising a number of single treatment facilities, which can be accessed one from the other without exiting the facility (i.e., does not involve widely separated locations or separate distinct practices). (FUN)

3.3.105 mV. Millivolt.

3.3.106 Nasal Cannula. Device consisting of two short tubes to be inserted into the nostrils to administer oxygen or other therapeutic gases. (MED)

3.3.107 Nebulizer. A device used for producing an aerosol of water and/or medication within inspired gas supply. (MED)

3.3.108 Negative Pressure. Pressure less than atmospheric. (MED)

3.3.109 Nitrogen. An element that, at atmospheric temperatures and pressures, exists as a clear, colorless, and tasteless gas; it comprises approximately four-fifths of the earth's atmosphere. (MED)

3.3.109.1 *Nitrogen NF.* Nitrogen complying as a minimum with nitrogen NF. (PIP)

3.3.110 Nitrogen Narcosis. A condition resembling alcoholic inebriation, which results from breathing nitrogen in the air under significant pressure. (HYP)

3.3.111 Nitrous Oxide. An inorganic compound, one of the oxides of nitrogen. It exists as a gas at atmospheric pressure and temperature, possesses a sweetish smell, and is used for inducing anesthesia when inhaled. The oxygen in the compound will be released under conditions of combustion, creating an oxygen-enriched atmosphere. (MED)

3.3.112 Noncombustible (Material). See 4.4.1.

3.3.113 Nonflammable. Not readily capable of burning with a flame and not liable to ignite and burn when exposed to flame.

3.3.114 Nonflammable Anesthetic Agent. Refers to those inhalation agents that, because of their vapor pressure at 37°C (98.6°F) and at atmospheric pressure, cannot attain flammable concentrations when mixed with air, oxygen, or mixtures of oxygen and nitrous oxide. (MED)

3.3.115* Nonflammable Medical Gas System. See 3.3.100, Medical Gas System, and Chapter 5.

3.3.116 Nonmedical Compressed Air. Air that is used for purposes other than patient care or medical devices that provide direct patient care. (MEC)

3.3.117 Nursing Home. A building or portion of a building used on a 24-hour basis for the housing and nursing care of four or more persons who, because of mental or physical incapacity, might be unable to provide for their own needs and safety without the assistance of another person. [*101*, 2015] (FUN)

3.3.118* Oxidizing Gas. A gas that supports combustion. (HYP)

3.3.119* Oxygen. A chemical element that, at normal atmospheric temperatures and pressures, exists as a colorless, odorless, and tasteless gas and comprises about 21 percent by volume of the earth's atmosphere. (MED)

3.3.119.1 *Gaseous Oxygen.* A colorless, odorless, tasteless, and nontoxic gas, comprising about 21 percent of normal air by volume, that is about 10 percent heavier than air; also the physical state of the element at atmospheric temperature and pressure. (MED)

3.3.119.2* *Liquid Oxygen.* Exists at cryogenic temperature, approximately −184.4°C (−300°F) at atmospheric pressure. It retains all of the properties of gaseous oxygen, but, in addition, when allowed to warm to room temperature at atmospheric pressure, it will evaporate and expand to fill a volume 860 times its liquid volume. (MED)

3.3.120* Oxygen Delivery Equipment. Any device used to transport and deliver an oxygen-enriched atmosphere to a patient. (MED)

3.3.121 Oxygen-Enriched Atmosphere (OEA). For the purposes of this code, an atmosphere in which the concentration of oxygen exceeds 23.5 percent by volume. (HYP)

3.3.122* Oxygen Hood. A device encapsulating a patient's head and used for a purpose similar to that of a mask. *(See also 3.3.95, Mask.)* (HYP)

3.3.123* Oxygen Toxicity (Hyperbaric). Physical impairment resulting from breathing gaseous mixtures containing oxygen-enriched atmospheres at elevated partial pressures for extended periods of time. (HYP)

3.3.124 Oxygen USP. Oxygen complying with Medical USP.

3.3.125 Patient Bed Location. The location of a patient sleeping bed, or the bed or procedure table of a critical care space. (ELS)

3.3.126 Patient-Care-Related Electrical Equipment. Electrical equipment appliance that is intended to be used for diagnostic, therapeutic, or monitoring purposes in a patient care vicinity. (MED)

3.3.127* Patient Care Space. Any space of a health care facility wherein patients are intended to be examined or treated. (FUN)

3.3.127.1* *Category 1 Space.* Space in which failure of equipment or a system is likely to cause major injury or death of patients, staff, or visitors. (FUN)

3.3.127.2* *Category 2 Space.* Space in which failure of equipment or a system is likely to cause minor injury to patients, staff, or visitors. (FUN)

3.3.127.3* *Category 3 Space.* Space in which the failure of equipment or a system is not likely to cause injury to patients, staff, or visitors but can cause discomfort. (FUN)

3.3.127.4* *Category 4 Space.* Space in which failure of equipment or a system is not likely to have a physical impact on patient care. (FUN)

3.3.128 Patient Care Vicinity. A space, within a location intended for the examination and treatment of patients, extending 1.8 m (6 ft) beyond the normal location of the bed, chair, table, treadmill, or other device that supports the patient during examination and treatment and extending vertically to 2.3 m (7 ft 6 in.) above the floor. (MED)

3.3.129 Patient Equipment Grounding Point. A jack or terminal that serves as the collection point for redundant grounding of electric appliances serving a patient care vicinity or for grounding other items in order to eliminate electromagnetic interference problems. (MED)

3.3.130* Patient Lead. Any deliberate electrical connection that can carry current between an appliance and a patient. (MED)

3.3.131 Patient Medical Gas. Piped gases such as oxygen, nitrous oxide, helium, carbon dioxide, and medical air that are used in the application of human respiration and the calibration of medical devices used for human respiration. (PIP)

3.3.132 Piped Distribution System. A pipeline network assembly of equipment that starts at and includes the source valve, warning systems (master, area, local alarms), bulk gas system signal actuating switch wiring, interconnecting piping, and all other components up to and including the station outlets/inlets. (PIP)

3.3.133 Piping. The tubing or conduit of the system. The three general classes of piping are main lines, risers, and branch (lateral) lines. (PIP)

3.3.133.1 *Branch (Lateral) Lines.* Those sections or portions of the piping system that serve a room or group of rooms on the same story of the facility. (PIP)

3.3.133.2 *Main Lines.* The piping that connects the source (pumps, receivers, etc.) to the risers or branches, or both. (PIP)

3.3.133.3 *Risers.* The vertical pipes connecting the system main line(s) with the branch lines on the various levels of the facility. (PIP)

3.3.134 Plug (Attachment Plug, Cap). A device that, by insertion in a receptacle, establishes connection between the conductors of the attached flexible cord and the conductors connected permanently to the receptacle. (MED)

3.3.135 Pressure.

3.3.135.1 *Absolute Pressure.* The total pressure in a system with reference to zero pressure. (HYP)

3.3.135.2 *Ambient Pressure.* Refers to total pressure of the environment referenced. (HYP)

3.3.135.3 *Gauge Pressure.* Refers to total pressure above (or below) atmospheric. (HYP)

3.3.135.4 *High Pressure.* A pressure exceeding 1.38 kPa (200 psi) gauge (215 psia). (MED)

3.3.135.5* *Operating Pressure.* The pressure that a particular piping system is set to operate at. (PIP)

3.3.135.6* *Partial Pressure.* The pressure, in absolute units, exerted by a particular gas in a gas mixture. (HYP)

3.3.135.7 *Positive Pressure.* Pressure greater than ambient atmospheric. (MED)

3.3.135.8 *Working Pressure (Rated).* The maximum rated operating pressure for a pipe, tube, or vessel based on its material, its allowable stress in tension, its outside diameter and wall thickness, the operating temperature, the joining method, and industry safety factors. (PIP)

3.3.136* **Pressure-Reducing Regulator.** A device that automatically reduces gas under high pressure to a usable lower working pressure. (MED)

3.3.137 Procedure Room. Where the proceduralist is using instrumentation that requires constant observation and control. (MED)

3.3.138 psia. Pounds per square inch absolute, a unit of pressure measurement with zero pressure as the base or reference pressure. (HYP)

3.3.139* **psig.** Pounds per square inch gauge, a unit of pressure measurement with atmospheric pressure as the base or reference pressure. (HYP)

3.3.140 Qualified Person. A person who, by possession of a recognized degree, certificate, professional standing, or skill, and who, by knowledge, training, and experience, has demonstrated the ability to perform the work. (HYP)

3.3.141 Reactance. The component of impedance contributed by inductance or capacitance. The unit of reactance is the ohm. (MED)

3.3.142 Receptacle. A receptacle is a contact device installed at the outlet for the connection of an attachment plug. A single receptacle is a single contact device with no other contact device on the same yoke. A multiple receptacle is two or more contact devices on the same yoke. [70, 2014] (ELS)

3.3.143 Reference Grounding Point. The ground bus of the panelboard or isolated power system panel supplying the patient care room. (MED)

3.3.144* **Remote.** A Level 3 source of supply that is accessed by exiting the single or multiple treatment facility. (PIP)

3.3.145 Reserve Supply. Where existing, that portion of the supply equipment that automatically supplies the system in the event of failure of the operating supply. The reserve supply only functions in an emergency and not as a normal operating procedure. (PIP)

3.3.146 Risk Categories.

3.3.146.1 *Category 1.* Activities, systems, or equipment whose failure is likely to cause major injury or death to patients, staff, or visitors.

3.3.146.2 *Category 2.* Activities, systems, or equipment whose failure is likely to cause minor injury to patients, staff, or visitors.

3.3.146.3 *Category 3.* Activities, systems, or equipment whose failure is not likely to cause injury to patients, staff, or visitors but can cause discomfort.

3.3.146.4 *Category 4.* Activities, systems, or equipment whose failure would have no impact on patient care.

3.3.147 Scavenging. Evacuation of exhaled mixtures of oxygen and nitrous oxide. (PIP)

3.3.148 Selected Receptacles. A minimal number of receptacles selected by the governing body of a facility as necessary to provide essential patient care and facility services during loss of normal power. (ELS)

3.3.149 Self-Extinguishing. A characteristic of a material such that, once the source of ignition is removed, the flame is quickly extinguished without the fuel or oxidizer being exhausted. (HYP)

3.3.150 Semipermanent Connection. A noninterchangeable connection, usually a D.I.S.S. connector, which is the termination of the pipeline and that is intended to be detached only for service. It is not the point at which the user makes connections or disconnections. (PIP)

3.3.151 Service Inlet. The pneumatic terminus of a Level 3 piped vacuum system. (PIP)

3.3.152 Service Outlet. The pneumatic terminus of a piped gas system for other than critical, continuous duty, nonflammable medical life support–type gases such as oxygen, nitrous oxide, or medical air. (PIP)

3.3.153* **Single Treatment Facility.** A diagnostic or treatment complex under a single management comprising a number of use points, but confined to a single contiguous group of use points (i.e., does not involve widely separated locations or separate distinct practices). (PIP)

3.3.154* Site of Intentional Expulsion. All points within 0.3 m (1 ft) of a point at which an oxygen-enriched atmosphere is intentionally vented to the atmosphere. (MED)

3.3.155 Space. A portion of the health care facility designated by the governing body that serves a specific purpose.

3.3.156 Standard Cubic Feet per Minute (SCFM). Volumetric flow rate of gas in units of standard cubic feet per minute. (PIP)

3.3.157 Station Inlet. An inlet point in a piped medical/surgical vacuum distribution system at which the user makes connections and disconnections. (PIP)

3.3.158 Station Outlet. An outlet point in a piped medical gas distribution system at which the user makes connections and disconnections. (PIP)

3.3.159 Supply Source.

    3.3.159.1 *Operating Supply.* The portion of the supply system that normally supplies the piping systems. The operating supply consists of a primary supply or a primary and secondary supply. (PIP)

    3.3.159.2 *Primary Supply.* That portion of the source equipment that actually supplies the system. (PIP)

    3.3.159.3 *Reserve Supply.* Where provided, that portion of the source equipment that automatically supplies the system in the event of failure of the primary and secondary operating supply. (PIP)

    3.3.159.4 *Secondary Supply.* Where provided, that portion of the source equipment that automatically supplies the system when the primary supply becomes exhausted. (PIP)

3.3.160* Surface-Mounted Medical Gas Rail Systems. A surface-mounted gas delivery system intended to provide ready access for two or more gases through a common delivery system to provide multiple gas station outlet locations within a single patient room or critical care area. (PIP)

3.3.161 Task Illumination. Provisions for the minimum lighting required to carry out necessary tasks in the areas described in Chapter 6, including safe access to supplies and equipment and access to exits. (ELS)

3.3.162 Terminal. The end of a flexible hose or tubing used in a manufactured assembly where the user is intended to make connection and disconnection. (PIP)

3.3.163 Touch Current. Leakage current flowing from the enclosure or from parts thereof, excluding patient connections, accessible to any operator or patient in normal use, through an external path other than the protective grounding (earth) conductor to earth or to another part of the enclosure. (MED)

3.3.164 Transfilling. The process of transferring a medical gas in gaseous or liquid state from one container or cylinder to another container or cylinder. (MED)

3.3.165 Tube.

    3.3.165.1* *Endotracheal Tube.* A tube for insertion through the mouth or nose into the upper portion of the trachea (windpipe). (MED)

3.3.166 Use Point. A location with any number of station outlets and inlets arranged for access by a practitioner during treatment of a patient. (PIP)

3.3.167 Vaporizer. A heat exchange unit designed to convert cryogenic liquid into the gaseous state. (PIP)

3.3.168 Ventilation. The mechanical or natural movement of air. (MEC)

3.3.169 Waste Anesthetic Gas Disposal (WAGD). The process of capturing and carrying away gases vented from the patient breathing circuit during the normal operation of gas anesthesia or analgesia equipment. (PIP)

3.3.170 Waste Water.

    3.3.170.1 *Black Waste Water.* Grossly contaminated and contain pathogenic, toxigenic, or other harmful agents.

    3.3.170.2 *Clear Waste Water.* Originates from a sanitary water source and does not pose substantial risk from dermal, ingestion, or inhalation exposure.

    3.3.170.3 *Gray Waste Water.* Contains significant contamination and has the potential to cause discomfort or sickness if contacted or consumed by humans. Can contain potentially unsafe levels of microorganisms or nutrients for microorganisms, as well as other organic or inorganic matter (chemical or biological).

3.3.171* Wet Procedure Locations. The area in a patient care space where a procedure is performed that is normally subject to wet conditions while patients are present, including standing fluids on the floor or drenching of the work area, either of which condition is intimate to the patient or staff. (FUN)

3.4 BICSI Definitions. These terms are defined in *The BICSI Information Transport Systems (ITS) Dictionary*. (HES)

3.4.1 Telecommunications Entrance Facility (EF). An entrance to a building for both public and private network service cables that includes the building entrance point and the entrance room or space at the point of demarcation between campus or utility service and building interior distribution of communications systems. (ELS)

3.4.2 Telecommunications Equipment Room (TER). An environmentally controlled centralized space for telecommunications equipment, typically including main or intermediate cross-connect equipment and cabling. (ELS)

3.4.3 Telecommunications Room (TR). An enclosed architectural space for housing telecommunications equipment, cable terminations, and cross-connect cabling, serving a floor or an area of a floor. (ELS)

Chapter 4 Fundamentals

4.1* Risk Categories. Activities, systems, or equipment shall be designed to meet Category 1 through Category 4 requirements, as detailed in this code.

4.1.1* Category 1. Activities, systems, or equipment whose failure is likely to cause major injury or death of patients, staff, or visitors shall be designed to meet Category 1 requirements, as detailed in this code.

4.1.2* Category 2. Activities, systems, or equipment whose failure is likely to cause minor injury of patients, staff, or visitors shall be designed to meet Category 2 requirements, as detailed in this code.

4.1.3 Category 3. Activities, systems, or equipment whose failure is not likely to cause injury to patients, staff, or visitors, but can cause discomfort, shall be designed to meet Category 3 requirements, as detailed in this code.

4.1.4 Category 4. Activities, systems, or equipment whose failure would have no impact on patient care shall be designed to meet Category 4 requirements, as detailed in this code.

4.2* Risk Assessment.

4.2.1 Categories shall be determined by following and documenting a defined risk assessment procedure.

4.2.2 A documented risk assessment shall not be required for Category 1.

4.3 Application. The Category definitions in Chapter 4 shall apply to Chapters 5 through 11.

4.4 Materials.

4.4.1* Noncombustible Material.

4.4.1.1 A material that complies with any of the following shall be considered a noncombustible material:

(1) A material that, in the form in which it is used and under the conditions anticipated, will not ignite, burn, support combustion, or release flammable vapors when subjected to fire or heat
(2) A material that is reported as passing ASTM E 136, *Standard Test Method for Behavior of Materials in a Vertical Tube Furnace at 750 Degrees C*
(3) A material that is reported as complying with the pass/fail criteria of ASTM E 136 when tested in accordance with the test method and procedure in ASTM E 2652, *Standard Test Method for Behavior of Materials in a Tube Furnace with a Cone-shaped Airflow Stabilizer, at 750 Degrees C*

4.4.1.2 Where the term *limited-combustible* is used in this code, it shall also include the term *noncombustible*. [*101:* 4.6.13]

4.4.2* Limited-Combustible Material. A material shall be considered a limited-combustible material where all the conditions of 4.4.2.1 and 4.4.2.2 and the conditions of either 4.4.2.3 or 4.4.2.4 are met:

4.4.2.1 The material shall not comply with the requirements for noncombustible material in accordance with 4.4.1.

4.4.2.2 The material, in the form in which it is used, shall exhibit a potential heat value not exceeding 3500 Btu/lb (8141 kJ/kg) where tested in accordance with NFPA 259, *Standard Test Method for Potential Heat of Building Materials.*

4.4.2.3 The material shall have the structural base of a noncombustible material with a surfacing not exceeding a thickness of ⅛ in. (3.2 mm) where the surfacing exhibits a flame spread index not greater than 50 when tested in accordance with ASTM E 84, *Standard Test Method for Surface Burning Characteristics of Building Materials,* or ANSI/UL 723, *Standard for Test for Surface Burning Characteristics of Building Materials.*

4.4.2.4 The material shall be composed of materials that, in the form and thickness used, exhibit neither a flame spread index greater than 25 nor evidence of continued progressive combustion when tested in accordance with ASTM E 84, *Standard Test Method for Surface Burning Characteristics of Building Materials,* or ANSI/UL 723, *Standard for Test for Surface Burning Characteristics of Building Materials,* and shall be of such composition that all surfaces that would be exposed by cutting through the material on any plane would exhibit neither a flame spread index greater than 25 nor exhibit evidence of continued progressive combustion when tested in accordance with ASTM E 84 or ANSI/UL 723.

4.4.2.5 Where the term *limited-combustible* is used in this *Code*, it shall also include the term *noncombustible*. [*101:*4.6.13.2]

Chapter 5 Gas and Vacuum Systems

5.1 Category 1 Piped Gas and Vacuum Systems.

5.1.1* Applicability.

5.1.1.1 These requirements shall apply to health care facilities that require Category 1 systems as referenced in Chapter 4.

5.1.1.2 Category 1 piped gas or piped vacuum system requirements shall be applied where any of the following criteria is met:

(1) General anesthesia or deep sedation is performed as defined in 3.3.61.1 and 3.3.61.2.
(2) The loss of the piped gas or piped vacuum systems is likely to cause major injury or death of patients, staff, or visitors.
(3) The facility piped gas or piped vacuum systems are intended for Category 1 patient care space per 3.3.127.1.

5.1.1.3* Where the terms *medical gas* or *medical support gas* occur, the provisions shall apply to all piped systems for oxygen, nitrous oxide, medical air, carbon dioxide, helium, nitrogen, instrument air, and mixtures thereof. Wherever the name of a specific gas service occurs, the provision shall apply only to that gas.

5.1.1.4 An existing system that is not in strict compliance with the provisions of this code shall be permitted to be continued in use as long as the authority having jurisdiction has determined that such use does not constitute a distinct hazard to life.

5.1.1.5 The following sections of this chapter shall apply to the operation, management, and maintenance of Category 1 medical gas and vacuum systems in both new and existing facilities:

(1) 5.1.2
(2) 5.1.3.1
(3) 5.1.3.2
(4) 5.1.3.3.4
(5) 5.1.3.6.2
(6) 5.1.3.8.4.2
(7) 5.1.14

5.1.2 Nature of Hazards of Gas and Vacuum Systems. Potential fire and explosion hazards associated with positive pressure gas central piping systems and medical–surgical vacuum systems shall be considered in the design, installation, testing, operation, and maintenance of these systems.

5.1.3* Category 1 Sources.

5.1.3.1 Central Supply System Identification and Labeling.

5.1.3.1.1* Containers, cylinders, and tanks shall be designed, fabricated, tested, and marked (stamped) in accordance with regulations of DOT, Transport Canada (TC) *Transportation of Dangerous Goods Regulations,* or the ASME *Boiler and Pressure Vessel Code,* "Rules for the Construction of Unfired Pressure Vessels," Section VIII. [**55:**7.1.5.1]

5.1.3.1.2* Cylinder contents shall be identified by attached labels or stencils naming the contents in accordance with the mandatory requirements of CGA C-7, *Guide to the Preparation of Precautionary Labeling and Marking of Compressed Gas Containers*.

5.1.3.1.3 Liquid containers shall have additional product identification visible from all directions with a minimum of 51 mm (2 in.) high letters such as a 360-degree wraparound tape for medical liquid containers.

5.1.3.1.4 Cryogenic liquid containers shall be provided with gas-specific outlet connections in accordance with the mandatory requirements of CGA V-5, *Diameter-Index Safety System (Noninterchangeable Low Pressure Connections for Medical Gas Applications)*, or CGA V-1, *Compressed Gas Association Standard for Compressed Gas Cylinder Valve Outlet and Inlet Connections*.

5.1.3.1.5 Cylinder and cryogenic liquid container outlet connections shall be affixed in such a manner as to be integral to the valve(s), unremovable with ordinary tools, or so designed as to render the attachment point unusable when removed.

5.1.3.1.6 The contents of cylinders and cryogenic liquid containers shall be verified prior to use.

5.1.3.1.7 Labels shall not be defaced, altered, or removed, and connecting fittings shall not be modified.

5.1.3.1.8 Locations containing positive pressure gases other than oxygen and medical air shall have their door(s) labeled as follows:

Positive Pressure Gases
NO Smoking or Open Flame
Room May Have Insufficient Oxygen
Open Door and Allow Room to Ventilate Before Entering

5.1.3.1.9 Locations containing central supply systems or cylinders containing only oxygen or medical air shall have their door(s) labeled as follows:

Medical Gases
NO Smoking or Open Flame

5.1.3.2 Central Supply System Operations.

5.1.3.2.1 The use of adapters or conversion fittings to adapt one gas-specific fitting to another shall be prohibited.

5.1.3.2.2 Cylinders and containers shall be handled in strict accordance with 11.6.2.

5.1.3.2.3 Only gas cylinders, reusable shipping containers, and their accessories shall be permitted to be stored in rooms containing central supply systems or gas cylinders.

5.1.3.2.4 No flammable materials, cylinders containing flammable gases, or containers containing flammable liquids shall be stored in rooms with gas cylinders.

5.1.3.2.5 If cylinders are wrapped when received, the wrappers shall be removed prior to storage.

5.1.3.2.6 Cylinders without correct markings or whose markings and gas-specific fittings do not match shall not be used.

5.1.3.2.7 Cryogenic liquid storage units intended to supply gas to the facility shall not be used to transfill other liquid storage vessels.

5.1.3.2.8 Care shall be exercised when handling cylinders that have been exposed to freezing temperatures or containers that contain cryogenic liquids to prevent injury to the skin.

5.1.3.2.9 Cylinders containing compressed gases and containers for volatile liquids shall be kept away from radiators, steam piping, and like sources of heat.

5.1.3.2.10 When cylinder valve protection caps are supplied, they shall be secured tightly in place unless the cylinder is connected for use.

5.1.3.2.11 Containers shall not be stored in a tightly closed space.

5.1.3.2.12 Cylinders in use and in storage shall be prevented from reaching temperatures in excess of 52°C (125°F).

5.1.3.2.13 Central supply systems for nitrous oxide and carbon dioxide using cylinders or portable containers shall be prevented from reaching temperatures lower than the recommendations of the central supply system's manufacturer, but shall never be lower than –7°C (20°F) or greater than 52°C (125°F).

5.1.3.3* Central Supply System Locations.

5.1.3.3.1 General. Central supply systems shall be located to meet the criteria in 5.1.3.3.1 through 5.1.3.3.1.10.

5.1.3.3.1.1 Any of the following systems shall be permitted to be located together in the same outdoor enclosure:

(1) Manifolds for gas cylinders *(see 5.1.3.5.11)*
(2) Manifolds for cryogenic liquid containers *(see 5.1.3.5.12)*
(3) Bulk cryogenic liquid systems *(see 5.1.3.5.14)*

5.1.3.3.1.2 Any of the following systems shall be permitted to be located together in the same indoor enclosure:

(1) Manifolds for gas cylinders *(see 5.1.3.5.11)*
(2) Manifolds for cryogenic liquid containers *(see 5.1.3.5.12)*
(3) In-building emergency reserves *(see 5.1.3.5.16)*
(4) Instrument air standby headers *(see 5.1.13.3.5.7)*

5.1.3.3.1.3 Any of the following systems shall be permitted to be located together in the same room:

(1) Medical air compressor supply sources *(see 5.1.3.6.3)*
(2) Medical–surgical vacuum sources *(see 5.1.3.7)*
(3) Waste anesthetic gas disposal (WAGD) sources *(see 5.1.3.8)*
(4) Instrument air compressor sources *(see 5.1.13.3.5)*
(5) Any other compressor, vacuum pump, or electrically powered machinery

5.1.3.3.1.4 Any system listed under 5.1.3.3.1.3 shall not be located in the same room with any system listed under 5.1.3.3.1.1 or 5.1.3.3.1.2, except instrument air reserve headers complying with 5.1.3.2.12 and 5.1.13.3.5.7 shall be permitted to be in the same room as an instrument air compressor.

5.1.3.3.1.5 Indoor locations for oxygen, nitrous oxide, and mixtures of these gases shall not communicate with the following:

(1) Areas involved in critical patient care
(2) Anesthetizing locations where moderate sedation, deep sedation, or general anesthesia is administered
(3) Locations storing flammables
(4) Rooms containing open electrical contacts or transformers
(5) Storage tanks for flammable or combustible liquids
(6) Engines
(7) Kitchens
(8) Areas with open flames

5.1.3.3.1.6 Central supply systems for oxygen with a total capacity connected and in storage of 566,335 L (20,000 ft^3) or more outside of the facility at standard temperature and pressure

2015 Edition

(STP) shall comply with NFPA 55, *Compressed Gases and Cryogenic Fluids Code*.

5.1.3.3.1.7 Central supply systems for nitrous oxide with a total capacity connected and in storage of 1451 kg (3200 lb) or more shall comply with the mandatory requirements of CGA G-8.1, *Standard for Nitrous Oxide Systems at Consumer Sites*.

5.1.3.3.1.8 Central supply systems for carbon dioxide using permanently installed containers with product capacities greater than 454 kg (1000 lb) shall comply with the mandatory requirements of CGA G-6.1, *Standard for Insulated Carbon Dioxide Systems at Consumer Sites*.

5.1.3.3.1.9 Central supply systems for carbon dioxide using permanently installed containers with product capacities of 454 kg (1000 lb) or less shall comply with the mandatory requirements of CGA G-6.5, *Standard for Small, Stationary, Insulated Carbon Dioxide Supply Systems*.

5.1.3.3.1.10* Central supply systems for bulk inert gases systems with a total capacity connected and in storage of 20,000 ft^3 or more of compressed gas or cryogenic fluid at standard temperature and pressure, shall comply with the mandatory requirements of CGA P-18, *Standard for Bulk Inert Gas Systems at Consumer Sites*.

5.1.3.3.2* Design and Construction. Locations for central supply systems and the storage of positive-pressure gases shall meet the following requirements:

(1) They shall be constructed with access to move cylinders, equipment, and so forth, in and out of the location on hand trucks complying with 11.4.3.1.1.
(2) They shall be provided with lockable doors or gates or otherwise able to be secured.
(3) If outdoors, they shall be provided with an enclosure (wall or fencing) constructed of noncombustible materials with a minimum of two entry/exits.
(4) If outdoors, bulk cryogenic liquid systems shall be provided with a minimum of two entry/exits.
(5) If indoors, they shall have interior finishes of noncombustible or limited-combustible materials.
(6)*If indoors, the room shall be separated from the rest of the building by walls and floors having a one-hour fire resistance rating with doors and other opening protectives having a ¾-hour fire protection rating.
(7)*They shall comply with *NFPA 70, National Electrical Code*, for ordinary locations.
(8) They shall be heated by indirect means (e.g., steam, hot water) if heat is required.
(9) They shall be provided with racks, chains, or other fastenings to secure all cylinders from falling, whether connected, unconnected, full, or empty.
(10)*They shall be supplied with electrical power compliant with the requirements for essential electrical systems as described in Chapter 6.
(11) They shall have racks, shelves, and supports, where provided, constructed of noncombustible materials or limited-combustible materials.
(12) They shall protect electrical devices from physical damage.
(13) They shall allow access by delivery vehicles and management of cylinders (e.g., proximity to loading docks, access to elevators, and passage of cylinders through public areas).
(14) They shall be designed to meet the operational requirements of 5.1.3.2 with regard to room temperature.

5.1.3.3.3 Ventilation.

5.1.3.3.3.1 Ventilation for Indoor Locations. Medical gas storage and transfilling room ventilation shall comply with 9.3.6.

5.1.3.3.3.2 Venting of Relief Valves. Indoor supply systems shall have all relief valves vented per 5.1.3.5.6.1(4) through 5.1.3.5.6.1(9).

5.1.3.3.3.3 Ventilation for Motor-Driven Equipment. The following source locations shall be adequately ventilated to prevent accumulation of heat:

(1) Medical air sources (*see 5.1.3.6*)
(2) Medical–surgical vacuum sources (*see 5.1.3.7*)
(3) Waste anesthetic gas disposal (WAGD) sources (*see 5.1.3.8.1*)
(4) Instrument air sources (*see 5.1.13.3.5*)

5.1.3.3.3.4 Ventilation for Outdoor Locations.

(1) Outdoor locations surrounded by impermeable walls, except fire barrier walls, shall have protected ventilation openings located at the base of each wall to allow free circulation of air within the enclosure.
(2) Walls that are shared with other enclosures or with buildings shall be permitted to not have openings.
(3) The fire barrier wall shall not have openings or penetrations, except conduit or piping shall be permitted, provided that the penetration is protected with a firestop system in accordance with the building code.

5.1.3.3.4 Storage.

5.1.3.3.4.1 Full or empty medical gas cylinders, when not connected, shall be stored in locations complying with 5.1.3.3.2 through 5.1.3.3.3 and shall be permitted to be in the same rooms or enclosures as their respective central supply systems.

5.1.3.3.4.2 Cylinders, whether full or empty, shall not be stored in enclosures containing motor-driven machinery, with the exception of cylinders intended for instrument air reserve headers complying with 5.1.13.3.5.7, which shall be permitted to be placed in the same location containing an instrument air compressor when it is the only motor-driven machinery located within the room. Only cylinders intended for instrument air reserve headers complying with 5.1.13.3.5.7 shall be permitted to be stored in enclosures containing instrument air compressors.

5.1.3.4 Control Equipment. For control equipment, as specified in 5.1.3.5.5, 5.1.3.5.6, and 5.1.3.5.8, that is physically remote from the supply system, the control equipment shall be installed within a secure enclosure to prevent unauthorized access in accordance with 5.1.3.3.2(2).

5.1.3.4.1 The enclosure shall provide enough space to perform maintenance and repair.

5.1.3.4.2 The location of the enclosure for control equipment other than for medical air shall not communicate with combustible or flammable materials.

5.1.3.5* Central Supply Systems. Central supply systems shall be permitted to consist of the following:

(1) Cylinder manifolds for gas cylinders per 5.1.3.5.11
(2) Manifolds for cryogenic liquid containers per 5.1.3.5.12
(3) Bulk cryogenic liquid systems per 5.1.3.5.14
(4) Medical air compressor systems per 5.1.3.6
(5) Medical–surgical vacuum producers per 5.1.3.7
(6) WAGD producers per 5.1.3.8
(7) Instrument air compressor systems per 5.1.13.3.5
(8) Proportioning systems for medical air USP per 5.1.3.6.3.14

5.1.3.5.1 General. Central supply systems shall be obtained from a supplier or manufacturer familiar with their proper construction and use and installed in accordance with the manufacturer's instructions.

5.1.3.5.2 Permitted Locations for Medical Gases. Central supply systems and medical gas outlets for oxygen, medical air, nitrous oxide, carbon dioxide, and all other patient medical gases shall be piped only into areas where the gases will be used under the direction of licensed medical professionals for purposes congruent with the following:

(1) Direct respiration by patients
(2) Clinical application of the gas to a patient, such as the use of an insufflator to inject carbon dioxide into patient body cavities during laparoscopic surgery and carbon dioxide used to purge heart-lung machine blood flow ways
(3) Medical device applications directly related to respiration
(4) Power for medical devices used directly on patients
(5) Calibration of medical devices intended for (1) through (4)

5.1.3.5.3 Support Gases. Central supply systems for support gases shall not be piped to, or used for, any purpose except medical support application.

5.1.3.5.4* Materials. Materials used in central supply systems shall meet the following requirements:

(1) In those portions of systems intended to handle oxygen at gauge pressures greater than 2413 kPa (350 psi), interconnecting hose shall contain no polymeric materials.
(2) In those portions of systems intended to handle oxygen or nitrous oxide material, construction shall be compatible with oxygen under the temperatures and pressures to which the components can be exposed in the containment and use of oxygen, nitrous oxide, mixtures of these gases, or mixtures containing more than 23.5 percent oxygen.
(3) If potentially exposed to cryogenic temperatures, materials shall be designed for low temperature service.
(4) If intended for outdoor installation, materials shall be installed per the manufacturer's requirements.

5.1.3.5.5 Final Line Pressure Regulators.

5.1.3.5.5.1 All positive pressure supply systems shall be provided with duplex line pressure regulators piped in parallel with the following characteristics:

(1) They shall be provided with isolation valves on the source side of each regulator.
(2) They shall be provided with isolation or check valves on the patient side of each regulator.
(3) A pressure indicator(s) shall be located downstream (patient or use side) of each regulator or immediately downstream of the isolating valves for the regulators.
(4) They shall be piped to allow either regulator to be serviced without interrupting supply.
(5) Each regulator shall be sized for 100 percent of the peak calculated demand.
(6) They shall be constructed of materials deemed suitable by the manufacturer.

5.1.3.5.5.2 The line pressure regulators required under 5.1.3.5.5.1, when used for bulk cryogenic liquid systems, shall be of a balanced design.

5.1.3.5.6 Relief Valves.

5.1.3.5.6.1 All pressure relief valves shall meet the following requirements:

(1) They shall be of brass, bronze, or stainless steel construction.
(2) They shall be designed for the specific gas service.
(3) They shall have a relief pressure setting not higher than the maximum allowable working pressure (MAWP) of the component with the lowest working pressure rating in the portion of the system being protected.
(4) They shall be vented to the outside of the building, except that relief valves for compressed air systems having less than 84,950 L (3000 ft^3) at STP shall be permitted to be diffused locally by means that will not restrict the flow.
(5) They shall have a vent discharge line that is not smaller than the size of the relief valve outlet.
(6) Where two or more relief valves discharge into a common vent line, its internal cross-sectional area shall be not less than the aggregate cross-sectional area of all relief valve vent discharge lines served.
(7) They shall not discharge into locations creating potential hazards.
(8) They shall have the discharge terminal turned down and screened to prevent the entry of rain, snow, or vermin.
(9) They shall be designed in accordance with ASME B31.3, *Pressure Process Piping*.

5.1.3.5.6.2 When vented to outdoors, materials and construction for relief valve discharge lines shall be the same as required for positive pressure gas distribution. *(See 5.1.10.1.)*

5.1.3.5.6.3 Central supply systems for positive pressure gases shall include one or more relief valves, all meeting the following requirements:

(1) They shall be located between each final line regulator and the source valve.
(2) They shall have a relief setting that is 50 percent above the normal system operating pressure, as indicated in Table 5.1.11.

5.1.3.5.6.4 When vented outside, relief valve vent lines shall be labeled in accordance with 5.1.11.1 in any manner that will distinguish them from the medical gas pipeline.

5.1.3.5.7 Auxiliary Source Connection. All source systems shall be provided with an auxiliary source connection point of the same size as the main line, which shall be located immediately on the patient side of the source valve.

5.1.3.5.7.1 The connection shall consist of a tee, a valve, and a removable plug or cap.

5.1.3.5.7.2 The auxiliary source connection valve shall be normally closed and secured.

5.1.3.5.8 Multiple Pressures. Where a single central supply system supplies separate piped distribution networks operating at different pressures, each piped distribution network shall comply with the following:

(1) Medical air compressor systems: 5.1.3.5.5 (pressure regulators) and 5.1.9.2.4(7) (master alarm)
(2) All central supply systems: 5.1.3.5.5 (pressure regulators), 5.1.3.5.6 (relief valves), 5.1.4.2 (source valve), and 5.1.9.2.4(7) (master alarm)

5.1.3.5.9 Local Signals.

5.1.3.5.9.1 The following systems shall have local signals located at the source equipment:

(1) Manifolds for gas cylinders without reserve supply *(see 5.1.3.5.11)*
(2) Manifolds for gas cylinders with reserve supply

(3) Manifolds for cryogenic liquid containers *(see 5.1.3.5.12)*
(4) Bulk cryogenic liquid systems *(see 5.1.3.5.14)*
(5) In-building emergency reserves *(see 5.1.3.5.16)*
(6) Instrument air headers *(see 5.1.3.5.10)*

5.1.3.5.9.2 The local signals shall meet the following requirements:

(1) Provision of visual indication only
(2) Labeling for the service and condition being monitored
(3) If intended for outdoor installation, be installed per manufacturer's requirements

5.1.3.5.10* Headers. In central supply systems using cylinders containing either gas or liquid, each header shall include the following:

(1)*Cylinder connections in the number required for the header's application
(2) Cylinder lead for each cylinder constructed of materials complying with 5.1.3.5.4 and provided with end fittings permanently attached to the cylinder lead complying with the mandatory requirements of CGA V-1, *Compressed Gas Association Standard for Compressed Gas Cylinder Valve Outlet and Inlet Connections* (ANSI B57.1)
(3) Filter of a material complying with 5.1.3.5.4 to prevent the intrusion of debris into the manifold controls
(4) Header shutoff valve downstream of the nearest cylinder connection, but upstream of the point at which the header connects to the central supply system
(5) Pressure indicator indicating the pressure of header contents
(6) Check valve to prevent backflow into the header and to allow service to the header
(7) If intended for gas cylinder service, a check valve at each connection for the cylinder lead in 5.1.3.5.10(2) to prevent loss of gas in the event of damage to the cylinder lead or operation of an individual cylinder relief valve
(8) If intended for gas cylinder service, a pressure regulator to reduce the cylinder pressure to an intermediate pressure to allow the proper operation of the primary and secondary headers
(9) If intended for service with cryogenic liquid containers, a pressure relief valve
(10) Vent valves, if fitted on a header, vented outside of the building per 5.1.3.5.6.1(5) through 5.1.3.5.6.1(9) and 5.1.3.5.6.2

5.1.3.5.11* Manifolds for Gas Cylinders.

5.1.3.5.11.1 The manifolds in this category shall be located in accordance with 5.1.3.3.1 and shall meet the following:

(1) If located outdoors, they shall be installed in an enclosure used only for this purpose and sited to comply with minimum distance requirements in NFPA 55.
(2) If located indoors, they shall be installed within a room used only for enclosure of such manifolds.

5.1.3.5.11.2 The manifold locations for this category shall be constructed in accordance with 5.1.3.3.2.

5.1.3.5.11.3 The manifold locations for this category shall be ventilated in accordance with 5.1.3.3.3.

5.1.3.5.11.4 The manifolds in this category shall consist of the following:

(1) Two equal headers in accordance with 5.1.3.5.10, each with a sufficient number of gas cylinder connections for an average day's supply, but not fewer than two connections, and with the headers connected to the final line pressure regulator assembly in such a manner that either header can supply the system
(2) Vent valves, if fitted on a header, vented outside of the building per 5.1.3.5.6.1(5) through (9) and 5.1.3.5.6.2
(3) Intermediate relief valve(s), piped to the outside in accordance with 5.1.3.5.6.1(5) through (9), that protects the piping between the header pressure regulator and the line pressure regulator assembly, and protects the line pressure regulators from overpressure in the event of a header regulator failure

5.1.3.5.11.5 The manifolds in this category shall include an automatic means of alternating the two headers to accomplish the following in normal operation:

(1) One header is the primary and the other is the secondary, with either being capable of either role.
(2) When the primary header is supplying the system, the secondary header is prevented from supplying the system.
(3) When the primary header is depleted, the secondary header automatically begins to supply the system.

5.1.3.5.11.6 The manifolds in this category shall have a local signal that visibly indicates the operating status of the equipment and shall activate an indicator at all master alarm panels when or at a predetermined set point before the secondary header begins to supply the system, indicating changeover has occurred or is about to occur.

5.1.3.5.11.7 If manifolds are located out of doors, they shall be installed per the manufacturer's requirements.

5.1.3.5.12* Manifolds for Cryogenic Liquid Containers.

5.1.3.5.12.1 Manifolds for cryogenic liquid containers shall be located in accordance with 5.1.3.3.1 and shall meet the following:

(1) If located outdoors, they shall be installed in an enclosure used only for the enclosure of such containers. *[See Figure A.5.1.3.5.14(a) for minimum siting distance requirements.]*
(2) If located indoors, they shall be installed within a room used only for the enclosure of such containers.

5.1.3.5.12.2 The manifolds in this category shall have their primary and secondary headers located in the same enclosure.

5.1.3.5.12.3 The reserve header shall be permitted to be located in the same enclosure as the primary and secondary headers or in another enclosure compliant with 5.1.3.5.12.1.

5.1.3.5.12.4 The manifolds in this category shall consist of the following:

(1) Two equal headers per 5.1.3.5.10, each having sufficient number of liquid container connections for an average day's supply, and with the headers connected to the final line pressure regulator assembly in such a manner that either header can supply the system
(2) Reserve header per 5.1.3.5.10 having sufficient number of gas cylinder connections for an average day's supply, but not fewer than three connections, and connected downstream of the primary/secondary headers and upstream of the final line pressure regulators
(3) Pressure relief installed downstream of the connection of the reserve header and upstream of the final line pressure regulating assembly and set at 50 percent above the nominal inlet pressure

5.1.3.5.12.5 The manifolds in this category shall include an automatic means of controlling the three headers to accomplish the following during normal operation:

(1) If provided with two liquid container headers, one cryogenic liquid header shall be the primary and the other shall be the secondary, with either being capable of either role.
(2) If provided with one liquid container header and one gas cylinder header (a hybrid arrangement), the liquid container header is the primary and the gas cylinder header is the secondary.
(3) When the primary header is supplying the system, the secondary header is prevented from supplying the system.
(4) When the primary header is depleted, the secondary header automatically begins to supply the system.

5.1.3.5.12.6 The manifolds in this category shall be equipped with a means to conserve the gas produced by evaporation of the cryogenic liquid in the secondary header (when so provided). This mechanism shall discharge the conserved gas into the system upstream of the final line regulator assembly.

5.1.3.5.12.7 The manifolds in this category shall include a manual or automatic means to place either header into the role of primary header and the other into the role of secondary header, except where a liquid/gas hybrid manifold is employed.

5.1.3.5.12.8 The manifolds in this category shall include a means to automatically activate the reserve header if for any reason the primary and secondary headers cannot supply the system.

5.1.3.5.12.9 The manifolds in this category shall have a local signal that visibly indicates the operating status of the equipment and activates an indicator at all master alarms under the following conditions:

(1) When or at a predetermined set point before the secondary header begins to supply the system, indicating changeover
(2) Where a hybrid arrangement is employed, when or at a predetermined set point before the secondary (cylinder) header contents fall to one day's average supply, indicating secondary low
(3) When or at a predetermined set point before the reserve header begins to supply the system, indicating reserve is in use
(4) When or at a predetermined set point before the reserve header contents fall to one day's average supply, indicating reserve low

5.1.3.5.13 Micro-Bulk or Small Bulk Cryogenic Liquid Systems.

5.1.3.5.13.1 Micro-bulk cryogenic liquid systems shall comply with the following requirements:

(1) If located indoors, be installed within a room used only for this purpose.
(2) If located outdoors, oxygen systems be sited to comply with the minimum distance requirements in NFPA 55.
(3) If located outdoors, nitrogen systems be sited to comply with the mandatory minimum distance requirements in CGA P-18, *Standard for Bulk Inert Gas Systems at Consumer Sites*.
(4) Be compliant with the mandatory requirements of CGA M-1, *Guide for Medical Gas Installations at Consumer Sites*.
(5) Be located in an enclosure constructed in accordance with 5.1.3.3.2(1) through 5.1.3.3.2(3) and 5.1.3.3.2(5), 5.1.3.3.2(8), and 5.1.3.3.2(9).
(6) Be located in an enclosure ventilated in accordance with 5.1.3.3.3.3.
(7) Be designed such that the items noted in 5.1.3.5.13.2 and items located in the trailer unloading area are readily visible to delivery personnel during filling operations.
(8) Be protected against overpressurization of the pressure vessel during filling operations.
(9) Not have a bottom fill valve.
(10) Be installed in accordance with 5.1.10.1 through 5.1.10.5.1.7.
(11) Be installed by personnel qualified to meet the mandatory requirements of CGA M-1, *Guide for Medical Gas Installations at Consumer Sites*, or ASSE 6015 *Professional Qualifications Standards for Bulk Medical Gas Systems Installers*.
(12) Be installed in compliance with Food and Drug Administration (FDA) Current Good Manufacturing Practices as found in 21 CFR 210 and 21 CFR 211.

5.1.3.5.13.2 A micro-bulk cryogenic liquid system with a primary and secondary supply shall have headers located in the same enclosure.

5.1.3.5.13.3* A micro-bulk cryogenic liquid system with a reserve header shall be permitted to be located in the same enclosure as the primary and secondary headers or in another enclosure compliant with 5.1.3.5.13.1.

5.1.3.5.13.4 A micro-bulk cryogenic liquid system shall consist of the following:

(1) Two equal headers each having sufficient capacity for an average day's supply, with either being capable of either role, consisting of one primary supply and one secondary supply, and with the headers connected to the final line pressure regulator assembly in such a manner that either header can supply the system and a reserve header, in accordance with 5.1.3.5.10, having sufficient number of gas cylinder connections for an average day's supply but not fewer than three, and connected downstream of the primary/secondary headers and upstream of the final line pressure regulators.
(2) One micro-bulk cryogenic liquid main header, having sufficient capacity for an average day's supply, one secondary supply consisting of a micro-bulk cryogenic liquid, liquid containers, or high-pressure cylinders and having sufficient capacity for an average day's supply, and a reserve header, in accordance with 5.1.3.5.10, having sufficient number of gas cylinder connections for an average day's supply but not fewer than three, and connected downstream of the primary/secondary headers and upstream of the final line pressure regulators.
(3) One micro-bulk cryogenic liquid main header, having sufficient capacity for an average day's supply, one reserve header consisting of either a micro-bulk cryogenic liquid supply or high-pressure cylinders in accordance with 5.1.3.5.10 connections for an average day's supply and connected downstream of the primary/secondary headers and upstream of the final line pressure regulators.

5.1.3.5.13.5 Conditions for the micro-bulk cryogenic system shall include the following:

(1) When the primary or main header is supplying the system, the secondary and reserve headers is prevented from supplying the system.
(2) When the primary or main header is depleted, the roles of primary or main, the secondary (when installed), and the

reserve headers alternate and will provide an operating cascade (primary-secondary-reserve) that automatically begins to supply the system.
(3) Capacity be determined after consideration of the customer usage requirements, delivery schedules, proximity of the facility to alternative supplies, and the emergency plan.
(4) Where there are two or more micro-bulk cryogenic liquid vessels of equal capacity, they are permitted to alternate in the roles of primary and secondary.
(5) A reserve supply sized for a greater than an average day's supply and the appropriate size of vessel or number of cylinders shall be determined after consideration of delivery schedules, proximity of the facility to alternative supplies, and the facility's emergency plan.
(6) At least two main vessel relief valves and rupture discs shall be installed downstream of a three-way (three-port) valve.
(7) A check valve shall be located in the primary supply piping upstream of the intersection with a secondary supply or reserve supply.
(8) A contents gauge shall be on each main vessel.
(9) A pressure relief shall be installed downstream of the connection of the reserve header and upstream of the final line pressure regulating assembly and set at 50 percent above the nominal inlet pressure.
(10) The manifolds in this category shall be equipped with a means to conserve the gas produced by evaporation of the cryogenic liquid in the secondary header (where so provided). This mechanism shall discharge the conserved gas into the system upstream of the final line regulator assembly.
(11) The manifolds for two equal headers shall include a manual or automatic means to place either header into the role as primary header and the other in the role of secondary header (where so provided).
(12) The manifolds for main supply with a secondary supply (where so provided) headers shall include a manual or automatic means to place the secondary header into the role as primary header during the filling of the main supply.
(13) The manifolds shall include a means to automatically actuate the reserve header if for any reason the primary and secondary (where so provided) headers cannot supply the system.
(14) Permanent anchors shall hold the components to the pad or flooring in accordance with the design requirements.

5.1.3.5.13.6 The micro-bulk cryogenic system in this category shall actuate a local signal and shall activate an indicator at all master alarms under the following conditions:

(1) When or at a predetermined set point before the main or primary supply reaches an average day's supply, indicating low contents
(2) If the secondary supply is a cryogenic vessel, when or at a predetermined set point before the secondary supply reaches an average day's supply, indicating low contents
(3) If the reserve supply is a cryogenic vessel, when or at a predetermined set point before the reserve supply reaches an average day's supply, indicating low contents
(4) Where there is more than one main supply vessel, when or at a predetermined set point before the secondary supply begins to supply the system, indicating changeover
(5) When or at a predetermined set point before the reserve supply begins to supply the system, indicating reserve is in use
(6) When or at a predetermined set point before the reserve supply contents fall to one day's average supply, indicating reserve low
(7) If the reserve is a cryogenic vessel, when or at a predetermined set point before the reserve internal pressure falls too low for the reserve to operate properly, indicating reserve failure

5.1.3.5.14* Bulk Cryogenic Liquid Systems.

5.1.3.5.14.1 Bulk cryogenic liquid storage systems shall be in accordance with NFPA 55, *Compressed Gases and Cryogenic Fluids Code*.

5.1.3.5.14.2 Bulk cryogenic liquid systems shall have the following protections:

(1) Be installed in accordance with NFPA 55, *Compressed Gases and Cryogenic Fluids Code*
(2) Meet the requirements of 5.1.3.3.2(1)
(3) Meet the requirements of 5.1.3.3.2(10)
(4) Meet the requirements of 5.1.3.3.2(12)
(5) Be installed meeting the requirements in 5.1.10.1 through 5.1.10.4.7
(6) Have a minimum work space clearance of 3 ft (1 m) around the storage container, vaporizer(s), and the cabinet opening or front side of the pressure regulating manifold for system maintenance and operation

5.1.3.5.14.3 Bulk cryogenic liquid sources shall include automatic means to provide the following functions:

(1) When the main supply is supplying the system, the reserve supply shall be prevented from supplying the system until the main supply is reduced to a level at or below the reserve activation pressure.
(2) When the main supply cannot supply the system, the reserve supply shall automatically begin to supply the system.
(3) Where there is more than one main supply vessel, the system shall operate as described in 5.1.3.5.12 for primary, secondary, and reserve operation.
(4) Where there are two or more cryogenic vessels, they shall be permitted to alternate (e.g., on a timed basis) in the roles of primary, secondary, and reserve, provided that an operating cascade (primary–secondary–reserve) as required in 5.1.3.5.12.5 is maintained at all times.
(5) Where a cryogenic vessel is used as the reserve, the reserve vessel shall include a means to conserve the gas produced by evaporation of the cryogenic liquid in the reserve vessel and to discharge the gas into the line upstream of the final line regulator assembly as required by 5.1.3.5.12.6.

5.1.3.5.14.4* The bulk systems shall have a local signal that visibly indicates the operating status of the equipment and an indicator at all master alarms under the following conditions:

(1) When or at a predetermined set point before the main supply reaches an average day's supply, indicating low contents
(2) When or at a predetermined set point before the reserve supply begins to supply the system, indicating reserve is in use
(3) When or at a predetermined set point before the reserve supply contents fall to one day's average supply, indicating reserve low
(4) If the reserve is a cryogenic vessel, when or at a predetermined set point before the reserve internal pressure falls too low for the reserve to operate properly, indicating reserve failure

(5) Where there is more than one main supply vessel, when or at a predetermined set point before the secondary vessel begins to supply the system, indicating changeover

5.1.3.5.15* Emergency Oxygen Supply Connection (EOSC). Emergency oxygen supply connections (EOSCs) shall be installed to allow connection of a temporary auxiliary source of supply for emergency or maintenance situations where either of the following conditions exist:

(1) The bulk cryogenic liquid central supply system or microbulk cryogenic liquid system is outside of and remote from the building that the oxygen supply serves, and there is no connected in-building oxygen reserve sufficient for an average day's supply. *(See 5.1.3.5.16 for requirements for such reserves.)*
(2) Multiple freestanding buildings are served from a single oxygen source such that damage to the interconnecting oxygen line could result in one or more buildings losing oxygen supply, in which case each building is required to be provided with a separate emergency connection.

5.1.3.5.15.1 EOSCs shall be located as follows:

(1) Located on the exterior of the building being served in a location accessible by emergency supply vehicles at all times in all weather conditions
(2) Connected to the main supply line immediately downstream of the main shutoff valve

5.1.3.5.15.2 EOSCs shall consist of the following:

(1) Physical protection to prevent unauthorized tampering
(2) Female DN (NPS) inlet for connection of the emergency oxygen source that is sized for 100 percent of the system demand at the emergency source gas pressure
(3) Manual shutoff valve to isolate the EOSC when not in use
(4) Two check valves, one downstream of the EOSC and one downstream of the main line shutoff valve, with both upstream from the tee connection for the two pipelines
(5) Relief valve sized to protect the downstream piping system and related equipment from exposure to pressures in excess of 50 percent higher than normal line pressure
(6) Any valves necessary to allow connection of an emergency supply of oxygen and isolation of the piping to the normal source of supply
(7) Minimum of 1 m (3 ft) of clearance around the EOSC for connection of temporary auxiliary source

5.1.3.5.16 In-Building Emergency Reserves (IBERs).

5.1.3.5.16.1 IBERs shall not be used as substitutes for the bulk gas reserve system that is required in 5.1.3.5.14.3.

5.1.3.5.16.2 When an IBER is provided inside the building as a substitute for the EOSC or for other purposes, it shall be located in accordance with 5.1.3.3 as follows:

(1) In a room or enclosure constructed per 5.1.3.3.2
(2) In a room or enclosure ventilated per 5.1.3.3.3

5.1.3.5.16.3 IBERs shall consist of either of the following:

(1) Gas cylinder header per 5.1.3.5.10 with sufficient cylinder connections to provide for at least an average day's supply with the appropriate number of connections being determined after consideration of the delivery schedule, the proximity of the facility to alternate supplies, and the facility's emergency plan
(2) Manifold for gas cylinders complying with 5.1.3.5.11

5.1.3.5.16.4 IBERs shall include a check valve in the main line placed on the distribution system side of the ordinary source's main line valve to prevent flow of gas from the emergency reserve to the ordinary source.

5.1.3.5.16.5 IBERs shall have a local signal that visibly indicates the operating status of the equipment and an alarm at all master alarms when or just before the reserve begins to serve the system.

5.1.3.6* Category 1 Medical Air Supply Systems.

5.1.3.6.1* Quality of Medical Air. Medical air shall be required to have the following characteristics:

(1) It shall be supplied from cylinders, bulk containers, or medical air compressor sources, or it shall be reconstituted from oxygen USP and oil-free, dry nitrogen NF.
(2) It shall meet the requirements of medical air USP.
(3) It shall have no detectable liquid hydrocarbons.
(4) It shall have less than 25 ppm gaseous hydrocarbons.
(5) It shall have equal to or less than 1 mg/m^3 (6.85 × 10^{-7} lb/yd^3) of permanent particulates sized 1 micron or larger in the air at normal atmospheric pressure.

5.1.3.6.2* Uses of Medical Air. Medical air sources shall be connected to the medical air distribution system only and shall be used only for air in the application of human respiration and calibration of medical devices for respiratory application.

5.1.3.6.3* Medical Air Compressor Sources.

5.1.3.6.3.1 Location. Medical air compressor systems shall be located per 5.1.3.3 as follows:

(1) Indoors in a dedicated mechanical equipment area, adequately ventilated and with any required utilities (e.g., electricity, drains, lighting)
(2) In a room ventilated per 5.1.3.3.3
(3) For air-cooled equipment, in a room designed to maintain the ambient temperature range as recommended by the manufacturer

5.1.3.6.3.2 Required Components. Medical air compressor systems shall consist of the following:

(1) Components complying with 5.1.3.6.3.4 through 5.1.3.6.3.8, arranged per 5.1.3.6.3.9
(2) Automatic means to prevent backflow from all on-cycle compressors through all off-cycle compressors
(3) Manual shutoff valve to isolate each compressor from the centrally piped system and from other compressors for maintenance or repair without loss of pressure in the system
(4) Intake filter–muffler(s) of the dry type
(5) Pressure relief valve(s) set at 50 percent above line pressure
(6) Piping and components between the compressor and the source shutoff valve that do not contribute to contaminant levels
(7) Except as defined in 5.1.3.6.3.2(1) through (6), materials and devices used between the medical air intake and the medical air source valve that are of any design or construction appropriate for the service as determined by the manufacturer

5.1.3.6.3.3 Air Drying Equipment. Medical air compressor systems shall preclude the condensation of water vapor in the piping distribution system by air drying equipment.

5.1.3.6.3.4 Compressors for Medical Air.

(A)* Compressors for medical air shall be designed to prevent the introduction of contaminants or liquid into the pipeline by any of the following methods:

(1) Elimination of oil anywhere in the compressor (e.g., liquid ring and permanently sealed bearing compressors)
(2) Reciprocating compressors provided with a separation of the oil-containing section from the compression chamber by at least two seals creating an area open to atmosphere that allows the following:
   (a) Direct and unobstructed visual inspection of the interconnecting shaft through vent and inspection openings no smaller than 1.5 shaft diameters in size
   (b) Confirmation by the facility operators of proper seal operation by direct visual inspection through the above-shaft opening, without disassembly of the compressor (e.g., extended head compressors with an atmospheric vent between the compression chamber and the crankcase)
(3) Rotating element compressors provided with a compression chamber free of oil that provide the following:
   (a) Separation of each oil-containing section from the compression chamber by at least one seal having atmospheric vents on each side with the vent closest to the oil-containing section supplied with a gravity drain to atmosphere
   (b) Unobstructed visualization of the atmospheric vent(s), closest to each oil-containing section, that is accessible for inspection without disassembling the compressor
   (c) Entry of the rotating shaft into each compression chamber at a point that is above atmospheric pressure
   (d) Confirmation by the facility operators of proper seal operation by direct visual inspection of the atmospheric vents

(B) For liquid ring compressors, service water and seal water shall be treated to control waterborne pathogens and chlorine from hyperchlorination from entering the medical air.

(C) Liquid ring compressors shall comply with the following:

(1) Service water and seal water of a quality recommended by the compressor manufacturer shall be used.
(2) Reserve medical air standby headers or a backup compressor shall be installed.
(3) When installed, the header shall comply with 5.1.3.5.10.
(4) When installed, the number of attached cylinders shall be sufficient for 1 hour normal operation.

(D) Compressors shall be constructed of materials deemed suitable by the manufacturer.

(E) Antivibration mountings shall be installed for compressors as required by equipment dynamics or location and in accordance with the manufacturer's recommendations.

(F) Flexible connectors shall connect the air compressors with their intake and outlet piping.

5.1.3.6.3.5 Aftercoolers.

(A) Aftercoolers, where required, shall be provided with individual condensate traps.

(B) The receiver shall not be used as an aftercooler or aftercooler trap.

(C) Aftercoolers shall be constructed of materials deemed suitable by the manufacturer.

(D) Antivibration mountings shall be installed for aftercoolers as required by equipment dynamics or location and in accordance with the manufacturer's recommendations.

5.1.3.6.3.6 Medical Air Receivers.
Receivers for medical air shall meet the following requirements:

(1) They shall be made of corrosion-resistant materials or otherwise be made corrosion resistant.
(2) They shall comply with Section VIII, "Unfired Pressure Vessels," of the ASME *Boiler and Pressure Vessel Code*.
(3) They shall be equipped with a pressure relief valve, automatic drain, manual drain, sight glass, and pressure indicator.
(4) They shall be of a capacity sufficient to prevent the compressors from short-cycling.

5.1.3.6.3.7 Medical Air Dryers.
Medical air dryers, where required, shall meet the following requirements:

(1) Be designed to provide air at a maximum dew point that is below the frost point [0°C (32°F)] at 345 kPa to 380 kPa (50 psi to 55 psi) at any level of demand
(2) Be sized for 100 percent of the system peak calculated demand at design conditions
(3) Be constructed of materials deemed suitable by the manufacturer
(4) Be provided with antivibration mountings installed as required by equipment dynamics or location and in accordance with the manufacturer's recommendations

5.1.3.6.3.8 Medical Air Filters.
Medical air filters shall meet the following requirements:

(1) Be appropriate for the intake air conditions
(2) Be located upstream (source side) of the final line regulators
(3) Be sized for 100 percent of the system peak calculated demand at design conditions and be rated for a minimum of 98 percent efficiency at 1 micron or greater
(4) Be equipped with a continuous visual indicator showing the status of the filter element life
(5) Be constructed of materials deemed suitable by the manufacturer

5.1.3.6.3.9 Piping Arrangement and Redundancies.

(A) Component arrangement shall be as follows:

(1) Components shall be arranged to allow service and a continuous supply of medical air in the event of a single fault failure.
(2) Component arrangement shall be permitted to vary as required by the technology(ies) employed, provided that an equal level of operating redundancy and medical air quality is maintained.

(B) Medical air compressors shall be sufficient to serve the peak calculated demand with the largest single compressor out of service. In no case shall there be fewer than two compressors.

(C) When aftercoolers are provided, they shall be arranged to meet either one of the following:

(1) Arranged as a duplex or multiplex set, sized to serve the peak calculated demand with the largest single aftercooler out of service, and provided with valves adequate,

to isolate any single aftercooler from the system without shutting down supply of medical air

(2) Arranged one per compressor, sized to handle the output of that compressor, and valved as appropriate to allow repair or replacement with that compressor out of service but without shutting down supply of medical air

(D)* A medical air receiver(s) shall be provided with proper valves to allow the flow of compressed air to enter and exit out of separate receiver ports during normal operation and allow the receiver to be bypassed during service without shutting down the supply of medical air.

(E) Dryers, filters, and regulators shall be at least duplexed, with each component sized to serve the peak calculated demand with the largest of each component out of service.

(F)* Dryers, filters, and regulators shall be provided with manual valves upstream and manual valves or check valves downstream to allow service to the components without shutting down the system in either one of the following ways:

(1) They shall be installed for each component, upstream and downstream of each component, allowing each to be individually isolated.
(2) They shall be installed upstream (source side) and downstream of components in series so as to create redundant parallel branches of components.

(G) A three-way valve (three-port), indexed to flow, full port shall be permitted to be used to isolate one branch or component for the purposes of 5.1.3.6.3.9(C), 5.1.3.6.3.9(D), 5.1.3.6.3.9(E), and 5.1.3.6.3.9(F).

(H) Under normal operation, only one aftercooler shall be open to airflow with the other aftercooler valved off.

(I) Under normal operation, only one dryer–filter(s)–regulator sequence shall be open to airflow with the other sequence valved off.

(J) If the relief valve required in 5.1.3.6.3.2(5) and 5.1.3.6.3.6(3) can be isolated from the system by the valve arrangement used to comply with 5.1.3.6.3.9(F), then a redundant relief valve(s) shall be installed in the parallel sequence.

(K) A DN8 (NPS ¼) valved sample port shall be provided downstream of the final line pressure regulators, dew point monitor, and carbon monoxide monitor and upstream of the source shutoff valve to allow for sampling of the medical air.

(L) Medical air source systems shall be provided with a source valve per 5.1.4.2.

(M) Where medical air piping systems at different operating pressures are required, the piping shall separate after the filters but shall be provided with separate line regulators, dew point monitors, relief valves, and source shutoff valves.

5.1.3.6.3.10 Electrical Power and Control.

(A) An additional compressor(s) shall automatically activate when the compressor(s) in operation is incapable of maintaining the required pressure.

(B) Automatic or manual alternation of compressors shall allow division of operating time. If automatic alternation of compressors is not provided, the facility staff shall arrange a schedule for manual alternation.

(C) Each compressor motor shall be provided with electrical components including, but not limited to, the following:

(1) Dedicated disconnect switch installed in the electrical circuit ahead of each motor starter
(2) Motor starting device
(3) Overload protection
(4) Where compressor systems having two or more compressors employ a control transformer or other voltage control power device, installation of at least two such devices
(5) Control circuits arranged in such a manner that the shutdown of one compressor does not interrupt the operation of another compressor
(6) Automatic restart function, such that the compressor(s) will restart after power interruption without manual intervention

(D) Electrical installation and wiring shall conform to the requirements of *NFPA 70, National Electrical Code.*

(E) Emergency electrical service for the compressors shall conform to the requirements of the essential electrical system as described in Chapter 6.

5.1.3.6.3.11 Compressor Intake.

(A) The medical air compressors shall draw their air from a source of clean air.

(B) The medical air intake shall be located a minimum of 7.6 m (25 ft) from ventilating system exhausts, fuel storage vents, combustion vents, plumbing vents, vacuum and WAGD discharges, or areas that can collect vehicular exhausts or other noxious fumes.

(C) The medical air intake shall be located a minimum of 6 m (20 ft) above ground level.

(D) The medical air intake shall be located a minimum of 3.0 m (10 ft) from any door, window, or other opening in the building.

(E) If an air source equal to or better than outside air (e.g., air already filtered for use in operating room ventilating systems) is available, it shall be permitted to be used for the medical air compressors with the following provisions:

(1) This alternate source of supply air shall be available on a continuous 24-hour-per-day, 7-day-per-week basis.
(2) Ventilating systems having fans with motors or drive belts located in the airstream shall not be used as a source of medical air intake.

(F) Compressor intake piping shall be permitted to be made of materials and use a joining technique as permitted under 5.1.10.2 and 5.1.10.3.

(G) Air intakes for separate compressors shall be permitted to be joined together to one common intake where the following conditions are met:

(1) The common intake is sized to minimize back pressure in accordance with the manufacturer's recommendations.
(2) Each compressor can be isolated by manual or check valve, blind flange, or tube cap to prevent open inlet piping when the compressor(s) is removed for service from the consequent backflow of room air into the other compressor(s).

(H) The end of the intake shall be turned down and screened or otherwise be protected against the entry of vermin, debris, or precipitation by screening fabricated or composed of a noncorroding material.

5.1.3.6.3.12 Operating Alarms and Local Signals. Medical air systems shall be monitored for conditions that can affect air quality during use or in the event of failure, based on the type of compressor(s) used in the system.

(A) A local alarm complying with 5.1.9.5 shall be provided for the medical air compressor source.

(B) Where liquid ring air compressors, compressors having water-cooled heads, or water-cooled aftercoolers are used, air receivers shall be equipped with a high water level sensor that shuts down the compressor system and activates a local alarm indicator. *[See 5.1.9.5.4(7).]*

(C) Where liquid ring compressors are used, each compressor shall have a liquid level sensor in each air–water separator that, when the liquid level is above the design level, shuts down its compressor and activates a local alarm indicator. *[See 5.1.9.5.4(8).]*

(D) Where nonliquid ring compressors compliant with 5.1.3.6.3.4(A)(1) are used, the air temperature at the immediate outlet of each compressor cylinder shall be monitored by a high-temperature sensor that shuts down that compressor and activates a local alarm indicator *[see 5.1.9.5.4(9)]*. The temperature setting shall be as recommended by the compressor manufacturer.

(E) Where compressors compliant with 5.1.3.6.3.4(A)(2) and (3) are used, the following requirements shall apply:

(1) The air temperature at the immediate outlet of each compressor chamber shall be monitored by a high-temperature sensor that shuts down that compressor and activates a local alarm indicator *(see 5.1.9.5.4)*, the temperature setting shall be as recommended by the compressor manufacturer.
(2) Coalescing filters with element change indicator shall be provided.
(3) Charcoal absorber shall be provided.
(4) Gaseous hydrocarbons shall be monitored on a quarterly basis.

(F) When the backup or lag compressor is running, a local alarm shall activate *[see 5.1.9.5.4(1)]*. This signal shall be manually reset.

5.1.3.6.3.13 Medical Air Quality Monitoring. Medical air quality shall be monitored downstream of the medical air regulators and upstream of the piping system as follows:

(1) Dew point shall be monitored and shall activate a local alarm and all master alarms when the dew point at system delivery pressure exceeds +2°C (+35°F).
(2) Carbon monoxide shall be monitored and shall activate a local alarm when the CO level exceeds 10 ppm. *[See 5.1.9.5.4(2).]*
(3) Dew point and carbon monoxide monitors shall activate their individual monitor's signal at the alarm panels where their signals are required when their power is lost.

5.1.3.6.3.14 Category 1 Medical Air Proportioning System.

(A) General.

(1) Medical air reconstituted from oxygen USP and nitrogen NF, produced using proportioning system(s), shall be required to meet the following:
  (a) The quality of medical air shall be in accordance with 5.1.3.6.1.
  (b) The system shall be capable of supplying this quality of medical air, per 5.1.3.6.1, over the entire range of flow.
  (c) The system shall produce medical air with an oxygen content of 19.5 percent to 23.5 percent.
  (d) The medical air shall be cleared for marketing by the FDA or approved by the FDA.
(2) The medical air proportioning system shall operate automatically.
(3) The mixture shall be analyzed continuously, and a recording capability shall be provided (e.g., via data port).
(4) The analyzing system specified in 5.1.3.6.3.14(A)(3) shall be a dedicated and an independent analyzer used to control the medical air proportioning system.
(5) If the mixture goes out of specification, an alarm shall be activated automatically, the primary medical air proportioning system shall be disconnected, and the reserve supply shall be activated.
(6) The system shall be arranged such that manual intervention is necessary to correct the composition of the mixture before reconnecting the medical air proportioning system to the health care facility pipeline system.
(7) If dedicated sources of oxygen USP and nitrogen NF supply the medical air proportioning system, reserve sources for the oxygen and nitrogen shall not be required.
(8) If dedicated sources of oxygen USP and nitrogen NF supply the medical air proportioning system, they shall not be used as the reserves for oxygen and nitrogen systems supplying the pipelines of the health care facility.
(9)*If the sources of oxygen USP and nitrogen NF that supply the medical air proportioning system are the same sources that supply the health care facility, engineering controls shall be provided to prevent cross contamination of oxygen and nitrogen supply lines, as provided in 5.1.3.5.8.
(10) A risk analysis and approval from the authority having jurisdiction shall be required.

(B) Location. The medical air proportioning system shall be located per 5.1.3.3 as follows:

(1) The medical air proportioning system's supply of oxygen USP and nitrogen NF shall be located per 5.1.3.3 and NFPA 55, as applicable.
(2) The mixing device and controls, analyzers, and receivers shall be located indoors within a room or area per 5.1.3.3.1.
(3) The indoor location shall include atmospheric monitoring for oxygen concentration.
(4) The indoor location shall be constructed with all required utilities (e.g., electricity, drains, lighting) per *NFPA 5000*.
(5) The indoor location shall be ventilated and heated per Chapter 9 and the manufacturer's recommendations.

(C) Required Components. The medical air proportioning system shall consist of the following:

(1) Supply of oxygen USP and supply of nitrogen NF as follows:
  (a) The supply lines shall be filtered to remove particulate entering the proportioning system.
  (b) The minimum safe supply gas temperature and recommended local signal shall be specified by the medical air proportioning system manufacturer.
(2) Mixing device with analyzers and engineering controls per manufacturer's recommendations to include, as a minimum, the following:
  (a) At least two oxygen analyzers capable of independently monitoring oxygen concentration

(b) Mechanism where each analyzer based upon nonconforming oxygen concentration is capable, directly or via other medical air proportioning system controls, of automatically shutting off the supply from the medical air proportioning system to the medical air piped distribution system and activating the reserve supply

(c) Mechanism where each analyzer, based upon nonconforming oxygen concentration, is capable, directly or via other proportioning system controls, of automatically shutting off the supply of oxygen and nitrogen to the proportioning system and activating the reserve supply

(d) Provision for manual resetting of the proportioning system after detection of nonconforming oxygen concentration and subsequent shutdown once conforming oxygen concentration is established, in order to re-establish flow to the medical air piping system

(e) Means of verifying the performance of the analyzers by reference to an air standard, with known traceable oxygen content

(3) Minimum of one recorder for recording the medical air proportioning system performance and air quality for a period of not less than 24 hours

(4) Continuous analysis of the mixture and a recording capability provided (e.g., via a data port)

(5) Mechanism for isolating the primary medical air proportioning system from the reserve supply and the medical air piping distribution system by employing sequential valves for redundancy

(6) Capability of the reserve supply to automatically activate if the primary supply is isolated

(7) Reserve supply of medical air USP sized, at minimum, for an average day's supply and consisting of one of the following:

(a) Additional medical air proportioning unit with a dedicated supply of oxygen USP and nitrogen NF

(b) Medical air compressor system per 5.1.3.5.11, with the exception of the allowance of a simplex medical air compressor system

(c) Medical air cylinder manifold per 5.1.3.5.11

(8) Receiver fitted with a pressure relief valve and pressure gauge as follows:

(a) The receiver shall be constructed of corrosion-resistant materials.

(b) The receiver, relief valves, and pressure gauges shall comply with ASME *Boiler and Pressure Vessel Code* and manufacturer's recommendations.

(9)*Warning systems per 5.1.9, including a local signal and master alarm that indicates nonconforming oxygen concentration per manufacturer's recommendations

(10) Final line pressure regulators complying with 5.1.3.5.5

(11) Pressure relief complying with 5.1.3.5.6

(12) Local signals complying with 5.1.3.5.9.2

5.1.3.7* Medical–Surgical Vacuum Supply Systems.

5.1.3.7.1 Medical–Surgical Vacuum Sources.

5.1.3.7.1.1 Medical–surgical vacuum sources shall be located per 5.1.3.3 as follows:

(1) Indoors in a dedicated mechanical equipment area, adequately ventilated and with any required utilities

(2) In a room ventilated per 5.1.3.3.3.3

(3) For air-cooled equipment, in a room designed to maintain the ambient temperature range as recommended by the equipment manufacturer

5.1.3.7.1.2 Medical–surgical vacuum sources shall consist of the following:

(1) Two or more vacuum pumps sufficient to serve the peak calculated demand with the largest single vacuum pump out of service

(2) Automatic means to prevent backflow from any on-cycle vacuum pumps through any off-cycle vacuum pumps

(3) Shutoff valve or other isolation means to isolate each vacuum pump from the centrally piped system and other vacuum pumps for maintenance or repair without loss of vacuum in the system

(4) Vacuum receiver

(5) Piping between the vacuum pump(s), discharge(s), receiver(s), and vacuum source shutoff valve in accordance with 5.1.10.2, except brass, galvanized, or black steel pipe, which is permitted to be used as recommended by the manufacturer

(6) Except as defined in 5.1.3.7.1.2(1) through (5), materials and devices used between the medical vacuum exhaust and the medical vacuum source that are permitted to be of any design or construction appropriate for the service as determined by the manufacturer

5.1.3.7.2 Vacuum Pumps.

5.1.3.7.2.1 Vacuum pumps shall be constructed of materials deemed suitable by the manufacturer.

5.1.3.7.2.2 Antivibration mountings shall be installed for vacuum pumps as required by equipment dynamics or location and in accordance with the manufacturer's recommendations.

5.1.3.7.2.3 Flexible connectors shall connect the vacuum pumps with their intake and outlet piping.

5.1.3.7.2.4 For liquid ring vacuum pumps, seal water shall be of a quality recommended by the vacuum pump manufacturer.

5.1.3.7.3 Vacuum Receivers. Receivers for vacuum shall meet the following requirements:

(1) They shall be made of materials deemed suitable by the manufacturer.

(2) They shall comply with Section VIII, "Unfired Pressure Vessels," of the ASME *Boiler and Pressure Vessel Code*.

(3) They shall be capable of withstanding a gauge pressure of 415 kPa (60 psi) and 760 mm (30 in.) gauge HgV.

(4) They shall be equipped with a manual drain.

(5) They shall be of a capacity based on the technology of the pumps.

5.1.3.7.4 Piping Arrangement and Redundancies.

5.1.3.7.4.1 Piping arrangement shall be as follows:

(1) Piping shall be arranged to allow service and a continuous supply of medical–surgical vacuum in the event of a single fault failure.

(2) Piping arrangement shall be permitted to vary based on the technology(ies) employed, provided that an equal level of operating redundancy is maintained.

(3) Where only one set of vacuum pumps is available for a combined medical–surgical vacuum system and an analysis, a research, or a teaching laboratory vacuum system, such laboratories shall be connected separately from the medical–surgical system directly to the receiver tank through its own isolation valve and fluid trap located at the receiver, and between the isolation valve and fluid trap, a scrubber shall be permitted to be installed.

5.1.3.7.4.2 The medical–surgical vacuum receiver(s) shall be serviceable without shutting down the medical–surgical vacuum system by any method to ensure continuation of service to the facility's medical–surgical pipeline distribution system.

5.1.3.7.4.3 Medical–surgical vacuum source systems shall be provided with a source shutoff valve per 5.1.4.2.

5.1.3.7.5 Electrical Power and Control.

5.1.3.7.5.1 Additional pumps shall automatically activate when the pump(s) in operation is incapable of adequately maintaining the required vacuum.

5.1.3.7.5.2 Automatic or manual alternation of pumps shall allow division of operating time. If automatic alternation of pumps is not provided, the facility staff shall arrange a schedule for manual alternation.

5.1.3.7.5.3 Each pump motor shall be provided with electrical components including, but not limited to, the following:

(1) Dedicated disconnect switch installed in the electrical circuit ahead of each motor starter
(2) Motor starting device
(3) Overload protection
(4) Where pump systems having two or more pumps employ a control transformer or other voltage control power device, at least two such devices
(5) Control circuits arranged in such a manner that the shutdown of one pump does not interrupt the operation of another pump
(6) Automatic restart function such that the pump(s) will restart after power interruption without manual intervention

5.1.3.7.5.4 Electrical installation and wiring shall conform to the requirements of *NFPA 70, National Electrical Code.*

5.1.3.7.5.5 Emergency electrical service for the pumps shall conform to the requirements of the essential electrical system as described in Chapter 6.

5.1.3.7.6 Medical–Surgical Vacuum Source Exhaust.

5.1.3.7.6.1 The medical–surgical vacuum pumps shall exhaust in a manner and location that minimizes the hazards of noise and contamination to the facility and its environment.

5.1.3.7.6.2 The exhaust shall be located as follows:

(1) Outdoors
(2) At least 7.5 m (25 ft) from any door, window, air intake, or other openings in buildings or places of public assembly
(3) At a level different from air intakes
(4) Where prevailing winds, adjacent buildings, topography, or other influences will not divert the exhaust into occupied areas or prevent dispersion of the exhaust

5.1.3.7.6.3 The end of the exhaust shall be turned down and screened or otherwise be protected against the entry of vermin, debris, or precipitation by screening fabricated or composed of a noncorroding material.

5.1.3.7.6.4 The exhaust shall be free of dips and loops that might trap condensate or oil or provided with a drip leg and valved drain at the bottom of the low point.

5.1.3.7.6.5 Vacuum exhausts from multiple pumps shall be permitted to be joined together to one common exhaust where the following conditions are met:

(1) The common exhaust is sized to minimize back pressure in accordance with the pump manufacturer's recommendations.
(2) Each pump can be isolated by manual or check valve, blind flange, or tube cap to prevent open exhaust piping when the pump(s) is removed for service from consequent flow of exhaust air into the room.

5.1.3.7.6.6 Vacuum exhaust piping shall be permitted to be made of materials and use a joining technique as permitted under 5.1.10.2 and 5.1.10.3.

5.1.3.7.7 Operating Alarms. Medical–surgical vacuum systems shall activate a local alarm when the backup or lag pump is running per 5.1.9.5. This signal shall be manually reset.

5.1.3.8* Waste Anesthetic Gas Disposal (WAGD).

5.1.3.8.1* Sources. WAGD sources shall be chosen in consultation with the medical staff having knowledge of the requirements to determine the type of system, number and placement of terminals, and other required safety and operating devices.

5.1.3.8.1.1 WAGD shall be permitted to be produced through the medical–surgical vacuum source, by a dedicated producer, or by venturi.

5.1.3.8.1.2 If WAGD is produced by the medical–surgical vacuum source, the following shall apply:

(1) The medical–surgical vacuum source shall comply with 5.1.3.7.
(2) The total concentration of oxidizers (oxygen and nitrous oxide) shall be maintained below 23.6 percent, or the vacuum pump shall comply with 5.1.3.8.2.1.
(3) The medical–surgical vacuum source shall be sized to accommodate the additional volume.

5.1.3.8.1.3 If WAGD is produced by a dedicated WAGD producer with a total power equal to or greater than 1 horsepower in total (both producers), the following shall apply:

(1) The WAGD source shall be located in accordance with 5.1.3.3.
(2) The WAGD source shall be located indoors in a dedicated mechanical equipment area with any required utilities.
(3) The WAGD source shall be ventilated per 5.1.3.3.3.3.
(4) For air-cooled equipment, the WAGD source shall be located to maintain the ambient temperature range as recommended by the manufacturer.
(5) The WAGD producers shall comply with 5.1.3.8.2.

5.1.3.8.1.4 If WAGD is produced by a dedicated WAGD producer with a total power less than 1 horsepower in total (both producers), the following shall apply:

(1) The WAGD source shall be permitted to be located near the inlet(s) served.
(2) For air-cooled equipment, the WAGD source shall be located to maintain the ambient temperature range as recommended by the manufacturer.

5.1.3.8.1.5 For liquid ring pumps in WAGD service, seal water shall be of a quality as recommended by the pump manufacturer.

5.1.3.8.1.6 The WAGD source shall consist of the following:

(1) Two or more WAGD producers sufficient to serve the peak calculated demand with the largest single WAGD producer out of service
(2) Automatic means to prevent backflow from any on-cycle WAGD producers through any off-cycle WAGD producers

(3) Shutoff valve to isolate each WAGD producer from the centrally piped system and other WAGD producers for maintenance or repair without loss of WAGD in the system
(4) Piping between the WAGD producers and the source shutoff valve compliant with 5.1.10.2, as recommended by the manufacturer
(5) Antivibration mountings installed for WAGD producers as required by equipment dynamics or location and in accordance with the manufacturer's recommendations
(6) Flexible connectors interconnecting the producers with their intake and outlet piping as required by equipment dynamics or location and in accordance with the WAGD producer manufacturer's recommendations

5.1.3.8.1.7 If WAGD is produced by a venturi, the following shall apply:

(1) The venturi shall not be user-adjustable (i.e., require the use of special tools).
(2) The venturi shall be driven using water, inert gas, instrument air, or other dedicated air source.
(3) Medical air shall not be used to power the venturi.

5.1.3.8.2 WAGD Producers.

5.1.3.8.2.1 Vacuum pumps dedicated for WAGD service shall be as follows:

(1) Compliant with 5.1.3.7.2
(2) Designed of materials and using lubricants and sealants that are inert in the presence of oxygen, nitrous oxide, and halogenated anesthetics

5.1.3.8.2.2 Vacuum producers (e.g., fans or blowers) designed for operation at vacuums below 130 mm (5 in.) HgV shall be as follows:

(1) Permitted to be made of any materials determined by the manufacturer as suitable for the service
(2) Provided with antivibration mountings as required by equipment dynamics or location and in accordance with the manufacturer's recommendation
(3) Connected with their intake and outlet piping through flexible connections
(4) Used only for WAGD service and not employed for other services
(5) Interconnected via piping, ductwork, and so forth, made of materials determined by the manufacturer as suitable to the service

5.1.3.8.3 WAGD Alarms.

5.1.3.8.3.1 When the WAGD system is served by a central source(s), a local alarm complying with 5.1.9.5 shall be provided for the WAGD source.

5.1.3.8.3.2 A WAGD source system shall activate a local alarm when the backup or lag producer is running.

5.1.3.8.4 Electrical Power and Control.

5.1.3.8.4.1 Additional producers shall automatically activate when the producer(s) in operation is incapable of maintaining the required vacuum.

5.1.3.8.4.2 Automatic or manual alternation of producers shall allow division of operating time. If automatic alternation of producers is not provided, the facility staff shall arrange a schedule for manual alternation.

5.1.3.8.4.3 Each producer motor shall be provided with electrical components including, but not limited to, the following:

(1) Dedicated disconnect switch installed in the electrical circuit ahead of each motor starter
(2) Motor starting device
(3) Overload protection
(4) Where WAGD systems having two or more producers employ a control transformer or other voltage control power device, at least two such devices
(5) Control circuits arranged in such a manner that the shutdown of one producer does not interrupt the operation of another producer
(6) Automatic restart function such that the pump(s) will restart after power interruption without manual intervention

5.1.3.8.4.4 Electrical installation and wiring shall conform to the requirements of *NFPA 70, National Electrical Code.*

5.1.3.8.4.5 Emergency electrical service for the producers shall conform to the requirements of the essential electrical system as described in Chapter 6.

5.1.3.8.5 WAGD Exhaust. The WAGD pumps shall exhaust in compliance with 5.1.3.7.6.

5.1.4* Valves.

5.1.4.1 General.

5.1.4.1.1 Gas and Vacuum Shutoff Valves. Shutoff valves shall be provided to isolate sections or portions of the piped distribution system for maintenance, repair, or planned future expansion need and to facilitate periodic testing.

5.1.4.1.2 Security. All valves, except valves in zone valve box assemblies, shall be secured by any of the following means:

(1) Located in secured areas
(2) Locked or latched in their operating position
(3) Located above ceilings, but remaining accessible and not obstructed

5.1.4.1.3 Labeling. All valves shall be labeled as to gas supplied and the area(s) controlled, in accordance with 5.1.11.2.

5.1.4.1.4 Accessibility.

(A) Zone valves shall be installed in valve boxes with removable covers large enough to allow manual operation of valves.

(B) Zone valves for use in certain areas, such as psychiatric or pediatric areas, shall be permitted to be secured with the approval of the authority having jurisdiction to prevent inappropriate access.

5.1.4.1.5 Flammable Gases. Valves for nonflammable medical gases shall not be installed with valves for flammable gases in the same zone valve box assembly with flammable gases.

5.1.4.1.6 Valve Types. New or replacement valves shall be permitted to be of any type as long as they meet the following conditions:

(1) They have a maximum pressure drop at intended maximum flow of 1.4 kPa (0.2 psig) in pressure service and 3.8 mm Hg (0.15 Hg) in vacuum service.
(2) They use a quarter turn to off.
(3) They are constructed of materials suitable for the service.
(4) They are provided with copper tube extensions by the manufacturer for brazing.
(5) They indicate to the operator if the valve is open or closed.

(6) They permit in-line serviceability.
(7) They are cleaned for oxygen service by the manufacturer if used for any positive pressure service.

5.1.4.2 Source Valve.

5.1.4.2.1 A shutoff valve shall be placed at the immediate connection of each source system to the piped distribution system to allow the entire source, including all accessory devices (e.g., air dryers, final line regulators), to be isolated from the facility.

5.1.4.2.2 The source valve shall be located in the immediate vicinity of the source equipment.

5.1.4.3* Main Line Valve.

5.1.4.3.1 A shutoff valve shall be provided in the main supply line inside of the buildings being served, except where one or more of the following conditions exist:

(1) The source and source valve are located inside the building served.
(2) The source system is physically mounted to the wall of the building served, and the pipeline enters the building in the immediate vicinity of the source valve.

5.1.4.3.2 The main line valve shall be located on the facility side of the source valve and outside of the source room, the enclosure, or where the main line first enters the building.

5.1.4.4 Riser Valve. Each riser supplied from the main line shall be provided with a shutoff valve in the riser adjacent to the main line.

5.1.4.5 Service Valves.

5.1.4.5.1 Service valves shall be installed to allow servicing or modification of lateral branch piping from a main or riser without shutting down the entire main, riser, or facility.

5.1.4.5.2 Only one service valve shall be required for each branch off of a riser, regardless of how many zone valve boxes are installed on that lateral.

5.1.4.5.3 Service valves shall be placed in the branch piping prior to any zone valve box assembly on that branch.

5.1.4.6 Zone Valves.

5.1.4.6.1 All station outlets/inlets shall be supplied through a zone valve as follows:

(1) The zone valve shall be placed such that a wall intervenes between the valve and outlets/inlets that it controls.
(2) The zone valve shall serve only outlets/inlets located on that same story.
(3) The zone valve shall not be located in a room with station outlets/inlets that it controls.

5.1.4.6.2 Zone valves shall be readily operable from a standing position in the corridor on the same floor they serve.

5.1.4.6.3 Zone valves shall be so arranged that shutting off the supply of medical gas or vacuum to one zone will not affect the supply of medical gas or vacuum to another zone or the rest of the system.

5.1.4.6.4 A pressure/vacuum indicator shall be provided on the station outlet/inlet side of each zone valve.

5.1.4.6.5 Zone valve boxes shall be installed where they are visible and accessible at all times.

5.1.4.6.6 Zone valve boxes shall not be installed behind normally open or normally closed doors or otherwise hidden from plain view.

5.1.4.6.7 Zone valve boxes shall not be located in closed or locked rooms, areas, or closets.

5.1.4.6.8 A zone valve shall be located immediately outside each vital life-support area, critical care area, and anesthetizing location of moderate sedation, deep sedation, or general anesthesia, in each medical gas or vacuum line, or both, and located so as to be readily accessible in an emergency.

5.1.4.6.9 All gas delivery columns, hose reels, ceiling tracks, control panels, pendants, booms, or other special installations shall be located downstream of the zone valve.

5.1.4.6.10 Zone valves shall be so arranged that shutting off the supply of gas to any one operating room or anesthetizing location will not affect the others.

5.1.4.7 In-Line Shutoff Valves. Optional in-line valves shall be permitted to be installed to isolate or shut off piping for servicing of individual rooms or areas.

5.1.4.8 Valves for Future Connections.

5.1.4.8.1 Future connection valves shall be labeled as to gas content.

5.1.4.8.2 Downstream piping shall be closed with a brazed cap with tubing allowance for cutting and rebrazing.

5.1.4.9 In-Line Check Valves. New or replacement check valves shall be as follows:

(1) They shall be of brass or bronze construction.
(2) They shall have brazed extensions.
(3) They shall have in-line serviceability.
(4) They shall not have threaded connections.
(5) They shall have threaded purge points of ⅛ in. NPT.

5.1.5* Station Outlets/Inlets.

5.1.5.1 Each station outlet/inlet for medical gases or vacuums shall be gas-specific, whether the outlet/inlet is threaded or is a noninterchangeable quick coupler.

5.1.5.2 Each station outlet shall consist of a primary and a secondary valve (or assembly).

5.1.5.3 Each station inlet shall consist of a primary valve (or assembly) and shall be permitted to include a secondary valve (or assembly).

5.1.5.4 The secondary valve (or assembly) shall close automatically to stop the flow of gas (or vacuum, if provided) when the primary valve (or assembly) is removed.

5.1.5.5 Each outlet/inlet shall be legibly identified in accordance with 5.1.11.3.

5.1.5.6 Threaded outlets/inlets shall be non-interchangeable connections complying with the mandatory requirements of CGA V-5, *Diameter-Index Safety System (Noninterchangeable Low Pressure Connections for Medical Gas Applications)*.

5.1.5.7 Each station outlet/inlet, including those mounted in columns, hose reels, ceiling tracks, or other special installations, shall be designed so that parts or components that are required to be gas-specific for compliance with 5.1.5.1 and 5.1.5.9 cannot be interchanged between the station outlet/inlet for different gases.

5.1.5.8 The use of common parts in outlets/inlets, such as springs, O-rings, fasteners, seals, and shutoff poppets, shall be permitted.

5.1.5.9 Components of a vacuum station inlet necessary for the maintenance of vacuum specificity shall be legibly marked to identify them as components or parts of a vacuum or suction system.

5.1.5.10 Components of inlets not specific to a vacuum shall not be required to be marked.

5.1.5.11 Factory-installed copper inlet tubes on station outlets extending no further than 205 mm (8 in.) from the body of the terminal shall be not less than DN8 (NPS ¼) (⅜ in. O.D.) size, with 8 mm (0.3 in.) minimum inside diameter.

5.1.5.12 Factory-installed copper outlet tubes on station inlets extending no further than 205 mm (8 in.) from the body of the terminal shall be not less than DN10 (NPS ⅜) (½ in. O.D.) size, with 10 mm (0.4 in.) minimum inside diameter.

5.1.5.13 Station outlets/inlets shall be permitted to be recessed or otherwise protected from damage.

5.1.5.14 When multiple wall outlets/inlets are installed, they shall be spaced to allow the simultaneous use of adjacent outlets/inlets with any of the various types of therapy equipment.

5.1.5.15 Station outlets in systems having nonstandard operating pressures shall meet the following additional requirements:

(1) They shall be gas-specific.
(2) They shall be pressure-specific where a single gas is piped at more than one operating pressure [e.g., a station outlet for oxygen at 550 kPa (80 psi) shall not accept an adapter for oxygen at 345 kPa (50 psi)].
(3) If operated at a pressure in excess of 550 kPa (80 psi), they shall be either D.I.S.S. connectors or comply with 5.1.5.15(4).
(4) If operated at a gauge pressure between 1380 kPa and 2070 kPa (200 psi and 300 psi), the station outlet shall be designed so as to prevent the removal of the adapter until the pressure has been relieved to prevent the adapter injuring the user or others when removed from the outlet.

5.1.5.16 WAGD networks shall provide a WAGD inlet in all locations where nitrous oxide or halogenated anesthetic gas is intended to be administered.

5.1.5.16.1 Station inlets for WAGD service shall have the following additional characteristics:

(1) They shall not be interchangeable with any other systems, including medical–surgical vacuum.
(2) Components necessary for the maintenance of WAGD specificity shall be legibly marked to identify them as components of a WAGD inlet.
(3) They shall be of a type appropriate for the flow and vacuum level required by the facility's gas anesthetic machines.
(4) They shall be located to avoid physical damage to the inlet.

5.1.6* Manufactured Assemblies.

5.1.6.1 Manufactured assemblies shall be pretested by the manufacturer prior to arrival at the installation site in accordance with the following:

(1) Initial blowdown test per 5.1.12.2.2
(2) Initial pressure test per 5.1.12.2.3
(3) Piping purge test per 5.1.12.2.5
(4) Standing pressure test per 5.1.12.2.6 or 5.1.12.2.7, except as permitted under 5.1.6.2

5.1.6.2 The standing pressure test under 5.1.6.1(4) shall be permitted to be performed by any testing method that will ensure a pressure decay of less than 1 percent in 24 hours.

5.1.6.3 The manufacturer of the assembly shall provide documentation certifying the performance and successful completion of the tests required in 5.1.6.1.

5.1.6.4 Manufactured assemblies employing flexible hose shall use hose and flexible connectors with a minimum burst gauge pressure of 6895 kPa (1000 psi).

5.1.6.5 Manufactured assemblies shall have a flame spread index of not greater than 200 when tested in accordance with ASTM E 84, *Standard Test Method for Surface Burning Characteristics of Building Materials*, or shall comply with the requirements for heat release in accordance with NFPA 286, *Standard Methods of Fire Tests for Evaluating Contribution of Wall and Ceiling Interior Finish to Room Fire Growth*, as described in Section 10.2 of NFPA *101, Life Safety Code*.

5.1.6.6 Manufactured assemblies employing flexible hose or tubing shall be attached to the pipelines using station outlets/inlets.

5.1.6.7 Manufactured assemblies employing hose or flexible connectors, where the station outlet/inlet attached to the piping is not fully and immediately accessible (i.e., cannot be manipulated without the removal of panels, doors, and so forth), shall have station outlets/inlets with the following additional characteristics:

(1) They shall be gas-specific connections with positive locking mechanisms that ensure the connector is firmly seated and cannot detach without intentional actuation of the release (e.g., D.I.S.S. connectors).
(2) In pressure gases, they shall be permitted to omit the secondary valve (or assembly) required in 5.1.5.2.
(3) In vacuum and WAGD, they shall be permitted to omit both primary and secondary valves (or assemblies) for minimum restriction to flow.
(4) They shall be provided with a second terminal at which the user connects and disconnects that complies with 5.1.5.

5.1.6.8 Station outlets/inlets installed in manufactured assemblies connected to the pipeline by brazing shall comply with 5.1.5.

5.1.6.9 The installation of manufactured assemblies shall be tested in accordance with 5.1.12.

5.1.7* Surface-Mounted Medical Gas Rails (MGR).

5.1.7.1 Medical gas rail (MGR) assemblies shall be permitted to be installed where multiple uses of medical gases and vacuum at a single patient location are required or anticipated.

5.1.7.2 MGR assemblies shall be entirely visible in the room, not passing into or through walls, partitions, and so forth.

5.1.7.3 MGR assemblies shall be made of materials with a melting point of at least 538°C (1000°F).

5.1.7.4 MGR assemblies shall be cleaned per 5.1.10.1.1.

5.1.7.5 Station outlets or inlets shall not be placed on the ends of MGR assemblies.

5.1.7.6 Openings for station outlets/inlets in the MGR shall be gas-specific.

5.1.7.7 Openings in the MGR not occupied by station outlets/inlets (e.g., for future use) shall be capped or plugged so that a special tool is required for removal (i.e., cannot be removed by a wrench, pliers, a screwdriver, or other common tool).

5.1.7.8 MGR assemblies shall connect to the pipeline through fittings that are brazed to the pipeline.

5.1.7.9* Where the pipeline and the MGR assembly are of dissimilar metals, the connections shall be plated or otherwise protected from interaction between the metals.

5.1.7.10 The installation of the MGR shall be tested in accordance with 5.1.12.

5.1.8 Pressure and Vacuum Indicators.

5.1.8.1 General.

5.1.8.1.1 Pressure indicators and manometers for medical gas piping systems shall be cleaned for oxygen service.

5.1.8.1.2 Gauges shall comply with ANSI/ASME B40.100, *Pressure Gauges and Gauge Attachments*.

5.1.8.1.3* The scale range of positive pressure analog indicators shall be such that the normal operating pressure is within the middle third of the total range [e.g., an indicator of 0 to 2070 kPa (0 to 300 psi) would have a lower third of 0 to 690 kPa (0 to 100 psig), a middle third of 690 kPa to 1380 kPa (100 psig to 200 psig), and a top third of 1380 kPa to 2070 kPa (200 psig to 300 psig)].

5.1.8.1.4 The accuracy of digital indicators shall be ±5 percent of the operating pressure at which they are used.

5.1.8.1.5 The scale range of vacuum indicators shall be 0 to 760 mm (0 to 30 in.) gauge HgV. Indicators with a normal range display shall indicate normal only above 300 mm (12 in.) gauge HgV.

5.1.8.1.6 Indicators adjacent to master alarm actuators and area alarms shall be labeled to identify the name of, or chemical symbol for, the particular piping system that they monitor.

5.1.8.2 Locations.

5.1.8.2.1 Pressure/vacuum indicators shall be readable from a standing position.

5.1.8.2.2 Pressure/vacuum indicators shall be provided at the following locations, as a minimum:

(1) Adjacent to the alarm-initiating device for source main line pressure and vacuum alarms in the master alarm system
(2) At or in area alarm panels to indicate the pressure/vacuum at the alarm-activating device for each system that is monitored by the panel
(3) On the station outlet/inlet side of zone valves

5.1.8.2.3 All pressure-sensing devices and main line pressure gauges downstream of the source valves shall be provided with a gas-specific demand check fitting to facilitate service testing or replacement.

5.1.8.2.3.1 Gas-specific demand check fittings shall not be required on zone valve pressure indicators.

5.1.8.2.4 Demand check fittings shall be provided for all monitors.

5.1.9* Category 1 Warning Systems.

5.1.9.1 General. All master, area, and local alarm systems used for medical gas and vacuum systems shall include the following:

(1) Separate visual indicators for each condition monitored, except as permitted in 5.1.9.5.2 for local alarms that are displayed on master alarm panels
(2) Visual indicators that remain in alarm until the situation that has caused the alarm is resolved
(3) Cancelable audible indication of each alarm condition that produces a sound with a minimum level of 80 dBA at 0.92 m (3 ft)
(4) Means to indicate a lamp or LED failure and audible failure
(5) Visual and audible indication that the communication with an alarm-initiating device is disconnected
(6) Labeling of each indicator, indicating the condition monitored
(7) Labeling of each alarm panel for its area of surveillance
(8) Reinitiation of the audible signal if another alarm condition occurs while the audible alarm is silenced
(9) Power for master, area alarms, sensors, and switches from the life safety branch of the essential electrical system as described in Chapter 6
(10) Power for local alarms, dew point sensors, and carbon monoxide sensors permitted to be from the same essential electrical branch as is used to power the air compressor system
(11) Where used for communications, wiring from switches or sensors that is supervised or protected as required by 517.30(C)(3) of *NFPA 70, National Electrical Code,* for life safety and critical branches circuits in which protection is any of the following types:
    (a) Conduit
    (b) Free air
    (c) Wire
    (d) Cable tray
    (e) Raceways
(12) Communication devices that do not use electrical wiring for signal transmission will be supervised such that failure of communication shall initiate an alarm.
(13) Assurance by the responsible authority of the facility that the labeling of alarms, where room numbers or designations are used, is accurate and up-to-date
(14) Provisions for automatic restart after a power loss of 10 seconds (e.g., during generator start-up) without giving false signals or requiring manual reset
(15) Alarm switches/sensors installed so as to be removable

5.1.9.2* Master Alarms. A master alarm system shall be provided to monitor the operation and condition of the source of supply, the reserve source (if any), and the pressure in the main lines of each medical gas and vacuum piping system.

5.1.9.2.1 The master alarm system shall consist of two or more alarm panels located in at least two separate locations, as follows:

(1) One master alarm panel shall be located in the office or work space of the on-site individual responsible for the maintenance of the medical gas and vacuum piping systems.

(2) In order to ensure continuous surveillance of the medical gas and vacuum systems while the facility is in operation, the second master alarm panel shall be located in an area of continuous observation (e.g., the telephone switchboard, security office, or other continuously staffed location).

5.1.9.2.2 A centralized computer system shall be permitted to be substituted for one of the master alarms required in 5.1.9.2.1 if the computer system complies with 5.1.9.3.

5.1.9.2.3 The master alarm panels required in 5.1.9.2.1 shall communicate directly to the alarm-initiating devices that they monitor.

5.1.9.2.3.1 If communication is achieved by wires, the following shall apply:

(A) Each of the two mandatory alarms shall be wired independently to the initiating device(s) for each signal.

(B) The wiring between each mandatory alarm(s) and the initiating device(s) shall not utilize common conductors that, if interrupted, would disable more than one signal.

(C) Each set of wires (in whatever number as required by the alarm) shall run to the initiating device(s) without interruption other than in-line splices necessary to complete the necessary length of wire.

(D) Where initiating devices are remote from the building and the wiring is to run underground in compliance with *NFPA 70*, the following exceptions shall be permitted to be used:

(1) Wiring from the initiating device and through the underground section shall be permitted to be run to a junction box located where the wiring first enters the building.
(2) A single set of wires complying with 5.1.9.2.3.1(B) and 5.1.9.2.3.1(C) for each signal shall be permitted to connect the initiating device and the junction box.
(3) Between the junction box and the two mandatory alarm panels, wiring shall comply with 5.1.9.2.3.1(A) through 5.1.9.2.3.1(C), 5.1.9.2.3.4, and 5.1.9.2.3.5 in all respects.

5.1.9.2.3.2 If communication is achieved by means other than wires, the following shall apply:

(A) Each of the mandatory alarms shall communicate independently to the initiating device(s) for each signal.

(B) The means of communication between each mandatory alarm(s) and the initiating device(s) shall not utilize a common communication device that, if interrupted, would disable more than one signal.

5.1.9.2.3.3 A single initiating device shall be permitted to actuate multiple master alarms.

5.1.9.2.3.4 The mandatory master alarm panels shall not be arranged such that failure of either panel would disable any signal on the other panel.

5.1.9.2.3.5 Where a relay is required to ensure correct operation of an initiating device, the control power for the relay shall not be such that disabling any master alarm panel would disable the relay.

5.1.9.2.3.6 Master alarm signals shall not be relayed from one master alarm panel to another.

5.1.9.2.3.7 Where multi-pole alarm relays are used to isolate the alarm-initiating signals to master alarm panels, the control power source for the relays shall be independent of any of the master alarm panels.

5.1.9.2.3.8 Multiple master alarms shall be permitted to monitor a single initiating device.

5.1.9.2.4 Master alarm panels for medical gas and vacuum systems shall each include the following signals:

(1) Alarm indication when, or just before, changeover occurs in a medical gas system that is supplied by a manifold or other alternating-type bulk system that has as a part of its normal operation a changeover from one portion of the operating supply to another
(2) Alarm indication for a bulk cryogenic liquid system when the main supply reaches an average day's supply, indicating low contents
(3) Alarm indication when, or just before, the changeover to the reserve supply occurs in a medical gas system that consists of one or more units that continuously supply the piping system while another unit remains as the reserve supply and operates only in the case of an emergency
(4) Alarm indication for cylinder reserve pressure low when the content of a cylinder reserve header is reduced below one average day's supply
(5) For bulk cryogenic liquid systems, an alarm when or at a predetermined set point before the reserve supply contents fall to one day's average supply, indicating reserve low
(6) Where a cryogenic liquid storage unit is used as a reserve for a bulk supply system, an alarm indication when the gas pressure available in the reserve unit is below that required for the medical gas system to function
(7) Alarm indication when the pressure in the main line of each separate medical gas system increases 20 percent or decreases 20 percent from the normal operating pressure
(8) Alarm indication when the medical–surgical vacuum pressure in the main line of each vacuum system drops to or below 300 mm (12 in.) gauge HgV
(9) Alarm indication(s) from the local alarm panel(s) as described in 5.1.9.5.2 to indicate when one or more of the conditions being monitored at a site is in alarm
(10) Medical air dew point high alarm from each compressor site to indicate when the line pressure dew point is greater than +2°C (+35°F)
(11) WAGD low alarm when the WAGD vacuum level or flow is below effective operating limits
(12) An instrument air dew point high alarm from each compressor site to indicate when the line pressure dew point is greater than −30°C (−22°F)
(13) Alarm indication if the primary or reserve production stops on a proportioning system

5.1.9.2.5 The alarm indications required in 5.1.9.2.4(7) and 5.1.9.2.4(8) shall originate from sensors installed in the main lines immediately downstream (on the patient or use side) of the source valves. Where it is necessary to install a main line valve in addition to a source valve *(see 5.1.4.3)*, the sensors shall be located downstream (on the patient or use side) of the main valve.

5.1.9.3 Master Alarms by Computer Systems. Computer systems used as substitute master alarms as required by 5.1.9.2.1(2) shall have the mechanical and electrical characteristics described in 5.1.9.3.1 and the programming characteristics described in 5.1.9.3.2.

5.1.9.3.1 Computer systems used to substitute for alarms shall have the following mechanical and electrical characteristics:

(1) The computer system shall be in continuous uninterrupted operation and provided with power supplies as needed to ensure such reliability.
(2) The computer system shall be continuously attended by responsible individuals or shall provide remote signaling of responsible parties (e.g., through pagers, telephone autodialers, or other such means).
(3) Where computer systems rely on signal interface devices (e.g., electronic interfaces, other alarm panels, 4 mA to 20 mA cards), such interfaces shall be supervised such that failure of the device(s) shall initiate an alarm(s).
(4) If the computer system does not power the signaling switches/sensors from the same power supply required in 5.1.9.3.1(1), the power supply for the signaling switches/sensors shall be powered from the life safety branch of the essential electrical system as described in Chapter 6.
(5) Computer systems shall be permitted to communicate directly to the sensors/switches in 5.1.9.2.3 in the same manner as an alarm panel if operation of another alarm panel(s) is not impaired.
(6) Communication from the computer system to the signaling switches or sensors shall be supervised such that failure of communication shall initiate an alarm.
(7) Computer systems shall be provided with an audio alert per 5.1.9.1(3), except the audio alert shall be permitted to be only as loud as needed to alert the system operator.
(8) The facility shall ensure compliance with 5.1.9.1(12).

5.1.9.3.2 The operating program for computer systems used to substitute for alarms shall include the following:

(1) The medical gas alarm shall be allocated the priority of a life safety signal.
(2) A medical gas alarm signal shall interrupt any other activity of a lesser priority to run the alarm algorithm(s).
(3) The alarm algorithm shall include activation of an audible alert, activation of any remote signaling protocol, and display of the specific condition in alarm.
(4) The alarm algorithm shall provide for compliance with 5.1.9.1(1), 5.1.9.1(2), 5.1.9.1(3), 5.1.9.1(4), 5.1.9.1(5), and 5.1.9.1(8).

5.1.9.4* Area Alarms. Area alarm panels shall be provided to monitor all medical gas, medical–surgical vacuum, and piped WAGD systems supplying the following:

(1) Anesthetizing locations where moderate sedation, deep sedation, or general anesthesia is administered
(2)*Critical care areas

5.1.9.4.1* Area alarms shall be located at a nurse's station or other similar location that will provide for surveillance.

5.1.9.4.2 Area alarm panels for medical gas systems shall indicate if the pressure in the lines in the area being monitored increases or decreases by 20 percent from the normal line pressure.

5.1.9.4.3 Area alarm panels for medical–surgical vacuum systems shall indicate if the vacuum in the area drops to or below 300 mm (12 in.) gauge HgV.

5.1.9.4.4 Alarm sensors for area alarms shall be located as follows:

(1)*Critical care areas shall have the alarm sensors installed on the patient or use side of each individual zone valve box assemblies.
(2)*Anesthetizing locations where moderate sedation, deep sedation, or general anesthesia is administered shall have the sensors installed either on the source side of any of the individual room zone valve box assemblies or on the patient or use side of each of the individual zone valve box assemblies.

5.1.9.4.5 Area alarm panels for medical gas systems shall provide visual and audible indication in the event a mismatch occurs between the transducer(s) and its associated circuit board(s).

5.1.9.5* Local Alarms. Local alarms shall be installed to monitor the function of the air compressor system(s), medical–surgical vacuum pump system(s), WAGD systems, instrument air systems, and proportioning systems.

5.1.9.5.1 The signals referenced in 5.1.9.5.4 shall be permitted to be located as follows:

(1) On or in the control panel(s) for the machinery being monitored
(2) Within a monitoring device (e.g., dew point monitor or carbon monoxide monitor)
(3) On a separate alarm panel(s)

5.1.9.5.2 The master alarm shall include at least one signal from the source equipment to indicate a problem with the source equipment at this location. This master alarm signal shall activate when any of the required local alarm signals for this source equipment activates.

5.1.9.5.3 If there is more than one medical air compressor system, instrument air compressor system, WAGD system, medical–surgical vacuum pump system, or proportioning system at different locations in the facility, or if the compressors and vacuum sources are in different locations in the facility, then it shall be necessary for each location to have separate alarms at the master panels.

5.1.9.5.4 The following functions shall be monitored at each local alarm site:

(1) Backup or lag compressor in operation, to indicate when the primary or lead air compressor is incapable of satisfying the demand of the requirements of the system, except when the medical air system consists of three or more compressors, in which case the backup or lag signal is permitted to energize when the last compressor has been signaled to start
(2) High carbon monoxide level, to indicate when the carbon monoxide level in the medical air system is 10 ppm or higher
(3) Medical air dew point high, to indicate when the line pressure dew point is greater than +2°C (+35°F)
(4) Backup or lag vacuum pump in operation, to indicate when the primary or lead vacuum pump is incapable of satisfying the demand of the requirements of the system, except when the vacuum pump system consists of three or more pumps, in which case the backup or lag signal is permitted to energize when the last pump has been signaled to start
(5) When a central dedicated WAGD producer is provided per 5.1.3.8.1.3, WAGD lag in use with the signal to be manually reset
(6) Instrument air dew point high, to indicate when the line pressure dew point is greater than -30°C (-22°F)
(7) For compressor systems using liquid ring compressors or compressors with water-cooled components, high water

in the receiver tank, to indicate when the water level in the receiver tank, has reached a level determined to be detrimental to the operation of the system

(8) For compressor systems using liquid ring compressors, high water in the separators
(9) For compressor systems using other than liquid ring compressors, high discharge air temperature
(10) Proportioning systems high/low indicator when the oxygen concentration is outside 19.5 percent to 23.5 percent oxygen
(11) Proportion systems reserve system in operation

5.1.10 Category 1 Distribution.

5.1.10.1 Piping Materials for Field-Installed Positive Pressure Medical Gas Systems.

5.1.10.1.1 Tubes, valves, fittings, station outlets, and other piping components in medical gas systems shall have been cleaned for oxygen service by the manufacturer prior to installation in accordance with the mandatory requirements of CGA G-4.1, *Cleaning Equipment for Oxygen Service*, except that fittings shall be permitted to be cleaned by a supplier or agency other than the manufacturer.

5.1.10.1.2 Each length of tube shall be delivered plugged or capped by the manufacturer and kept sealed until prepared for installation.

5.1.10.1.3 Fittings, valves, and other components shall be delivered sealed and labeled and kept sealed until prepared for installation.

5.1.10.1.4* Tubes shall be hard-drawn seamless copper in accordance with ASTM B 819, *Standard Specification for Seamless Copper Tube for Medical Gas Systems*, medical gas tube, Type L, except Type K shall be used where operating pressures are above a gauge pressure of 1275 kPa (185 psi) and the pipe sizes are larger than DN80 [NPS 3 (3⅛ in. O.D.)].

5.1.10.1.5 ASTM B 819, *Standard Specification for Seamless Copper Tube for Medical Gas Systems*, medical gas tube shall be identified by the manufacturer's markings "OXY," "MED," "OXY/MED," "OXY/ACR," or "ACR/MED" in blue (Type L) or green (Type K).

5.1.10.1.6 The installer shall furnish documentation certifying that all installed piping materials comply with the requirements of 5.1.10.1.1.

5.1.10.2 Piping Materials for Field-Installed Medical–Surgical Vacuum and WAGD Systems.

5.1.10.2.1 Tubes for Vacuum. Piping for vacuum systems shall be constructed of any of the following:

(1) Hard-drawn seamless copper tube in accordance with the following:
 (a) ASTM B 88, *Standard Specification for Seamless Copper Water Tube*, copper tube (Type K, Type L, or Type M)
 (b) ASTM B 280, *Standard Specification for Seamless Copper Tubing for Air Conditioning and Refrigeration Field Service*, copper ACR tube
 (c) ASTM B 819, *Standard Specification for Seamless Copper Tube for Medical Gas Systems*, copper medical gas tubing (Type K or Type L)
(2) Stainless steel tube in accordance with the following:
 (a) ASTM A 269 TP304L or 316L, *Standard Specification for Seamless and Welded Austenitic Stainless Steel Tubing for General Service*
 (b) ASTM A 312 TP304L or 316L, *Standard Specification for Seamless and Welded Austenitic Stainless Steel Pipes*
 (c) A312 TP 304L/316L, Sch. 5S pipe, and A403 WP304L/316L, Sch. 5S fittings

5.1.10.2.2 Vacuum Tube Marking Where Required.

5.1.10.2.2.1 If copper vacuum tubing is installed along with any medical gas tubing, the vacuum tubing shall, prior to installation, be prominently labeled or otherwise identified to preclude using materials or installation procedures in the medical gas system that are not suitable for oxygen service.

5.1.10.2.2.2 If medical gas tube in accordance with ASTM B 819, *Standard Specification for Seamless Copper Tube for Medical Gas Systems*, is used for vacuum piping, such special marking shall not be required, provided that the vacuum piping installation meets all other requirements for medical gas piping, including the prohibition of flux on copper-to-copper joints and the use of a nitrogen purge while brazing.

5.1.10.2.3 WAGD System Piping. WAGD systems shall be piped as follows:

(1) Using materials compliant with 5.1.10.2.1 or 5.1.10.2.2
(2) In systems operated under 130 mm (5 in.) HgV maximum vacuum only, using any noncorroding tube or ductwork

5.1.10.2.3.1 If joined to the vacuum piping, WAGD system piping shall be connected at a minimum distance of 1.5 m (5 ft) from any vacuum inlet.

5.1.10.3 Joints.

5.1.10.3.1* Positive pressure patient gas systems, medical support gas systems, vacuum systems, and WAGD systems shall have all turns, offsets, and other changes in direction made using fittings or techniques appropriate to any of the following acceptable joining methods:

(1) Brazing, as described in 5.1.10.4
(2) Welding, as described in 5.1.10.5
(3) Memory metal fittings, as described in 5.1.10.6
(4) Axially swaged, elastic preload fittings, as described in 5.1.10.7
(5) Threaded, as described under 5.1.10.8

5.1.10.3.2 Vacuum systems and WAGD systems shall be permitted to have branch connections made using mechanically formed, drilled, and extruded tee-branch connections that are formed in accordance with the tool manufacturer's instructions. Such branch connections shall be joined by brazing, as described in 5.1.10.4.

5.1.10.3.3 WAGD systems designed for operation below 130 mm (5 in.) HgV shall be permitted to be joined using any method that will result in a leak-free network when tested per 5.1.12.3.2.

5.1.10.4 Brazed Joints.

5.1.10.4.1 General Requirements.

5.1.10.4.1.1 Fittings shall be wrought copper capillary fittings complying with ASME B16.22, *Wrought Copper and Copper Alloy Solder-Joint Pressure Fittings*, or brazed fittings complying with ANSI/ASME B16.50, *Wrought Copper and Copper Alloy Braze-Joint Pressure Fittings*.

5.1.10.4.1.2 Cast copper alloy fittings shall not be permitted.

5.1.10.4.1.3 Brazed joints shall be made using a brazing alloy that exhibits a melting temperature in excess of 538°C (1000°F) to retain the integrity of the piping system in the event of fire exposure.

5.1.10.4.1.4 Brazed tube joints shall be the socket type.

5.1.10.4.1.5 Filler metals shall bond with and be metallurgically compatible with the base metals being joined.

5.1.10.4.1.6 Filler metals shall comply with ANSI/AWS A5.8, *Specification for Filler Metals for Brazing and Braze Welding*.

5.1.10.4.1.7 Copper-to-copper joints shall be brazed using a copper–phosphorus or copper–phosphorus–silver brazing filler metal (BCuP series) without flux.

5.1.10.4.1.8 Brazing performed between bulk cryogenic liquid vessels and their vaporizers (i.e., subject to cryogenic exposure) shall be permitted to be brazed using BAg brazing alloy with flux by a brazer qualified to the mandatory requirements of CGA M-1, *Guide for Medical Gas Installations at Consumer Sites*.

5.1.10.4.1.9 Joints to be brazed in place shall be accessible for necessary preparation, assembly, heating, filler application, cooling, cleaning, and inspection.

5.1.10.4.1.10 Braze joints shall be continuously purged with nitrogen NF.

5.1.10.4.2 Cutting Tube Ends.

5.1.10.4.2.1 Tube ends shall be cut square using a sharp tubing cutter to avoid deforming the tube.

5.1.10.4.2.2 The cutting wheels on tubing cutters shall be free from grease, oil, or other lubricant not suitable for oxygen service.

5.1.10.4.2.3 The cut ends of the tube shall be permitted to be rolled smooth or deburred with a sharp, clean deburring tool, taking care to prevent chips from entering the tube.

5.1.10.4.3 Cleaning Joints for Brazing.

5.1.10.4.3.1 The interior surfaces of tubes, fittings, and other components that are cleaned for oxygen service shall be stored and handled to avoid contamination prior to assembly and brazing.

5.1.10.4.3.2 The exterior surfaces of tube ends shall be cleaned prior to brazing to remove any surface oxides.

5.1.10.4.3.3 When cleaning the exterior surfaces of tube ends, no matter shall be allowed to enter the tube.

5.1.10.4.3.4 If the interior surfaces of fitting sockets become contaminated prior to brazing, they shall be recleaned for oxygen in accordance with 5.1.10.4.3.10 and be cleaned for brazing with a clean, oil-free wire brush.

5.1.10.4.3.5 Clean, nonshedding, abrasive pads shall be used to clean the exterior surfaces of the tube ends.

5.1.10.4.3.6 The use of steel wool or sand cloth shall be prohibited.

5.1.10.4.3.7 The cleaning process shall not result in grooving of the surfaces to be joined.

5.1.10.4.3.8 After being abraded, the surfaces shall be wiped using a clean, lint-free white cloth.

5.1.10.4.3.9 Tubes, fittings, valves, and other components shall be visually examined internally before being joined to verify that they have not become contaminated for oxygen service and that they are free of obstructions or debris.

5.1.10.4.3.10 The interior surfaces of tube ends, fittings, and other components that were cleaned for oxygen service by the manufacturer, but that became contaminated prior to being installed, shall be permitted to be recleaned on-site by the installer by thoroughly scrubbing the interior surfaces with a clean, hot water–alkaline solution, such as sodium carbonate or trisodium phosphate, using a solution of 450 g (1 lb) of sodium carbonate or trisodium phosphate to 11 L (3 gal) of potable water, and thoroughly rinsing them with clean, hot, potable water.

5.1.10.4.3.11 Other aqueous cleaning solutions shall be permitted to be used for on-site recleaning permitted in 5.1.10.4.3.10, provided that they are as recommended in the mandatory requirements of CGA G-4.1, *Cleaning Equipment for Oxygen Service*, and are listed in CGA O2-DIR, *Directory of Cleaning Agents for Oxygen Service*.

5.1.10.4.3.12 Material that has become contaminated internally and is not clean for oxygen service shall not be installed.

5.1.10.4.3.13 Joints shall be brazed within 8 hours after the surfaces are cleaned for brazing.

5.1.10.4.4 Brazing Dissimilar Metals.

5.1.10.4.4.1 Flux shall only be used when brazing dissimilar metals, such as copper and bronze or brass, using a silver (BAg series) brazing filler metal.

5.1.10.4.4.2 Surfaces shall be cleaned for brazing in accordance with 5.1.10.4.3.

5.1.10.4.4.3 Flux shall be applied sparingly to minimize contamination of the inside of the tube with flux.

5.1.10.4.4.4 The flux shall be applied and worked over the cleaned surfaces to be brazed using a stiff bristle brush to ensure complete coverage and wetting of the surfaces with flux.

5.1.10.4.4.5 Where possible, short sections of copper tube shall be brazed onto the noncopper component, and the interior of the subassembly shall be cleaned of flux prior to installation in the piping system.

5.1.10.4.4.6 On joints DN20 (NPS ¾) (⅞ in. O.D.) size and smaller, flux-coated brazing rods shall be permitted to be used in lieu of applying flux to the surfaces being joined.

5.1.10.4.5* Nitrogen Purge.

5.1.10.4.5.1 When brazing, joints shall be continuously purged with oil-free, dry nitrogen NF to prevent the formation of copper oxide on the inside surfaces of the joint.

5.1.10.4.5.2 The source of the purge gas shall be monitored, and the installer shall be audibly alerted when the source content is low.

5.1.10.4.5.3 The purge gas flow rate shall be controlled by the use of a pressure regulator and flowmeter, or combination thereof.

5.1.10.4.5.4 Pressure regulators alone shall not be used to control purge gas flow rates.

5.1.10.4.5.5 In order to ensure that all ambient air has been removed from the pipeline prior to brazing, an oxygen analyzer shall be used to verify the effectiveness of the purge. The oxygen analyzer shall read below 1 percent oxygen concentration before brazing begins.

5.1.10.4.5.6 During and after installation, openings in the piping system shall be kept sealed to maintain a nitrogen atmosphere within the piping to prevent debris or other contaminants from entering the system.

5.1.10.4.5.7 While a joint is being brazed, a discharge opening shall be provided on the opposite side of the joint from where the purge gas is being introduced.

5.1.10.4.5.8 The flow of purge gas shall be maintained until the joint is cool to the touch.

5.1.10.4.5.9 After the joint has cooled, the purge discharge opening shall be sealed to prevent contamination of the inside of the tube and maintain the nitrogen atmosphere within the piping system.

5.1.10.4.5.10 The final brazed connection of new piping to an existing pipeline containing the system gas shall be permitted to be made without the use of a nitrogen purge.

5.1.10.4.5.11 After a final brazed connection in a positive pressure medical gas pipeline is made without a nitrogen purge, an outlet in the immediate downstream zone of the affected portion(s) of both the new and existing piping shall be tested in accordance with the final tie-in test in 5.1.12.3.9.

5.1.10.4.5.12* When using the autogenous orbital welding process, joints shall be continuously purged inside and outside with inert gas(es) in accordance with the qualified welding procedure.

5.1.10.4.6 Assembling and Heating Brazed Joints.

5.1.10.4.6.1 Tube ends shall be inserted into the socket, either fully or to a mechanically limited depth that is not less than the minimum cup depth (overlap) specified by ANSI/ASME B16.50, *Wrought Copper and Copper Alloy Braze-Joint Pressure Fittings*.

5.1.10.4.6.2 Where flux is permitted, the joint shall be heated slowly until the flux has liquefied.

5.1.10.4.6.3 After flux is liquefied, or where flux is not permitted to be used, the joint shall be heated quickly to the brazing temperature, taking care not to overheat the joint.

5.1.10.4.6.4 Techniques for heating the joint, applying the brazing filler metal, and making horizontal, vertical, and large-diameter joints shall be as stated in sections on applying heat and brazing and horizontal and vertical joints in Chapter VII, "Brazed Joints," in the CDA *Copper Tube Handbook*.

5.1.10.4.7 Inspection of Brazed Joints.

5.1.10.4.7.1 After brazing, the outside of all joints shall be cleaned by washing with water and a wire brush to remove any residue and allow clear visual inspection of the joint.

5.1.10.4.7.2 Where flux has been used, the wash water shall be hot.

5.1.10.4.7.3 Each brazed joint shall be visually inspected after cleaning the outside surfaces.

5.1.10.4.7.4 Joints exhibiting the following conditions shall not be permitted:

(1) Flux or flux residue (when flux or flux-coated BAg series rods are used with dissimilar metals)
(2) Base metal melting or erosion
(3) Unmelted filler metal
(4) Failure of the filler metal to be clearly visible all the way around the joint at the interface between the socket and the tube
(5) Cracks in the tube or component
(6) Cracks in the braze filler metal
(7) Failure of the joint to hold the test pressure under the installer-performed initial pressure test *(see 5.1.12.2.3)* and standing pressure test *(see 5.1.12.2.6 or 5.1.12.2.7)*

5.1.10.4.7.5 Brazed joints that are identified as defective under the conditions of 5.1.10.4.7.4(2) or (5) shall be replaced.

5.1.10.4.7.6 Brazed joints that are identified as defective under the conditions of 5.1.10.4.7.4(1), (3), (4), (6), or (7) shall be permitted to be repaired, except that no joint shall be reheated more than once before being replaced.

5.1.10.5 Welded Joints.

5.1.10.5.1 Gas Tungsten Arc Welding (GTAW) for Copper and Stainless Tube.

5.1.10.5.1.1 Welded joints for medical gas and medical–surgical vacuum systems shall be permitted to be made using a gas tungsten arc welding (GTAW) autogenous orbital procedure.

5.1.10.5.1.2 The GTAW autogenous orbital procedure and the welder qualification procedure shall be qualified in accordance with Section IX, "Welding and Brazing Qualifications," of the ASME *Boiler and Pressure Vessel Code*.

5.1.10.5.1.3 Welder qualification procedures shall include a bend test and a tensile test in accordance with Section IX, "Welding and Brazing Qualifications," of the ASME *Boiler and Pressure Vessel Code* on each tube size diameter.

5.1.10.5.1.4 Each welder shall qualify to a welding procedure specification (WPS) for each tube diameter.

5.1.10.5.1.5* GTAW autogenous orbital welded joints shall be purged during welding with a commercially available mixture of 75 percent helium (±5 percent) and 25 percent argon (±5 percent).

5.1.10.5.1.6 The shield gas shall be as required in 5.1.10.5.1.5.

5.1.10.5.1.7 Test coupons shall be welded and inspected, as a minimum, at start of work and every 4 hours thereafter, or when the machine is idle for more than 30 minutes, and at the end of the work period.

5.1.10.5.1.8 Test coupons shall be inspected on the I.D. and O.D. by a qualified quality control inspector.

5.1.10.5.1.9 Test coupons shall also be welded at change of operator, weld head, welding power supply, or gas source.

5.1.10.5.1.10 All production welds shall be visually inspected on the O.D. by the operator, and any obvious weld failures shall be cut out and re-welded.

5.1.10.5.2 Welding for Stainless Tube.

5.1.10.5.2.1 Stainless tube shall be welded using metal inert gas (MIG) welding, tungsten inert gas (TIG) welding, or other welding techniques suited to joining stainless tube.

5.1.10.5.2.2 Welders shall be qualified to Section IX, "Welding and Brazing Qualifications," of the ASME *Boiler and Pressure Vessel Code*.

5.1.10.6 Memory Metal Fittings.

5.1.10.6.1 Memory metal fittings having a temperature rating not less than 538°C (1000°F) and a pressure rating not less than 2070 kPa (300 psi) shall be permitted to be used to join copper or stainless steel tube.

5.1.10.6.2 Memory metal fittings shall be installed by qualified technicians in accordance with the manufacturer's instructions.

5.1.10.7 Axially Swaged Fittings.

5.1.10.7.1 Axially swaged, elastic strain preload fittings providing metal-to-metal seals, having a temperature rating not less than 538°C (1000°F) and a pressure rating not less than 2070 kPa (300 psi), and that, when complete, are permanent and nonseparable shall be permitted to be used to join copper or stainless steel tube.

5.1.10.7.2 Axially swaged, elastic strain preload fittings shall be installed by qualified technicians in accordance with the manufacturer's instructions.

5.1.10.8 Threaded Fittings. Threaded fittings shall meet the following criteria:

(1) They shall be limited to connections for pressure and vacuum indicators, alarm devices, gas-specific demand check fittings, and source equipment on the source side of the source valve.
(2) They shall be tapered pipe threads complying with ASME B1.20.1, *Pipe Threads, General Purpose, Inch.*
(3)*They shall be made up with polytetrafluoroethylene (PTFE) tape or other thread sealant recommended for oxygen service, with sealant applied to the male threads only and care taken to ensure sealant does not enter the pipe.

5.1.10.9 Special Fittings.

5.1.10.9.1 Listed or approved metallic gas tube fittings that, when made up, provide a permanent joint having the mechanical, thermal, and sealing integrity of a brazed joint shall be permitted to be used.

5.1.10.9.2 Dielectric Fittings. Dielectric fittings that comply with the following shall be permitted only where required by the manufacturer of special medical equipment to electrically isolate the equipment from the system distribution piping:

(1) They shall be of brass or copper construction with an appropriate dielectric.
(2) They shall be permitted to be a union.
(3) They shall be clean for oxygen where used for medical gases and medical support gases.

5.1.10.10 Prohibited Joints. The following joints shall be prohibited throughout medical gas and vacuum distribution pipeline systems:

(1) Flared and compression-type connections, including connections to station outlets and inlets, alarm devices, and other components
(2) Other straight-threaded connections, including unions
(3) Pipe-crimping tools used to permanently stop the flow of medical gas and vacuum piping
(4) Removable and nonremovable push-fit fittings that employ a quick assembly push fit connector

5.1.10.11 Installation of Piping and Equipment.

5.1.10.11.1 Pipe Sizing.

5.1.10.11.1.1 Piping systems shall be designed and sized to deliver the required flow rates at the utilization pressures.

5.1.10.11.1.2 Mains and branches in medical gas piping systems shall be not less than DN15 (NPS ½) (⅝ in. O.D.) size.

5.1.10.11.1.3 Mains and branches in medical–surgical vacuum systems shall be not less than DN20 (NPS ¾) (⅞ in. O.D.) size.

5.1.10.11.1.4 Drops to individual station outlets and inlets shall be not less than DN15 (NPS ½) (⅝ in. O.D.) size.

5.1.10.11.1.5 Runouts to alarm panels and connecting tubing for gauges and alarm devices shall be permitted to be DN8 (NPS ¼) (⅜ in. O.D.) size.

5.1.10.11.2 Protection of Piping. Piping shall be protected against freezing, corrosion, and physical damage.

5.1.10.11.2.1 Piping exposed in corridors and other areas where subject to physical damage from the movement of carts, stretchers, portable equipment, or vehicles shall be protected.

5.1.10.11.2.2 Piping underground within buildings or embedded in concrete floors or walls shall be installed in a continuous conduit.

5.1.10.11.3 Location of Piping.

5.1.10.11.3.1 Piping risers shall be permitted to be installed in pipe shafts if protected from physical damage, effects of excessive heat, corrosion, or contact with oil.

5.1.10.11.3.2 Piping shall not be installed in kitchens, elevator shafts, elevator machine rooms, areas with open flames, electrical service equipment over 600 volts, and areas prohibited under *NFPA 70, National Electrical Code*, except for the following locations:

(1) Room locations for medical air compressor supply systems and medical–surgical vacuum pump supply systems
(2) Room locations for secondary distribution circuit panels and breakers having a maximum voltage rating of 600 volts

5.1.10.11.3.3 Medical gas piping shall be permitted to be installed in the same service trench or tunnel with fuel gas lines, fuel oil lines, electrical lines, steam lines, and similar utilities, provided that the space is ventilated (naturally or mechanically) and the ambient temperature around the medical gas piping is limited to 54°C (130°F) maximum.

5.1.10.11.3.4 Medical gas piping shall not be located where subject to contact with oil, including a possible flooding area in the case of a major oil leak.

5.1.10.11.4 Pipe Support.

5.1.10.11.4.1 Piping shall be supported from the building structure.

5.1.10.11.4.2 Hangers and supports shall comply with and be installed in accordance with MSS SP-58, *Pipe Hangers and Supports — Materials, Design, Manufacture, Selection, Application, and Installation.*

5.1.10.11.4.3 Supports for copper tube shall be sized for copper tube.

5.1.10.11.4.4 In potentially damp locations, copper tube hangers or supports that are in contact with the tube shall be plastic-coated or otherwise be electrically insulated from the tube by a material that will not absorb moisture.

5.1.10.11.4.5 Maximum support spacing shall be in accordance with Table 5.1.10.11.4.5.

Table 5.1.10.11.4.5 Maximum Pipe Support Spacing

Pipe Size	Hanger Spacing mm	Hanger Spacing ft
DN8 (NPS ¼) (⅜ in. O.D.)	1520	5
DN10 (NPS ⅜) (½ in. O.D.)	1830	6
DN15 (NPS ½) (⅝ in. O.D.)	1830	6
DN20 (NPS ¾) (⅞ in. O.D.)	2130	7
DN25 (NPS 1) (1⅛ in. O.D.)	2440	8
DN32 (NPS 1¼) (1⅜ in. O.D.)	2740	9
DN40 (NPS 1½) (1⅝ in. O.D.) and larger	3050	10
Vertical risers, all sizes, every floor, but not to exceed	4570	15

5.1.10.11.4.6 Where required, medical gas and vacuum piping shall be seismically restrained against earthquakes in accordance with the applicable building code.

5.1.10.11.5 Underground Piping Outside of Buildings.

5.1.10.11.5.1 Buried piping outside of buildings shall be installed below the local level of frost penetration.

5.1.10.11.5.2 The installation procedure for underground piping shall protect the piping from physical damage while being backfilled.

5.1.10.11.5.3 If underground piping is protected by a conduit, cover, or other enclosure, the following requirements shall be met:

(1) Access shall be provided at the joints for visual inspection and leak testing.
(2) The conduit, cover, or enclosure shall be self-draining and not retain groundwater in prolonged contact with the pipe.

5.1.10.11.5.4 Buried piping that will be subject to surface loads shall be buried at a depth that will protect the piping or its enclosure from excessive stresses.

5.1.10.11.5.5 The minimum backfilled cover above the top of the pipe or its enclosure for buried piping outside of buildings shall be 900 mm (36 in.), except that the minimum cover shall be permitted to be reduced to 450 mm (18 in.) where there is no potential for damage from surface loads or surface conditions.

5.1.10.11.5.6 Trenches shall be excavated so that the pipe or its enclosure has firm, substantially continuous bearing on the bottom of the trench.

5.1.10.11.5.7 Backfill shall be clean, free from material that can damage the pipe, and compacted.

5.1.10.11.5.8 A continuous tape or marker placed immediately above the pipe or its enclosure shall clearly identify the pipeline by specific name.

5.1.10.11.5.9 A continuous warning means shall also be provided above the pipeline at approximately one-half the depth of burial.

5.1.10.11.5.10 Where underground piping is installed through a wall sleeve, the outdoor end of the sleeve shall be sealed to prevent the entrance of groundwater into the building.

5.1.10.11.6 Hose and Flexible Connectors.

5.1.10.11.6.1 Hose and flexible connectors, both metallic and nonmetallic, shall be no longer than necessary and shall not penetrate or be concealed in walls, floors, ceilings, or partitions.

5.1.10.11.6.2 Flexible connectors, metallic or nonmetallic, shall have a minimum burst pressure, with a gauge pressure of 6895 kPa (1000 psi).

5.1.10.11.6.3 Metallic flexible joints shall be permitted in the pipeline where required for expansion joints, seismic protection, thermal expansion, or vibration control and shall be as follows:

(1) For all wetted surfaces, made of bronze, copper, or stainless steel
(2) Cleaned at the factory for oxygen service and received on the job site with certification of cleanliness
(3) Suitable for service at 2070 kPa (300 psig) or above and able to withstand temperatures of 538°C (1000°F)
(4) Provided with brazing extensions to allow brazing into the pipeline per 5.1.10.4
(5) Supported with pipe hangers and supports as required for their additional weight

5.1.10.11.7 Prohibited System Interconnections.

5.1.10.11.7.1 Two or more medical gas or vacuum piping systems shall not be interconnected for installation, testing, or any other reason, except as permitted by 5.1.10.11.7.2.

5.1.10.11.7.2 Medical gas and vacuum systems with the same contents shall be permitted to be interconnected with an inline valve installed between the systems.

5.1.10.11.7.3 Leak testing shall be accomplished by separately charging and testing each individual piping system.

5.1.10.11.8 Manufacturer's Instructions.

5.1.10.11.8.1 The installation of individual components shall be made in accordance with the instructions of the manufacturer.

5.1.10.11.8.2 Manufacturer's instructions shall include directions and information deemed by the manufacturer to be adequate for attaining proper operation, testing, and maintenance of the medical gas and vacuum systems.

5.1.10.11.8.3 Copies of the manufacturer's instructions shall be left with the system owner.

5.1.10.11.9 Changes in System Use.

5.1.10.11.9.1 Where a positive pressure medical gas piping distribution system originally used or constructed for use at one pressure and for one gas is converted for operation at another pressure or for another gas, all provisions of 5.1.10 shall apply as if the system were new.

5.1.10.11.9.2 A vacuum system shall not be permitted to be converted for use as a gas system.

5.1.10.11.10 Qualification of Installers.

5.1.10.11.10.1 The installation of medical gas and vacuum systems shall be made by qualified, competent technicians who are experienced in performing such installations, including all personnel who actually install the piping system.

5.1.10.11.10.2 Installers of medical gas and vacuum piped distribution systems, all appurtenant piping supporting pump and compressor source systems, and appurtenant piping supporting source gas manifold systems not including permanently installed bulk source systems, shall be certified in accordance with ASSE 6010, *Professional Qualification Standard for Medical Gas Systems Installers.*

5.1.10.11.10.3 Installers of medical gas and vacuum systems shall not use their certification to oversee installation by non-certified personnel.

5.1.10.11.10.4 Brazing shall be performed by individuals who are qualified in accordance with the provisions of 5.1.10.11.11.

5.1.10.11.10.5 Prior to any installation work, the installer of medical gas and vacuum piping shall provide and maintain documentation on the job site for the qualification of brazing procedures and individual brazers that is required under 5.1.10.11.11.

5.1.10.11.10.6 Health care organization personnel shall be permitted to install piping systems if all of the requirements of 5.1.10.11.10 are met during the installation.

5.1.10.11.11 Qualification of Brazing Procedures and Brazing.

5.1.10.11.11.1 Brazing procedures and brazer performance for the installation of medical gas and vacuum piping shall be qualified in accordance with either Section IX, "Welding and Brazing Qualifications," of the ASME *Boiler and Pressure Vessel Code*, or AWS B2.2, *Standard for Brazing Procedure and Performance Qualification*, both as modified by 5.1.10.11.11.2 through 5.1.10.11.11.5.

5.1.10.11.11.2 Brazers shall be qualified by visual examination of the test coupon followed by sectioning.

5.1.10.11.11.3 The brazing procedure specification shall address cleaning, joint clearance, overlap, internal purge gas, purge gas flow rate, and filler metal.

5.1.10.11.11.4 The brazing procedure qualification record and the record of brazer performance qualification shall document filler metal used, cleaning, joint clearance, overlap, internal purge gas and flow rate during brazing of coupon, and absence of internal oxidation in the completed coupon.

5.1.10.11.11.5 Brazing procedures qualified by a technically competent group or agency shall be permitted under the following conditions:

(1) The brazing procedure specification and the procedure qualification records meet the requirements of this code.
(2) The employer obtains a copy of both the brazing procedure specification and the supporting qualification records from the group or agency and signs and dates these records, thereby accepting responsibility for the qualifications that were performed by the group or agency.
(3) The employer qualifies at least one brazer following each brazing procedure specification used.

5.1.10.11.11.6 An employer shall be permitted to accept brazer qualification records of a previous employer under the following conditions:

(1) The brazer has been qualified following the same or an equivalent procedure that the new employer uses.
(2) The new employer obtains a copy of the record of brazer performance qualification tests from the previous employer and signs and dates these records, thereby accepting responsibility for the qualifications performed by the previous employer.

5.1.10.11.11.7 Performance qualifications of brazers shall remain in effect indefinitely, unless the brazer does not braze with the qualified procedure for a period exceeding 6 months or there is a specific reason to question the ability of the brazer.

5.1.10.11.12 Breaching or Penetrating Medical Gas Piping.

5.1.10.11.12.1 Positive pressure patient medical gas piping and medical support gas piping shall not be breached or penetrated by any means or process that will result in residual copper particles or other debris remaining in the piping or affect the oxygen-clean interior of the piping.

5.1.10.11.12.2 The breaching or penetrating process shall ensure that any debris created by the process remains contained within the work area.

5.1.11* Labeling, Identification, and Operating Pressure. Color and pressure requirements shall be in accordance with Table 5.1.11.

5.1.11.1 Pipe Labeling.

5.1.11.1.1 Piping shall be labeled by stenciling or adhesive markers that identify the patient medical gas, the support gas, or the vacuum system and include the following:

(1) Name of the gas or vacuum system or the chemical symbol per Table 5.1.11
(2) Gas or vacuum system color code per Table 5.1.11
(3) Where positive pressure gas piping systems operate at pressures other than the standard gauge pressure in Table 5.1.11, the operating pressure in addition to the name of the gas

5.1.11.1.2 Pipe labels shall be located as follows:

(1) At intervals of not more than 6.1 m (20 ft)
(2) At least once in or above every room
(3) On both sides of walls or partitions penetrated by the piping
(4) At least once in every story height traversed by risers

5.1.11.1.3 Medical gas piping shall not be painted.

5.1.11.2 Shutoff Valves.

5.1.11.2.1 Shutoff valves shall be identified with the following:

(1) Name or chemical symbol for the specific medical gas or vacuum system
(2) Room or areas served
(3) Caution to not close or open the valve except in emergency

5.1.11.2.2 Where positive pressure gas piping systems operate at pressures other than the standard gauge pressure of 345 kPa to 380 kPa (50 psi to 55 psi) or a gauge pressure of 1100 kPa to 1275 kPa (160 psi to 185 psi) for nitrogen or instrument air, the valve identification shall also include the nonstandard operating pressure.

5.1.11.2.3 Source valves shall be labeled in substance as follows:

SOURCE VALVE

FOR THE (SOURCE NAME).

5.1.11.2.4 Main line valves shall be labeled in substance as follows:

MAIN LINE VALVE FOR THE (GAS/VACUUM NAME) SERVING (NAME OF THE BUILDING).

Table 5.1.11 Standard Designation Colors and Operating Pressures for Gas and Vacuum Systems

Gas Service	Abbreviated Name	Colors (Background/Text)	Standard Gauge Pressure	
			kPa	psi
Medical air	Med air	Yellow/black	345–380	50–55
Carbon dioxide	CO_2	Gray/black or gray/white	345–380	50–55
Helium	He	Brown/white	345–380	50–55
Nitrogen	N_2	Black/white	1100–1275	160–185
Nitrous oxide	N_2O	Blue/white	345–380	50–55
Oxygen	O_2	Green/white or white/green	345–380	50–55
Oxygen/carbon dioxide mixtures	$O_2/CO_2 n\%$ (n = % of CO_2)	Green/white	345–380	50–55
Medical–surgical vacuum	Med vac	White/black	380 mm to 760 mm (15 in. to 30 in.) HgV	
Waste anesthetic gas disposal	WAGD	Violet/white	Varies with system type	
Other mixtures	Gas A%/Gas B%	Colors as above Major gas for background/minor gas for text	None	
Nonmedical air (Category 3 gas-powered device)		Yellow and white diagonal stripe/black	None	
Nonmedical and Category 3 vacuum		White and black diagonal stripe/black boxed	None	
Laboratory air		Yellow and white checkerboard/black	None	
Laboratory vacuum		White and black checkerboard/black boxed	None	
Instrument air		Red/white	1100–1275	160–185

5.1.11.2.5 The riser valve(s) shall be labeled in substance as follows:

RISER FOR THE (GAS/VACUUM NAME) SERVING (NAME OF THE AREA/BUILDING SERVED BY THE PARTICULAR RISER).

5.1.11.2.6 The service valve(s) shall be labeled in substance as follows:

SERVICE VALVE FOR THE (GAS/VACUUM NAME) SERVING (NAME OF THE AREA/BUILDING SERVED BY THE PARTICULAR VALVE).

5.1.11.2.7* Zone valve box assemblies shall be labeled outside of the valve box as to the areas that they control as follows:

ZONE VALVES FOR THE (GAS/VACUUM NAME) SERVING (NAME OF AREA SERVED BY THE PARTICULAR VALVE)

5.1.11.3 Station Outlets and Inlets.

5.1.11.3.1 Station outlets and inlets shall be identified as to the name or chemical symbol for the specific medical gas or vacuum provided.

5.1.11.3.1.1 In sleep labs, where the outlet is downstream of a flow control device, the station outlet identification shall include a warning not to use the outlet for ventilating patients.

5.1.11.3.2 Where medical gas systems operate at pressures other than the standard gauge pressure of 345 kPa to 380 kPa (50 psi to 55 psi) or a gauge pressure of 1100 kPa to 1275 kPa (160 psi to 185 psi) for nitrogen, the station outlet identification shall include the nonstandard operating pressure in addition to the name of the gas.

5.1.11.4 Alarm Panels.

5.1.11.4.1 Labeling of alarm panels for each indicator shall indicate the condition monitored and its area of surveillance.

5.1.11.4.2* Area alarm panels shall be identified with the following:

(1) Name or chemical symbol of the specific medical gas or system being monitored
(2) Area(s) monitored by the alarm panel

5.1.12* Performance Criteria and Testing — Category 1 (Gases, Medical–Surgical Vacuum, and WAGD).

5.1.12.1 General.

5.1.12.1.1 Inspection and testing shall be performed on all new piped gas systems, additions, renovations, temporary installations, or repaired systems to ensure, by a documented procedure, that all applicable provisions of this document have been adhered to and system integrity has been achieved or maintained.

5.1.12.1.2 Inspection and testing shall include all components of the system, or portions thereof, including, but not limited to, gas bulk source(s); manifolds; compressed air source systems (e.g., compressors, dryers, filters, regulators); source alarms and monitoring safeguards; master alarms; pipelines; isolation valves; area alarms; zone valves; and station inlets (vacuum) and outlets (pressure gases).

5.1.12.1.3 All systems that are breached and components that are subject to additions, renovations, or replacement (e.g., new gas sources: bulk, manifolds, compressors, dryers, alarms) shall be inspected and tested.

5.1.12.1.4 Systems shall be deemed breached at the point of pipeline intrusion by physical separation or by system component removal, replacement, or addition.

5.1.12.1.5 Breached portions of the systems subject to inspection and testing shall be confined to only the specific altered zone and components in the immediate zone or area that is located upstream for vacuum systems and downstream for pressure gases at the point or area of intrusion.

5.1.12.1.6 The inspection and testing reports shall be submitted directly to the party that contracted for the testing, who shall submit the report through channels to the responsible facility authority and any others that are required.

5.1.12.1.7 Reports shall contain detailed listings of all findings and results.

5.1.12.1.8 The responsible facility authority shall review these inspection and testing records prior to the use of all systems to ensure that all findings and results of the inspection and testing have been successfully completed.

5.1.12.1.9 All documentation pertaining to inspections and testing shall be maintained on-site within the facility.

5.1.12.1.10 Before piping systems are initially put into use, the facility authority shall be responsible for ascertaining that the gas/vacuum delivered at the outlet/inlet is that shown on the outlet/inlet label and that the proper connecting fittings are installed for the specific gas/vacuum service.

5.1.12.1.11 Acceptance of the verifier's report shall be permitted to satisfy the requirements in 5.1.12.1.10.

5.1.12.1.12 The removal of components within a source system for repair and reinstallation, or the replacement of components like for like, shall be treated as new work for the purposes of testing whenever such work involves cutting or brazing new piping, or both.

5.1.12.1.12.1 Where no piping is changed, functional testing shall be performed as follows:

(1) To verify the function of the replaced device
(2) To ensure no other equipment in the system has been adversely impacted

5.1.12.1.12.2 Where no piping is changed, in addition to tests of general function required by 5.1.12.1.12.1, testing shall be performed as follows:

(1) Pressure gas sources shall be tested for compliance with 5.1.12.3.14.2 as applicable to the equipment type.
(2) Medical air and instrument air sources shall be tested to 5.1.12.3.14.3.
(3) Vacuum and WAGD systems shall be tested to 5.1.12.3.14.5.
(4) Alarm systems shall be tested to 5.1.12.3.5.2 and 5.1.12.3.5.3.
(5) All affected components shall be tested as appropriate to that specific component (e.g., a replaced dew point monitor would be tested to 5.1.3.6.3.13).

5.1.12.1.13 The rated accuracy of pressure and vacuum indicators used for testing shall be 1 percent (full scale) or better.

5.1.12.2 Installer-Performed Tests.

5.1.12.2.1 General.

5.1.12.2.1.1 The tests required by 5.1.12.2 shall be performed and documented by the installer prior to the tests listed in 5.1.12.3.

5.1.12.2.1.2 The test gas shall be oil-free, dry nitrogen NF.

5.1.12.2.1.3 Where manufactured assemblies are to be installed, the tests required by 5.1.12.2 shall be performed as follows:

(1) After completion of the distribution piping, but before the standing pressure test
(2) Prior to installation of manufactured assemblies supplied through flexible hose or flexible tubing
(3) At all station outlets/inlets on installed manufactured assemblies supplied through copper tubing

5.1.12.2.2 Initial Piping Blowdown.
Piping in medical gas and vacuum distribution systems shall be blown clear by means of oil-free, dry nitrogen NF after installation of the distribution piping but before installation of station outlet/inlet rough-in assemblies and other system components (e.g., pressure/vacuum alarm devices, pressure/vacuum indicators, pressure relief valves, manifolds, source equipment).

5.1.12.2.3 Initial Pressure Test.

5.1.12.2.3.1 Each section of the piping in medical gas and vacuum systems shall be pressure tested.

5.1.12.2.3.2 Initial pressure tests shall be conducted as follows:

(1) After blowdown of the distribution piping
(2) After installation of station outlet/inlet rough-in assemblies
(3) Prior to the installation of components of the distribution piping system that would be damaged by the test pressure (e.g., pressure/vacuum alarm devices, pressure/vacuum indicators, line pressure relief valves)

5.1.12.2.3.3 The source shutoff valve shall remain closed during the tests specified in 5.1.12.2.3.

5.1.12.2.3.4 The test pressure for pressure gases and vacuum systems shall be 1.5 times the system operating pressure but not less than a gauge pressure of 1035 kPa (150 psi).

5.1.12.2.3.5* The test pressure shall be maintained until each joint has been examined for leakage by means of a leak detector that is safe for use with oxygen and does not contain ammonia.

5.1.12.2.3.6 Leaks, if any, shall be located, repaired (if permitted), replaced (if required), and retested.

5.1.12.2.4 Initial Cross-Connection Test. It shall be determined that no cross-connections exist between the various medical gas and vacuum piping systems.

5.1.12.2.4.1 All piping systems shall be reduced to atmospheric pressure.

5.1.12.2.4.2 Sources of test gas shall be disconnected from all piping systems, except for the one system being tested.

5.1.12.2.4.3 The system under test shall be charged with oil-free, dry nitrogen NF to a gauge pressure of 345 kPa (50 psi).

5.1.12.2.4.4 After the installation of the individual faceplates with appropriate adapters matching outlet/inlet labels, each individual outlet/inlet in each installed medical gas and vacuum piping system shall be checked to determine that the test gas is being dispensed only from the piping system being tested.

5.1.12.2.4.5 The cross-connection test referenced in 5.1.12.2.4 shall be repeated for each installed medical gas and vacuum piping system.

5.1.12.2.4.6 The proper labeling and identification of system outlets/inlets shall be confirmed during these tests.

5.1.12.2.5 Initial Piping Purge Test. The outlets in each medical gas piping system shall be purged to remove any particulate matter from the distribution piping.

5.1.12.2.5.1 Using appropriate adapters, each outlet shall be purged with an intermittent high-volume flow of test gas until the purge produces no discoloration in a clean white cloth.

5.1.12.2.5.2 The purging required in 5.1.12.2.5.1 shall be started at the closest outlet/inlet to the zone valve and continue to the furthest outlet/inlet within the zone.

5.1.12.2.6 Standing Pressure Test for Positive Pressure Medical Gas Piping. After successful completion of the initial pressure tests under 5.1.12.2.3, medical gas distribution piping shall be subject to a standing pressure test.

5.1.12.2.6.1 Tests shall be conducted after the final installation of station outlet valve bodies, faceplates, and other distribution system components (e.g., pressure alarm devices, pressure indicators, line pressure relief valves, manufactured assemblies, hose).

5.1.12.2.6.2 The source valve shall be closed during this test.

5.1.12.2.6.3 The piping systems shall be subjected to a 24-hour standing pressure test using oil-free, dry nitrogen NF.

5.1.12.2.6.4 Test pressures shall be 20 percent above the normal system operating line pressure.

5.1.12.2.6.5* At the conclusion of the tests, there shall be no change in the test pressure except that attributed to specific changes in ambient temperature.

5.1.12.2.6.6 Leaks, if any, shall be located, repaired (if permitted) or replaced (if required), and retested.

5.1.12.2.6.7 The 24-hour standing pressure test of the positive pressure system shall be witnessed by the authority having jurisdiction or its designee. A form indicating that this test has been performed and witnessed shall be provided to the verifier at the start of the tests required in 5.1.12.3.

5.1.12.2.7 Standing Vacuum Test for Vacuum Piping. After successful completion of the initial pressure tests under 5.1.12.2.3, vacuum distribution piping shall be subjected to a standing vacuum test.

5.1.12.2.7.1 Tests shall be conducted after installation of all components of the vacuum system.

5.1.12.2.7.2 The piping systems shall be subjected to a 24-hour standing vacuum test.

5.1.12.2.7.3 Test pressure shall be between 300 mm (12 in.) HgV and full vacuum.

5.1.12.2.7.4 During the test, the source of test vacuum shall be disconnected from the piping system.

5.1.12.2.7.5* At the conclusion of the test, there shall be no change in the vacuum other than that attributed to changes of ambient temperature.

5.1.12.2.7.6 The 24-hour standing pressure test of the vacuum system shall be witnessed by the authority having jurisdiction or its designee. A form indicating that this test has been performed and witnessed shall be provided to the verifier at the start of the tests required in 5.1.12.3.

5.1.12.2.7.7 Leaks, if any, shall be located, repaired (if permitted) or replaced (if required), and retested.

5.1.12.3 System Verification.

5.1.12.3.1 General.

5.1.12.3.1.1 Verification tests shall be performed only after all tests required in 5.1.12.2, Installer Performed Tests, have been completed.

5.1.12.3.1.2 The test gas shall be oil-free, dry nitrogen NF or the system gas where permitted.

5.1.12.3.1.3 Testing shall be conducted by a party technically competent and experienced in the field of medical gas and vacuum pipeline testing and meeting the requirements of ASSE 6030, *Professional Qualifications Standard for Medical Gas Systems Verifiers.*

5.1.12.3.1.4 Testing shall be performed by a party other than the installing contractor.

5.1.12.3.1.5 When systems have not been installed by in-house personnel, testing shall be permitted by personnel of that organization who meet the requirements of 5.1.12.3.1.3.

5.1.12.3.1.6 All tests required under 5.1.12.3 shall be performed after installation of any manufactured assemblies supplied through tubing or flexible hose.

5.1.12.3.1.7 Where there are multiple possible connection points for terminals, each possible position shall be tested independently.

5.1.12.3.1.8 The gas of system designation shall be permitted to be used for all tests, regardless of the size of the system, which include the following:

(1) Standing pressure *(see 5.1.12.3.2)*
(2) Cross-connection *(see 5.1.12.3.3)*
(3) Alarms *(see 5.1.12.3.5)*
(4) Piping purge *(see 5.1.12.3.6)*
(5) Piping particulates *(see 5.1.12.3.7)*

5.1.12.3.2* Standing Pressure Test. Piping systems shall be subjected to a 10-minute standing pressure test at operating line pressure using the following procedure:

(1) After the system is filled with nitrogen or source gas, the source valve and all zone valves shall be closed.
(2) The piping system shall show no decrease in pressure after 10 minutes.
(3) Any leaks found shall be located, repaired, and retested per 5.1.12.2.6.

5.1.12.3.3 Cross-Connection Test. After the closing of walls and completion of the requirements of 5.1.12.2, it shall be determined that no cross-connection of piping systems exists by either of the methods detailed in 5.1.12.3.3.1 or 5.1.12.3.3.2.

5.1.12.3.3.1 Individual Pressurization Method.

(A) All medical gas and vacuum piping systems shall be reduced to atmospheric pressure.

(B) All sources of test gas from all of the medical gas and vacuum systems, with the exception of the one system to be checked, shall be disconnected.

(C) The system being checked shall be pressurized to a gauge pressure of 345 kPa (50 psi).

(D) With adapters matching outlet labels, each individual station outlet/inlet of all medical gas and vacuum systems installed shall be checked to determine that test gas is being dispensed only from the outlets/inlets of the piping system being tested.

(E) The source of test gas shall be disconnected, and the system tested reduced to atmospheric pressure.

(F) Proceed to test each additional piping system until all medical gas and vacuum piping systems are free of cross-connections.

5.1.12.3.3.2 Pressure Differential Method.

(A) The pressure in all medical gas systems shall be reduced to atmospheric.

(B) The test gas pressure in all medical gas piping systems shall be increased to the values indicated in Table 5.1.12.3.3.2(B), simultaneously maintaining these nominal pressures throughout the test.

(C) Systems with nonstandard operating pressures shall be tested at a gauge pressure of at least 70 kPa (10 psi) higher or lower than any other system being tested.

Table 5.1.12.3.3.2(B) Alternate Test Pressures

Medical Gas	Pressure (Gauge)
Gas mixtures	140 kPa (20 psi)
Nitrogen/instrument air	210 kPa (30 psi)
Nitrous oxide	275 kPa (40 psi)
Oxygen	345 kPa (50 psi)
Medical air	415 kPa (60 psi)
Systems at nonstandard pressures	70 kPa (10 psi) greater or less than any other system HgV vacuum
Vacuum	510 mm (20 in.) HgV
WAGD	380 mm (15 in.) HgV (if so designed)

(D) Any vacuum systems shall be in operation so that these vacuum systems are tested at the same time the medical gas systems are tested.

(E) Following the adjustment of pressures in accordance with 5.1.12.3.3.2(B) and 5.1.12.3.3.2(C), each station outlet for each medical gas system shall be tested using the gas-specific connection for each system with test gauge attached to verify that the correct test pressure/vacuum is present at each outlet/inlet of each system as listed in Table 5.1.12.3.3.2(B).

(F) Each test gauge used in performing this test shall be calibrated with the pressure indicator used for the line pressure regulator used to provide the source pressure.

(G) Each station outlet shall be identified by label (and color marking, if used), and the pressure indicated on the test gauge shall be that listed in Table 5.1.12.3.3.2(B) for the system being tested.

5.1.12.3.4 Valve Test. Valves installed in each medical gas and vacuum piping system shall be tested to verify proper operation and rooms or areas of control.

5.1.12.3.4.1 Records shall be made listing the rooms or areas controlled by each valve for each gas.

5.1.12.3.4.2 The information shall be utilized to assist and verify the proper labeling of the valves.

5.1.12.3.5 Alarm Test.

5.1.12.3.5.1 General.

(A) All warning systems for each medical gas and vacuum system(s) shall be tested to ensure that all components function properly prior to placing the system in service.

(B) Permanent records of these tests shall be maintained.

(C) Warning systems that are part of an addition to an existing piping system shall be tested prior to the connection of the new piping to the existing system.

(D) Tests of warning systems for new installations (initial tests) shall be performed after the cross-connection testing (see 5.1.12.3.3), but before purging the piping (see 5.1.12.3.6) and performing the remaining verification tests. (See 5.1.12.3.7 through 5.1.12.3.14.)

(E) Initial tests of warning systems that can be included in an addition or extension to an existing piping system shall be completed before connection of the addition to the existing system.

(F) Test gases for the initial tests shall be oil-free, dry nitrogen NF, the gas of system designation, or operating vacuum.

(G) Where computer systems are used as substitutes for a required alarm panel as permitted under 5.1.9.2.2, the computer system shall be included in the alarm tests as modified in 5.1.9.3.

5.1.12.3.5.2 Master Alarms.

(A) The master alarm system tests shall be performed for each of the medical gas and vacuum piping systems.

(B) Permanent records of these tests shall be maintained with those required under 5.1.12.1.7.

(C) The audible and noncancelable visual signals of 5.1.9.1 shall indicate if the pressure in the main line increases or decreases 20 percent from the normal operating pressure.

(D) The operation of all master alarm signals referenced in 5.1.9.2.4 shall be verified.

5.1.12.3.5.3 Area Alarms. The warning signals for all medical gas piping systems shall be tested to verify an alarm condition if the pressure in the piping system increases or decreases 20 percent from the normal operating pressure for positive pressure gases, or when the vacuum system(s) drops below a gauge pressure of 300 mm (12 in.) HgV.

5.1.12.3.6 Piping Purge Test. In order to remove any traces of particulate matter deposited in the pipelines as a result of construction, a heavy, intermittent purging of the pipeline shall be done.

5.1.12.3.6.1 The appropriate adapter shall be obtained from the facility or manufacturer, and high purge rates of at least 225 Nl/min (8 SCFM) shall be put on each outlet.

5.1.12.3.6.2 After the purge is started, it shall be rapidly interrupted several times until the purge produces no discoloration in a white cloth loosely held over the adapter during the purge.

5.1.12.3.6.3 In order to avoid possible damage to the outlet and its components, this test shall not be conducted using any implement other than the proper adapter.

5.1.12.3.6.4* No pronounced or objectionable odor shall be discernible from any positive pressure outlet.

5.1.12.3.7 Piping Particulate Test. For each positive pressure gas system, the cleanliness of the piping system shall be verified.

5.1.12.3.7.1 A minimum of 1000 L (35 ft^3) of gas shall be filtered through a clean, white 0.45 micron filter at a minimum flow rate of 100 Nl/min (3.5 SCFM).

5.1.12.3.7.2 Twenty-five percent of the zones shall be tested at the outlet most remote from the source.

5.1.12.3.7.3 The filter shall accrue no more than 0.001 g (1 mg) of matter from any outlet tested.

5.1.12.3.7.4 If any outlet fails this test, the most remote outlet in every zone shall be tested.

5.1.12.3.7.5 The test shall be performed with the use of oil-free, dry nitrogen NF.

5.1.12.3.8* Verifier Piping Purity Test. For each medical gas system, the purity of the piping system shall be verified in accordance with 5.1.12.3.8.

5.1.12.3.8.1 These tests shall be performed with oil-free, dry nitrogen NF or the system gas.

5.1.12.3.8.2 The outlet most remote from the source shall be tested for total non-methane hydrocarbons and compared to the source gas.

5.1.12.3.8.3 If the system gas is used as the source gas, it shall be tested at the source equipment.

5.1.12.3.8.4 The difference between the two tests shall in no case exceed 5 ppm of total non-methane hydrocarbons.

5.1.12.3.8.5 The difference between the two tests shall in no case exceed 5 ppm halogenated hydrocarbons.

5.1.12.3.8.6 The moisture concentration of the outlet test shall not exceed 500 ppm or an equivalent pressure dew point of -12°C (10°F) at a gauge pressure of 345 kPa (50 psi).

5.1.12.3.9 Final Tie-In Test.

5.1.12.3.9.1 Prior to the connection of any work or any extension or addition to an existing piping system, the tests in 5.1.12.3.1 through 5.1.12.3.8 shall be successfully performed on the new work.

5.1.12.3.9.2 Each joint in the final connection between the new work and the existing system shall be leak-tested with the gas of system designation at the normal operating pressure by means of a leak detectant that is safe for use with oxygen and does not contain ammonia.

5.1.12.3.9.3 Vacuum joints shall be tested using an ultrasonic leak detector or other means that will allow detection of leaks in an active vacuum system.

5.1.12.3.9.4 For pressure gases, immediately after the final brazed connection is made and leak-tested, an outlet in the new piping and an outlet in the existing piping that are immediately downstream from the point or area of intrusion shall be purged in accordance with the applicable requirements of 5.1.12.3.6.

5.1.12.3.9.5 Before the new work is used for patient care, positive pressure gases shall be tested for operational pressure and gas concentration in accordance with 5.1.12.3.10 and 5.1.12.3.11.

5.1.12.3.9.6 Permanent records of these tests shall be maintained in accordance with 5.1.14.4.

5.1.12.3.10 Operational Pressure Test. Operational pressure tests shall be performed at each station outlet/inlet or terminal where the user makes connections and disconnections.

5.1.12.3.10.1 Tests shall be performed with the gas of system designation or the operating vacuum.

5.1.12.3.10.2 All gas outlets with a gauge pressure of 345 kPa (50 psi), including, but not limited to, oxygen, nitrous oxide, medical air, and carbon dioxide, shall deliver 100 SLPM (3.5 SCFM) with a pressure drop of not more than 35 kPa (5 psi) and static pressure of 345 kPa to 380 kPa (50 psi to 55 psi).

5.1.12.3.10.3 Support gas outlets shall deliver 140 SLPM (5.0 SCFM) with a pressure drop of not more than 35 kPa (5 psi) gauge and static pressure of 1100 kPa to 1275 kPa (160 psi to 185 psi) gauge.

5.1.12.3.10.4 Medical–surgical vacuum inlets shall draw 85 Nl/min (3 SCFM) without reducing the vacuum pressure below 300 mm (12 in.) gauge HgV at any adjacent station inlet.

5.1.12.3.10.5 Oxygen and medical air outlets serving critical care areas shall allow a transient flow rate of 170 SLPM (6 SCFM) for 3 seconds.

5.1.12.3.10.6* Where outlets are being fed with non-standard line pressure, volume, or gas content, for clinical reasons, they shall be labeled in accordance with 5.1.11.

5.1.12.3.11 Medical Gas Concentration Test. After purging each system with the gas of system designation, the following shall be performed:

(1) Each pressure gas source and outlet shall be analyzed for concentration of gas, by volume.
(2) Analysis shall be conducted with instruments designed to measure the specific gas dispensed.
(3)*Allowable concentrations shall be as indicated in Table 5.1.12.3.11.

5.1.12.3.12 Medical Air Purity Test for Compressor Sources.

5.1.12.3.12.1 The medical air source shall be analyzed for concentration of contaminants by volume prior to the source valve being opened.

Table 5.1.12.3.11 Gas Concentrations

Medical Gas	Concentration
Oxygen	≥99% oxygen
Nitrous oxide	≥99% nitrous oxide
Nitrogen	≤1% oxygen or ≥99% nitrogen
Medical air	19.5%–23.5% oxygen
Other gases	As specified by ±1%, unless otherwise specified

5.1.12.3.12.2 A sample(s) shall be taken for the air system test at the system sample port.

5.1.12.3.12.3 The test results shall not exceed the parameters in Table 5.1.12.3.12.3.

Table 5.1.12.3.12.3 Contaminant Parameters for Medical Air

Parameter	Limit Value
Pressure dew point	2°C (35°F)
Carbon monoxide	10 ppm
Carbon dioxide	500 ppm
Gaseous hydrocarbons	25 ppm (as methane)
Halogenated hydrocarbons	2 ppm

5.1.12.3.13 Labeling. The presence and correctness of labeling required by this code for all components (e.g., station outlets/inlets, shutoff valves, and alarm panels) shall be verified.

5.1.12.3.14 Source Equipment Verification.

5.1.12.3.14.1 General. Source equipment verification shall be performed following the installation of the interconnecting pipelines, accessories, and source equipment.

5.1.12.3.14.2 Gas Supply Sources.

(A) The system apparatus shall be tested for proper function, including the changeover from primary to secondary supply (with its changeover signal) and the operation of the reserve (with its reserve-in-use signal), before the system is put into service.

(B) If the system has an actuating switch and signal to monitor the contents of the reserve, its function shall be tested before the system is put into service.

(C) If the system has an actuating switch and signal to monitor the pressure of the reserve unit, its function shall be tested before the system is put into service.

(D) Testing of the bulk supply signal and the master signal panel installations shall be arranged with the owner or the organization responsible for the operation and maintenance of the supply system for the testing of the bulk supply signals to ensure proper identification and activation of the master signal panels so that the facility can monitor the status of that supply system.

(E) The tests required in 5.1.12.3.14.2(D) shall also be conducted when the storage units are changed or replaced.

5.1.12.3.14.3 Medical Air Compressor Systems.

(A) Tests of the medical air compressor system shall include the purity test for air quality, and the test of the alarm sensors after calibration and setup per the manufacturer's instructions, as well as lead-lag controls.

(B) Tests shall be conducted at the sample port of the medical air system.

(C) The operation of the system control sensors, such as dew point, air temperature, and all other air quality monitoring sensors and controls, shall be checked for proper operation and function before the system is put into service.

(D) The quality of medical air as delivered by the compressor air supply shall be verified after installation of new components prior to use by patients.

(E) The air quality tests in 5.1.12.3.14.3(D) shall be conducted after the medical air source system has been operating normally but with the source valve closed under a simulated load for an elapsed time of at least 12 hours.

(F) The aggregate run time on the compressors shall not be used to determine the elapsed time.

(G) Loading shall be simulated by continuously venting air at approximately 25 percent of the rated system capacity.

(H) A demand of approximately 25 percent of the rated compressor capacity shall be created to cause the compressors to cycle on and off continuously and the dryers to operate for the 24-hour period.

5.1.12.3.14.4 Proportioning Systems for Medical Air USP.

(A) The system apparatus shall be tested for proper function, including the changeover from primary to secondary (if applicable) and operation of the reserve, before the system is put into service.

(B) Tests shall include the purity of the air quality and test of the alarm sensors after calibration and setup per the manufacturer's instructions.

(C) Tests shall be conducted at the sample port of the proportioning system.

(D) The operation of the control sensors and all quality monitoring sensors and controls shall be checked for proper operation and function before the system is put into service.

5.1.12.3.14.5 Medical–Surgical Vacuum Systems. The proper functioning of the medical–surgical vacuum source system(s) shall be tested before it is put into service.

5.1.13 Category 1 Support Gases.

5.1.13.1* Applicability.

5.1.13.1.1 Support gases are any gases that are used primarily for powering equipment used in patient care procedures (typical support gases are nitrogen and instrument air). Support gas applications require delivery at pressures, cleanliness, or purities specific to their intended function(s) (e.g., to operate medical–surgical tools). Support gases shall be permitted to be piped into areas intended for any medical support purpose and, if appropriate to the procedures, to be piped into laboratories.

5.1.13.1.2 Support gas sources shall be permitted to be used for many general utility uses (e.g., to remove excess moisture from instruments before further processing, or to operate gas-driven booms, boom brakes, pendants, or similar applications). Requirements for general utility systems will be found in Chapter 9.

5.1.13.1.3 Support gas systems shall not convey oxidizing gases other than air or gases intended for patient or staff respiration.

5.1.13.2 Nature of Hazards. Design, installation, and operation of support gas systems shall consider all hazards involved with any pressurized gas except those associated with oxidizing gases and hazards associated with the elevated pressures typical of these systems.

5.1.13.3 Sources.

5.1.13.3.1 Support gases shall be permitted to be supplied from the same sources as patient care gases. Where this is done, they shall be treated as the patient care gas and not as a support gas *(see 5.1.1 through 5.1.12 and 5.1.14)*.

5.1.13.3.2 Sources for support gases delivered from cylinders shall comply with 5.1.3.3 through 5.1.3.5.10.

5.1.13.3.3 Sources for support gases delivered from containers shall comply with 5.1.3.3 through 5.1.3.5.11 except 5.1.3.5.10.

5.1.13.3.4 Sources for support gases delivered from bulk sources shall comply with 5.1.3.3 through 5.1.3.5.11 except 5.1.3.5.10.

5.1.13.3.5* Instrument Air Supply Systems.

5.1.13.3.5.1 Quality of Instrument Air. The quality of instrument air shall be as follows:

(1) Compliant with ANSI/ISA S-7.0.01, *Quality Standard for Instrument Air*
(2) Filtered to 0.01 micron
(3) Free of liquids (e.g., water, hydrocarbons, solvents)
(4) Free of hydrocarbon vapors
(5) Dry to a dew point of −40°C (−40°F)

5.1.13.3.5.2 Instrument air supply systems shall be located per 5.1.3.3 as follows:

(1) Indoors, in a dedicated mechanical equipment area that is adequately ventilated and with any required utilities
(2) In a room ventilated per 5.1.3.3.3.3
(3) For air-cooled equipment, in a room designed to maintain the ambient temperature range as recommended by the equipment manufacturer

5.1.13.3.5.3 Instrument air sources shall provide air with the following characteristics:

(1) A gauge pressure not less than 1380 kPa (200 psi) at the compressor
(2) The quality of instrument air, as described in 5.1.13.3.5.1

5.1.13.3.5.4 Instrument air sources shall be of either of the following formats:

(1) At least two compressors
(2) One compressor and a standby header complying with 5.1.3.5.10

5.1.13.3.5.5 Instrument air sources shall include the components specified in 5.1.3.6.3.2, 5.1.3.6.3.5, 5.1.3.6.3.6, and 5.1.3.6.3.7 [except 5.1.3.6.3.7(1)].

5.1.13.3.5.6 Instrument air compressors shall be permitted to be of any type capable of not less than a gauge pressure of 1380 kPa (200 psi) output pressure and of providing air meeting the definition of instrument air in 5.1.13.3.5.1.

5.1.13.3.5.7 Instrument Air Standby Headers. Where instrument air systems are provided with a standby header, the header shall meet the following requirements:

(1) It shall comply with 5.1.3.5.10, except that the number of attached cylinders shall be sufficient for 1 hour of normal operation.
(2) It shall use connectors as for medical air in the mandatory requirements of CGAV-1, *Compressed Gas Association Standard for Compressed Gas Cylinder Valve Outlet and Inlet Connections* (ANSI B57.1).
(3) It shall enter the system upstream of the final line filters.
(4) It shall automatically serve the system in the event of a failure of the compressor.

5.1.13.3.5.8* Intake Air. Intake air for instrument air compressors shall be permitted to be drawn from outside, from ducted air, or from the equipment location.

5.1.13.3.5.9 Instrument Air Filters. Instrument air sources shall be provided with filtration sized for 100 percent of the system peak calculated demand at design conditions and with the following elements and characteristics:

(1) Activated carbon filters located upstream (source side) of the final line filters
(2) Line filters located upstream (source side) of the final line regulators and downstream of the carbon filters rated for a minimum of 98 percent efficiency at 0.01 micron
(3) Equipped with a continuous visual indicator showing the status of the line filter element life
(4) Constructed of materials deemed suitable by the manufacturer
(5) Filters combining the functions in (1) to (4) in a single unit shall be permitted to be used

5.1.13.3.5.10 Instrument Air Accessories. Accessories used for instrument air sources shall comply with the following subparagraphs:

(1) 5.1.3.6.3.5 for aftercoolers
(2) 5.1.3.6.3.6 for air receivers
(3) 5.1.3.6.3.7 for air dryers
(4) 5.1.3.5.5 for air regulators

5.1.13.3.5.11 Instrument Air Piping Arrangement and Redundancies. Instrument air sources shall comply with 5.1.3.6.3.9, except for the following:

(1) Systems employing a standby header shall be permitted to have simplex aftercoolers and dryers.
(2) Systems employing a standby header shall not require a three-valve receiver bypass.
(3) Standby headers, where provided, shall be isolated from the compressor by a check valve to prevent backflow through the compressor.

5.1.13.3.5.12 Instrument Air Monitoring and Alarms.

(A) Instrument air sources shall include the following alarms:

(1) Local alarm that activates when or just before the backup compressor (if provided) activates, indicating that the lag compressor is in operation and must be manually reset
(2) Local alarm and alarms at all master alarm panels that activate when the dew point at system pressure exceeds −30°C (−22°F), indicating a high dew point

(B) For sources with standby headers, the following additional conditions shall activate a local alarm at the compressor site, a local signal at the header location, and alarms at all master alarm panels:

(1) Alarm that activates when or just before the reserve begins to supply the system, indicating reserve in use
(2) Alarm that activates when or just before the reserve falls below an average hour's supply, indicating reserve is low

5.1.13.9.5.13 Electrical Power and Control.

(1) When multiple compressors are used, an additional compressor(s) shall automatically activate when the compressor(s) in operation is incapable of maintaining the required pressure.
(2) When multiple compressors are used, automatic or manual alternation of compressors shall allow division of operating time. If automatic alternation of compressors is not provided, the facility staff shall arrange a schedule for manual alternation.
(3) Each compressor motor shall be provided with electrical components including, but not limited to, the following:
   (a) Dedicated disconnect switch installed in the electrical circuit ahead of each motor starter
   (b) Motor starting device
   (c) Overload protection
   (d) Where compressor systems having two or more compressors employ a control transformer or other voltage control power device, installation of at least two such devices
   (e) Control circuits arranged in such a manner that the shutdown of one compressor does not interrupt the operation of another compressor
   (f) Automatic restart function such that the compressor(s) will restart after power interruption without manual intervention
(4) Electrical installation and wiring shall conform to the requirements of *NFPA 70, National Electrical Code*.
(5) Emergency electrical service for the compressors shall conform to the requirements of the essential electrical system, as described in Chapter 6.

5.1.13.4 Valves. Requirements for support gas valves shall be in accordance with 5.1.4.1.1 through 5.1.4.8.

5.1.13.5 Outlets. Requirements for support gas outlets shall be in accordance with 5.1.5.1, 5.1.5.2, 5.1.5.4 through 5.1.5.8, 5.1.5.11, and 5.1.5.13 through 5.1.5.15.

5.1.13.6 Manufactured Assemblies. Requirements for support gases in manufactured assemblies shall be in accordance with 5.1.6.1 through 5.1.6.9.

5.1.13.7 Pressure Indicators. Requirements for support gas pressure indicators shall be in accordance with 5.1.8.1.1 through 5.1.8.1.4, 5.1.8.1.6, and 5.1.8.2.

5.1.13.8 Warning Systems.

5.1.13.8.1 General requirements for support gas warning systems shall be in accordance with 5.1.9.1.

5.1.13.8.2 Master alarm requirements for support gas shall be in accordance with 5.1.9.2.

5.1.13.8.3 Area alarm requirements for support gas shall be in accordance with 5.1.9.4.

5.1.13.8.4 Local alarm requirements for support gas shall be in accordance with 5.1.9.5.

5.1.13.9 Distribution. Requirements for support gas piping distribution shall be in accordance with 5.1.10.1, 5.1.10.3, 5.1.10.4, 5.1.10.4.1 through 5.1.10.4.6, 5.1.10.9, 5.1.10.9(1), 5.1.10.9(2), 5.1.10.9(3), and 5.1.10.11.

5.1.13.10 Labeling and Identification. Requirements for support gas labeling shall be in accordance with 5.1.11.1 through 5.1.11.4.

5.1.13.11 Performance Testing. Requirements for support gas performance testing shall be in accordance with 5.1.12, with the following exceptions:

(1) The piping purity test *(see 5.1.12.3.8)* shall be permitted to be omitted.
(2) The medical gas concentration test *(see 5.1.12.3.11)* shall be permitted to be omitted.

5.1.14* Category 1 Operation and Management.

5.1.14.1 Special Precautions — Patient Gas, Vacuum, WAGD, and Medical Support Gas Systems.

5.1.14.1.1* Piping systems shall not be used for the distribution of flammable anesthetic gases.

5.1.14.1.2 Piping systems shall not be used as a grounding electrode.

5.1.14.1.3* Liquid or debris shall not be introduced into the medical–surgical vacuum or WAGD systems for disposal.

5.1.14.1.4* The medical–surgical vacuum and WAGD systems shall not be used for nonmedical applications (e.g., vacuum steam condensate return).

5.1.14.2 Maintenance of Medical Gas, Vacuum, WAGD, and Medical Support Gas Systems.

5.1.14.2.1* General. Health care facilities with installed medical gas, vacuum, WAGD, or medical support gas systems, or combinations thereof, shall develop and document periodic maintenance programs for these systems and their subcomponents as appropriate to the equipment installed.

5.1.14.2.2 Maintenance Programs.

5.1.14.2.2.1 Inventories. Inventories of medical gas, vacuum, WAGD, and medical support gas systems shall include at least all source subsystems, control valves, alarms, manufactured assemblies containing patient gases, and outlets.

5.1.14.2.2.2* Inspection Schedules. Scheduled inspections for equipment and procedures shall be established through the risk assessment of the facility and developed with consideration of the original equipment manufacturer recommendations and other recommendations as required by the authority having jurisdiction.

5.1.14.2.2.3 Inspection Procedures. The facility shall be permitted to use any inspection procedure(s) or testing methods established through its own risk assessment.

5.1.14.2.2.4 Maintenance Schedules. Scheduled maintenance for equipment and procedures shall be established through the risk assessment of the facility and developed with consideration of the original equipment manufacturer recommendations and other recommendations as required by the authority having jurisdiction.

5.1.14.2.2.5 Qualifications.

(A) Persons maintaining these systems shall be qualified to perform these operations.

(B) Appropriate qualification shall be demonstrated by any of the following:

(1) A documented training program acceptable to the health care facility by which such persons are employed or contracted to work with specific equipment as installed in that facility
(2) Credentialing to the requirements of ASSE 6040, *Professional Qualification Standard for Medical Gas Maintenance Personnel*, and technically competent on the specific equipment as installed in that facility.
(3) Credentialing to the requirements of ASSE 6030, *Professional Qualification Standard for Medical Gas Systems Verifiers*, and technically competent on the specific equipment as installed in that facility.

5.1.14.2.3* Inspection and Testing Operations.

5.1.14.2.3.1 Manufactured Assemblies Employing Flexible Connection(s) Between the User Terminal and the Piping System.

(A) Nonstationary booms and articulating assemblies, other than head walls utilizing flexible connectors, shall be tested for leaks, per manufacturer's recommendations, every 18 months or at a duration as determined by a risk assessment.

(B) The system pressure to nonstationary booms and articulating arms shall be maintained at operating pressure until each joint has been examined for leakage by effective means of leak detection that is safe for use with oxygen.

(C) Safe working condition of the flexible assemblies shall be confirmed.

(D) D.I.S.S. connectors internal to the boom and assemblies shall be checked for leakage.

(E) Leaks, if any, shall be repaired (if permitted), or the components replaced (if required), and the equipment retested prior to placing the equipment back into service.

(F) Additional testing of nonstationary booms or articulating arms shall be performed at intervals defined by documented performance data.

5.1.14.3 Medical Gas and Vacuum Systems Information and Warning Signs.

5.1.14.3.1 The gas content of medical gas and vacuum piping systems shall be labeled in accordance with 5.1.11.1.

5.1.14.3.2 Labels for shutoff valves shall be in accordance with 5.1.11.2 and updated when modifications are made changing the areas served.

5.1.14.3.3 Station inlets and outlets shall be identified in accordance with 5.1.11.3.

5.1.14.3.4 Alarm panel labeling shall be in accordance with 5.1.11.4 and updated when modifications are made changing the areas served.

5.1.14.4 Medical Gas and Vacuum Systems Maintenance and Record Keeping.

5.1.14.4.1 Permanent records of all tests required by 5.1.12.3.1 through 5.1.12.3.14 shall be maintained in the organization's files.

5.1.14.4.2 The supplier of the bulk cryogenic liquid system shall, upon request, provide documentation of vaporizer(s) sizing criteria to the facility.

5.1.14.4.3 An annual review of bulk system capacity shall be conducted to ensure the source system has sufficient capacity.

5.1.14.4.4 Central supply systems for nonflammable medical gases shall conform to the following:

(1) They shall be inspected annually.
(2) They shall be maintained by a qualified representative of the equipment owner.
(3) A record of the annual inspection shall be available for review by the authority having jurisdiction.

5.1.14.4.5 A periodic testing procedure for nonflammable medical gas and vacuum and related alarm systems shall be implemented.

5.1.14.4.6 Whenever modifications are made that breach the pipeline, any necessary installer and verification test specified in 5.1.12 shall be conducted on the downstream portions of the medical gas piping system.

5.1.14.4.7 Procedures, as specified, shall be established for the following:

(1) Maintenance program for the medical air compressor supply system in accordance with the manufacturer's recommendations
(2) Facility testing and calibration procedure that ensures carbon monoxide monitors are calibrated at least annually or more often if recommended by the manufacturer
(3) Maintenance program for both the medical–surgical vacuum piping system and the secondary equipment attached to medical–surgical vacuum station inlets to ensure the continued good performance of the entire medical–surgical vacuum system
(4) Maintenance program for the WAGD system to ensure performance

5.1.14.4.8 Audible and visual alarm indicators shall meet the following requirements:

(1) They shall be periodically tested to determine that they are functioning properly.
(2) Records of the test shall be maintained until the next test is performed.

5.1.14.4.9 Medical–surgical vacuum station inlet terminal performance, as required in 5.1.12.3.10.4, shall be tested as follows:

(1) On a regular preventive maintenance schedule as determined by the facility maintenance staff
(2) Based on flow of free air (Nl/min or SCFM) into a station inlet while simultaneously checking the vacuum level

5.2 Category 2 Piped Gas and Vacuum Systems.

5.2.1* Applicability.

5.2.1.1 These requirements shall apply to health care facilities that qualify for Category 2 systems as referenced in Chapter 4.

5.2.1.2 Category 2 piped gas or piped vacuum system requirements shall be permitted when all of the following criteria are met:

(1) Only moderate sedation; minimal sedation, as defined in 3.3.61.3 and 3.3.61.4; or no sedation is performed. Deep sedation and general anesthesia shall not be permitted.
(2) The loss of the piped gas or piped vacuum systems is likely to cause minor injury to patients, staff, or visitors.
(3) The facility piped gas or piped vacuum systems are intended for Category 2 patient care space per 3.3.127.2.

5.2.1.3 The following subsections of this chapter shall apply to the operation, management, and maintenance of Category 2 medical gas and vacuum systems in both new and existing health care facilities:

(1) 5.1.3.6.2
(2) 5.1.3.8.4.2
(3) 5.1.10.11.7.1
(4) 5.2.3.1
(5) 5.2.3.2
(6) 5.2.3.3
(7) 5.2.3.5(2)
(8) 5.2.3.6(2)
(9) 5.2.3.7(2)
(10) 5.2.13
(11) 5.2.14

5.2.2 Nature of Hazards of Gas and Vacuum Systems. The requirement of 5.1.2 shall apply to the nature of hazards of gas and vacuum systems.

5.2.3 Category 2 Sources.

5.2.3.1 Central Supply System Identification and Labeling. Category 2 systems shall comply with 5.1.3.1.

5.2.3.2 Central Supply Operations. Category 2 systems shall comply with 5.1.3.2.

5.2.3.3 Central Supply System Locations. Category 2 systems shall comply with 5.1.3.3.

5.2.3.4 Central Supply Systems. Category 2 systems shall comply with 5.1.3.5.

5.2.3.5 Category 2 Medical Air Supply Systems. Category 2 systems shall comply with 5.1.3.6, except as follows:

(1) Medical air compressors, dryers, aftercoolers, filters, and regulators shall be permitted to be simplex.
(2) The facility staff shall develop their emergency plan to deal with the loss of medical air.

5.2.3.6 Category 2 Medical–Surgical Vacuum. Category 2 systems shall comply with 5.1.3.7, except as follows:

(1) Medical–surgical vacuum systems shall be permitted to be simplex.
(2) The facility staff shall develop their emergency plan to deal with the loss of medical–surgical vacuum.

5.2.3.7 Category 2 WAGD. Category 2 systems shall comply with 5.1.3.8, except as follows:

(1) Medical WAGD pumps shall be permitted to be simplex.
(2) The facility staff shall develop their emergency plan to deal with the loss of WAGD.

5.2.3.8 Instrument Air Supply Systems. Category 2 systems shall comply with 5.1.13.3.5.

5.2.4 Valves. Category 2 systems shall comply with 5.1.4.

5.2.5 Station Outlets and Inlets. Category 2 systems shall comply with 5.1.5.

5.2.6 Manufactured Assemblies. Category 2 systems shall comply with 5.1.6.

5.2.7 Surface-Mounted Medical Gas Rails. Category 2 systems shall comply with 5.1.7.

5.2.8 Pressure and Vacuum Indicators. Category 2 systems shall comply with 5.1.8.

5.2.9 Warning Systems (Category 2). Warning systems associated with Category 2 systems shall provide the master, area, and local alarm functions of a Category 1 system as required in 5.1.9, except as follows:

(1) Warning systems shall be permitted to be a single alarm panel.
(2) The alarm panel shall be located in an area of continuous surveillance while the facility is in operation.
(3) Pressure and vacuum switches/sensors shall be mounted at the source equipment with a pressure indicator at the master alarm panel.

5.2.10 Category 2 Distribution. Level 2 systems shall comply with 5.1.10.

5.2.11 Labeling and Identification. Category 2 systems shall comply with 5.1.11.

5.2.12 Performance Criteria and Testing — Category 2 (Gas, Medical–Surgical Vacuum, and WAGD). Category 2 systems shall comply with 5.1.12.

5.2.13 Category 2 Support Gases. Category 2 systems shall comply with 5.1.13.

5.2.14* Category 2 Operation and Management. Category 2 systems shall comply with 5.1.14.

5.3* Category 3 Piped Gas and Vacuum Systems.

5.3.1* Applicability. These requirements shall apply to health care facilities that qualify to install Category 3 systems as defined in Chapter 4.

5.3.1.1 The following sections of this chapter shall apply to the operation, management, and maintenance of the medical gas and vacuum systems in both new and existing Category 3 health care facilities:

(1) 5.3.1.2(1)
(2) 5.3.1.6
(3) 5.3.2
(4) 5.3.4.1(4)
(5) 5.3.3.3.2
(6) 5.3.3.3.3
(7) 5.3.3.3.4
(8) 5.3.3.3.5
(9) 5.3.3.3.6
(10) 5.3.3.3.7
(11) 5.3.3.4(4)
(12) 5.3.14

5.3.1.2 Category 3 piped gas and vacuum systems shall be permitted when all of the following criteria are met:

(1)*Only moderate sedation; minimal sedation, as defined in 3.3.61.3 and 3.3.61.4; or no sedation is performed. Deep sedation and general anesthesia shall not be permitted.

(2) The loss of the piped gas and vacuum systems is not likely to cause injury to patients, staff, or visitors, but can cause discomfort.
(3) The facility piped gas and vacuum systems are intended for Category 3 or Category 4 patient care rooms per 3.3.127.3 and 3.3.127.4.

5.3.1.3 Where the term *medical gas* occurs, the provisions shall apply to all piped systems for oxygen, nitrous oxide, medical air, carbon dioxide, helium, air, and mixtures thereof. Wherever the name of a specific gas service occurs, the provision shall apply only to that gas.

5.3.1.4 Where the term *medical support gas* occurs, the provisions shall apply to all piped systems for nitrogen and dental air. Wherever the name of a specific gas service occurs, the provision shall apply only to that gas.

5.3.1.5 Wherever the term *vacuum* occurs, the provisions shall apply to all piped systems for medical–surgical vacuum, waste anesthetic gas disposal (WAGD), and dental vacuum. Wherever the name of a specific vacuum service occurs, the provision shall apply only to that vacuum service.

5.3.1.6 An existing system that is not in strict compliance with the requirements of this code shall be permitted to continue in use as long as the authority having jurisdiction has determined that such use does not constitute a distinct hazard to life.

5.3.2 Nature of Hazards of Gas and Vacuum Systems. Potential fire and explosion hazards associated with Category 3 gas and dental vacuum systems shall be considered in the design, installation, testing, operation, and maintenance of the systems.

5.3.3 Sources. Category 3 systems shall comply with 5.2.3, except as included in 5.3.3.1 through 5.3.3.11.

5.3.3.1 Central Supply System Identification and Labeling. Category 3 systems shall comply with 5.2.3.1.

5.3.3.2 Central Supply Operations. Category 3 systems shall comply with 5.2.3.2.

5.3.3.3 Central Supply System Locations. Category 3 systems shall comply with 5.2.3.3.

5.3.3.3.1 Ventilation for motor-driven equipment, including dental air sources and dental vacuum sources, shall comply with 5.2.3.3.

5.3.3.3.2 Enclosures shall serve no purpose other than to contain the medical gas source equipment, except that nitrogen source equipment and dental air cylinders in 5.3.3.6.1 shall be permitted in the enclosure.

5.3.3.3.3 Dental air compressors, dental vacuum pumps, and other equipment shall not be located in enclosures for medical gas cylinders.

5.3.3.3.4 Dental air compressors shall be installed in a designated mechanical equipment area, heated and ventilated in accordance with 5.2.3.3, and have required utilities (e.g., electrical power, drains, lighting).

5.3.3.3.5 Where nitrogen or dental air in cylinders is used, the cylinders shall be permitted to be located in a dental air compressor equipment room.

5.3.3.3.6 Nitrogen and dental air cylinders shall be permitted to be located in enclosures for medical gases.

5.3.3.3.7 Cylinders in service and in storage shall be individually secured and located to prevent falling or being knocked over.

5.3.3.4 Central Supply Systems. Category 3 systems, including dental air sources and dental vacuum sources shall comply with 5.2.3.4 except as follows:

(1) The central supply system's final line regulators shall be permitted to be simplex.
(2) For a single treatment facility, the central supply system shall contain a minimum of two equal headers, of one or more cylinders, with each header containing a minimum of an average day's supply.
(3) Where the central supply system is remote from the building being served, the manifold in this category shall include an automatic means of alternating the primary and secondary headers.
(4) Where the central supply system is not remote, the manifold in this category shall include a manual or automatic means of alternating the primary and secondary headers.
(5) Where the central supply system serves multiple treatment facilities, the manifold in this category shall include an automatic means of alternating the primary and secondary headers.
(6) For dental applications, flexible connectors of other than all-metal construction that connect manifolds to the gas distribution piping shall not exceed 1.52 m (5 ft) in length and shall not penetrate walls, floors, ceilings, or partitions.
(7) Pressure relief valve discharge that will not create an oxygen deficient atmosphere hazard shall be permitted to exhaust inside the manifold room.

5.3.3.5 Category 3 Medical Air Supply Systems. Category 3 medical air supply systems shall comply with 5.2.3.5.

5.3.3.6* Dental Air Supply Systems. Dental air supply systems shall comply with 5.3.3.6.1 or 5.3.3.6.2.

5.3.3.6.1 Dental Air Compressor Supply Systems.

5.3.3.6.1.1 General. Category 3 dental air compressor supply systems shall include the following:

(1) Disconnect switch(es)
(2) Motor starting device(s)
(3) Motor overload protection device(s)
(4) One or more compressors
(5) CFor single, duplex, or multiple compressor systems, means for activation/deactivation of each individual compressor
(6) When multiple compressors are used, manual or automatic means to alternate individual compressors
(7) When multiple compressors are used, manual or automatic means to activate the additional unit(s) should the in-service unit(s) be incapable of maintaining adequate pressure
(8) Intake filter–muffler(s) of the dry type
(9) Receiver(s) with a manual or automatic drain
(10) Shutoff valves
(11) Compressor discharge check valve(s) (for multiple compressors)
(12) Air dryer(s) that maintains a minimum of 40 percent relative humidity at operating pressure and temperature
(13) In-line final particulate/coalescing filters rated at 0.01 μ, with filter status indicator to ensure the delivery of dental air with a maximum allowable 0.05 ppm liquid oil

(14) Pressure regulator(s)
(15) Pressure relief valve
(16) Pressure indicator
(17) Moisture indicator

5.3.3.6.1.2 Receivers. Receivers shall have the following:

(1) The capacity to prevent short cycling of the compressor(s)
(2) Compliance with Section VIII, "Unfired Pressure Vessels," of the ASME *Boiler and Pressure Vessel Code*

5.3.3.6.1.3* Moisture Indicator. Moisture indicators shall have the following:

(1) A location in the active airstream prior to, or after, the receiver and upstream of any system pressure regulators
(2) The ability to indicate (e.g., by color change, digital readout, or other method understood by the user) when the relative humidity of the dental air exceeds 40 percent at line pressure and temperature

5.3.3.6.1.4 Pressure Relief Valve Discharge. Pressure relief valves for dental air systems having less than 84,950 L (3000 ft^3) at STP shall be permitted to discharge locally indoors in a safe manner that will not restrict the flow.

5.3.3.6.1.5* Source of Dental Air Compressor Intake. Dental air sources for a compressor(s) shall meet the following requirements:

(1) If the intake is located inside the building, it shall be located within a space where no chemical-based materials are stored or used.
(2) If the intake is located inside the building, it shall be located in a space that is not used for patient medical treatment.
(3) If the intake is located inside the building, it shall not be taken from a room or space in which there is an open or semi-open discharge from a Category 3 vacuum system.
(4) If the intake is located outside the building, it shall be drawn from locations where no contamination from vacuum exhaust discharges or particulate matter is anticipated.

5.3.3.6.2 Dental Air Cylinder Supply Systems.

5.3.3.6.2.1 Quality of Dental Air Cylinder. Dental air cylinders shall meet or exceed the quality grade requirements of industrial air.

5.3.3.6.2.2 Dental air cylinders shall be permitted to be installed in enclosures for Category 3 medical gases or in a mechanical room.

5.3.3.6.2.3 Dental air cylinder source equipment shall include the following:

(1) One or more cylinders of dental air, each providing at least an average day's supply
(2) A manifold if primary and secondary cylinders are provided
(3) A line pressure regulating valve
(4) A check valve downstream from the pressure regulating valve
(5) A pressure relief valve set at 50 percent above the normal line pressure and located downstream from the check valve

5.3.3.6.2.4 Mechanical means shall be provided to ensure that the dental air cylinder gas source equipment is connected to the correct gas distribution piping system.

5.3.3.6.2.5 Threaded connections to manifolds shall comply with the mandatory requirements of CGA V-5, *Diameter-Index Safety System (Noninterchangeable Low Pressure Connections for Medical Gas Applications)*.

5.3.3.6.2.6 Flexible connectors shall have a gauge pressure rating not less than 6895 kPa (1000 psi).

5.3.3.6.2.7 Flexible connectors of other than all-metal construction that connect manifolds to the gas distribution piping shall not exceed 1.52 m (5 ft) in length and shall not penetrate walls, floors, ceilings, or partitions.

5.3.3.6.2.8 Pressure relief valves for dental air cylinder systems having less than 84,950 L (3000 ft^3) at STP shall be permitted to discharge locally indoors in a safe manner that will not restrict the flow.

5.3.3.7 Instrument Air Supply Systems. A Category 3 instrument air supply system, if used, shall comply with 5.2.3.8, except that instrument air supply system compressors, dryers, aftercoolers, filters, and regulators shall be permitted to be simplex.

5.3.3.8* Nitrogen Supply System. Nitrogen source equipment shall be permitted to be installed in enclosures for Category 3 medical gases or in a mechanical room.

5.3.3.8.1 Nitrogen source equipment shall include the following:

(1) One or more cylinders of nitrogen NF, each providing at least an average day's supply
(2) A manifold, if primary and secondary cylinders are provided
(3) A line pressure regulating valve
(4) A check valve downstream from the pressure regulating valve
(5) A pressure relief valve set at 50 percent above the normal line pressure and located downstream from the check valve
(6) A pressure relief valve discharge piped to outdoors at a point that will not create a probable hazard and that is turned down to prevent the entry of rain or snow

5.3.3.8.2 Flexible connectors of other than all-metal construction that connect manifolds to the gas distribution piping shall not exceed 1.52 m (5 ft) in length and shall not penetrate walls, floors, ceilings, or partitions.

5.3.3.9 Medical–Surgical Vacuum. Category 3 medical–surgical vacuum systems, if used, shall comply with 5.2.3.5.

5.3.3.10 Dental Vacuum Supply Systems.

5.3.3.10.1 Category 3 Dental Vacuum Supply Systems.

5.3.3.10.1.1 Category 3 vacuum sources shall include the following:

(1) A vacuum pump(s) suited for wet or dry service as intended in the system design
(2) If intended for wet service, properly vented liquid/air separator

5.3.3.10.1.2 Category 3 vacuum source equipment shall be obtained from, and be installed under the supervision of, the manufacturer(s) or supplier(s) who is familiar with its installation, operation, and use.

5.3.3.10.1.3* Drainage from Vacuum Equipment. Drainage from vacuum equipment shall include the following:

(1) Liquids drained from a Category 3 vacuum source shall discharge indirectly to a sanitary drainage system through an approved air gap to a trapped and vented drain.
(2) The clear air gap between a vacuum drain outlet or indirect drain pipe, and the flood category rim of an indirect waste receptor or other point of disposal, shall be not less than twice the diameter of the effective opening of the drain served, but not less than 25.4 mm (1 in.), unless the local plumbing code requires a larger air gap.
(3) Where the drainage is from a waste holding tank on the suction side of the vacuum source, the following requirements shall be met:
   (a) A check valve shall be installed in the drain line from the holding tank between the tank and any vent lines.
   (b) The trap in the building drainage system shall be the deep-seal type that is conventionally vented within the plumbing system.
   (c) An additional vent shall be installed between the holding tank drain check valve and the drain trap, on the inlet side of the trap, to close and seal the check valve while the holding tank is operating under vacuum and collecting waste.
   (d) The additional vent described in 5.3.3.10.1.3(3)(c) shall be permitted to be connected to the plumbing system vents, unless a drain pump system with a positive pressure discharge is installed, in which case 5.3.3.10.1.3(4) shall apply.
   (e) Both of the vents in 5.3.3.10.1.3(3)(c) and 5.3.3.10.1.3(3)(d) shall extend vertically to not less than 152 mm (6 in.) above the top of the holding tank before turning horizontal.
   (f) Outdoor vents shall be protected against the entry of insects, vermin, debris, and precipitation.
   (g) The trap and drain branch shall be not less than two pipe sizes larger than the waste pipe from the separator, but not less than DN50 (NPS 2).
   (h) The trap seal shall be not less than 100 mm (4 in.) deep.
   (i) The vent for the vacuum check valve shall be not less than the size of the check valve.
   (j) The vent for the trap shall be not less than one-half the size of the trap and drain branch.
(4)*Where the drainage is from a waste holding tank on the suction side of the vacuum source and a positive discharge pump drain system is in place, the following requirements shall be met:
   (a) The pump shall drain indirectly to the plumbing system through an air gap equal to the diameter of the discharge pipe but not less than 25.4 mm (1 in.) above the rim.
   (b) A check valve shall be installed in the drain line from the holding tank to the drain.
   (c) The trap in the building drainage system shall be the deep seal type that is conventionally vented within the plumbing system.
   (d) The trap and drain branch shall be not less than two pipe sizes larger than the waste pipe from the separator, but not less than DN40 (NPS 1½).
   (e) The trap seal shall be at least two times the exhaust back pressure in the separator but not less than 100 mm (4 in.) deep.
(5) Where the drainage is at a positive pressure from an air/waste separator on the discharge side of the vacuum source, the following requirements shall be met:
   (a) Where there is a positive pressure discharge from a vacuum pump, it shall be required to drain through an air/waste separator.
   (b) Discharge shall be either of the following:
      i. Direct into a trap in the building drainage system that is the deep-seal type and is conventionally vented within the plumbing system
      ii. Indirect to the plumbing system through an air gap equal to the diameter of the discharge pipe, but not less than 25.4 mm (1 in.) above the rim
   (c) The trap vent shall extend vertically to not less than 152 mm (6 in.) above the top of the separator before turning horizontal.
   (d) Outdoor vents shall be protected against the entry of insects, vermin, debris, and precipitation.
   (e) The trap and drain branch shall be two pipe sizes larger than the waste pipe from the separator, but not less than DN40 (NPS 1½).
   (f) The air/waste separator vent shall be the full size of the separator vent connection.
   (g) The separator vent shall be separate from the building vent piping.
(6) The indirect drainage from vacuum equipment shall discharge to the sanitary drainage system through an approved air gap without causing overflow or splatter on building surfaces.
(7) None of the requirements within this chapter for drainage in Category 3 dental vacuum systems shall supersede provisions of the local plumbing code.

5.3.3.10.1.4 Vacuum Exhaust. The exhaust from Category 3 vacuum sources shall comply with the following:

(1) It shall be piped to the outside through a separate vent system.
(2) The exhaust point shall be chosen to minimize the hazards of noise.
(3) The exhaust point shall be remote from any door, window, or other opening into the building.
(4) The exhaust point shall be located at a different elevation than air intakes.
(5) The exhaust point shall not be located where affected by prevailing winds, adjacent buildings, topography, or other obstacles to the rapid dispersion of the exhaust gases.
(6) The exhaust point shall be protected against the entry of insects, vermin, debris, and precipitation.
(7) The exhaust piping shall be sized to prevent back pressure greater than the pump manufacturer's recommendations.
(8)*Where multiple pumps exhaust through a common pipe, each pump shall be fitted with a check valve or a manual isolation valve or shall be arranged to allow capping the individual pump exhausts when a pump is removed for service.
(9) Where multiple pumps exhaust through a common pipe, piping shall be arranged following the pump manufacturer's recommendations.

5.3.3.11 Waste Anesthetic Gas Disposal (WAGD). Category 3 systems shall comply with 5.2.3.7.

5.3.4 Valves.

5.3.4.1 Emergency Shutoff Valves. Category 3 systems shall comply with 5.2.4, except as follows:

(1)*Where a central Category 3 medical gas supply is remote from a single treatment facility, the main supply line shall be provided with an emergency shutoff valve so located in the single treatment facility to be accessible from all use-point locations in an emergency.

(2) Where a central Category 3 medical gas supply system supplies two treatment facilities, each facility shall be provided with an emergency shutoff valve so located in the treatment facility to be accessible from all use-point locations in an emergency.

(3) Emergency shutoff valves shall be labeled to indicate the gas they control and shall shut off only the gas to the treatment facility that they serve.

(4) A remotely activated shutoff valve at a supply manifold shall not be used for emergency shutoff. For clinical purposes, such a remote valve actuator shall not fail-closed in the event of a loss of electric power. Where remote actuators are the type that fail-open, it shall be mandatory that cylinder shut-off valves be closed whenever the system is not in use.

5.3.5* Station Outlets and Inlets. Category 3 systems shall comply with 5.2.5.

5.3.6 Manufactured Assemblies. Category 3 systems shall comply with 5.2.6.

5.3.7 Surface-Mounted Medical Gas Rails. Category 3 systems shall comply with 5.2.7.

5.3.8 Pressure and Vacuum Indicators. Category 3 systems shall comply with 5.2.8.

5.3.9 Category 3 Warning Systems. Category 3 warning systems shall comply with 5.2.9 except as follows:

(1) Warning systems shall be permitted to be a single alarm panel.
(2) The alarm panel shall be located in an area of continuous surveillance while the facility is in operation.
(3) Pressure and vacuum switches/sensors shall be mounted at the source equipment with a pressure indicator at the master alarm panel.
(4) Warning systems for medical gas systems shall provide the following alarms:
   (a) Oxygen main line pressure low
   (b) Oxygen main line pressure high
   (c) Oxygen changeover to secondary bank or about to changeover (if automatic)
   (d) Nitrous oxide main line pressure low
   (e) Nitrous oxide main line pressure high
   (f) Nitrous oxide changeover to secondary bank or about to changeover (if automatic)
(5) Audible and noncancelable alarm visual signals shall indicate if the pressure in the main line increases or decreases 20 percent from the normal operating pressure.
(6) Visual indications shall remain until the situation that caused the alarm is resolved.
(7) Pressure switches/sensors shall be installed downstream of any emergency shutoff valves and any other shutoff valves in the system and shall cause an alarm for the medical gas if the pressure decreases or increases 20 percent from the normal operating pressure.
(8) A cancelable audible indication of each alarm condition that produces a sound at the alarm panel shall reinitiate the audible signal if another alarm condition occurs while the audible signal is silenced.

5.3.10 Distribution. Category 3 systems shall comply with 5.1.10, except as follows:

(1) Dental air and dental vacuum shall comply with 5.1.10.2.1, except the tubing shall be permitted to be annealed (soft temper).
(2) Dental vacuum tubing shall be permitted to be:
   (a) PVC plastic pipe shall be Schedule 40 or Schedule 80, complying with ASTM D 1785, *Standard Specification for Poly (Vinyl Chloride) (PVC) Plastic Pipe, Schedules 40, 80, and 120*.
   (b) PVC plastic fittings shall be Schedule 40 or Schedule 80 to match the pipe, complying with ASTM D 2466, *Standard Specification for Poly (Vinyl Chloride) (PVC) Plastic Pipe Fittings, Schedule 40*; or ASTM D 2467, *Standard Specification Poly (Vinyl Chloride) (PVC) Plastic Pipe Fittings, Schedule 80*.
   (c) Joints in PVC plastic piping shall be solvent-cemented in accordance with ASTM D 2672, *Standard Specification for Joints for IPS PVC Pipe Using Solvent Cement*.
   (d) CPVC IPS plastic pipe shall be Schedule 40 or Schedule 80, complying with ASTM F 441, *Standard Specification for Chlorinated Poly (Vinyl Chloride) (CPVC) Plastic Pipe*, Schedules 40 and 80.
   (e) CPVC IPS plastic fittings shall be Schedule 40 or Schedule 80 to match the pipe, complying with ASTM F 438, *Standard Specification for Socket-Type Chlorinated Poly (Vinyl Chlorinated) (CPVC) Plastic Pipe Fittings, Schedule 40*; or ASTM F 439, *Standard Specification for Chlorinated Poly (Vinyl Chlorinated) (CPVC) Plastic Pipe Fittings, Schedule 80*.
   (f) CPVC CTS plastic pipe and fittings ½ in. through 2 in. size shall be SDR 11, complying with ASTM D 2846, *Standard Specification for Chlorinated Poly (Vinyl Chloride) (CPVC) Plastic Hot- and Cold-Water Distribution Systems*.
   (g) Solvent cement for joints in CPVC plastic piping shall comply with ASTM F 493, *Solvent Cements for CPVC Pipe and Fittings*.
(3) Dental air and dental vacuum fittings shall be permitted to be:
   (a) Soldered complying with ASME B16.22, *Wrought Copper and Copper Alloy Solder-Joint Pressure Fittings*
   (b) Flared fittings complying with ASME B16.26, *Cast Copper Alloy Fittings for Flared Copper Tubes*
   (c) Compression fittings (¾ in. maximum size)
(4) Soldered joints in Category 3 dental air supply piping shall be made in accordance with ASTM B 828, *Standard Practice for Making Capillary Joints by Soldering of Copper and Copper Alloy Tube and Fittings*, using a "lead-free" solder filler metal containing not more than 0.2 percent lead by volume that complies with ASTM B 32, *Standard Specification for Solder Metal*.
(5) Where required, gas and vacuum equipment and piping shall be seismically restrained against earthquakes in accordance with the applicable building code.
(6) Gas and vacuum piping systems shall be designed and sized to deliver the required flow rates at the utilized pressures.

5.3.10.1 Installation of Vacuum Piping.

5.3.10.1.1 Pipe Sizing. Piping systems shall be designed and sized to draw the required flow rates at the utilization vacuums.

5.3.10.1.2 Protection of Piping.

5.3.10.1.2.1 Piping shall be protected against freezing, corrosion, and physical damage.

5.3.10.1.2.2 Piping exposed in corridors and other locations where subject to physical damage from the movement of carts, stretchers, beds, portable equipment, or vehicles, shall be protected.

5.3.10.1.3 Copper Pipe Support. Pipe support for copper tube shall be in accordance with Table 5.3.10.1.3.

Table 5.3.10.1.3 Maximum Copper Tube Support Spacing

Pipe Size	Hanger Spacing	
	mm	ft
DN8 (NPS ¼) (⅜ in. O.D.)	1520	5
DN10 (NPS ⅜) (½ in. O.D.)	1830	6
DN15 (NPS ½) (⅝ in. O.D.)	1830	6
DN20 (NPS ¾) (⅞ in. O.D.)	2130	7
DN25 (NPS 1) (1⅛ in. O.D.)	2440	8
DN32 (NPS 1¼) (1⅜ in. O.D.)	2740	9
DN40 (NPS 1½) (1⅝ in. O.D.) and larger	3050	10
Vertical risers, all sizes, every floor, but not to exceed	4570	15

5.3.10.1.4 Plastic Pipe Support. The maximum support spacing for plastic pipe shall be in accordance with Table 5.3.10.1.4.

Table 5.3.10.1.4 Maximum Plastic Pipe Support Spacing

Pipe Size	Hanger Spacing	
	mm	ft
DN15 (NPS ½) (⅝ in. O.D.)	1220	4.00
DN20 (NPS ¾) (⅞ in. O.D.)	1220	4.00
DN25 (NPS 1) (1⅛ in. O.D.)	1320	4.33
DN32 (NPS 1¼) (1⅜ in. O.D.)	1320	4.33
DN40 (NPS 1½) (1⅝ in. O.D.)	1420	4.66
DN50 (NPS 2) (2⅜ in. O.D.)	1420	4.66
DN65 (NPS 2½) (2⅞ in. O.D.) and larger	1520	5.00
Vertical risers, all sizes, every floor, but not to exceed	3040	10.00

5.3.10.1.5 Underground Piping Outside of Buildings. Buried piping outside of buildings shall be in accordance with 5.1.10.11.5.

5.3.10.1.6 Underground Piping Within Buildings. Underground piping within buildings shall be in accordance with the following:

(1) The installation procedure for underground piping shall prevent physical damage to the piping while being backfilled
(2) If the underground piping is protected by a conduit, cover, or other enclosure, access shall be provided at the joints during construction for visual inspection and leak testing

5.3.10.1.7 Piping Within Floor Slabs.

5.3.10.1.7.1 Copper Category 3 vacuum piping that is installed within floor slabs shall be enclosed in a conduit, flexible plastic tubing, or other means to prevent contact between the copper tubing and concrete.

5.3.10.1.7.2 Plastic Category 3 vacuum piping shall be permitted to contact concrete.

5.3.10.1.7.3 During construction, access shall be provided at all joints for visual inspection and leak testing.

5.3.10.1.7.4 Care shall be taken to protect plastic piping from damage from vibrators while wet concrete is being consolidated.

5.3.10.1.8 Plastic Pipe Installation.

5.3.10.1.8.1 Horizontal piping in Category 3 dental vacuum systems shall be sloped a minimum of 7 mm per 3.05 m (¼ in. per 10 ft) toward the vacuum source equipment.

5.3.10.1.8.2 Horizontal piping shall include no sags or low points that will permit fluids or debris to accumulate.

5.3.10.1.9 Valves in Vacuum Systems. Shutoff valves shall be permitted to be installed in Category 3 vacuum piping.

5.3.11 Labeling, Pressure, and Identification. Category 3 systems shall comply with 5.2.11.

5.3.12 Performance Criteria and Testing. Category 3 systems for medical gas, medical support gas, medical surgical vacuum, WAGD, dental air and dental vacuum shall comply with 5.2.12, except as follows:

5.3.12.1 General.

5.3.12.1.1 The initial tests required by 5.3.12.1 shall be performed prior to the final tests required by 5.3.12.2.10.

5.3.12.1.2 Initial tests shall be conducted by one or more of the following, who shall be experienced in the installation, operation, and testing of Category 3 medical support gas, dental vacuum, dental air and dental vacuum supply systems:

(1) Installer
(2) Representative of the system supplier
(3) Representative of the system manufacturer
(4) ASSE 6030 medical gas system's verifier

5.3.12.1.3 The test gas for Category 3 copper piping supply systems shall be oil-free, dry nitrogen NF or the system gas.

5.3.12.1.4 Where manufactured assemblies are to be installed, the initial tests required under 5.3.12.2 shall be performed as follows:

(1) After completion of the distribution piping
(2) Prior to installation or connection of manufactured assemblies having internal tubing or hose.
(3) At all outlets and inlets on manufactured assemblies having internal copper tubing

5.3.12.2 Category 3 Dental Vacuum Supply Systems.

5.3.12.2.1 Blowdown. Piping in Category 3 gas-powered device supply systems shall be blown clear using oil-free, dry nitrogen NF as follows:

(1) After installation of the distribution piping
(2) After installation of outlet shutoff valves
(3) Before connection to the use points
(4) Before installation of system components (e.g., pressure indicators, pressure relief valves, manifolds, source equipment)

5.3.12.2.2 Initial Pressure Test for Copper Piping Systems.

5.3.12.2.3 Each section of the piping in Category 3 gas powered device supply systems, copper vacuum systems, shall be pressure tested using oil-free, dry nitrogen NF or the system gas.

5.3.12.2.4 Initial pressure tests shall be conducted as follows:

(1) After blowdown of the distribution piping
(2) After installation of outlet and inlet shutoff valves station outlets and inlets
(3) Prior to the installation of components of the distribution piping system that would be damaged by the test pressure (e.g., pressure/vacuum indicators, line pressure relief valves)
(4) The source shutoff valves for the piping systems shall remain closed during the tests, unless being used for the pressure test gas
(5) With test pressure 1.5 times the system operating pressure but not less than a gauge pressure of 1035 kPa (150 psi)
(6)*With test pressure maintained until each joint is examined for leakage by means of a detectant that is safe for use with oxygen and that does not contain ammonia
(7) With leaks, if any, located, repaired (if permitted), or replaced (if required) by the installer and retested

5.3.12.2.5 Initial Leak Test for Plastic Vacuum Piping Systems. Initial leak tests shall be conducted as follows:

(1) Each section of the piping in Category 3 vacuum systems with plastic piping shall be leak tested using a test vacuum or the vacuum source equipment.
(2) If installed, the vacuum source shutoff valves for the piping systems shall remain closed during the tests, unless being used for the leak test vacuum source.
(3) The leak test vacuum shall be a minimum of 300 mm (12 in.) HgV.
(4) The test vacuum shall be maintained until each joint has been examined for leakage. An ultrasonic leak detector shall be permitted to be used.
(5) Leaks, if any, shall be located, repaired, or replaced (if required) by the installer and retested.

5.3.12.2.6 Initial Cross-Connection Test for Copper Piping Systems. Initial cross-connection tests for copper piping systems shall be conducted as follows:

(1) Tests shall be conducted to determine that no cross-connections exist between the Category 3 copper piping systems and Category 3 copper vacuum piping systems.
(2) The piping systems shall be at atmospheric pressure.
(3) The test gas shall be oil-free, dry nitrogen NF or dental air.
(4) The source of test gas shall be connected only to the piping system being tested.
(5) The piping system being tested shall be pressurized to a gauge pressure of 345 kPa (50 psi).
(6) The individual system gas outlet and vacuum inlet in each installed gas-powered device and copper vacuum or copper piping system shall be checked to determine that the test gas pressure is present only at the piping system being tested.
(7) The cross-connection test shall be repeated for each installed Category 3 piping system for gas-powered devices and for vacuum with copper piping.
(8) The proper labeling and identification of system outlets/inlets shall be confirmed during the tests.

5.3.12.2.7 Initial Cross-Connection Test for Plastic Vacuum Piping Systems. Initial cross-connection tests for plastic vacuum piping systems shall be conducted as follows:

(1) Tests shall be conducted to determine that no cross connections exist between any Category 3 plastic vacuum piping systems or Category 3 copper piping systems
(2) The vacuum source shutoff valves for the vacuum piping systems shall remain closed during the tests, unless they are being used for the cross-connection test vacuum source.
(3) The cross-connection test vacuum shall be a minimum of 300 mm (12 in.) HgV.
(4) The source of test vacuum shall be connected only to the vacuum piping system being tested.
(5) The individual gas-powered device system gas outlets and vacuum system inlets shall be checked to determine that the test vacuum is only present at the vacuum piping system being tested.
(6) The cross-connection tests shall be repeated for each installed vacuum system with plastic piping.
(7) The proper labeling and identification of system outlets/inlets shall be confirmed during the tests.

5.3.12.2.8 Initial Piping Purge Test for Dental Air and Nitrogen Supply Systems. Initial piping purge tests for dental air and nitrogen supply systems shall be conducted as follows:

(1) The outlets in each Category 3 dental air and nitrogen supply piping system shall be purged to remove any particulate matter from the distribution piping.
(2) The test gas shall be oil-free, dry nitrogen NF or the system gas.
(3) Each outlet shall be purged with an intermittent high-volume flow of test gas until the purge produces no discoloration in a clean white cloth.
(4) The purging shall be started at the furthest outlet in the system and proceed toward the source equipment.

5.3.12.2.9 Initial Standing Pressure Test for Dental Air and Nitrogen Supply Systems. After successful completion of the initial pressure tests under 5.3.12.2.6, Category 3 gas-powered device distribution piping shall be subjected to a standing pressure test, which includes the following:

(1) Tests shall be conducted after the installation of outlet valves and other distribution system components (e.g., pressure indicators and line pressure relief valves).
(2) The source valve shall be closed unless the source gas is being used for the test.
(3) The piping systems shall be subjected to a 24-hour standing pressure test using oil-free, dry nitrogen NF or the system gas.
(4) Test pressures shall be 20 percent above the normal system operating line pressure.
(5) At the conclusion of the tests, there shall be no change in the test pressure greater than a gauge pressure of 35 kPa (5 psi).
(6) Leaks, if any, shall be located, repaired (unless prohibited), or replaced (if required) by the installer and retested.

5.3.12.2.10 Final Testing of Category 3 Dental Air Supply Systems, Nitrogen Supply Systems, and Vacuum Systems. Final testing of dental air supply systems, nitrogen supply systems, and vacuum systems shall be conducted as follows:

(1) Final testing of gas-powered device systems and vacuum systems shall be performed only after all initial tests required by 5.3.12.2.1 through 5.3.12.2.9 have been performed.

(2) The final tests required by 5.3.12.2.11 through 5.3.12.2.15 shall be performed by one or more of the following, who shall be experienced with the installation, operation, and testing of Category 3 gas-powered device supply systems and vacuum systems:

  (a) Installer
  (b) Representative of the system supplier
  (c) Representative of the system manufacturer
  (d) ASSE 6030 medical gas system's verifier

(3) The test gas shall be oil-free, dry nitrogen NF or the system gas or vacuum.

5.3.12.2.11 Final Standing Pressure Test (Category 3 Dental Air and Nitrogen). Each gas-powered device piping system shall be subjected to a 10-minute standing pressure test at operating line pressure using the following procedures:

(1) After the system is filled with oil-free, dry nitrogen NF or the system gas, the source valve shall be closed.
(2) The piping system downstream of the valve shall show no decrease in pressure after 10 minutes.
(3) Any leaks found shall be located, repaired (unless prohibited) or replaced (if required) by the installer, and retested.

5.3.12.2.12 Final Standing Vacuum Test (Category 3 Vacuum Systems). Each Category 3 vacuum piping system shall be subjected to a 10-minute standing vacuum test at operating line vacuum using the following procedures:

(1) After the system has stabilized at the operating line vacuum, the source valve and any zone valves shall be closed.
(2) The piping system upstream of the valves shall show no decrease in vacuum after 10 minutes.
(3) Leaks, if any, shall be located, repaired (unless prohibited) or replaced (if required) by the installer, and retested.

5.3.12.2.13 Final Cross-Connection Test (Category 3 Gas-Powered Devices and Vacuum and Scavenging Systems). After closing of walls and completion of the requirements of 5.3.12.2, it shall be determined that no cross-connections exist between the piping systems for Category 3 gas-powered devices and vacuum and scavenging systems using the following method:

(1) Test each piping system independently, starting with the vacuum systems first, and check that the test vacuum is present only at inlets of the system being tested.
(2) Reduce all piping systems to atmospheric pressure.
(3) Operate the Category 3 vacuum or scavenging system being tested at the normal system vacuum, using the source equipment.
(4) Test each gas outlet and vacuum inlet using appropriate adapters to verify that vacuum is present only at the vacuum inlets in the system being tested, and not at any gas outlets or inlets.
(5) Shut down the vacuum source equipment and slowly break the vacuum in the vacuum piping system, increasing its pressure to atmospheric.
(6) Test each Category 3 vacuum system until all are determined to be free of cross-connections.
(7) Using oil-free, dry nitrogen NF or the system gas, pressurize the gas piping system to a gauge pressure of 345 kPa (50 psi).
(8) Test each gas-powered device gas outlet using appropriate adapters to verify that the test gas pressure is present only at the outlets in the gas-powered device system being tested.
(9) After it has been determined that a gas-powered device piping system is free of cross-connections, disconnect the source of test gas and reduce the piping to atmospheric pressure.
(10) Proceed to test each gas-powered device piping system until all are determined to be free of cross-connections.

5.3.12.2.14 Final Piping Purge Test (Category 3 Gas-Powered Devices). To remove any traces of particulate matter deposited in the pipelines as a result of construction, a heavy, intermittent purging of each gas-powered device pipeline shall be done.

(1) The appropriate adapter shall be obtained from the facility or manufacturer, and high purge rates shall be put on each outlet.
(2) After the purge is started, it shall be rapidly interrupted several times until the purge produces no discoloration in a white cloth loosely held over the adapter during the purge.
(3) To avoid possible damage to the outlet and its components, the test shall not be conducted using any implement other than the correct adapter.

5.3.12.2.15 Final Tie-In Test (Category 3 Dental Air, Nitrogen, and Vacuum Systems).

(1) Prior to the connection of any new piping in extensions or additions to an existing piping system, the final tests in 5.3.12.2 shall be successfully performed on the new work.
(2) Each joint in the final connection between the new work and the existing system shall be leak-tested, with the gas of system designation or vacuum at the normal operating pressure or vacuum, by means of a leak detectant that is safe for use with oxygen and does not contain ammonia.
(3) For gas piping, immediately after a final connection is made and leak-tested, the specific altered zone and components in the immediate zone or area that is downstream from the point or area of intrusion shall be purged per 5.3.12.2.14.

5.3.12.2.16 Source Equipment Testing (Category 3 Dental Air, Nitrogen, and Vacuum Systems). Source equipment testing shall be conducted as follows:

(1) Source equipment checks shall be performed following the installation of the interconnecting pipelines, accessories, and source equipment.
(2) Where the source equipment and system gas or vacuum is used for testing of the distribution piping, the source equipment shall be checked out and placed in operation prior to testing the distribution piping.
(3) The source equipment shall be checked out and placed in operation according to the manufacturer's instructions.

5.3.13 Reserved.

5.3.14 Operation and Management of Category 3 Systems. Category 3 systems shall comply with 5.2.14.

Chapter 6   Electrical Systems

6.1* Applicability.

6.1.1 This chapter shall apply to new health care facilities as specified in Section 1.3.

6.1.2 The following paragraphs of this chapter shall apply to new and existing health care facilities:

(1) 6.3.2.2.4.2
(2) 6.3.2.2.6.1
(3) 6.3.2.2.6.2(F)
(4) 6.3.2.2.8.5(B)(2) and (3)
(5) 6.3.2.2.8.7
(6) 6.3.4
(7) 6.4.1.1.18.7
(8) 6.4.2.2.6.2(C)
(9) 6.4.2.2.6.3
(10) 6.4.4
(11) 6.5.4

6.1.3 Paragraph 6.3.2.2.2.3 shall apply only to existing facilities.

6.2 Nature of Hazards.

6.2.1* Fire and Explosions.

6.2.2 Shock. (Reserved)

6.2.3 Thermal. (Reserved)

6.3 Electrical System.

6.3.1 Sources. Each health care appliance requiring electrical line power for operation shall be supported by power sources that provide power adequate for each service.

6.3.1.1 Power/Utility Company. (Reserved)

6.3.1.2 On-Site Generator Set. (Reserved)

6.3.2 Distribution.

6.3.2.1 Electrical Installation. Installation shall be in accordance with *NFPA 70, National Electrical Code*.

6.3.2.1.1* Distribution system arrangements shall be designed to minimize interruptions to the electrical systems due to internal failures by the use of adequately rated equipment.

6.3.2.2 All Patient Care Rooms.

6.3.2.2.1* Branch circuit wiring 600 V or less shall comply with the requirements in 6.3.2.2.1.1 through 6.3.2.2.1.4.

6.3.2.2.1.1* Circuits.

(A) Branch circuits serving a given patient bed location shall be fed from not more than one normal branch-circuit distribution panel.

(B) When required, branch circuits serving a given patient bed location shall be permitted to be fed from more than one critical branch-circuit distribution panel.

6.3.2.2.1.2 Category 1 Spaces. Category 1 spaces shall be served by circuits from a critical branch panel(s) served from a single automatic transfer switch and a minimum of one circuit served by the normal power distribution system or by a system originating from a second critical branch automatic transfer switch.

6.3.2.2.1.3 Access to Overcurrent Protective Devices.

(A) Only authorized personnel shall have access to overcurrent protective devices serving Category 1 and Category 2 spaces.

(B) Overcurrent protective devices serving Category 1 and Category 2 spaces shall not be permitted to be located in public access spaces.

(C) Where used in locations such as in Category 1 spaces, isolated power panels shall be permitted in those locations.

6.3.2.2.1.4 Special-Purpose Outlets. Branch circuits serving only special-purpose outlets or receptacles (e.g., portable X-ray receptacles) shall not be required to conform to the requirements of 6.3.2.2.1.2.

6.3.2.2.2 Grounding requirements shall comply with the requirements in 6.3.2.2.2.1 through 6.3.2.2.2.4.

6.3.2.2.2.1 Grounding Circuitry Integrity. Grounding circuits and conductors in patient care spaces shall be installed in such a way that the continuity of other parts of those circuits cannot be interrupted nor the resistance raised above an acceptable level by the installation, removal, and replacement of any installed equipment, including power receptacles.

6.3.2.2.2.2 Reliability of Grounding. The grounding conductor shall conform to *NFPA 70, National Electrical Code*.

6.3.2.2.2.3 Separate Grounding Conductor. When existing construction does not have a separate grounding conductor, the continued use of the system shall be permitted, provided that it meets the performance requirements in 6.3.3.1.

6.3.2.2.2.4 Metal Receptacle Boxes. Where metal receptacle boxes are used, the performance of the connection between the receptacle grounding terminal and the metal box shall be equivalent to the performance provided by copper wire no smaller than 12 AWG.

6.3.2.2.3* Grounding Interconnects. In patient care spaces supplied by the normal distribution system and any branch of the essential electrical system, the grounding system of the normal distribution system and that of the essential electrical system shall be interconnected.

6.3.2.2.4 Protection Against Ground Faults.

6.3.2.2.4.1* Equipment Protection. The main and downstream ground-fault protective devices (where required) shall be coordinated as required in 6.3.2.5.

6.3.2.2.4.2 Personnel Protection. If used, ground-fault circuit interrupters (GFCIs) shall be listed.

6.3.2.2.5 Low-voltage wiring shall comply with either of the following:

(1) Fixed systems of 30 V (dc or ac rms) or less shall be permitted to be ungrounded, provided that the insulation between each ungrounded conductor and the primary circuit, which is supplied from a conventionally grounded distribution system, is the same protection as required for the primary voltage.
(2) A grounded low-voltage system shall be permitted, provided that load currents are not carried in the grounding conductors.

6.3.2.2.6 Receptacles.

6.3.2.2.6.1* Types of Receptacles.

(A) Each power receptacle shall provide at least one separate, highly dependable grounding pole capable of maintaining low-contact resistance with its mating plug, despite electrical and mechanical abuse. The grounding terminal of each receptacle shall be connected to the reference grounding point by means of an insulated copper equipment grounding conductor.

(B) Special receptacles, such as the following, shall be permitted:

(1) Four-pole units providing an extra pole for redundant grounding or ground continuity monitoring
(2) Locking-type receptacles
(3) Where required for reduction of electrical noise on the grounding circuit, receptacles in which the grounding terminals are purposely insulated from the receptacle yoke

(C) All single, duplex, or quadruplex type receptacles, or any combination thereof, located at patent bed locations in Category 1 spaces shall be listed hospital grade.

6.3.2.2.6.2 Minimum Number of Receptacles. The number of receptacles shall be determined by the intended use of the spaces in accordance with 6.3.2.2.6.2(A) through 6.3.2.2.6.2(F).

(A) Receptacles for Patient Bed Locations in Category 2 Spaces. Each patient bed location shall be provided with a minimum of eight receptacles. They shall be permitted to be of the locking or nonlocking type, single, duplex, or quadruplex type, or any combination of the three. All receptacles shall be listed hospital grade.

(B) Receptacles for Patient Bed Locations in Category 1 Spaces. Each patient bed location shall be provided with a minimum of 14 receptacles. They shall be permitted to be of the locking or nonlocking type, single, duplex, or quadruplex type, or any combination of the three. All receptacles shall be listed hospital grade.

(C) Receptacles for Operating Rooms. Operating rooms shall be provided with a minimum of 36 receptacles. They shall be permitted to be of the locking or nonlocking type, single, duplex, or quadruplex type, or any combination of the three. All receptacles shall be listed hospital grade.

(D) Receptacles for Bathrooms or Toilets. Receptacles shall not be required in bathrooms or toilet rooms.

(E) Receptacles for Special Rooms. Receptacles shall not be required in rooms where medical requirements mandate otherwise (e.g., certain psychiatric, pediatric, or hydrotherapy rooms).

(F) Designated Pediatric Locations. Receptacles that are located within the patient rooms, bathrooms, playrooms, and activity rooms of pediatric units or spaces with similar risk as determined by the governing body, other than nurseries, shall be listed tamper-resistant or shall employ a listed tamper-resistant cover.

6.3.2.2.6.3 Polarity of Receptacles. Each receptacle shall be wired in accordance with *NFPA 70, National Electrical Code*, to ensure correct polarity.

6.3.2.2.6.4 Other Services Receptacles. Receptacles provided for other services having different voltages, frequencies, or types on the same premises shall be of such design that attachment plugs and caps used in such receptacles cannot be connected to circuits of a different voltage, frequency, or type, but shall be interchangeable within each classification and rating required for two-wire, 125-V, single-phase ac service.

6.3.2.2.7 Special Grounding.

6.3.2.2.7.1* Use of Isolated Ground Receptacles.

(A) An isolated ground receptacle, if used, shall not defeat the purposes of the safety features of the grounding systems detailed herein.

(B) An isolated ground receptacle shall not be installed within a patient care vicinity.

6.3.2.2.7.2 Patient Equipment Grounding Point. A patient equipment grounding point comprising one or more grounding terminals or jacks shall be permitted in an accessible location in the patient care vicinity.

6.3.2.2.7.3* Special Grounding in Patient Care Rooms. In addition to the grounding required to meet the performance requirements of 6.3.3.1, additional grounding shall be permitted where special circumstances so dictate.

6.3.2.2.8 Wet Procedure Locations.

6.3.2.2.8.1* Wet procedure locations shall be provided with special protection against electric shock.

6.3.2.2.8.2 This special protection shall be provided as follows:

(1) Power distribution system that inherently limits the possible ground-fault current due to a first fault to a low value, without interrupting the power supply
(2) Power distribution system in which the power supply is interrupted if the ground-fault current does, in fact, exceed the trip value of a Class A GFCI

6.3.2.2.8.3 Patient beds, toilets, bidets, and wash basins shall not be required to be considered wet procedure locations.

6.3.2.2.8.4* Operating rooms shall be considered to be a wet procedure location, unless a risk assessment conducted by the health care governing body determines otherwise.

6.3.2.2.8.5 In existing construction, the requirements of 6.3.2.2.8.1 shall not be required when a written inspection procedure, acceptable to the authority having jurisdiction, is performed by a designated individual at the hospital to indicate that equipment grounding conductors for 120-V, single-phase, 15-A and 20-A receptacles; equipment connected by cord and plug; and fixed electrical equipment are installed and maintained in accordance with *NFPA 70, National Electrical Code*, and the applicable performance requirements of this chapter.

(A) The procedure shall include electrical continuity tests of all required equipment, grounding conductors, and their connections.

(B) Fixed receptacles, equipment connected by cord and plug, and fixed electrical equipment shall be tested as follows:

(1) When first installed
(2) Where there is evidence of damage
(3) After any repairs

6.3.2.2.8.6 The use of an isolated power system (IPS) shall be permitted as a protective means capable of limiting ground-fault current without power interruption. When installed, such a power system shall conform to the requirements of 6.3.2.6.

6.3.2.2.8.7* Operating rooms defined as wet procedure locations shall be protected by either isolated power or ground-fault circuit interrupters.

6.3.2.2.8.8 Where GFCI protection is used in an operating room, one of the following shall apply:

(1) Each receptacle shall be an individual GFCI device.
(2) Each receptacle shall be individually protected by a single GFCI device.

6.3.2.2.9 Isolated Power.

6.3.2.2.9.1 An isolated power system shall not be required to be installed in any patient care space, except as specified in 6.3.2.2.8.

6.3.2.2.9.2 The system shall be permitted to be installed where it conforms to the performance requirements specified in 6.3.2.6.

6.3.2.2.10 Essential Electrical Systems (EES).

6.3.2.2.10.1 Category 1 spaces shall be served only by a Type 1 EES.

6.3.2.2.10.2 Category 2 spaces shall be served by a Type 1 or Type 2 EES.

6.3.2.2.10.3 A Type 1 EES serving a Category 1 space shall be permitted to serve Category 2 spaces in the same facility.

6.3.2.2.10.4 Category 3 or Category 4 spaces shall not be required to be served by an EES.

6.3.2.2.11 Battery-Powered Lighting Units.

6.3.2.2.11.1 One or more battery-powered lighting units shall be provided within locations where deep sedation and general anesthesia is administered.

6.3.2.2.11.2 The lighting level of each unit shall be sufficient to terminate procedures intended to be performed within the operating room.

6.3.2.2.11.3 The sensor for units shall be wired to the branch circuit(s) serving general lighting within the room.

6.3.2.2.11.4 Units shall be capable of providing lighting for 1½ hours.

6.3.2.2.11.5 Units shall be tested monthly for 30 seconds, and annually for 30 minutes.

6.3.2.3 Laboratories. Outlets with two to four receptacles, or an equivalent power strip, shall be installed every 0.5 m to 1.0 m (1.6 ft to 3.3 ft) in instrument usage areas, and either installation shall be at least 80 mm (3.15 in.) above the countertop.

6.3.2.4 Other Nonpatient Areas. (Reserved)

6.3.2.5 Ground-Fault Protection.

6.3.2.5.1 Applicability. The requirements of 6.3.2.5.2 shall apply to hospitals and other buildings housing Category 1 spaces or utilizing life-support equipment and buildings that provide essential utilities or services for the operation of Category 1 spaces or electrical life-support equipment.

6.3.2.5.2 When ground-fault protection is provided for operation of the service or feeder disconnecting means, an additional step of ground-fault protection shall be provided in the next level of feeder downstream toward the load.

6.3.2.5.3 Ground-fault protection for operation of the service and feeder disconnecting means shall be fully selective such that the downstream device and not the upstream device shall open for downstream ground faults.

6.3.2.6* Isolated Power Systems.

6.3.2.6.1 Isolation Transformer.

6.3.2.6.1.1 The isolation transformer shall be listed and approved for the purpose.

6.3.2.6.1.2 The primary winding shall be connected to a power source so that it is not energized with more than 600 V (nominal).

(A) If present, the neutral of the primary winding shall be grounded in an approved manner.

(B) If an electrostatic shield is present, it shall be connected to the reference grounding point.

6.3.2.6.1.3 Wiring of isolated power systems shall be in accordance with 517.160 of *NFPA 70, National Electrical Code*.

6.3.2.6.2 Impedance of Isolated Wiring.

6.3.2.6.2.1* The impedance (capacitive and resistive) to ground of either conductor of an isolated system shall exceed 200,000 ohms when installed. The installation at this point shall include receptacles but is not required to include lighting fixtures or components of fixtures. This value shall be determined by energizing the system and connecting a low-impedance ac milliammeter (0 to 1 mA scale) between the reference grounding point and either conductor in sequence. This test shall be permitted to be performed with the line isolation monitor (see 6.3.2.6.3) connected, provided that the connection between the line isolation monitor and the reference grounding point is open at the time of the test. After the test is made, the milliammeter shall be removed and the grounding connection of the line isolation monitor shall be restored. When the installation is completed, including permanently connected fixtures, the reading of the meter on the line isolation monitor, which corresponds to the unloaded line condition, shall be made. This meter reading shall be recorded as a reference for subsequent line impedance evaluation. This test shall be conducted with no phase conductors grounded.

6.3.2.6.2.2 An approved capacitance suppressor shall be permitted to be used to improve the impedance of the permanently installed isolated system; however, the resistive impedance to ground of each isolated conductor of the system shall be at least 1 megohm prior to the connection of the suppression equipment. Capacitance suppressors shall be installed so as to prevent inadvertent disconnection during normal use.

6.3.2.6.3 Line Isolation Monitor.

6.3.2.6.3.1* In addition to the usual control and protective devices, each isolated power system shall be provided with an approved, continually operating line isolation monitor that indicates possible leakage or fault currents from either isolated conductor to ground.

6.3.2.6.3.2 The monitor shall be designed such that a green signal lamp, conspicuously visible in the area where the line isolation monitor is utilized, remains lighted when the system is adequately isolated from ground; and an adjacent red signal lamp and an audible warning signal (remote if desired) shall be energized when the total hazard current (consisting of possible resistive and capacitive leakage currents) from either isolated conductor to ground reaches a threshold value of 5.0 mA under normal line voltage conditions. The line isolation monitor shall not alarm for a fault hazard current of less than 3.7 mA.

6.3.2.6.3.3* The line isolation monitor shall comply with either of the following:

(1) It shall have sufficient internal impedance such that, when properly connected to the isolated system, the maxi-

mum internal current that will flow through the line isolation monitor, when any point of the isolated system is grounded, shall be 1 mA.

(2) It shall be permitted to be of the low-impedance type such that the current through the line isolation monitor, when any point of the isolated system is grounded, will not exceed twice the alarm threshold value for a period not exceeding 5 milliseconds.

6.3.2.6.3.4* An ammeter connected to indicate the total hazard current of the system (contribution of the fault hazard current plus monitor hazard current) shall be mounted in a plainly visible place on the line isolation monitor with the "alarm on" zone (total hazard current = 5.0 mA) at approximately the center of the scale. A line isolation monitor shall be located in the operating room.

6.3.2.6.3.5 Means shall be provided for shutting off the audible alarm while leaving the red warning lamp activated. When the fault is corrected and the green signal lamp is reactivated, the audible alarm-silencing circuit shall reset automatically, or an audible or distinctive visual signal shall indicate that the audible alarm is silenced.

6.3.2.6.3.6 A reliable test switch shall be mounted on the line isolation monitor to test its capability to operate (i.e., cause the alarms to operate and the meter to indicate in the "alarm on" zone). This switch shall transfer the grounding connection of the line isolation monitor from the reference grounding point to a test impedance arrangement connected across the isolated line; the test impedance(s) shall be of the appropriate magnitude to produce a meter reading corresponding to the rated total hazard current at the nominal line voltage, or to a lesser alarm hazard current if the line isolation monitor is so rated. The operation of this switch shall break the grounding connection of the line isolation monitor to the reference grounding point before transferring this grounding connector to the test impedance(s), so that making this test will not add to the hazard of a system in actual use; nor will the test include the effect of the line-to-ground stray impedance of the system. The test switch shall be of a self-restoring type.

6.3.2.6.3.7 The line isolation monitor shall not generate energy of sufficient amplitude or frequency, as measured by a physiological monitor with a gain of at least 10^4 with a source impedance of 1000 ohms connected to the balanced differential input of the monitor, to create interference or artifact on human physiological signals. The output voltage from the amplifier shall not exceed 30 mV when the gain is 10^4. The impedance of 1000 ohms shall be connected to the ends of typical unshielded electrode leads that are a normal part of the cable assembly furnished with physiological monitors. A 60 Hz notch filter shall be used to reduce ambient interference, as is typical in physiological monitor design.

6.3.2.6.4 Identification of Conductors for Isolated (Ungrounded) Systems. The isolated conductors shall be identified in accordance with 517.160(A)(5) of *NFPA 70, National Electrical Code.*

6.3.3 Performance Criteria and Testing.

6.3.3.1 Grounding System in Patient Care Spaces.

6.3.3.1.1* Grounding System Testing. The effectiveness of the grounding system shall be determined by voltage measurements and impedance measurements.

6.3.3.1.1.1 For new construction, the effectiveness of the grounding system shall be evaluated before acceptance.

6.3.3.1.1.2 Small wall-mounted conductive surfaces not likely to become energized, such as surface-mounted towel and soap dispensers, mirrors, and so forth, shall not be required to be intentionally grounded or tested.

6.3.3.1.1.3 Large metal conductive surfaces not likely to become energized, such as windows, door frames, and drains, shall not be required to be intentionally grounded or periodically tested.

6.3.3.1.1.4* Whenever the electrical system has been altered or replaced, that portion of the system shall be tested.

6.3.3.1.2 Reference Point. The voltage and impedance measurements shall be taken with respect to a reference point, which shall be one of the following:

(1) Reference grounding point *(see Chapter 3)*
(2) Grounding point, in or near the room under test, that is electrically remote from receptacles (e.g., an all-metal cold-water pipe)
(3) Grounding contact of a receptacle that is powered from a different branch circuit from the receptacle under test

6.3.3.1.3* Voltage Measurements.

6.3.3.1.3.1 The voltage measurements shall be made under no-fault conditions between a reference point and exposed fixed electrical equipment with conductive surfaces in a patient care vicinity.

6.3.3.1.3.2 The voltage measurements shall be made with an accuracy of ±20 percent.

6.3.3.1.3.3 Voltage measurements for faceplates of wiring devices shall not be required.

6.3.3.1.4* Impedance Measurements. The impedance measurement shall be made with an accuracy of ±20 percent.

6.3.3.1.4.1 For new construction, the impedance measurement shall be made between the reference point and the grounding contact of 10 percent of all receptacles within the patient care vicinity.

6.3.3.1.4.2 The impedance measurement shall be the ratio of voltage developed (either 60 Hz or dc) between the point under test and the reference point to the current applied between these two points.

6.3.3.1.5 Test Equipment. Electrical safety test instruments shall be tested periodically, but not less than annually, for acceptable performance.

6.3.3.1.5.1 Voltage measurements specified in 6.3.3.1.3 shall be made with an instrument having an input resistance of 1000 ohms ±10 percent at frequencies of 1000 Hz or less.

6.3.3.1.5.2 The voltage across the terminals (or between any terminal and ground) of resistance-measuring instruments used in occupied patient care rooms shall not exceed 500 mV rms or 1.4 dc or peak to peak.

6.3.3.1.6 Criteria for Acceptability for New Construction.

6.3.3.1.6.1 The voltage limit shall be 20 mV.

6.3.3.1.6.2 The impedance limit shall be 0.2 ohm for systems containing isolated ground receptacles and 0.1 ohm for all others.

6.3.3.2 Receptacle Testing in Patient Care Spaces.

6.3.3.2.1 The physical integrity of each receptacle shall be confirmed by visual inspection.

6.3.3.2.2 The continuity of the grounding circuit in each electrical receptacle shall be verified.

6.3.3.2.3 Correct polarity of the hot and neutral connections in each electrical receptacle shall be confirmed.

6.3.3.2.4 The retention force of the grounding blade of each electrical receptacle (except locking-type receptacles) shall be not less than 115 g (4 oz).

6.3.3.3 Isolated Power Systems.

6.3.3.3.1 Patient Care Spaces. If installed, the isolated power system shall be tested in accordance with 6.3.3.3.2.

6.3.3.3.2 Line Isolation Monitor Tests. The line isolation monitor (LIM) circuit shall be tested after installation, and prior to being placed in service, by successively grounding each line of the energized distribution system through a resistor whose value is 200 × V (ohms), where V equals measured line voltage. The visual and audible alarms *(see 6.3.2.6.3.2)* shall be activated.

6.3.3.4 Ground-Fault Protection Testing. When equipment ground-fault protection is first installed, each level shall be performance-tested to ensure compliance with 6.3.2.5.

6.3.4* Administration of Electrical System.

6.3.4.1 Maintenance and Testing of Electrical System.

6.3.4.1.1 Where hospital-grade receptacles are required at patient bed locations and in locations where deep sedation or general anesthesia is administered, testing shall be performed after initial installation, replacement, or servicing of the device.

6.3.4.1.2 Additional testing of receptacles in patient care spaces shall be performed at intervals defined by documented performance data.

6.3.4.1.3 Receptacles not listed as hospital-grade, at patient bed locations and in locations where deep sedation or general anesthesia is administered, shall be tested at intervals not exceeding 12 months.

6.3.4.1.4 The LIM circuit shall be tested at intervals of not more than 1 month by actuating the LIM test switch *(see 6.3.2.6.3.6)*. For a LIM circuit with automated self-test and self-calibration capabilities, this test shall be performed at intervals of not more than 12 months. Actuation of the test switch shall activate both visual and audible alarm indicators.

6.3.4.1.5 After any repair or renovation to an electrical distribution system, the LIM circuit shall be tested in accordance with 6.3.3.3.2.

6.3.4.2 Record Keeping.

6.3.4.2.1* General.

6.3.4.2.1.1 A record shall be maintained of the tests required by this chapter and associated repairs or modification.

6.3.4.2.1.2 At a minimum, the record shall contain the date, the rooms or areas tested, and an indication of which items have met, or have failed to meet, the performance requirements of this chapter.

6.3.4.2.2 Isolated Power System (Where Installed). A permanent record shall be kept of the results of each of the tests.

6.4 Essential Electrical System Requirements — Type 1.

6.4.1 Sources (Type 1 EES).

6.4.1.1 On-Site Generator Set.

6.4.1.1.1* Design Considerations. Dual sources of normal power shall not constitute an alternate source of power as described in this chapter.

6.4.1.1.1.1 Distribution system arrangements shall be designed to minimize interruptions to the electrical systems due to internal failures by the use of adequately rated equipment.

6.4.1.1.1.2 The following factors shall be considered in the design of the distribution system:

(1) Abnormal voltages, such as single phasing of three-phase utilization equipment; switching or lightning surges, or both; voltage reductions; and so forth
(2) Capability of achieving the fastest possible restoration of any given circuit(s) after clearing a fault
(3) Effects of future changes, such as increased loading or supply capacity, or both
(4) Stability and power capability of the prime mover during and after abnormal conditions
(5)*Sequence reconnection of loads to avoid large current inrushes that trip overcurrent devices or overload the generator(s)
(6) Bypass arrangements to allow testing and maintenance of system components that could not otherwise be maintained without disruption of important hospital functions
(7) Effects of any harmonic currents on neutral conductors and equipment

6.4.1.1.2 Current-sensing devices, phase and ground, shall be selected to minimize the extent of interruption to the electrical system due to abnormal current caused by overload or short circuits, or both.

6.4.1.1.3 Generator load-shed circuits designed for the purpose of load reduction or for load priority systems shall not shed life safety branch loads, critical branch loads serving critical care areas, medical air compressors, medical–surgical vacuum pumps, fire pumps, the pressure maintenance (jockey) pump(s) for water-based fire protection systems, generator fuel pumps, or other generator accessories.

6.4.1.1.4 Essential electrical systems shall have a minimum of the following two independent sources of power: a normal source generally supplying the entire electrical system and one or more alternate sources for use when the normal source is interrupted.

6.4.1.1.5 Where the normal source consists of generating units on the premises, the alternate source shall be either another generating set or an external utility service.

6.4.1.1.6 General. Generator sets installed as an alternate source of power for essential electrical systems shall be designed to meet the requirements of such service.

6.4.1.1.6.1 Type 1 and Type 2 essential electrical system power sources shall be classified as Type 10, Class X, Level 1 generator sets per NFPA 110, *Standard for Emergency and Standby Power Systems*.

6.4.1.1.6.2 Type 3 essential electrical system power sources shall be classified as Type 10, Class X, Level 2 generator sets per NFPA 110, *Standard for Emergency and Standby Power Systems*.

6.4.1.1.7 Fuel Cell Systems. Fuel cell systems shall be permitted to serve as the alternate source for all or part of an essential electrical system, provided the following conditions apply:

6.4.1.1.7.1 Installation shall comply with NFPA 853, *Standard for Installation of Stationary Fuel Cell Power Systems*.

6.4.1.1.7.2 N+1 units shall be provided where N units have sufficient capacity to supply the demand load of the portion of the system served.

6.4.1.1.7.3* System shall be able to assume loads within 10 seconds of loss of normal power source.

6.4.1.1.7.4 System shall have a continuing source of fuel supply, together with sufficient on-site fuel storage for the essential system type.

6.4.1.1.7.5 A connection shall be provided for a portable diesel generator to supply life safety and critical portions of the distribution system (if present).

6.4.1.1.8 Uses for Essential Electrical System.

6.4.1.1.8.1 The generating equipment used shall be either reserved exclusively for such service or normally used for other purposes of peak demand control, internal voltage control, load relief for the external utility, or cogeneration. If normally used for such other purposes, two or more sets shall be installed, such that the maximum actual demand likely to be produced by the connected load of the life safety and critical branches, as well as medical air compressors, medical–surgical vacuum pumps, electrically operated fire pumps, jockey pumps, fuel pumps, and generator accessories, shall be met by a multiple generator system, with the largest generator set out of service (not available). The alternate source of emergency power for illumination and identification of means of egress shall be the essential electrical system. The alternate power source for fire protection signaling systems shall be the essential electrical system.

6.4.1.1.8.2 A single generator set that operates the essential electrical system shall be permitted to be part of the system supplying the other purposes as specified in 6.4.1.1.8.1, provided that any such use will not decrease the mean period between service overhauls to less than 3 years.

6.4.1.1.8.3* Optional loads shall be permitted to be served by the essential electrical system generating equipment. Optional loads shall be served by their own transfer means, such that these loads shall not be transferred onto the generating equipment if the transfer will overload the generating equipment and shall be shed upon a generating equipment overload. Use of the generating equipment to serve optional loads shall not constitute "other purposes" as described in 6.4.1.1.8.1 and, therefore, shall not require multiple generator sets.

6.4.1.1.8.4 Where optional loads include contiguous or same-site facilities not covered in this code, provisions shall be made to meet the requirements of NFPA *101, Life Safety Code*; Article 700 of *NFPA 70, National Electrical Code*; and other applicable NFPA requirements for emergency egress under load-shed conditions.

6.4.1.1.9 Work Space or Room.

6.4.1.1.9.1 The EPS shall be installed in a separate room for Level 1 installations. EPSS equipment shall be permitted to be installed in this room. [110:7.2.1]

(A) The room shall have a minimum 2-hour fire rating or be located in an adequate enclosure located outside the building capable of resisting the entrance of snow or rain at a maximum wind velocity required by local building codes. [110:7.2.1.1]

(B) The rooms, enclosures, or separate buildings housing Level 1 or Level 2 EPSS equipment shall be designed and located to minimize damage from flooding, including that caused by the following:

(1) Flooding resulting from fire fighting
(2) Sewer water backup
(3) Other disasters or occurrences

[110:7.2.3]

6.4.1.1.9.2 The EPS equipment shall be installed in a location that permits ready accessibility and a minimum of 0.9 m (36 in.) from the skid rails' outermost point in the direction of access for inspection, repair, maintenance, cleaning, or replacement. This requirement shall not apply to units in outdoor housings. [110: 7.2.6]

6.4.1.1.10* Capacity and Rating. The generator set(s) shall have the capacity and rating to meet the maximum actual demand likely to be produced by the connected load of the essential electrical system(s).

6.4.1.1.11 Load Pickup. The energy converters shall have the required capacity and response to pick up and carry the load within the time specified in Table 4.1(b) of NFPA 110, *Standard for Emergency and Standby Power Systems*, after loss of primary power.

6.4.1.1.12 Maintenance of Temperature. The EPS shall be heated as necessary to maintain the water jacket and battery temperature determined by the EPS manufacturer for cold start and load acceptance for the type of EPSS. [110:5.3.1]

6.4.1.1.13* Heating, Cooling, and Ventilating. With the EPS running at rated load, ventilation airflow shall be provided to limit the maximum air temperature in the EPS room or the enclosure housing the unit to the maximum ambient air temperature required by the EPS manufacturer. [110:7.7.1]

6.4.1.1.13.1 Consideration shall be given to all the heat emitted to the EPS equipment room by the energy converter, uninsulated or insulated exhaust pipes, and other heat-producing equipment. [110:7.7.1.1]

6.4.1.1.13.2 Air shall be supplied to the EPS equipment for combustion. [110:7.7.2]

(A) For EPS supplying Level 1 EPSS, ventilation air shall be supplied directly from a source outside the building by an exterior wall opening or from a source outside the building by a 2-hour fire-rated air transfer system. [110:7.7.2.1]

(B) For EPS supplying Level 1 EPSS, discharge air shall be directed outside the building by an exterior wall opening or to an exterior opening by a 2-hour fire-rated air transfer system. [110:7.7.2.2]

(C) Fire dampers, shutters, or other self-closing devices shall not be permitted in ventilation openings or ductwork for supply or return/discharge air to EPS equipment for Level 1 EPSS. [110:7.7.2.3]

6.4.1.1.13.3 Ventilation air supply shall be from outdoors or from a source outside of the building by an exterior wall opening or from a source outside the building by a 2-hour fire-rated air transfer system. [110:7.7.3]

6.4.1.1.13.4 Ventilation air shall be provided to supply and discharge cooling air for radiator cooling of the EPS when running at rated load. [**110:**7.7.4]

(A) Ventilation air supply and discharge for radiator-cooled EPS shall have a maximum static restriction of 125 Pa (0.5 in. of water column) in the discharge duct at the radiator outlet. [**110:**7.7.4.1]

(B) Radiator air discharge shall be ducted outdoors or to an exterior opening by a 2-hour rated air transfer system. [**110:**7.7.4.2]

6.4.1.1.13.5 Motor-operated dampers, when used, shall be spring operated to open and motor closed. Fire dampers, shutters, or other self-closing devices shall not be permitted in ventilation openings or ductwork for supply or return/discharge air to EPS equipment for Level 1 EPSS. [**110:**7.7.5]

6.4.1.1.13.6 The ambient air temperature in the EPS equipment room or outdoor housing containing Level 1 rotating equipment shall be not less than 4.5°C (40°F). [**110:** 5.3.5]

6.4.1.1.13.7 Units housed outdoors shall be heated as specified in 5.3.1 [of NFPA 110, *Standard for Emergency and Standby Power Systems*]. [**110:**7.7.7]

6.4.1.1.13.8 Design of the heating, cooling, and ventilation system for the EPS equipment room shall include provision for factors including, but not limited to, the following:

(1) Heat
(2) Cold
(3) Dust
(4) Humidity
(5) Snow and ice accumulations around housings
(6) Louvers
(7) Remote radiator fans
(8) Prevailing winds blowing against radiator fan discharge air

[**110:** 7.7.7]

6.4.1.1.14 Cranking Batteries. Internal combustion engine cranking batteries shall be in accordance with the battery requirements of NFPA 110, *Standard for Emergency and Standby Power Systems*.

6.4.1.1.15 Compressed Air Starting Devices. Other types of stored energy starting systems (except pyrotechnic) shall be permitted to be used where recommended by the manufacturer of the prime mover and subject to approval of the authority having jurisdiction, under the following conditions:

(1) Where two complete periods of cranking cycles are completed without replacement of the stored energy
(2) Where a means for automatic restoration from the emergency source of the stored energy is provided
(3) Where the stored energy system has the cranking capacity specified in 5.6.4.2.1 of NFPA 110, *Standard for Emergency and Standby Power Systems*
(4) Where the stored energy system has a "black start" capability in addition to normal discharge capability

[**110:**5.6.4.1.2]

6.4.1.1.16 Fuel Supply. The fuel supply for the generator set shall comply with Sections 5.5 and 7.9 of NFPA 110, *Standard for Emergency and Standby Power Systems*.

6.4.1.1.17 Requirements for Safety Devices.

6.4.1.1.17.1 Internal Combustion Engines. Internal combustion engines serving generator sets shall be equipped with the following:

(1) Sensor device plus visual warning device to indicate a water-jacket temperature below that required in 6.4.1.1.12
(2) Sensor devices plus visual pre-alarm warning device to indicate the following:
    (a) High engine temperature (above manufacturer's recommended safe operating temperature range)
    (b) Low lubricating oil pressure (below manufacturer's recommended safe operating range)
    (c) Low water coolant level
(3) Automatic engine shutdown device plus visual device to indicate that a shutdown took place due to the following:
    (a) Overcrank (failed to start)
    (b) Overspeed
    (c) Low lubricating oil pressure
    (d) Excessive engine temperature
(4) Common audible alarm device to warn that one or more of the pre-alarm or alarm conditions exist

6.4.1.1.17.2 Safety indications and shutdowns shall be in accordance with Table 6.4.1.1.17.2.

6.4.1.1.18 Alarm Annunciator. A remote annunciator that is storage battery powered shall be provided to operate outside of the generating room in a location readily observed by operating personnel at a regular work station *(see 700.12 of NFPA 70, National Electrical Code)*. The annunciator shall be hard-wired to indicate alarm conditions of the emergency or auxiliary power source as follows:

(1) Individual visual signals shall indicate the following:
    (a) When the emergency or auxiliary power source is operating to supply power to load
    (b) When the battery charger is malfunctioning
(2) Individual visual signals plus a common audible signal to warn of an engine-generator alarm condition shall indicate the following:
    (a) Low lubricating oil pressure
    (b) Low water temperature (below that required in 6.4.1.1.12)
    (c) Excessive water temperature
    (d) Low fuel when the main fuel storage tank contains less than a 4-hour operating supply
    (e) Overcrank (failed to start)
    (f) Overspeed

6.4.1.1.18.1* A remote, common audible alarm shall be provided as specified in 6.4.1.1.18.6. [**110:**5.6.6]

6.4.1.1.18.2 The following annunciation shall be provided at a minimum:

(1) For Level 1 EPS, local annunciation and facility remote annunciation, or local annunciation and network remote annunciation
(2) For Level 2 EPS, local annunciation

[**110:**5.6.6.2]

6.4.1.1.18.3 For the purposes of defining the types of annunciation in 6.4.1.1.18.2, the following shall apply:

(1) Local annunciation is located on the equipment itself or within the same equipment room.
(2) Facility remote annunciation is located on site but not within the room where the equipment is located.
(3) Network remote annunciation is located off site.

[**110:**5.6.6.3]

Table 6.4.1.1.17.2 Safety Indications and Shutdowns

Indicator Function (at Battery Voltage)	Level 1 CV	Level 1 S	Level 1 RA
(a) Overcrank	X	X	X
(b) Low water temperature	X	—	X
(c) High engine temperature pre-alarm	X	—	X
(d) High engine temperature	X	X	X
(e) Low lube oil pressure pre-alarm	X	—	X
(f) Low lube oil pressure	X	X	X
(g) Overspeed	X	X	X
(h) Low fuel main tank	X	—	X
(i) Low coolant level	X	O	X
(j) EPS supplying load	X	—	—
(k) Control switch not in automatic position	X	—	X
(l) High battery voltage	X	—	—
(m) Low cranking voltage	X	—	X
(n) Low voltage in battery	X	—	—
(o) Battery charger ac failure	X	—	—
(p) Lamp test	X	—	—
(q) Contacts for local and remote common alarm	X	—	X
(r) Audible alarm-silencing switch	—	—	X
(s) Low starting air pressure	X	—	—
(t) Low starting hydraulic pressure	X	—	—
(u) Air shutdown damper when used	X	X	X
(v) Remote emergency stop	—	X	—

CV: Control panel–mounted visual. S: Shutdown of EPS indication. RA: Remote audible. X: Required. O: Optional.

Notes:
(1) Item (p) shall be provided, but a separate remote audible signal shall not be required when the regular work site in 5.6.6 of NFPA 110, *Standard for Emergency and Standby Power Systems*, is staffed 24 hours a day.
(2) Item (b) is not required for combustion turbines.
(3) Item (r) or (s) is required only where used as a starting method.
(4) Item (j): EPS ac ammeter shall be permitted for this function.
(5) All required CV functions shall be visually annunciated by a remote, common visual indicator.
(6) All required functions indicated in the RA column shall be annunciated by a remote, common audible alarm as required in 5.6.5.2(4) of NFPA 110.
(7) Item (i) requires a low gas pressure alarm on gaseous systems.
(8) Item (b) must be set at 11°C (20°F) below the regulated temperature determined by the EPS manufacturer, as required in 5.3.1 of NFPA 110.

6.4.1.1.18.4 An alarm-silencing means shall be provided, and the panel shall include repetitive alarm circuitry so that, after the audible alarm has been silenced, it reactivates after the fault condition has been cleared and has to be restored to its normal position to be silenced again. [**110:** 5.6.6.4]

6.4.1.1.18.5 In lieu of the requirement of 5.6.6.4 of NFPA 110, a manual alarm-silencing means shall be permitted that silences the audible alarm after the occurrence of the alarm condition, provided such means do not inhibit any subsequent alarms from sounding the audible alarm again without further manual action. [**110:** 5.6.6.5]

6.4.1.1.18.6 Individual alarm indication to annunciate any of the conditions listed in Table 6.4.1.1.17.2 shall have the following characteristics:

(1) It shall be battery powered.
(2) It shall be visually indicated.
(3) It shall have additional contacts or circuits for a common audible alarm that signals locally and remotely when any of the itemized conditions occurs.
(4) It shall have a lamp test switch(es) to test the operation of all alarm lamps.

6.4.1.1.18.7 A centralized computer system (e.g., building automation system) shall not be permitted to be substituted for the alarm annunciator in 6.4.1.1.18 but shall be permitted to be used to supplement the alarm annunciator.

6.4.1.2 Battery. Battery systems shall meet all requirements of Article 700 of *NFPA 70, National Electrical Code*.

6.4.2* Distribution (Type 1 EES).

6.4.2.1 General Requirements.

6.4.2.1.1 Electrical characteristics of the transfer switches shall be suitable for the operation of all functions and equipment they are intended to supply.

6.4.2.1.2* Coordination.

6.4.2.1.2.1 Overcurrent protective devices serving the essential electrical system shall be coordinated for the period of time that a fault's duration extends beyond 0.1 second.

6.4.2.1.2.2 Coordination shall not be required as follows:

(1) Between transformer primary and secondary overcurrent protective devices, where only one overcurrent protective device or set of overcurrent protective devices exists on the transformer secondary
(2) Between overcurrent protective devices of the same size (ampere rating) in series

6.4.2.1.3 Switch Rating. The rating of the transfer switches shall be adequate for switching all classes of loads to be served and for withstanding the effects of available fault currents without contact welding

6.4.2.1.4 Automatic Transfer Switch. Transfer of all loads shall be accomplished using an automatic transfer switch(es). Each automatic transfer switch of 600 V or less shall be listed for the purpose and approved for emergency electrical service *(see NFPA 70, National Electrical Code, Article 700.3)* as a complete assembly.

6.4.2.1.5 Automatic Transfer Switch Features.

6.4.2.1.5.1 Source Monitoring.

(A)* Undervoltage-sensing devices shall be provided to monitor all ungrounded lines of the primary source of power as follows:

(1) When the voltage on any phase falls below the minimum operating voltage of any load to be served, the transfer switch shall automatically initiate engine start and the process of transfer to the emergency power supply (EPS).
(2)*When the voltage on all phases of the primary source returns to within specified limits for a designated period of time, the process of transfer back to primary power shall be initiated.

[**110**:6.2.2.1]

(B) Both voltage-sensing and frequency-sensing equipment shall be provided to monitor one ungrounded line of the EPS. [**110**:6.2.2.2]

(C) Transfer to the EPS shall be inhibited until the voltage and frequency are within a specified range to handle loads to be served. [**110**:6.2.2.3]

(D) Sensing equipment shall not be required in the transfer switch, provided it is included with the engine control panel. [**110**:6.2.2.3.1]

(E) Frequency-sensing equipment shall not be required for monitoring the public utility source where used as an EPS, as permitted by 5.1.3 of NFPA 110. [**110**:6.2.2.3.2]

6.4.2.1.5.2 Interlocking. Mechanical interlocking or an approved alternate method shall prevent the inadvertent interconnection of the primary power supply and the EPS, or any two separate sources of power. [**110**:6.2.3]

6.4.2.1.5.3* Manual Operation. Instruction and equipment shall be provided for safe manual nonelectric transfer in the event the transfer switch malfunctions. [**110**:6.2.4]

6.4.2.1.5.4* Time Delay on Starting of EPS. A time-delay device shall be provided to delay starting of the EPS. The timer shall prevent nuisance starting of the EPS and possible subsequent load transfer in the event of harmless momentary power dips and interruptions of the primary source. [**110**:6.2.5]

6.4.2.1.5.5 Time Delay at Engine Control Panel. Time delays shall be permitted to be located at the engine control panel in lieu of in the transfer switches. [**110**:6.2.6]

6.4.2.1.5.6 Time Delay on Transfer to EPS. An adjustable time-delay device shall be provided to delay transfer and sequence load transfer to the EPS to avoid excessive voltage drop when the transfer switch is installed for Level 1 use. [**110**:6.2.7]

(A) Time Delay Commencement. The time delay shall commence when proper EPS voltage and frequency are achieved. [**110**:6.2.7.1]

(B) Time Delay at Engine Control Panel. Time delays shall be permitted to be located at the engine control panel in lieu of in the transfer switches. [**110**:6.2.7.2]

6.4.2.1.5.7* Time Delay on Retransfer to Primary Source. An adjustable time-delay device with automatic bypass shall be provided to delay retransfer from the EPS to the primary source of power, and allow the primary source to stabilize before retransfer of the load. [**110**:6.2.8]

6.4.2.1.5.8 Time Delay Bypass If EPS Fails. The time delay shall be automatically bypassed if the EPS fails. [**110**:6.2.9]

(A) The transfer switch shall be permitted to be programmed for a manually initiated retransfer to the primary source to provide for a planned momentary interruption of the load. [**110**:6.2.9.1]

(B) If used, the arrangement in 6.2.9.1 of NFPA 110 shall be provided with a bypass feature to allow automatic retransfer in the event that the EPS fails and the primary source is available. [**110**:6.2.9.2]

6.4.2.1.5.9 Time Delay on Engine Shutdown. A minimum time delay of 5 minutes shall be provided for unloaded running of the EPS prior to shutdown to allow for engine cooldown. [**110**:6.2.10]

(A) The minimum 5-minute delay shall not be required on small (15 kW or less) air-cooled prime movers. [**110**:6.2.10.1]

(B) A time-delay device shall not be required, provided it is included with the engine control panel, or if a utility feeder is used as an EPS. [**110**:6.2.10.2]

6.4.2.1.5.10 Engine Generator Exercising Timer. A program timing device shall be provided to exercise the EPS as described in Chapter 8 of NFPA 110. [**110**:6.2.11]

(A) Transfer switches shall transfer the connected load to the EPS and immediately return to primary power automatically in case of the EPS failure. [**110**:6.2.11.1]

(B) Exercising timers shall be permitted to be located at the engine control panel in lieu of in the transfer switches. [**110**:6.2.11.2]

(C) A program timing device shall not be required in health care facilities that provide scheduled testing in accordance with NFPA 99, *Health Care Facilities Code*. [**110**:6.2.11.3]

6.4.2.1.5.11 Test Switch. A test means shall be provided on each automatic transfer switch (ATS) that simulates failure of the primary power source and then transfers the load to the EPS. [**110**:6.2.12]

6.4.2.1.5.12* Indication of Transfer Switch Position. Two pilot lights with identification nameplates or other approved position indicators shall be provided to indicate the transfer switch position. [**110**:6.2.13]

6.4.2.1.5.13 Motor Load Transfer. Provisions shall be included to reduce currents resulting from motor load transfer if such currents could damage EPSS equipment or cause nuisance tripping of EPSS overcurrent protective devices. [**110:**6.2.14]

6.4.2.1.5.14* Isolation of Neutral Conductors. Provisions shall be included for ensuring continuity, transfer, and isolation of the primary and the EPS neutral conductors wherever they are separately grounded to achieve ground-fault sensing. [**110:**6.2.15]

6.4.2.1.5.15* Nonautomatic Transfer Switch Features. Switching devices shall be mechanically held and shall be operated by direct manual or electrical remote manual control. [**110:**6.2.16]

(A) Interlocking. Reliable mechanical interlocking, or an approved alternate method, shall prevent the inadvertent interconnection of the primary power source and the EPS. [**110:**6.2.16.1]

(B) Indication of Switch Position. Two pilot lights with identification nameplates, or other approved position indicators, shall be provided to indicate the switch position. [**110:**6.2.16.2]

6.4.2.1.6 Nonautomatic Transfer Device Classification. Nonautomatic transfer devices of 600 V or less shall be listed for the purpose and approved.

6.4.2.1.7 Nonautomatic Transfer Device Features.

6.4.2.1.7.1 General. Switching devices shall be mechanically held and shall be operated by direct manual or electrical remote manual control. [**110:**6.2.16]

6.4.2.1.7.2 Interlocking. Reliable mechanical interlocking, or an approved alternate method, shall prevent the inadvertent interconnection of the primary power source and the EPS. [**110:**6.2.16.1]

6.4.2.1.7.3 Indication of Switch Position. Two pilot lights with identification nameplates, or other approved position indicators, shall be provided to indicate the switch position. [**110:**6.2.16.2]

6.4.2.1.8 Bypass and Isolating Transfer Switches. Bypass-isolation switches shall be permitted for bypassing and isolating the transfer switch and installed in accordance with 6.4.2, 6.4.3, and 6.4.4 of NFPA 110. [**110:**6.4.1]

6.4.2.1.8.1 Bypass-Isolation Switch Rating. The bypass-isolation switch shall have a continuous current rating and a current rating compatible with that of the associated transfer switch. [**110:**6.4.2]

6.4.2.1.8.2 Bypass-Isolation Switch Classification. Each bypass-isolation switch shall be listed for emergency electrical service as a completely factory-assembled and factory-tested apparatus. [**110:**6.4.3]

6.4.2.1.8.3* Operation. With the transfer switch isolated or disconnected, the bypass-isolation switch shall be designed so it can function as an independent nonautomatic transfer switch and allow the load to be connected to either power source. [**110:**6.4.4]

6.4.2.1.8.4 Reconnection of Transfer Switch. Reconnection of the transfer switch shall be possible without a load interruption greater than the maximum time, in seconds, specified by the type of system. [**110:**6.4.5]

6.4.2.2 Branches.

6.4.2.2.1* General.

6.4.2.2.1.1 The essential electrical system shall be divided into the following three branches:

(1) Life safety
(2) Critical
(3) Equipment

6.4.2.2.1.2 The division between the branches shall occur at transfer switches where more than one transfer switch is required.

6.4.2.2.1.3 Each branch shall be arranged for connection, within time limits specified in this chapter, to an alternate source of power following a loss of the normal source.

6.4.2.2.1.4 The number of transfer switches to be used shall be based upon reliability, design, and load considerations.

(A) Each branch of the essential electrical system shall have one or more transfer switches.

(B) One transfer switch shall be permitted to serve one or more branches in a facility with a continuous load on the switch of 150 kVA (120 kW) or less.

6.4.2.2.1.5 For the purposes of this code, the provisions for emergency systems in Article 700 of *NFPA 70, National Electrical Code*, shall be applied only to the life safety branch.

6.4.2.2.1.6 The following portions of Article 700 of *NFPA 70* shall be amended as follows:

(A) 700.4 shall not apply.

(B) 700.10(D)(1) through (3) shall not apply.

(C) 700.17 Branch Circuits for Emergency Lighting. Branch circuits that supply emergency lighting shall be installed to provide service from a source complying with 700.12 when the normal supply for lighting is interrupted or where single circuits supply luminaries containing secondary batteries.

(D) 700.28 shall not apply.

6.4.2.2.2 Feeders from Alternate Source.

6.4.2.2.2.1 A single feeder supplied by a local or remote alternate source shall be permitted to supply the essential electrical system to the point at which the life safety, critical, and equipment branches are separated.

6.4.2.2.2.2 Installation of the transfer equipment shall be permitted at other than the location of the alternate source.

6.4.2.2.3 Life Safety Branch.

6.4.2.2.3.1 The life safety branch shall be limited to circuits essential to life safety.

6.4.2.2.3.2 The life safety branch shall supply power as follows:

(1) Illumination of means of egress in accordance with NFPA *101, Life Safety Code*
(2) Exit signs and exit directional signs in accordance with NFPA *101, Life Safety Code*
(3)*Hospital communications systems, where used for issuing instruction during emergency conditions
(4) Generator set location as follows:
   (a) Task illumination
   (b) Battery charger for emergency battery-powered lighting unit(s)
   (c) Select receptacles at the generator set location and essential electrical system transfer switch locations
(5) Elevator cab lighting, control, communications, and signal systems
(6) Electrically powered doors used for building egress
(7) Fire alarms and auxiliary functions of fire alarm combination systems complying with *NFPA 72, National Fire Alarm and Signaling Code*

6.4.2.2.3.3 Alarm and alerting systems (other than fire alarm systems) shall be connected to the life safety branch or critical branch.

6.4.2.2.3.4 Loads dedicated to a specific generator, including the fuel transfer pump(s), ventilation fans, electrically operated louvers, controls, cooling system, and other generator accessories essential for generator operation, shall be connected to the life safety branch or the output terminals of the generator with overcurrent protective devices.

6.4.2.2.3.5 No functions other than those in 6.4.2.2.3.2, 6.4.2.2.3.3, and 6.4.2.2.3.4 shall be connected to the life safety branch, except as specifically permitted in 6.4.2.2.3.

6.4.2.2.4* Critical Branch.

6.4.2.2.4.1 The critical branch shall be permitted to be subdivided into two or more branches.

6.4.2.2.4.2 The critical branch shall supply power for task illumination, fixed equipment, select receptacles, and select power circuits serving the following spaces and functions related to patient care:

(1) Critical care spaces that utilize anesthetizing gases, task illumination, select receptacles, and fixed equipment
(2) Isolated power systems in special environments
(3) Task illumination and select receptacles in the following:
   (a) Patient care spaces, including infant nurseries, selected acute nursing areas, psychiatric bed areas (omit receptacles), and ward treatment rooms
   (b) Medication preparation spaces
   (c) Pharmacy dispensing spaces
   (d) Nurses' stations (unless adequately lighted by corridor luminaires)
(4) Additional specialized patient care task illumination and receptacles, where needed
(5) Nurse call systems
(6) Blood, bone, and tissue banks
(7)*Telephone equipment rooms and closets
(8) Task illumination, select receptacles, and select power circuits for the following areas:
   (a) General care beds with at least one duplex receptacle per patient bedroom, and task illumination as required by the governing body of the health care facility
   (b) Angiographic labs
   (c) Cardiac catheterization labs
   (d) Coronary care units
   (e) Hemodialysis rooms or areas
   (f) Emergency room treatment areas (select)
   (g) Human physiology labs
   (h) Intensive care units
   (i) Postoperative recovery rooms (select)
(9) Additional task illumination, receptacles, and select power circuits needed for effective facility operation, including single-phase fractional horsepower motors, which are permitted to be connected to the critical branch

6.4.2.2.5 Equipment Branch.

6.4.2.2.5.1 General. The equipment branch shall be connected to equipment described in 6.4.2.2.5.3 through 6.4.2.2.5.4.

6.4.2.2.5.2 Connection to Alternate Power Source.

(A) The equipment branch shall be installed and connected to the alternate power source, such that equipment described in 6.4.2.2.5.3 is automatically restored to operation at appropriate time-lag intervals following the energizing of the life safety and critical branches.

(B) The arrangement of the connection to the alternate power source shall also provide for the subsequent connection of equipment described in 6.4.2.2.5.4.

6.4.2.2.5.3* Equipment for Delayed-Automatic Connection.

(A) The following equipment shall be permitted to be arranged for delayed-automatic connection to the alternate power source:

(1) Central suction systems serving medical and surgical functions, including controls, with such suction systems permitted to be placed on the critical branch
(2) Sump pumps and other equipment required to operate for the safety of major apparatus, including associated control systems and alarms
(3) Compressed air systems serving medical and surgical functions, including controls, with such air systems permitted to be placed on the critical branch
(4) Smoke control and stair pressurization systems
(5) Kitchen hood supply or exhaust systems, or both, if required to operate during a fire in or under the hood
(6) Supply, return, and exhaust ventilating systems for the following:
   (a) Airborne infectious/isolation rooms
   (b) Protective environment rooms
   (c) Exhaust fans for laboratory fume hoods
   (d) Nuclear medicine areas where radioactive material is used
   (e) Ethylene oxide evacuation
   (f) Anesthetic evacuation

(B) Where delayed-automatic connection is not appropriate, the ventilation systems specified in 6.4.2.2.5.3(A)(6) shall be permitted to be placed on the critical branch.

6.4.2.2.5.4* Equipment for Delayed-Automatic or Manual Connection. The following equipment shall be permitted to be arranged for either delayed-automatic or manual connection to the alternate power source (also see A.6.4.2.2.5.3):

(1) Heating equipment used to provide heating for operating, delivery, labor, recovery, intensive care, coronary care, nurseries, infection/isolation rooms, emergency treatment spaces, and general patient rooms; and pressure maintenance (jockey or make-up) pump(s) for water-based fire protection systems
(2)*Heating of general patient rooms during disruption of the normal source shall not be required under any of the following conditions:
   (a) Outside design temperature is higher than -6.7°C (+20°F)
   (b) Outside design temperature is lower than -6.7°C (+20°F), where a selected room(s) is provided for the needs of all confined patients [then only such room(s) need be heated].
(3) Elevator(s) selected to provide service to patient, surgical, obstetrical, and ground floors during interruption of normal power
(4) Supply, return, and exhaust ventilating systems for surgical and obstetrical delivery suites, intensive care, coronary care, nurseries, and emergency treatment spaces
(5) Hyperbaric facilities
(6) Hypobaric facilities
(7) Autoclaving equipment, which is permitted to be arranged for either automatic or manual connection to the alternate source
(8) Controls for equipment listed in 6.4.2.2.4
(9)*Other selected equipment

6.4.2.2.6 Wiring Requirements.

6.4.2.2.6.1* Separation from Other Circuits. The life safety branch and critical branch shall be kept independent of all other wiring and equipment.

6.4.2.2.6.2 Receptacles. The requirements for receptacles shall comply with 6.4.2.2.6.2(A), 6.4.2.2.6.2(B), and 6.4.2.2.6.2(C).

(A) The number of receptacles on a single branch circuit for areas described in 6.4.2.2.4.2(8) shall be minimized to limit the effects of a branch-circuit outage.

(B) Branch-circuit overcurrent devices shall be readily accessible to authorized personnel.

(C)* The electrical receptacles or the cover plates for the electrical receptacles supplied from the life safety and critical branches shall have a distinctive color or marking so as to be readily identifiable.

6.4.2.2.6.3 Switches. Switches of all types shall be permitted in the lighting circuits connected to the essential electrical system that do not serve as the illumination of egress as required by NFPA *101, Life Safety Code.*

6.4.2.2.6.4 Mechanical Protection of the Life Safety and Critical Branches. The wiring of the life safety and critical branches shall be mechanically protected by *raceways*, as defined in NFPA *70, National Electrical Code.*

6.4.2.2.6.5 Flexible power cords of appliances or other utilization equipment connected to the life safety and critical branches shall not be required to be enclosed in raceways.

6.4.2.2.6.6 Secondary circuits of transformer-powered communication or signaling systems shall not be required to be enclosed in raceways unless otherwise specified by Chapters 7 or 8 of *NFPA 70, National Electrical Code.*

6.4.3 Performance Criteria and Testing (Type 1 EES).

6.4.3.1 Source. The life safety and critical branches shall be installed and connected to the alternate power source specified in 6.4.1.1.4 and 6.4.1.1.5 so that all functions specified herein for the life safety and critical branches are automatically restored to operation within 10 seconds after interruption of the normal source.

6.4.3.2 Transfer Switches.

6.4.3.2.1 All ac-powered support and accessory equipment necessary to the operation of the EPS shall be supplied from the load side of the automatic transfer switch(es), or the output terminals of the EPS, ahead of the main EPS overcurrent protection, as necessary, to ensure continuity of the EPSS operation and performance. [**110:**7.12.5]

6.4.3.2.2 The essential electrical system shall be served by the normal power source, except when the normal power source is interrupted or drops below a predetermined voltage level. Settings of the sensors shall be determined by careful study of the voltage requirements of the load.

6.4.3.2.3 Failure of the normal source shall automatically start the alternate source generator after a short delay, as described in 6.4.2.1.5.4. When the alternate power source has attained a voltage and frequency that satisfies minimum operating requirements of the essential electrical system, the load shall be connected automatically to the alternate power source.

6.4.3.2.4 Upon connection of the alternate power source, the loads comprising the life safety and critical branches shall be automatically re-energized. The load comprising the equipment system shall be connected either automatically after a time delay, as described in 6.4.2.1.5.6, or nonautomatically and in such a sequential manner as not to overload the generator.

6.4.3.2.5 When the normal power source is restored, and after a time delay, as described in 6.4.2.1.5.7, the automatic transfer switches shall disconnect the alternate source of power and connect the loads to the normal power source. The alternate power source generator set shall continue to run unloaded for a preset time delay, as described in 6.4.2.1.5.9.

6.4.3.2.6 If the emergency power source fails and the normal power source has been restored, retransfer to the normal source of power shall be immediate, bypassing the retransfer delay timer.

6.4.3.2.7 If the emergency power source fails during a test, provisions shall be made to immediately retransfer to the normal source.

6.4.3.2.8 Nonautomatic transfer switching devices shall be restored to the normal power source as soon as possible after the return of the normal source or at the discretion of the operator.

6.4.4 Administration (Type 1 EES).

6.4.4.1 Maintenance and Testing of Essential Electrical System.

6.4.4.1.1 Maintenance and Testing of Alternate Power Source and Transfer Switches.

6.4.4.1.1.1 Maintenance of Alternate Power Source. The generator set or other alternate power source and associated equipment, including all appurtenance parts, shall be so maintained as to be capable of supplying service within the shortest time practicable and within the 10-second interval specified in 6.4.1.1.11 and 6.4.3.1.

6.4.4.1.1.2 The 10-second criterion shall not apply during the monthly testing of an essential electrical system. If the 10-second criterion is not met during the monthly test, a process shall be provided to annually confirm the capability of the life safety and critical branches to comply with 6.4.3.1.

6.4.4.1.1.3 Maintenance shall be performed in accordance with NFPA 110, *Standard for Emergency and Standby Power Systems*, Chapter 8.

6.4.4.1.1.4 Inspection and Testing. Criteria, conditions, and personnel requirements shall be in accordance with 6.4.4.1.1.4(A) through 6.4.4.1.1.4(C).

(A)* **Test Criteria.** Generator sets shall be tested 12 times a year, with testing intervals of not less than 20 days nor more than 40 days. Generator sets serving essential electrical systems shall be tested in accordance with NFPA 110, *Standard for Emergency and Standby Power Systems*, Chapter 8.

(B) Test Conditions. The scheduled test under load conditions shall include a complete simulated cold start and appropriate automatic and manual transfer of all essential electrical system loads.

(C) Test Personnel. The scheduled tests shall be conducted by competent personnel to keep the machines ready to function and, in addition, serve to detect causes of malfunction and to train personnel in operating procedures.

6.4.4.1.2 Maintenance and Testing of Circuitry.

6.4.4.1.2.1* Circuit Breakers. Main and feeder circuit breakers shall be inspected annually, and a program for periodically exercising the components shall be established according to manufacturer's recommendations.

6.4.4.1.2.2 Insulation Resistance. The resistance readings of main feeder insulation shall be taken prior to acceptance and whenever damage is suspected.

6.4.4.1.3 Maintenance of Batteries. Batteries for on-site generators shall be maintained in accordance with NFPA 110, *Standard for Emergency and Standby Power Systems*.

6.4.4.2 Record Keeping. A written record of inspection, performance, exercising period, and repairs shall be regularly maintained and available for inspection by the authority having jurisdiction.

6.5 Essential Electrical System Requirements — Type 2.

6.5.1 Sources (Type 2 EES). The requirements for sources for Type 2 essential electrical systems shall conform to those listed in 6.4.1.

6.5.2 Distribution (Type 2 EES).

6.5.2.1 General. The distribution requirements for Type 2 essential electrical systems shall conform to those listed in 6.4.2.1.

6.5.2.1.1* Coordination.

6.5.2.1.1.1 Overcurrent protective devices serving the essential electrical system shall be coordinated for the period of time that a fault's duration extends beyond 0.1 second.

6.5.2.1.1.2 Coordination shall not be required as follows:

(1) Between transformer primary and secondary overcurrent protective devices, where only one overcurrent protective device or set of overcurrent protective devices exists on the transformer secondary
(2) Between overcurrent protective devices of the same size (ampere rating) in series

6.5.2.2 Specific Requirements.

6.5.2.2.1* General.

6.5.2.2.1.1 The number of transfer switches to be used shall be based upon reliability, design, and load considerations.

6.5.2.2.1.2 The essential electrical system shall be divided into the following two branches:

(1) Life safety branch
(2) Equipment branch

6.5.2.2.1.3 Each branch of the essential electrical system shall have one or more transfer switches.

6.5.2.2.1.4 For the purposes of this code, Article 700 shall only be applied to the life safety branch.

6.5.2.2.1.5 The following portions of Article 700 of *NFPA 70* shall be amended as follows:

(A) 700.4 shall not apply.

(B) 700.10(D)(1) through (3) shall not apply.

(C) 700.17 Branch Circuits for Emergency Lighting. Branch circuits that supply emergency lighting shall be installed to provide service from a source complying with 700.12 when the normal supply for lighting is interrupted or where single circuits supply luminaires containing secondary batteries.

(D) 700.28 shall not apply.

6.5.2.2.1.6 One transfer switch shall be permitted to serve one or more branches in a facility with a continuous load on the switch of 150 kVA (120 kW) or less.

6.5.2.2.2 Life Safety Branch.

6.5.2.2.2.1 The life safety branch shall supply power as follows:

(1) Illumination of means of egress in accordance with NFPA *101, Life Safety Code*
(2) Exit signs and exit directional signs in accordance with NFPA *101, Life Safety Code*
(3) Alarm and alerting systems, including the following:
   (a) Fire alarms
   (b) Alarms required for systems used for the piping of nonflammable medical gases as specified in Chapter 5
(4)*Communications systems, where used for issuing instructions during emergency conditions
(5) Sufficient lighting in dining and recreation areas to provide illumination to exit ways at a minimum of 5 ft-candles
(6) Task illumination and select receptacles at the generator set location
(7) Elevator cab lighting, control, communications, and signal systems

6.5.2.2.2.2 No functions, other than those listed in 6.5.2.2.2.1(1) through 6.5.2.2.2.1(7), shall be connected to the life safety.

6.5.2.2.3 Equipment Branch.

6.5.2.2.3.1 General.

(A) The equipment branch shall be installed and connected to the alternate power source such that equipment listed in 6.5.2.2.3.2 is automatically restored to operation at appropriate time-lag intervals following the restoration of the life safety branch to operation.

(B) The equipment branch arrangement shall also provide for the additional connection of equipment listed in 6.5.2.2.3.3.

6.5.2.2.3.2 AC Equipment for Nondelayed-Automatic Connection. Generator accessories including, but not limited to, the transfer fuel pump, electrically operated louvers, and other generator accessories essential for generator operation shall be arranged for automatic connection to the alternate power source.

6.5.2.2.3.3 Delayed-Automatic Connections to Equipment Branch. The following equipment shall be permitted to be connected to the equipment branch and shall be arranged for delayed-automatic connection to the alternate power source:

(1) Task illumination and select receptacles in the following:
   (a) Patient care spaces
   (b) Medication preparation spaces
   (c) Pharmacy dispensing spaces
   (d) Nurses' stations (unless adequately lighted by corridor luminaires)
(2) Supply, return, and exhaust ventilating systems for airborne infectious isolation rooms
(3) Sump pumps and other equipment required to operate for the safety of major apparatus and associated control systems and alarms
(4) Smoke control and stair pressurization systems
(5) Kitchen hood supply or exhaust systems, or both, if required to operate during a fire in or under the hood
(6) Nurse call systems

6.5.2.2.3.4* Delayed-Automatic or Manual Connections to Equipment Branch. The equipment in 6.5.2.2.3.4(A) and 6.5.2.2.3.4(B) shall be permitted to be connected to the equipment branch and shall be arranged for either delayed-automatic or manual connection to the alternate power source.

(A) Heating Equipment to Provide Heating for General Patient Rooms. Heating of general patient rooms during disruption of the normal source shall not be required under any of the following conditions:

(1)*The outside design temperature is higher than −6.7°C (+20°F).
(2) The outside design temperature is lower than −6.7°C (+20°F) and, where a selected room(s) is provided for the needs of all confined patients, then only such room(s) need be heated.
(3) The facility is served by a dual source of normal power. See A.6.4.1.1.1 for more information.

(B)* Elevator Service. In instances where interruptions of power would result in elevators stopping between floors, throw-over facilities shall be provided to allow the temporary operation of any elevator for the release of passengers.

(C) Optional Connections to the Equipment Branch. Additional illumination, receptacles, and equipment shall be permitted to be connected only to the equipment branch.

(D) Multiple Systems. Where one switch serves multiple systems as permitted in 6.5.2.2, transfer for all loads shall be nondelayed automatic.

6.5.2.2.4 Wiring Requirements.

6.5.2.2.4.1* Separation from Other Circuits. The life safety and equipment branches shall be kept entirely independent of all other wiring and equipment.

6.5.2.2.4.2* Receptacles. The electrical receptacles or the cover plates for the electrical receptacles supplied from the life safety and equipment branches shall have a distinctive color or marking so as to be readily identifiable.

6.5.3 Performance Criteria and Testing (Type 2 EES).

6.5.3.1 Source. The life safety and equipment branches shall be installed and connected to the alternate source of power specified in 6.4.1.1.4 and 6.4.1.1.5 so that all functions specified herein for the life safety and equipment branches are automatically restored to operation within 10 seconds after interruption of the normal source.

6.5.3.2 Transfer Switches. The essential electrical system shall be served by the normal power source until the normal power source is interrupted or drops below a predetermined voltage level. Settings of the sensors shall be determined by careful study of the voltage requirements of the load.

6.5.3.2.1 Failure of the normal source shall automatically start the alternate source generator after a short delay, as described in 6.4.2.1.5.4. When the alternate power source has attained a voltage and frequency that satisfies minimum operating requirements of the essential electrical system, the load shall be connected automatically to the alternate power source.

6.5.3.2.2 All ac-powered support and accessory equipment necessary to the operation of the EPS shall be supplied from the load side of the automatic transfer switch(es), or the output terminals of the EPS, ahead of the main EPS overcurrent protection to ensure continuity of the EPSS operation and performance. [110:7.12.5]

6.5.3.2.3 Upon connection of the alternate power source, the loads comprising the life safety and equipment branches shall be automatically re-energized. The loads comprising the equipment branch shall be connected either automatically after a time delay, as described in 6.4.2.1.5.6, or nonautomatically and in such a sequential manner as not to overload the generator.

6.5.3.2.4 When the normal power source is restored, and after a time delay as described in 6.4.2.1.5.7, the automatic transfer switches shall disconnect the alternate source of power and connect the loads to the normal power source. The alternate power source generator set shall continue to run unloaded for a preset time delay as described in 6.4.2.1.5.9.

6.5.3.2.5 If the emergency power source fails and the normal power source has been restored, retransfer to the normal source of power shall be immediate, bypassing the retransfer delay timer.

6.5.3.2.6 If the emergency power source fails during a test, provisions shall be made to immediately retransfer to the normal source.

6.5.3.2.7 Nonautomatic transfer switching devices shall be restored to the normal power source as soon as possible after the return of the normal source or at the discretion of the operator.

6.5.4 Administration (Type 2 EES).

6.5.4.1 Maintenance and Testing of Essential Electrical System.

6.5.4.1.1 Maintenance and Testing of Alternate Power Source and Transfer Switches.

6.5.4.1.1.1 Maintenance of Alternate Power Source. The generator set or other alternate power source and associated equipment, including all appurtenance parts, shall be so maintained as to be capable of supplying service within the shortest time practicable and within the 10-second interval specified in 6.4.1.1.8 and 6.4.3.1.

6.5.4.1.1.2 Inspection and Testing. Generator sets shall be inspected and tested in accordance with 6.4.4.1.1.4.

6.5.4.1.2 Maintenance and Testing of Circuitry. Circuitry shall be maintained and tested in accordance with 6.4.4.1.2.

6.5.4.1.3 Maintenance of Batteries. Batteries shall be maintained in accordance with 6.4.4.1.3.

6.5.4.2 Record Keeping. A written record of inspection, performance, exercising period, and repairs shall be regularly maintained and available for inspection by the authority having jurisdiction.

Chapter 7 Information Technology and Communications Systems

7.1* Applicability. This chapter shall apply to information technology and communications systems in all health care facilities that provide services to human beings.

7.2 Reserved.

7.3 Category 1 Systems.

7.3.1 Information Technology and Communications Systems Infrastructure.

7.3.1.1 Premises Distribution System (Fiber and Copper).

7.3.1.1.1 Cables and installation shall be in compliance with *NFPA 70, National Electrical Code*, and TIA/EIA 568-B.

7.3.1.1.2 Distribution system cable labeling, record keeping, and alphanumeric schemes shall be in accordance with TIA/EIA 606-A.

7.3.1.2* **Telecommunications Systems' Spaces and Pathways.**

7.3.1.2.1 Entrance Facility (EF).

7.3.1.2.1.1 General. The entrance facility (EF) location shall be permitted to be combined with the telecommunications equipment room (TER).

7.3.1.2.1.2 Not less than two physically separated service entrance pathways into this location shall be required.

7.3.1.2.1.3 Remote Primary Data Center.

(A) In a facility where the primary data center is located remotely, two EFs and redundant telecommunications service entrances shall be provided.

(B)* Electronic storage with a minimum capacity to store all inpatient records shall be provided at the building.

7.3.1.2.1.4 Location Requirements and Restrictions.

(A) The EF shall be permitted to be located with the telecommunications equipment room (TER).

(B) Where the EF is combined with the TER, the space and electrical power and cabling shall be added to the TER to accommodate the telecommunications service provider's space and access requirements.

(C)* The EF shall be dedicated to low-voltage communication systems.

(D) Electrical equipment or fixtures (e.g., transformers, panelboards, conduit, wiring) that are not directly related to the support of the EF shall not be installed in or pass through the EF.

(E) Mechanical equipment and fixtures (e.g., water or drainage piping of any kind, ductwork, pneumatic tubing) that are not directly related to the support of the EF shall not be installed in, pass through, or enter the EF.

(F)* The EF shall be located not less than 3.66 m (12 ft) from any permanent source of electromagnetic interference.

(G) The EF shall be located in an area not subject to flooding.

(H) The EF shall be as close as practicable to the building communications service entrance point.

7.3.1.2.1.5 Working Space (Reserved).

7.3.1.2.1.6 Security. Access to EFs shall be restricted and controlled.

7.3.1.2.1.7 Power Requirements.

(A) Circuits serving the EF shall be dedicated to serving the EF.

(B) Circuits serving equipment in the EF shall be connected to the critical branch of the essential electrical system.

(C) A minimum of one duplex receptacle served from normal power shall be provided on one wall of the EF for service and maintenance.

7.3.1.2.1.8 Environmental Requirements.

(A) Temperature and humidity in the EF shall be controlled in accordance with the manufacturer's equipment requirements.

(B)* HVAC systems serving the EF shall be connected to the equipment branch of the essential electrical system.

(C)* A positive pressure differential with respect to surrounding areas shall be provided.

(D) Sprinklers shall be provided with wire cages or shall be recessed to prevent accidental operation.

7.3.1.2.1.9 Other Requirements (Reserved).

7.3.1.2.2 Telecommunications Equipment Room (TER). Each facility shall have at least one TER space that meets the minimum requirements of this chapter.

7.3.1.2.2.1 General.

(A) The telecommunications equipment room (TER) houses the main networking equipment and shall be permitted to also house application servers and data storage devices that serve the health care facility.

(B) Central equipment for other communications systems shall be permitted to be housed in the TER.

7.3.1.2.2.2* The entrance facility (EF) shall be permitted to be combined with the TER space.

7.3.1.2.2.3 Reserved.

7.3.1.2.2.4 Location Requirements and Restrictions.

(A) Electrical equipment or fixtures (e.g., transformers, panelboards, conduit, wiring) that are not directly related to the support of the TER shall not be installed in, pass through, or enter the TER.

(B) Any mechanical equipment or fixtures (e.g., water or drainage piping of any kind, ductwork, pneumatic tubing) not directly related to the support of the TER shall not be installed in, pass through, or enter the TER.

(C) The TER shall be located in a nonsterile area of the facility.

(D) In geographic areas prone to hurricanes or tornados, the TER shall be located away from exterior curtain walls to prevent wind and water damage.

(E)* The TER shall be located not less than 3.66 m (12 ft) from any permanent source of electromagnetic interference.

(F) The TER shall be located or designed to avoid vibration from mechanical equipment or other sources.

7.3.1.2.2.5 Working Space. Working space about communications cabinets, racks, or other equipment shall be in accordance with 110.26(A) of *NFPA 70, National Electrical Code.*

7.3.1.2.2.6 Security. Access to the TER shall be restricted and controlled.

7.3.1.2.2.7 Power Requirements.

(A) Circuits serving the TER and the equipment within the TER shall be dedicated to serving the TER.

(B) Circuits serving fire alarms, medical gas alarms, elevator communications, and communications systems used for issuing instructions during emergency conditions (e.g., fire fighter's phone system) shall be connected to the life safety branch of the essential electrical system.

(C) Circuits serving other communications equipment in the TER shall be connected to the essential electrical system.

(D) A minimum of one duplex outlet shall be provided on each wall and shall be connected to normal power for service and maintenance.

(E) Consideration shall be given to the reliability of power supply to the HVAC equipment because of its important function within the TER.

7.3.1.2.2.8 Environmental Requirements.

(A) Temperature and humidity in the TER shall be controlled in accordance with the manufacturer's equipment requirements.

(B) HVAC systems serving the TER shall be connected to the equipment branch of the essential electrical system.

(C) A positive pressure differential with respect to surrounding areas shall be provided.

7.3.1.2.2.9 Other Requirements (Reserved).

7.3.1.2.3 Telecommunications Room (TR).

7.3.1.2.3.1 General. A telecommunications room (TR) houses telecommunications equipment, cable terminations, and cross-connect cabling.

7.3.1.2.3.2 Sufficient TRs shall be provided so that any data or communications outlet in the building can be reached without exceeding 90 m (292 ft) maximum pathway distance from the termination point in the TR to the outlet.

7.3.1.2.3.3 Reserved.

7.3.1.2.3.4 Location Requirements and Restrictions.

(A) Switchboards, panelboards, transformers, and similar electrical equipment that are not directly related to the support of the TR shall not be installed in the TR.

(B) Any mechanical equipment or fixtures (e.g., water or drainage piping of any kind, ductwork, pneumatic tubing) not directly related to the support of the TR shall not be installed in, pass through, or enter the TR.

(C) In geographic areas prone to hurricanes or tornados, TRs shall be located away from exterior curtain walls to prevent wind and water damage.

(D)* The TR shall be located a minimum of 3.66 m (12 ft) from any permanent source of electromagnetic interference.

(E) A minimum of one TR shall be on each floor of the facility.

(F) A TR shall serve a maximum of 1858 m^2 (20,000 ft^2) of usable space on a single floor.

7.3.1.2.3.5 Working Space. Working space about communications cabinets, racks, or other equipment shall be in accordance with 110.26(A) of *NFPA 70, National Electrical Code.*

7.3.1.2.3.6 Security. Access to TRs shall be restricted and controlled.

7.3.1.2.3.7 Power Requirements.

(A) Circuits serving the TR and the equipment within the TR shall be dedicated to serving the TR.

(B) Circuits serving the TR shall be connected to the critical branch of the essential electrical system.

(C) A minimum of one duplex receptacle shall be provided in each TR and shall be connected to normal power for service and maintenance.

7.3.1.2.3.8 Environmental Requirements.

(A) Temperature and humidity in the TR shall be controlled in accordance with the manufacturer's equipment requirements.

(B) Sprinklers shall be provided with wire cages or shall be recessed to prevent accidental discharge.

7.3.1.2.3.9 Other Requirements. Dropped ceilings shall not be installed in the TR.

7.3.1.2.4 Cabling Pathways and Raceway Requirements.

7.3.1.2.4.1 Backbone Distribution. Redundant pathways shall be provided between the EF and TER.

7.3.1.2.4.2 Conduits shall be provided for cabling in inaccessible ceiling spaces.

7.3.1.2.5 Outside Plant (OSP) Infrastructure.

7.3.1.2.5.1 General. Outside plant (OSP) infrastructure shall consist of the conduits, vaults, and other pathways and cabling used to connect buildings on a campus and to provide services from off-campus service providers.

7.3.1.2.5.2 Pathways.

(A) Dual telecommunications service entrance pathways shall be provided to the TEF.

(B) Service entrance pathways shall be a minimum of 6.1 m (20 ft) apart.

(C) Underground conduits for technology systems shall be a minimum of 0.61 m (2 ft) from underground steam and water piping if crossing perpendicularly, and a minimum of 1.83 m (6 ft) if parallel.

(D) Underground conduits for technology systems shall be a minimum of 0.61 m (2 ft) below grade.

7.3.1.3 Antennas. (Reserved)

7.3.2 Voice, Data, Communications, and Cable Television Systems.

7.3.2.1 Voice/Telecommunications. (Reserved)

7.3.2.2 Local Area Networks (LANS). (Reserved)

7.3.2.3 Wireless Local Area Network (LAN) Systems and Public Wifi Hot Spots. (Reserved)

7.3.2.4 Wireless Voice Systems and In-Building Cellular Networks. (Reserved)

7.3.2.5 UHF, VHF, 800 MHz, and 900 MHz Radio Communication Systems. (Reserved)

7.3.2.6 Cable Television. (Reserved)

7.3.3 Other Communications Systems.

7.3.3.1 Nurse Call Systems.

7.3.3.1.1* General. The nurse call systems shall communicate patient and staff calls for assistance and information in health care facilities.

7.3.3.1.1.1 The nurse call systems shall be the audiovisual type or tone visual type and listed for the purpose.

7.3.3.1.1.2 The recognized standard for a listed nurse call system shall be ANSI/UL 1069, *Safety Standard for Hospital Signaling and Nurse Call Equipment.*

7.3.3.1.1.3* The nurse call system shall provide event notifications for one or more of the following: medical device alarms, staff emergency calls, code calls, and staff or patient requests for help or assistance.

7.3.3.1.1.4 Primary notification of nurse call events shall be provided by a listed nurse call system in accordance with 7.3.3.1.8.

7.3.3.1.1.5* Supplemental features shall be permitted to include call notification to alphanumeric pagers and other wireless devices carried by health care facility staff.

7.3.3.1.2 Patient Area Call Stations. The locations of call stations and calling devices shall be in accordance with the requirements set forth in the FGI Guidelines and as required by state and local codes.

7.3.3.1.2.1* Each patient bed location shall be provided with a call station.

7.3.3.1.2.2* A single call station that provides two-way voice communications shall not serve more than two adjacent beds with calling devices.

7.3.3.1.2.3* Call stations at patient bed locations shall be permitted to provide supplemental signaling of medical device alarms.

7.3.3.1.2.4 When provided, supplemental signaling of a medical device alarm shall be in accordance with 7.3.3.1.8.

7.3.3.1.2.5 Bath stations shall be provided at each inpatient toilet, bath, shower, or sitz bath and shall be accessible to a patient lying on the floor.

7.3.3.1.2.6 A pull cord shall be permitted to enable access by a patient lying on the floor.

7.3.3.1.3 Staff Emergency Call. The locations of staff emergency call stations shall be in accordance with the requirements set forth in the FGI Guidelines, and as required by state and local codes.

7.3.3.1.3.1* A staff emergency call shall be turned off only at the station, room, or space from where it originates.

7.3.3.1.4* Code Call. The nurse call system shall include provisions to summon assistance from medical emergency resuscitation teams, in locations set forth in the FGI Guidelines and as required by state and local codes.

7.3.3.1.4.1* A code call shall be turned off only at the station, room, or space from where it originates.

7.3.3.1.5 Call stations located in areas where patients are under constant visual surveillance, such as pre-op, recovery, and emergency units shall be permitted to be limited to the staff emergency call and the code call, and two-way communication with the patient bed location shall not be required.

7.3.3.1.6 Nurse call system provisions shall be provided for geriatric, Alzheimer's, and other dementia units where:

(1) All call stations shall have tamper-resistant fasteners.
(2) Provisions shall be made for the removal or covering of call buttons and outlets.
(3) Call cords or pull strings in excess of 152 mm (6 in.) shall not be permitted.

7.3.3.1.7 Nurse call system provisions shall not be required in psychiatric units, except for psychiatric seclusion ante/exam rooms where staff emergency call stations shall be provided:

(1) Call stations shall have tamper-resistant fasteners.
(2) Provisions shall be made for the removal or covering of call buttons and outlets.
(3) Call cords or pull strings shall not be permitted.
(4) Control to limit unauthorized use shall be permitted.

7.3.3.1.8 Notification Signals. The nurse call system shall annunciate each call visibly and audibly to all areas to where calls need to be directed and as required by state and local codes.

7.3.3.1.8.1 Notification signals for a code call and staff emergency call shall be individually identifiable and distinct from all other nurse call signals.

7.3.3.1.8.2 Activation of a call station including patient station, bath station, staff emergency station, and code call station shall activate the following notification signals:

(1) Visual signal in the corridor at the patient room door or care space
(2) Visual signals at corridor intersections where individual patient room door or care space signals are not directly visible from the associated nursing station
(3) Visible and audible signals at the nurse master station and associated duty stations
(4) Visible signals at the calling station from which the call originates
(5) A visual or aural signal indication at each audio calling station to indicate voice circuit operation

7.3.3.2 Patient Tracking. (Reserved)

7.3.3.3 Equipment and Asset Tracking. (Reserved)

7.3.3.4 Staff and Visitor Tracking. (Reserved)

7.3.3.5 Wireless Phone and Paging Integration. (Reserved)

7.3.3.6 Patient and Equipment Monitoring Systems. (Reserved)

7.3.3.7 Clinical Information Systems. (Reserved)

7.3.3.8 Pharmacy. (Reserved)

7.3.3.9 Material Management Information Systems. (Reserved)

7.3.3.10 Electronic Medical Records and Dictation Systems. (Reserved)

7.3.3.11 Medical Imaging Systems. (Reserved)

7.3.3.12 Archiving Systems. (Reserved)

7.3.4 Security Systems.

7.3.4.1 Internet Protocol (IP) Security Cameras Systems. (Reserved)

7.3.4.2 Digital Video Recording. (Reserved)

7.3.4.3 Intrusion Detection Systems. (Reserved)

7.3.4.4 Sitewide Monitoring. (Reserved)

7.3.4.5 Access Control Systems. (Reserved)

7.3.4.6 ID Badging Systems Integrated with Point of Sales Systems. (Reserved)

7.3.4.7 Threat Protection Systems. (Reserved)

7.3.4.8 Parking Access Systems. (Reserved)

7.4 Category 2 Systems.

7.4.1 Information Technology and Communications Systems Infrastructure.

7.4.1.1 Requirements for information technology and communications systems infrastructure shall be in accordance with 7.3.1, except as specified in 7.4.1.1.1 and 7.4.1.1.2.

7.4.1.1.1 HVAC systems serving the TEF, the TER, and TRs shall be connected to the essential electrical system.

7.4.1.1.2 Redundant pathways and cabling for the backbone distribution system shall not be required.

7.4.2 Voice, Data, Communications, and Cable Television Systems.

7.4.2.1 Voice/Telecommunications. (Reserved)

7.4.2.2 Local Area Networks (LANS). (Reserved)

7.4.2.3 Wireless Local Area Network (LAN) Systems and Public Wifi Hot Spots. (Reserved)

7.4.2.4 Wireless Voice Systems and In-Building Cellular Networks. (Reserved)

7.4.2.5 Cable Television. (Reserved)

7.4.3 Other Communications Systems.

7.4.3.1 Nurse Call Systems.

7.4.3.1.1 General. The nurse call system shall be in accordance with the requirements in 7.3.3.1.

7.4.3.1.2 Provisions for medical device alarms and code calls shall not be required. *(See 7.3.3.1.1.3.)*

7.4.3.2 Patient Tracking. (Reserved)

7.4.3.3 Equipment and Asset Tracking. (Reserved)

7.4.3.4 Staff and Visitor Tracking. (Reserved)

7.4.3.5 Wireless Phone and Paging Integration. (Reserved)

7.4.3.6 Patient and Equipment Monitoring Systems. (Reserved)

7.4.3.7 Material Management Information Systems. (Reserved)

7.4.3.8 Electronic Medical Records and Dictation Systems. (Reserved)

7.4.3.9 Medical Imaging Systems. (Reserved)

7.4.3.10 Archiving Systems. (Reserved)

7.4.4 Security Systems.

7.4.4.1 Internet Protocol (IP) Security Cameras Systems. (Reserved)

7.4.4.2 Digital Video Recording. (Reserved)

7.4.4.3 Intrusion Detection Systems. (Reserved)

7.4.4.4 Sitewide Monitoring. (Reserved)

7.4.4.5 Access Control Systems. (Reserved)

7.4.4.6 ID Badging Systems Integrated with Point of Sales Systems. (Reserved)

7.4.4.7 Threat Protection Systems. (Reserved)

7.4.4.8 Parking Access Systems. (Reserved)

7.5 Category 3 Systems.

7.5.1 Information Technology and Communications Systems Infrastructure.

7.5.1.1 Requirements for information technology and communications systems infrastructure shall be in accordance with 7.3.1, with exceptions as noted in 7.5.1.1.1 through 7.5.1.1.4.

7.5.1.1.1 Dual service entrance pathways into the EF are not required.

7.5.1.1.2 Power circuits serving equipment in the EF, the TER, and TRs shall not be required to be connected to the essential electrical system.

7.5.1.1.3 HVAC systems serving the EF, the ER, and TRs shall not be required to be connected to the essential electrical system.

7.5.1.1.4 Redundant pathways and cabling for the backbone distribution system shall not be required.

7.5.2 Voice, Data, Communications, and Cable Television Systems.

7.5.2.1 Voice/Telecommunications. (Reserved)

7.5.2.2 Local Area Networks (LANS). (Reserved)

7.5.2.3 Cable Television. (Reserved)

7.5.3 Other Communications Systems.

7.5.3.1 Nurse Call Systems. (Reserved)

7.5.3.2 Electronic Medical Records and Dictation Systems. (Reserved)

7.5.3.3 Medical Imaging Systems. (Reserved)

7.5.3.4 Archiving Systems. (Reserved)

7.5.4 Security Systems.

7.5.4.1 Internet Protocol (IP) Security Cameras Systems. (Reserved)

7.5.4.2 Digital Video Recording. (Reserved)

7.5.4.3 Intrusion Detection Systems. (Reserved)

7.5.4.4 Access Control Systems. (Reserved)

Chapter 8 Plumbing

8.1 Applicability.

8.1.1 This chapter shall apply to construction of new health care facilities, except as noted in 8.1.2 and 8.1.3.

8.1.2 This chapter shall also apply to the altered, renovated, or modernized portions of existing systems or individual components.

8.1.3 Existing construction or equipment shall be permitted to be continued in use when such use does not constitute a distinct hazard to life.

8.2 System Category Criteria. The health care facility's governing body that has the responsibility for the building system components as identified in this chapter shall designate, in accordance with the function of each space, building system categories in accordance with Sections 4.1 and 4.2.

8.2.1* The category of risk applied to each plumbing system serving a space shall be independent of the category of risk applied to other systems serving that same space.

8.3 General Requirements.

8.3.1 Potable Water. Potable water systems shall comply with applicable plumbing codes and FGI Guidelines.

8.3.2 Nonpotable Water. Nonpotable water systems shall comply with applicable plumbing codes and FGI Guidelines.

8.3.3* Water Heating. Maximum hot water temperatures shall comply with applicable plumbing codes and FGI Guidelines.

8.3.4 Water Conditioning. Water shall be treated or heated to control pathogens in the water.

8.3.5 Nonmedical Compressed Air.

8.3.5.1 Nonmedical air compressors shall be listed or approved.

8.3.5.2 Nonmedical compressed air shall not be used for powering medical instruments or for human respiration.

8.3.5.3 Nonmedical compressed air shall meet the quality and pressure requirements of the equipment connected to the system.

8.3.6 Special Use Water Systems. When special use water systems are required, application of FGI Guidelines or other appropriate publicly reviewed nationally published standards shall be followed.

8.3.7 Grease Interceptors.

8.3.7.1 Sizing for grease interceptors shall be permitted in accordance with local plumbing codes or an engineered calculation factoring meals served per day.

8.3.8 Fixtures. Plumbing fixtures shall be in accordance with the FGI Guidelines.

8.3.9 Black Waste Water. Black waste water shall be discharged to a sanitary sewer or private on-site waste treatment system as permitted by applicable plumbing codes.

8.3.10 Gray Waste Water.

8.3.10.1 Gray waste water shall be permitted to be stored on-site and used for nonpotable water systems as permitted by applicable plumbing codes.

8.3.10.2 Gray waste water shall not be used for any system that aerosolizes the water in a breathing zone or has direct contact with humans.

8.3.10.3 Excess gray waste water shall be discharged to a sanitary sewer or private on-site waste treatment system as permitted by applicable plumbing codes.

8.3.11 Clear Waste Water.

8.3.11.1 Clear waste water shall be permitted to be stored on-site and used for nonpotable water systems as permitted by applicable plumbing codes.

8.3.11.2 Clear waste water that has been treated to potable water standards shall be permitted to be used as nonpotable water.

8.3.11.3 Clear waste water that has not been treated to potable water standards shall not be used for any system that aerosolizes the water in a breathing zone or has direct contact with humans.

8.3.11.4 Excess clear waste water shall be discharged to a storm sewer, held in detention ponds, or recharged into the water table as permitted by applicable plumbing codes.

8.3.12 Drainage Systems. Drainage systems shall comply with applicable plumbing codes and FGI Guidelines.

8.4 Category 1. (Reserved)

8.5 Category 2. (Reserved)

8.6 Category 3. (Reserved)

Chapter 9   Heating, Ventilation, and Air Conditioning (HVAC)

9.1 Applicability.

9.1.1 This chapter shall apply to construction of new health care facilities, except as noted in 9.1.2 and 9.1.3.

9.1.2 This chapter shall also apply to the altered, renovated, or modernized portions of existing systems or individual components.

9.1.3 Existing construction or equipment shall be permitted to be continued in use when such use does not constitute a distinct hazard to life.

9.2 System Category Criteria. The health care facility's governing body that has the responsibility for the building system components as identified in this chapter shall designate, in accordance with the function of each space, building system categories in accordance with Sections 4.1 and 4.2.

9.2.1* The category of risk applied to each HVAC system serving a space shall be independent of the category of risk applied to other systems serving that same space.

9.3 General.

9.3.1 Heating, Cooling, Ventilating, and Process Systems.

9.3.1.1 Heating, cooling, ventilating, and process systems serving spaces or providing health care functions covered by this code or listed within ASHRAE 170, *Ventilation of Health Care Facilities*, shall be provided in accordance with ASHRAE 170, and as amended by Sections 9.4 through 9.6.

9.3.1.2 Laboratories shall comply with NFPA 45, *Standard on Fire Protection for Laboratories Using Chemicals*.

9.3.1.3* Windowless Anesthetizing Locations. Anesthetizing locations shall not be required to have a smoke purge system.

9.3.2 Energy Conservation. Heating, cooling, and ventilating systems serving spaces or providing health care functions covered by this code shall comply with ASHRAE 90.1, *Energy Standard for Buildings Except Low-Rise Residential Buildings*, or another locally adopted energy code.

9.3.3 Commissioning.

9.3.3.1 Heating, cooling, ventilating, and process systems serving spaces or providing health care functions covered by this code shall be commissioned in accordance with ASHRAE 90.1, *Energy Standard for Buildings Except Low-Rise Residential Buildings*.

9.3.3.2* Commissioning shall follow any publically reviewed document acceptable to the authority having jurisdiction.

9.3.4 Piping. Heating, cooling, ventilating, and process systems serving spaces or providing health care functions covered by this code shall utilize piping systems complying with applicable mechanical codes.

9.3.5 Ductwork. Heating, cooling, ventilating, and process systems serving spaces or providing health care functions covered by this code shall utilize ductwork systems complying with NFPA 90A, *Standard for the Installation of Air-Conditioning and Ventilating Systems*, or applicable mechanical codes.

9.3.6 Medical Gas Storage or Transfilling.

9.3.6.1 All gases, other than medical gases, shall be provided with ventilation per NFPA 55, *Compressed Gases and Cryogenic Fluids Code*.

9.3.6.2 Outdoor storage/installations for medical gases and cryogenic fluids shall be provided with ventilation per NFPA 55, *Compressed Gases and Cryogenic Fluids Code.*

9.3.6.3* Medical gases and cryogenic fluids that are in use per Chapter 11 shall not require special ventilation.

9.3.6.4 Transfilling area shall be provided with ventilation in accordance with NFPA 55, *Compressed Gases and Cryogenic Fluids Code.*

9.3.6.5 Indoor storage or manifold areas and storage or manifold buildings for medical gases and cryogenic fluids shall be provided with natural ventilation or mechanical exhaust ventilation in accordance with 9.3.6.5.1 through 9.3.6.8.

9.3.6.5.1* For the purposes of this section the volume of fluid (gas and liquid) to be used in determining the ventilation requirements shall be the volume of the stored fluid when expanded to standard temperature and pressure (STP) of either the largest single vessel in the enclosed space or of the entire volume of the connected vessels that are on a common manifold in the enclosed space, whichever is larger.

9.3.6.5.2 Natural Ventilation.

9.3.6.5.2.1 Natural ventilation shall consist of two nonclosable louvered openings, each having an aggregate free opening area of at least 155 cm^2/35 L (24 in.^{2}/1000 ft^3) of the fluid designed to be stored in the space and in no case less than 465 cm^2 (72 in.^{2}).

9.3.6.5.2.2 One opening shall be located within 30 cm (1 ft) of the floor, and one shall be located within 30 cm (1 ft) of the ceiling.

9.3.6.5.2.3 The openings shall be located to ensure cross ventilation.

9.3.6.5.2.4 Natural ventilation openings shall be directly to the outside atmosphere without ductwork.

9.3.6.5.2.5 Mechanical ventilation shall be provided if natural ventilation requirements cannot be met.

9.3.6.5.3 Mechanical Ventilation.

9.3.6.5.3.1 Mechanical exhaust to maintain a negative pressure in the space shall be provided continuously, unless an alternative design is approved by the authority having jurisdiction.

9.3.6.5.3.2 Mechanical exhaust shall be at a rate of 1 L/sec of airflow for each 300 L (1 cfm per 5 ft^3 of fluid) designed to be stored in the space and not less than 24 L/sec (50 cfm) nor more than 235 L/sec (500 cfm).

9.3.6.5.3.3 Mechanical exhaust inlets shall be unobstructed and shall draw air from within 300 mm (1 ft) of the floor and adjacent to the cylinder or containers.

9.3.6.5.3.4 Mechanical exhaust air fans shall be supplied with electrical power from the essential electrical system.

9.3.6.5.3.5 Dedicated exhaust systems shall not be required, provided that the system does not connect to spaces that contain combustible or flammable materials.

9.3.6.5.3.6 The exhaust duct material shall be noncombustible.

9.3.6.5.3.7 A means of make-up air shall be provided according to one of the following:

(1) Air shall be permitted via noncombustible ductwork to be transferred from adjacent spaces, from outside the building, or from spaces that do not contain combustible or flammable materials.

(2) Air shall be permitted to be transferred from a corridor under the door up to the greater of 24 L/sec (50 cfm) or 15 percent of the room exhaust in accordance with NFPA 90A, *Standard for the Installation of Air-Conditioning and Ventilating Systems.*

(3) Supply air shall be permitted to be provided from any building ventilation system that does not contain flammable or combustible vapors.

9.3.6.6 Discharge from the natural and mechanical ventilation systems shall be sited by a minimum separation distance in accordance with NFPA 55, *Compressed Gases and Cryogenic Fluids Code.*

9.3.6.7 A storage room shall maintain a temperature not greater than 52°C (125°F).

9.3.6.8 A transfer or manifold room shall maintain a temperature not greater than 52°C (125°F) and not less than −7°C (20°F).

9.3.7 Waste Gas.

9.3.7.1 Removal of excess anesthetic gases from the anesthesia circuit shall be accomplished by waste anesthetic gas disposal (WAGD), as described in Chapter 5, or by an active or passive scavenging ventilation system.

9.3.7.1.1 Active Systems. A dedicated exhaust system with an exhaust fan shall be provided to interconnect all of the anesthesia gas circuits to provide sufficient airflow and negative pressure in the gas disposal tubing so that cross contamination does not occur in the other circuits connected to the system.

9.3.7.1.2 Passive Systems.

9.3.7.1.2.1 A dedicated exhaust system with an exhaust fan shall be provided to exhaust snorkels at all of the anesthesia gas circuits to provide sufficient airflow to capture the gases, vapors, and particles expelled from the gas disposal tubing.

9.3.7.1.2.2 The snorkel shall include a minimum 25.4 mm (1 in.) diameter tubing connected to the exhaust system.

9.3.7.2 All the exhausted air shall be vented to the external atmosphere.

9.3.7.3 The excess anesthetic gases shall be deposited into the exhaust stream either at the exhaust grille or further downstream in the exhaust duct.

9.3.8 Medical Plume Evacuation. Plumes from medical procedures, including the use of lasers, shall be captured by one of the following methods:

(1) Direct connection to an unfiltered dedicated exhaust system that discharges outside the building
(2) HEPA filtering and direct connection to a return or exhaust duct
(3) Chemical and thermal sterilization and return to the space

9.3.9 Emergency Power System Room. Heating, cooling, and ventilating of the emergency power system shall be in accordance with NFPA 110, *Standard for Emergency and Standby Power Systems.*

9.3.10 Ventilation During Construction. Ventilation during construction shall comply with the FGI Guidelines.

9.4 Category 1. (Reserved)

9.5 Category 2. (Reserved)

9.6 Category 3. (Reserved)

Chapter 10 Electrical Equipment

10.1* Applicability.

10.1.1 This chapter shall apply to the performance, maintenance, and testing of electrical equipment in health care facilities, as specified in Section 1.3.

10.1.2 Experimental or research apparatus built to order or under development shall be used under qualified supervision and shall have a degree of safety that is equivalent to that described herein or that has been deemed acceptable by the facility.

10.1.3* Reserved.

10.2 Performance Criteria and Testing for Patient Care–Related Electrical Appliances and Equipment.

10.2.1 Permanently Connected — Fixed Equipment. Patient-connected electric appliances shall be grounded to the equipment grounding bus in the distribution panel by an insulated grounding conductor run with the power conductors.

10.2.2 Cord- and Plug-Connected — Portable Equipment.

10.2.2.1 Grounding of Appliances.

10.2.2.1.1 All cord-connected electrically powered appliances that are not double insulated and are used in the patient care vicinity shall be provided with a three-wire power cord and a three-pin grounding-type plug.

10.2.2.1.2 Double-insulated appliances shall be permitted to have two conductor cords and shall be rated as Class II devices.

10.2.2.2 Attachment Plugs. Attachment plugs listed for the purpose shall be used on all cord-connected appliances.

10.2.2.3 Construction and Use. The attachment plug shall be a two-pole, three-wire grounding type.

10.2.2.3.1 Appliances supplied by other than 120-V single-phase systems shall use the grounding-type plug (cap) appropriate for the particular power system.

10.2.2.3.2 The grounding prong of the plug shall be the first to be connected to, and the last to be disconnected from, the receptacle.

10.2.2.3.3 If screw terminals are used, the stranded conductor shall be twisted to prevent stray strands, but the bundle shall not be tinned after twisting.

10.2.2.3.4 If the conductor is not twisted, it shall be attached by an approved terminal lug.

10.2.2.3.5 The power cord conductors shall be arranged so that the conductors are not under tension in the plug.

10.2.2.3.6 The grounding conductor shall be the last one to disconnect when a failure of the plug's strain relief allows the energized conductors to be disrupted.

10.2.2.3.7 Strain Relief. Strain relief shall be provided.

10.2.2.3.7.1 The strain relief shall not cause thinning of the conductor insulation.

10.2.2.3.7.2 The strain relief of replaceable plugs shall be capable of being disassembled.

10.2.2.3.7.3 Plugs shall be permitted to be integrally molded onto the cord jacket if the design is listed for the purpose.

10.2.2.3.8 Testing. The wiring of each cord assembly shall be tested for continuity and polarity at the time of manufacture, when assembled into an appliance, and when repaired.

10.2.3 Power Cords.

10.2.3.1 Material and Gauge.

10.2.3.1.1 The flexible cord, including the grounding conductor, shall be of a type suitable for the particular application; shall be listed for use at a voltage equal to or greater than the rated power line voltage of the appliance; and shall have an ampacity, as given in Table 400.5(A) of *NFPA 70, National Electrical Code,* equal to or greater than the current rating of the device.

10.2.3.1.2 "Hard Service" (SO, ST, or STO), "Junior Hard Service" (SJO, SJT, or SJTO), or equivalent listed flexible cord shall be used, except where an appliance with a cord of another designation has been listed for the purpose.

10.2.3.2 Grounding Conductor.

10.2.3.2.1 Each electric appliance shall be provided with a grounding conductor in its power cord.

10.2.3.2.2 The grounding conductor shall be not smaller than 18 AWG.

10.2.3.2.3 The grounding conductor of cords longer than 4.6 m (15 ft) shall be not smaller than 16 AWG.

10.2.3.2.4* A grounding conductor in the power cord shall not be required for double-insulated appliances, but a functional ground conductor (functional earth conductor) shall be permitted.

10.2.3.3 Detachable Power Cords.

10.2.3.3.1 A detachable power cord shall be permitted if an accidental disconnection would not present an unacceptable hazard or if a mechanism that reliably prevents inadvertent disconnection is used.

10.2.3.3.2 Detachable power cords shall be designed so that the grounding conductor is the first to be connected and the last to be disconnected.

10.2.3.3.3 The cord set to the appliance shall be listed for the purpose.

10.2.3.4 Connection to Circuit and Color Codes.

10.2.3.4.1 Power cords, regardless of whether intended for use on grounded or isolated power systems, shall be connected in accordance with the conventions of a grounded system.

10.2.3.4.2 The circuit conductors in the cord shall be connected to the plug and the wiring in the appliance so that any of the following devices, when used in the primary circuit, are connected to the ungrounded conductor:

(1) Center contact of an Edison base lampholder
(2) Solitary fuseholder
(3) Single-pole, overcurrent protective device
(4) Any other single-pole, current-interrupting device

10.2.3.4.3 A second fuseholder or other overcurrent protective device provided in the appliance shall be permitted to be placed in the grounded side of the line.

10.2.3.5 Cord Strain Relief.

10.2.3.5.1 Cord strain relief shall be provided at the attachment of the power cord to the appliance so that mechanical stress, either pull, twist, or bend, is not transmitted to internal connections.

10.2.3.5.2 A strain relief molded onto the cord shall be bonded to the jacket and shall be of compatible material.

> Paragraph 10.2.3.6(5) was deleted by a tentative interim amendment (TIA). See page 1.

10.2.3.6 Multiple Outlet Connection. Two or more power receptacles supplied by a flexible cord shall be permitted to be used to supply power to plug-connected components of a movable equipment assembly that is rack-, table-, pedestal-, or cart-mounted, provided that all of the following conditions are met:

(1) The receptacles are permanently attached to the equipment assembly.
(2)*The sum of the ampacity of all appliances connected to the outlets does not exceed 75 percent of the ampacity of the flexible cord supplying the outlets.
(3) The ampacity of the flexible cord is in accordance with *NFPA 70, National Electrical Code.*
(4)*The electrical and mechanical integrity of the assembly is regularly verified and documented.

10.2.4 Adapters and Extension Cords.

10.2.4.1 Three-prong to two-prong adapters shall not be permitted.

10.2.4.2 Adapters and extension cords meeting the requirements of 10.2.4.2.1 through 10.2.4.2.3 shall be permitted.

10.2.4.2.1 All adapters shall be listed for the purpose.

10.2.4.2.2 Attachment plugs and fittings shall be listed for the purpose.

10.2.4.2.3 The cabling shall comply with 10.2.3.

10.2.5 Leakage Current — Fixed Equipment. The leakage current flowing through the ground conductor of the power supply connection to ground of permanently wired appliances installed in general or critical care areas shall not exceed 10.0 mA (ac or dc) with all grounds lifted.

10.2.6* Touch Current — Portable Equipment. The touch current for cord connected equipment shall not exceed 500 μA with normal polarity and the ground wire disconnected (if a ground wire is provided).

10.3 Testing Requirements — Fixed and Portable.

10.3.1* Physical Integrity. The physical integrity of the power cord assembly composed of the power cord, attachment plug, and cord-strain relief shall be confirmed by visual inspection.

10.3.2* Resistance.

10.3.2.1 For appliances that are used in the patient care vicinity, the resistance between the appliance chassis, or any exposed conductive surface of the appliance, and the ground pin of the attachment plug shall be less than 0.50 ohm under the following conditions:

(1) The cord shall be flexed at its connection to the attachment plug or connector.
(2) The cord shall be flexed at its connection to the strain relief on the chassis.

10.3.2.2 The requirement of 10.3.2.1 shall not apply to accessible metal parts that achieve separation from main parts by double insulation or metallic screening or that are unlikely to become energized (e.g., escutcheons or nameplates, small screws).

10.3.3* Leakage Current Tests.

10.3.3.1 General.

10.3.3.1.1 The requirements in 10.3.3.2 through 10.3.3.4 shall apply to all tests.

10.3.3.1.2 Tests shall be performed with the power switch ON and OFF.

10.3.3.2 Resistance Test. The resistance tests of 10.3.2 shall be conducted before undertaking any leakage current measurements.

10.3.3.3* Techniques of Measurement. The test shall not be made on the load side of an isolated power system or separable isolation transformer.

10.3.3.4 Leakage and Touch Current Limits. The leakage and touch current limits in 10.2.5 and 10.2.6 shall be followed.

10.3.4 Leakage Current — Fixed Equipment.

10.3.4.1 Permanently wired appliances in the patient care vicinity shall be tested prior to installation while the equipment is temporarily insulated from ground.

10.3.5 Touch Current — Portable Equipment.

10.3.5.1 If multiple devices are connected together and one power cord supplies power, the touch current shall be measured as an assembly.

10.3.5.2 When multiple devices are connected together and more than one power cord supplies power, the devices shall be separated into groups according to their power supply cord, and the touch current shall be measured independently for each group as an assembly.

10.3.5.3 Touch Leakage Test Procedure. Measurements shall be made using the circuit, such as the one illustrated in Figure 10.3.5.3, with the appliance ground broken in two modes of appliance operation as follows:

(1) Power plug connected normally with the appliance on
(2) Power plug connected normally with the appliance off (if equipped with an on/off switch)

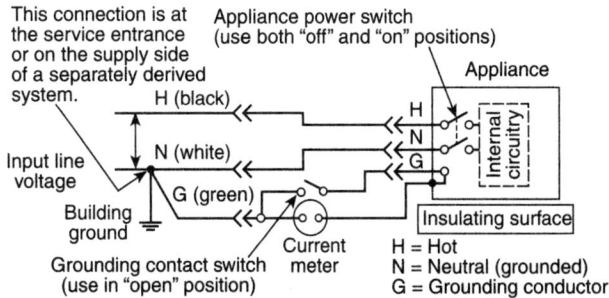

FIGURE 10.3.5.3 Example Test Circuit for Measuring Touch Leakage Current.

10.3.5.3.1 If the appliance has fixed redundant grounding (e.g., permanently fastened to the grounding system), the touch leakage current test shall be conducted with the redundant grounding intact.

10.3.6* Lead Leakage Current Tests and Limits — Portable Equipment.

10.3.6.1 The leakage current between all patient leads connected together and ground shall be measured with the power plug connected normally and the device on.

10.3.6.2 An acceptable test configuration shall be as illustrated in Figure 10.3.6.2.

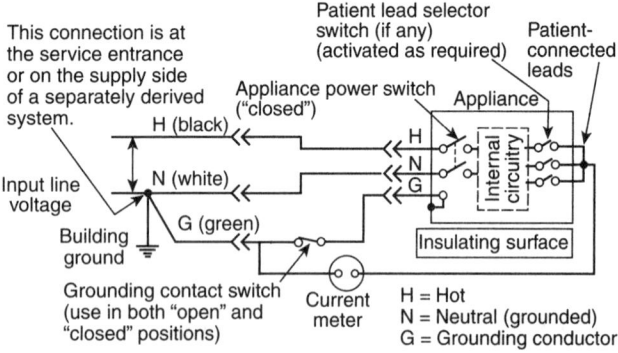

FIGURE 10.3.6.2 Test Circuit for Measuring Leakage Current Between Patient Leads and Ground — Nonisolated.

10.3.6.3 The leakage current shall not exceed 100 µA for ground wire closed and 500 µA ac for ground wire open.

10.4 Nonpatient Electrical Appliances and Equipment.

10.4.1 Permanently Connected — Fixed. (Reserved)

10.4.2 Cord- and Plug-Connected — Portable Equipment in Patient Care Vicinity.

10.4.2.1 Nonpatient care–related electrical equipment, including facility- or patient-owned appliances that are used in the patient care vicinity and will, in normal use, contact patients, shall be visually inspected by the patient's care staff or other personnel.

10.4.2.2 Any equipment that appears not to be in proper working order or in a worn condition shall be removed from service or reported to the appropriate maintenance staff.

10.4.2.3 Household or office appliances not commonly equipped with grounding conductors in their power cords shall be permitted, provided that they are not located within the patient care vicinity. Double-insulated appliances shall be permitted in the patient care vicinity.

10.5 Administration.

10.5.1 Responsibilities of Governing Body. (Reserved)

10.5.2 Policies.

10.5.2.1 Testing Intervals.

10.5.2.1.1 The facility shall establish policies and protocols to identify what patient care-related electrical equipment requires periodic inspection and, where applicable, the type of test and intervals of testing.

10.5.2.1.2 All patient care–related electrical equipment used in the patient care vicinity shall be tested in accordance with 10.3.5.3 or 10.3.6 before being put into service for the first time and after any repair or modification that might have compromised electrical safety.

10.5.2.2 Protection of Patients with Direct Electrical Pathways to the Heart. Only equipment that is specifically designed for the purpose [i.e., provided with suitable isolated patient leads or connections (cardiac floating, also known as CF, according to IEC 60601-1, *Medical Electrical Equipment — Part 1: General Requirements for Basic Safety and Essential Performance*)] shall be connected directly to electrically conductive pathways to a patient's heart.

10.5.2.3 Adapters and Extension Cords.

10.5.2.3.1 Adapters and extension cords meeting the requirements of 10.2.4 shall be permitted to be used.

10.5.2.3.2 Three-to-two-prong adapters shall not be permitted.

10.5.2.3.3 The wiring shall be tested for all of the following:

(1) Physical integrity
(2) Polarity
(3) Continuity of grounding at the time of assembly and periodically thereafter

10.5.2.4 Devices Likely to Be Used During Defibrillation. Devices that are critical to patient safety and that are likely to be attached to the patient when a defibrillator is used (such as ECG monitors) shall be rated as "defibrillator proof."

10.5.2.5* System Demonstration. Any system consisting of several electric appliances shall be demonstrated to comply with this code as a complete system.

10.5.2.6 Electrical Equipment Systems. Purchase contracts for electrical equipment systems, such as nurse call and signaling that consist of interconnected elements, shall require all of the following:

(1) The elements are intended to function together.
(2) The manufacturers provide documentation for such interconnection.
(3) The systems are installed by personnel qualified to do such installations.

10.5.2.7 Appliances Not Provided by the Facility. Policies shall be established for the control of appliances not supplied by the facility.

10.5.3 Servicing and Maintenance of Equipment.

10.5.3.1 The manufacturer of the appliance shall furnish documents containing at least a technical description, instructions for use, and a means of contacting the manufacturer.

10.5.3.2 The documents specified in 10.5.3.1 shall include the following, where applicable:

(1) Illustrations that show the location of controls
(2) Explanation of the function of each control
(3) Illustrations of proper connection to the patient or other equipment, or both
(4) Step-by-step procedures for testing and proper use of the appliance
(5) Safety considerations in use and servicing of the appliance

(6) Precautions to be taken if the appliance is used on a patient simultaneously with other electric appliances
(7) Schematics, wiring diagrams, mechanical layouts, parts lists, and other pertinent data for the appliance
(8) Instructions for cleaning, disinfection, or sterilization
(9) Utility supply requirements (electrical, gas, ventilation, heating, cooling, and so forth)
(10) Explanation of figures, symbols, and abbreviations on the appliance
(11) Technical performance specifications
(12) Instructions for unpacking, inspection, installation, adjustment, and alignment
(13) Preventive and corrective maintenance, inspection and repair procedures

10.5.4 Administration of Oxygen Therapy.

10.5.4.1 Electrical Equipment in Oxygen-Enriched Atmospheres. Appliances, or a part(s) of an appliance or a system (e.g., pillow speaker, remote control, pulse oximeter probe), to be used in the site of intentional expulsion shall comply with one of the following:

(1) They shall be listed for use in oxygen-enriched atmospheres.
(2) They shall be sealed so as to prevent an oxygen-enriched atmosphere from reaching electrical components, with sealing material of the type that will still seal even after repeated exposure to water, oxygen, mechanical vibration, and heating from the external circuitry.
(3) They shall be ventilated so as to limit the oxygen concentration surrounding electrical components to below 23.5 percent by volume.
(4) They shall have both of the following characteristics:
  (a) No hot surfaces over 300°C (573°F), except for small (less than 2 W) hermetically sealed heating elements, such as light bulbs
  (b) No exposed switching or sparking points of electrical energy that fall to the right of the curve for the appropriate type of circuit illustrated in Figure 10.5.4.1(a) through Figure 10.5.4.1(f), with the dc (or peak ac) open-circuit voltage and short-circuit current required to be used

10.5.4.2 When only the remote control or signal leads of a device are to be used in the site of intentional expulsion, only the control or signal leads shall be required to comply with 10.5.4.1.

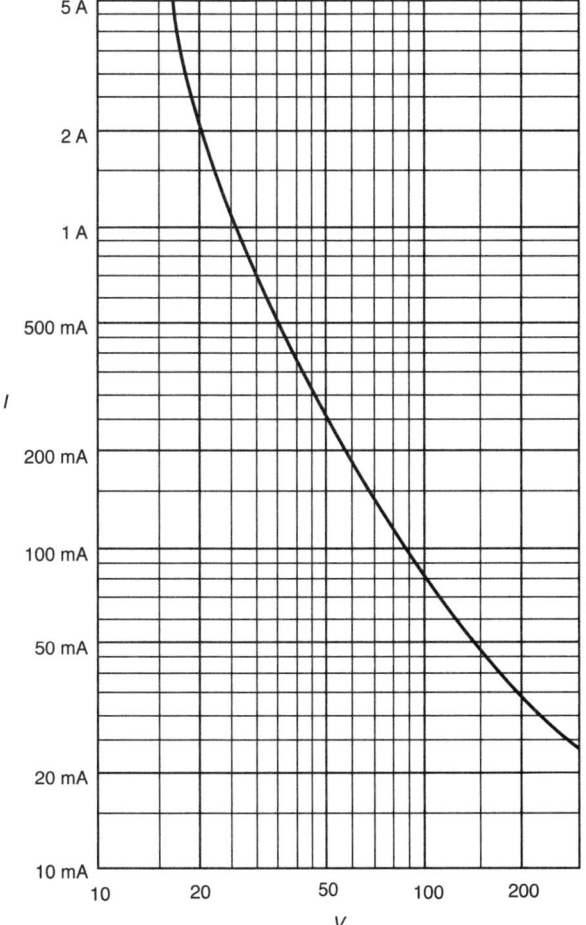

FIGURE 10.5.4.1(a) Resistance Circuits (L < 1 mH): Minimum Igniting Currents, Applicable to All Circuits Containing Cadmium, Zinc, or Magnesium.

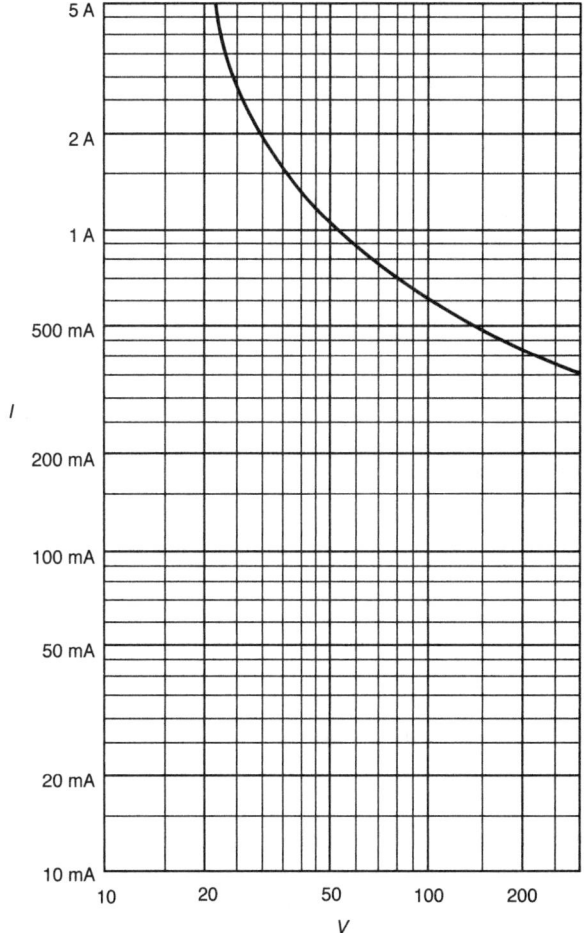

FIGURE 10.5.4.1(b) Resistance Circuits (L < 1 mH): Minimum Igniting Currents, Applicable to Circuits Where Cadmium, Zinc, or Magnesium Can Be Excluded.

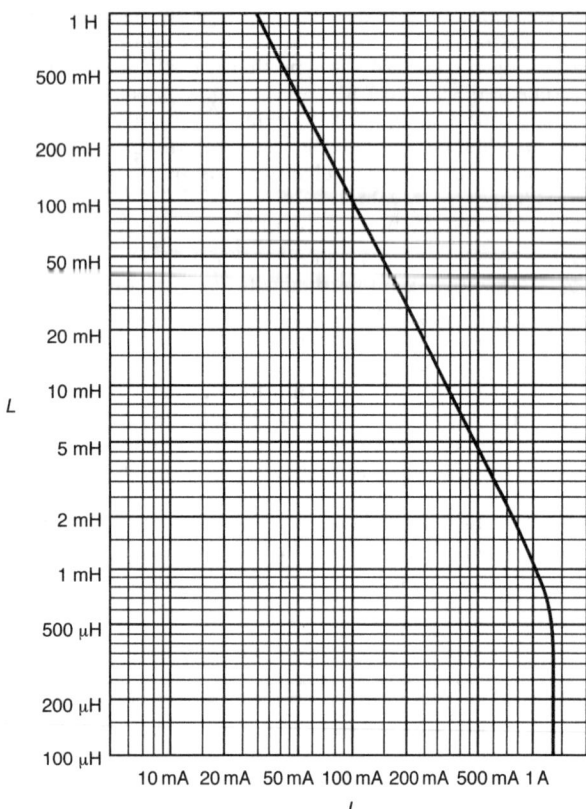

FIGURE 10.5.4.1(c) Inductance Circuits (L > 1 mH): Minimum Igniting Currents at 24 V, Applicable to All Circuits Containing Cadmium, Zinc, or Magnesium.

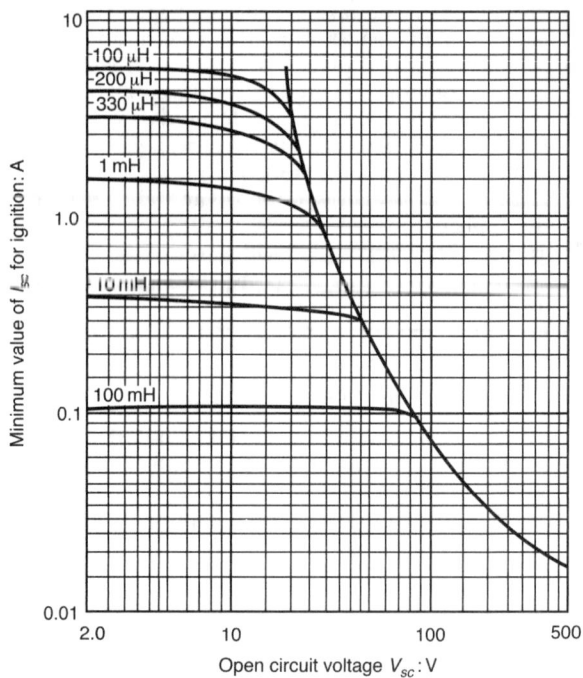

FIGURE 10.5.4.1(d) Inductance Circuits (L > 1 mH): Minimum Igniting Currents for Various Voltages, Applicable to All Circuits Containing Cadmium, Zinc, or Magnesium.

10.5.4.3 Subparagraphs 10.5.4.1 and 10.5.4.2 shall not apply to small (less than 2 W), hermetically sealed heating elements such as light bulbs.

10.5.4.4 Electrical equipment sold with the intent to be used in oxygen-enriched atmospheres shall be listed for use in oxygen-enriched atmospheres.

10.5.4.5* Electrical equipment used within oxygen delivery equipment shall be listed for use in oxygen-enriched atmospheres in accordance with ANSI/AAMI ES 60601-1, *Medical Electrical Equipment.*

10.5.4.6* High-energy-delivering probes (such as defibrillator paddles) or other electrical devices that do not comply with 10.5.4.1 and 10.5.4.2, that are deemed essential to the care of an individual patient, and that must be used within an administration site or within oxygen-delivery equipment shall be permitted.

10.5.5 Laboratory.

10.5.5.1* The laboratory shall establish policies and protocols for the type of test and intervals of testing for appliances.

10.5.6 Record Keeping — Patient Care Appliances.

10.5.6.1 Instruction Manuals.

10.5.6.1.1 Instruction and maintenance manuals shall be accessible to the group responsible for the maintenance of the appliance.

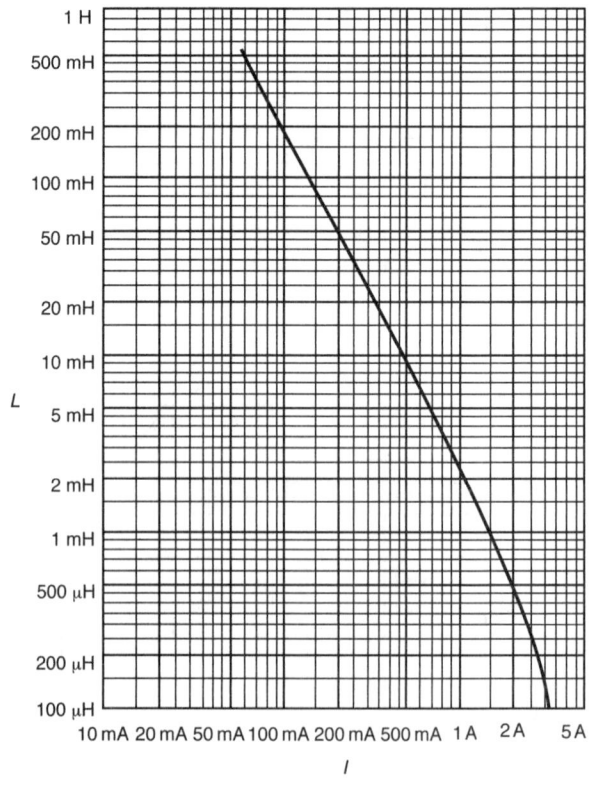

FIGURE 10.5.4.1(e) Inductance Circuits (L > 1 mH): Minimum Igniting Currents at 24 V, Applicable Only to Circuits Where Cadmium, Zinc, or Magnesium Can Be Excluded.

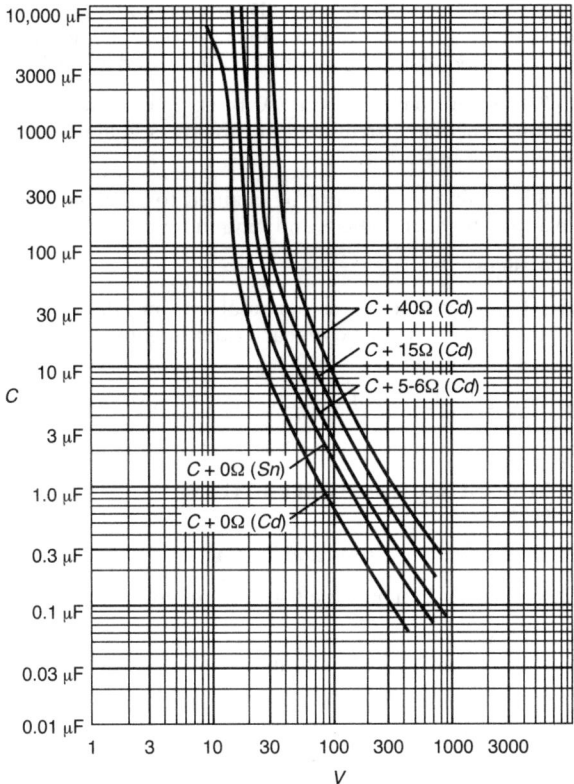

FIGURE 10.5.4.1(f) Capacitance Circuits Minimum Ignition Voltages. (The curves correspond to values of current-limiting resistance as indicated. The curve marked *Sn* is applicable only where cadmium, zinc, or magnesium can be excluded.)

10.5.6.1.2 Instruction and user maintenance manuals shall be accessible to the user.

10.5.6.1.3 Any safety labels and condensed operating instructions on an appliance shall be maintained in legible condition.

10.5.6.2* Documentation.

10.5.6.2.1 A record shall be maintained of the tests required by this chapter and associated repairs or modifications.

10.5.6.2.2 At a minimum, the record shall contain all of the following:

(1) Date
(2) Unique identification of the equipment tested
(3) Indication of which items have met or have failed to meet the performance requirements of 10.5.6.2

10.5.6.3 Records Retention. The records shall be maintained and kept for a period of time in accordance with a health care facility's record retention policy.

10.5.7 Use. (Reserved)

10.5.8 Qualification and Training of Personnel.

10.5.8.1* Personnel concerned for the application or maintenance of electric appliances shall be trained on the risks associated with their use.

10.5.8.1.1 The health care facilities shall provide programs of continuing education for its personnel.

10.5.8.1.2 Continuing education programs shall include periodic review of manufacturers' safety guidelines and usage requirements for electrosurgical units and similar appliances.

10.5.8.2 Personnel involved in the use of energy-delivering devices including, but not limited to, electrosurgical, surgical laser, and fiber-optic devices shall receive periodic training in fire prevention and suppression.

10.5.8.3* Equipment shall be serviced by qualified personnel only.

Chapter 11 Gas Equipment

11.1 Applicability.

11.1.1 This chapter shall apply to the performance, maintenance, and testing of gas equipment in health care facilities, as specified in Section 1.3.

11.1.2* This chapter shall apply to the use, at normal atmospheric pressure, of all of the following:

(1) Nonflammable medical gases
(2) Vapors and aerosols
(3) Equipment required for the administration of 11.1.2(1) and 11.1.2(2)

11.1.3 When used in this chapter, the term *oxygen* shall be intended to mean 100 percent oxygen as well as mixtures of oxygen and air.

11.1.4* This chapter shall not apply to special atmospheres, such as those encountered in hyperbaric chambers.

11.1.5* Reserved.

11.2 Cylinder and Container Source.

11.2.1 Cylinders and containers shall comply with 5.1.3.1.

11.2.2 Cylinder valve outlet connections shall conform to CGA V-1, *Standard for Compressed Gas Cylinder Valve Outlet and Inlet Connections* (ANSI B57.1) (includes Pin-Index Safety System for medical gases). (See 5.1.3.1.4.)

11.2.3 When low pressure threaded connections are employed, they shall be in accordance with CGA V-5, *Diameter-Index Safety System (Noninterchangeable Low Pressure Connections for Medical Gas Applications)*, for noninterchangeable, low pressure connections for medical gases, air, and suction.

11.2.4 Low pressure quick coupler connections shall be noninterchangeable between gas services.

11.2.5 Pressure reducing regulators and gauges intended for use in high pressure service shall be listed for such service.

11.2.6 Pressure reducing regulators shall be used on high-pressure cylinders to reduce the cylinder pressure to operating pressures.

11.2.7 Approved pressure reducing regulators or other gas-flow control devices shall be used to reduce the cylinder pressure of every cylinder used for medical purposes. All such devices shall have connections so designed that they attach only to cylinders of gas for which they are designated.

11.2.8* Equipment that could allow the intermixing of different gases, either through defects in the mechanism or through error in manipulation in any portion of the high pressure side of any system in which these gases might flow, shall not be used for joining cylinders containing compressed gases.

11.2.9 Cylinder valve outlet connections for oxygen shall be Connection No. 540 or Connection No. 870 as described in CGA V-1, *Standard for Compressed Gas Cylinder Valve Outlet and Inlet Connections* (ANSI B57.1).

11.2.10 Cylinder valve outlet connections for nitrous oxide shall be Connection No. 326 or Connection No. 910 as described in CGA V-1, *Standard for Compressed Gas Cylinder Valve Outlet and Inlet Connections* (ANSI B57.1).

11.3 Cylinder and Container Storage Requirements.

11.3.1* Storage for nonflammable gases equal to or greater than 85 m³ (3000 ft³) at STP shall comply with 5.1.3.3.2 and 5.1.3.3.3.

11.3.2* Storage for nonflammable gases greater than 8.5 m³ (300 ft³), but less than 85 m³ (3000 ft³), at STP shall comply with the requirements in 11.3.2.1 through 11.3.2.8.

11.3.2.1 Storage locations shall be outdoors in an enclosure or within an enclosed interior space of noncombustible or limited-combustible construction, with doors (or gates outdoors) that can be secured against unauthorized entry.

11.3.2.2 Oxidizing gases, such as oxygen and nitrous oxide, shall not be stored with any flammable gas, liquid, or vapor.

11.3.2.3 Oxidizing gases such as oxygen and nitrous oxide shall be separated from combustibles or flammable materials by one of the following:

(1) Minimum distance of 6.1 m (20 ft)
(2) Minimum distance of 1.5 m (5 ft) if the entire storage location is protected by an automatic sprinkler system designed in accordance with NFPA 13, *Standard for the Installation of Sprinkler Systems*
(3) A gas cabinet constructed per NFPA 30, *Flammable and Combustible Liquids Code*, or NFPA 55, *Compressed Gases and Cryogenics Fluids Code*, if the entire storage location is protected by an automatic sprinkler system designed in accordance with NFPA 13

11.3.2.4 Gas cylinder and cryogenic liquid container storage shall comply with 5.1.3.3.2 and 5.1.3.3.3.

11.3.2.5 Cylinder and container storage locations shall comply with 5.1.3.2.12 with respect to temperature limitations.

11.3.2.6 Cylinder or container restraints shall comply with 11.6.2.3.

11.3.2.7 Smoking, open flames, electric heating elements, and other sources of ignition shall be prohibited within storage locations and within 6.1 m (20 ft) of outside storage locations.

11.3.2.8 Cylinder valve protection caps shall comply with 11.6.2.3.

11.3.3 Storage for nonflammable gases with a total volume equal to or less than 8.5 m³ (300 ft³) shall comply with the requirements in 11.3.3.1 and 11.3.3.2.

11.3.3.1 Individual cylinder storage associated with patient care areas, not to exceed 2100 m² (22,500 ft²) of floor area, shall not be required to be stored in enclosures.

11.3.3.2 Precautions in handling cylinders specified in 11.3.3.1 shall be in accordance with 11.6.2.

11.3.3.3 When small-size (A, B, D, or E) cylinders are in use, they shall be attached to a cylinder stand or to medical equipment designed to receive and hold compressed gas cylinders.

11.3.3.4 Individual small-size (A, B, D, or E) cylinders available for immediate use in patient care areas shall not be considered to be in storage.

11.3.3.5 Cylinders shall not be chained to portable or movable apparatus such as beds and oxygen tents.

11.3.4 Signs.

11.3.4.1 A precautionary sign, readable from a distance of 1.5 m (5 ft), shall be displayed on each door or gate of the storage room or enclosure.

11.3.4.2 The sign shall include the following wording as a minimum:

CAUTION:
OXIDIZING GAS(ES) STORED WITHIN
NO SMOKING

11.4 Performance Criteria and Testing.

11.4.1 Portable Patient Care Gas Equipment.

11.4.1.1* Anesthetic apparatus shall be subject to approval by the authority having jurisdiction.

11.4.1.2* Each yoke on anesthetic apparatus constructed to allow attachment of a small cylinder equipped with a flush-type valve shall have two pins installed as specified in CGA V-1, *Standard for Compressed Gas Cylinder Valve Outlet and Inlet Connections* (ANSI B57.1).

11.4.1.3 Testing.

11.4.1.3.1 Interventions requiring testing shall include, but not be limited to, the following:

(1) Alteration of pipeline hose or pipeline fittings
(2) Alteration of internal piping
(3) Adjustment of selector switches or flush valves
(4) Replacement or repair of flowmeters or bobbins

11.4.1.3.2 After any adjustment or repair involving use of tools, or any modification of the gas piping supply connections or the pneumatic power supply connections for the anesthesia ventilator, or other pneumatically powered device, if one is present, and before use on patients, the gas anesthesia apparatus shall be tested at the final common path to the patient to determine that oxygen, and only oxygen, is delivered from the oxygen flowmeters and the oxygen flush valve, if any.

11.4.1.3.3 Before the gas anesthesia apparatus is returned to service, each fitting and connection shall be checked to verify its proper indexing to the respective gas service involved.

11.4.1.3.4 Before the gas anesthesia apparatus is returned to service, an oxygen analyzer, or a similar device, shall be used to verify the oxygen concentration.

11.4.1.4* Yoke-type connections between anesthesia apparatus and flush-type cylinder valves (commonly used with anesthetic gas cylinders) shall be of the Connection No. 860 type in accordance with CGA V-1, *Compressed Gas Cylinder Valve Outlet and Inlet Connections* (ANSI B57.1).

11.4.2 Apparatus for Administering Respiratory Therapy.

11.4.2.1 Oxygen-delivery equipment intended to rest on the floor shall be equipped with a base designed to render the entire assembly stable during storage, transport, and use. If casters are used, they shall conform to Class C of U.S. Government Commercial Standard 223-59, *Casters, Wheels, and Glides for Hospital Equipment*.

11.4.2.2 Oxygen enclosures of rigid materials shall be fabricated of noncombustible materials.

11.4.2.3 Equipment supplied from cylinders or containers shall be designed and constructed for service at full cylinder or container pressure or constructed for use with, or equipped with, pressure reducing regulators.

11.4.2.4 Humidification or reservoir jars containing liquid to be dispersed into a gas stream shall be made of transparent or translucent material, shall be impervious to contained solutions and medications, and shall allow observation of the liquid level and consistency.

11.4.2.5 Humidifiers and nebulizers shall be equipped with provisions for overpressure relief or alarm if the flow becomes obstructed.

11.4.2.6 Humidifiers and nebulizers shall be incapable of tipping or shall be mounted so that any tipping or alteration from the vertical shall not interfere with function or accuracy.

11.4.3 Nonpatient Gas Equipment.

11.4.3.1 Carts and Hand Trucks.

11.4.3.1.1 Construction. Carts and hand trucks for cylinders and containers shall be constructed for the intended purpose, be self-supporting, and be provided with appropriate chains or stays to retain cylinders or containers.

11.4.3.2* Medical Devices. Medical devices not for patient care and requiring oxygen USP shall meet the following:

(1) Be listed for the intended purpose by the United States Food and Drug Administration
(2) Be under the direction of a licensed medical professional, if connected to the piped distribution system
(3) Not be permanently attached to the piped distribution system
(4) Be installed and used per the manufacturer's instructions
(5) Be equipped with a backflow prevention device

11.5 Administration.

11.5.1 Policies.

> Subsection 11.5.1.1 was revised by a tentative interim amendment (TIA). See page 1.

11.5.1.1 Elimination of Sources of Ignition.

11.5.1.1.1 Smoking materials (e.g., matches, cigarettes, lighters, lighter fluid, tobacco in any form) shall be removed from patients receiving respiratory therapy.

11.5.1.1.2* When a nasal cannula and its associated supply tubing are delivering oxygen outside of a patient care room, no sources of open flame shall be permitted in the site of intentional expulsion.

11.5.1.1.3* When any other oxygen delivery equipment not specified in 11.5.1.1.2 is in use, no sources of open flame shall be permitted in the area of administration.

11.5.1.1.4* Solid fuel-burning appliances shall not be permitted in the area of administration.

11.5.1.1.5* Sparking toys shall not be permitted in any patient care room.

11.5.1.1.6 Nonmedical appliances that have hot surfaces or sparking mechanisms shall not be permitted within oxygen-delivery equipment or within the site of intentional expulsion.

11.5.1.2 Misuse of Flammable Substances.

11.5.1.2.1 Flammable or combustible aerosols or vapors, such as alcohol, shall not be used in oxygen-enriched atmospheres.

11.5.1.2.2 Oil, grease, or other flammable substances shall not be used on/in oxygen equipment.

11.5.1.2.3 Flammable and combustible liquids shall not be permitted within the site of intentional expulsion.

11.5.1.3 Servicing and Maintenance of Equipment.

11.5.1.3.1 Defective equipment shall be immediately removed from service.

11.5.1.3.2 Areas designated for the servicing of oxygen equipment shall be clean and free of oil, grease, or other flammable substances.

11.5.1.3.3* A scheduled preventive maintenance program shall be followed.

11.5.2 Gases in Cylinders and Liquefied Gases in Containers.

11.5.2.1 Qualification and Training of Personnel.

11.5.2.1.1* Personnel concerned with the application and maintenance of medical gases and others who handle medical gases and the cylinders that contain the medical gases shall be trained on the risks associated with their handling and use.

11.5.2.1.2 Health care facilities shall provide programs of continuing education for their personnel.

11.5.2.1.3 Continuing education programs shall include periodic review of safety guidelines and usage requirements for medical gases and their cylinders.

11.5.2.1.4 Equipment shall be serviced only by personnel trained in the maintenance and operation of the equipment.

11.5.2.1.5 If a bulk cryogenic system is present, the supplier shall provide annual training on its operation.

11.5.2.2 Transfilling Cylinders.

11.5.2.2.1 Mixing of compressed gases in cylinders shall be prohibited.

11.5.2.2.2* Transfilling of gaseous oxygen from one cylinder to another shall be in accordance with the mandatory requirements in CGA P-2.5, *Transfilling of High Pressure Gaseous Oxygen to Be Used for Respiration*.

11.5.2.2.3 Transfilling of any gases from one cylinder to another in the patient care vicinity shall be prohibited.

11.5.2.3 Transfilling Liquid Oxygen. Transfilling of liquid oxygen shall comply with 11.5.2.3.1 or 11.5.2.3.2, as applicable.

11.5.2.3.1 Transfilling to liquid oxygen base reservoir containers or to liquid oxygen portable containers over 344.74 kPa (50 psi) shall include the following:

(1) A designated area separated from any portion of a facility wherein patients are housed, examined, or treated by a fire barrier of 1 hour fire-resistive construction.
(2) The area is mechanically ventilated, is sprinklered, and has ceramic or concrete flooring.

(3) The area is posted with signs indicating that transfilling is occurring and that smoking in the immediate area is not permitted.
(4) The individual transfilling the container(s) has been properly trained in the transfilling procedures.

11.5.2.3.2* Where transfilling to liquid oxygen portable containers at 344.74 kPa (50 psi) and under, the following conditions shall be met:

(1) The area is well ventilated and has noncombustible flooring.
(2) The area is posted with signs indicating that smoking in the area is not permitted.
(3) The individual transfilling the liquid oxygen portable container has been properly trained in the transfilling procedure.
(4) The mandatory requirements of CGA P-2.6, *Transfilling of Low-Pressure Liquid Oxygen to Be Used for Respiration*, are met.

11.5.2.4* Filling Cylinders from Oxygen Concentrators. Filling cylinders from oxygen concentrators, including in the patient care vicinity, shall be in accordance with the manufacturer's instructions, not to exceed the limits in 11.5.2.4.1 through 11.5.2.4.4.

11.5.2.4.1 The cylinder contents shall not exceed 700 L (25 ft^3).

11.5.2.4.2 The flow shall not exceed 5 L/min (0.2 ft^3/min).

11.5.2.4.3 The pressure shall not exceed the DOT rating of the cylinder or 20,700 kPa (3000 psi), whichever is less.

11.5.2.4.4 The cylinders shall be in accordance with DOT requirements or those of the applicable regulatory agency.

11.5.2.5 Ambulatory Patients. Ambulatory patients on oxygen therapy shall be permitted access to all flame- and smoke-free areas within the health care facility.

11.5.3 Use (Including Information and Warning Signs).

11.5.3.1 Labeling.

11.5.3.1.1 Equipment listed for use in oxygen-enriched atmospheres shall be so labeled.

11.5.3.1.2 Oxygen-metering equipment and pressure reducing regulators shall be conspicuously labeled as follows:

OXYGEN — USE NO OIL

11.5.3.1.3 Flowmeters, pressure reducing regulators, and oxygen-dispensing apparatus shall be clearly and permanently labeled, designating the gas or mixture of gases for which they are intended.

11.5.3.1.4 Apparatus whose calibration or function is dependent on gas density shall be labeled as to the proper supply gas gauge pressure (kPa/psi) for which it is intended.

11.5.3.1.5 Oxygen-metering equipment, pressure reducing regulators, humidifiers, and nebulizers shall be labeled with the name of the manufacturer or supplier.

11.5.3.1.6 Cylinders and containers shall be labeled in accordance with CGA C-7, *Guide to the Preparation of Precautionary Labeling and Marking of Compressed Gas Containers*. Color coding shall not be utilized as a primary method of determining cylinder or container content.

11.5.3.1.7 All labeling shall be durable and withstand cleansing or disinfection.

11.5.3.2* Signs.

11.5.3.2.1 In health care facilities where smoking is not prohibited, precautionary signs readable from a distance of 1.5 m (5 ft) shall be conspicuously displayed wherever supplemental oxygen is in use and in aisles and walkways leading to such an area.

11.5.3.2.2 The signs shall be attached to adjacent doorways or to building walls or be supported by other appropriate means.

11.5.3.2.3 In health care facilities where smoking is prohibited and signs are prominently (strategically) placed at all major entrances, secondary signs with no smoking language shall not be required.

11.5.3.2.4 The nonsmoking policies shall be strictly enforced.

11.5.3.3 Transportation, Storage, and Use of Equipment.

11.5.3.3.1 Flow-control valves on administering equipment shall be closed prior to connection and when not in use.

11.5.3.3.2 Apparatus shall not be stored or transported with liquid agents in reservoirs.

11.5.3.3.3 Care shall be taken in attaching connections from gas services to equipment and from equipment to patients.

11.5.3.3.4 Fixed or adjustable orifice mechanisms, metering valves, pressure reducing regulators, and gauges shall not be connected directly to high pressure cylinders, unless specifically listed for such use and provided with appropriate safety devices.

11.5.3.3.5 Equipment shall only be serviced by qualified personnel.

11.6 Operation and Management of Cylinders.

11.6.1 Administration. Administrative authorities of health care organizations shall provide policies and procedures for safe practices.

11.6.1.1 Purchase specifications shall include the following:

(1) Specifications for cylinders
(2) Marking of cylinders, regulators, and valves
(3) Proper connections on the cylinders supplied to the facility

11.6.1.2 Training procedures shall include the following:

(1) Maintenance programs in accordance with the manufacturer's recommendations for the piped gas system
(2) Use and transport of equipment and the proper handling of cylinders, containers, hand trucks, supports, and valve protection caps
(3) Verification of gas content and mechanical connection specificity of each cylinder or container prior to placing it into service

11.6.1.3 Policies for enforcement shall include the following:

(1) Regulations for the storage and handling of cylinders and containers of oxygen and nitrous oxide
(2) Prompt evaluation of all signal warnings and all necessary measures taken to re-establish the proper functions of the medical gas and vacuum systems
(3) Organizational capability and resources to cope with a complete loss of any medical gas or vacuum system
(4) Successful completion of all tests required in 5.1.12.3 prior to the use of any medical gas or vacuum piping system for patient care
(5) Locations intended for the delivery vehicle delivering cryogenic liquid to bulk cryogenic liquid systems to remain open and not be used for any other purpose (e.g., vehicle parking, storage of trash containers)

11.6.2 Special Precautions for Handling Oxygen Cylinders and Manifolds. Handling of oxygen cylinders and manifolds shall be based on CGA G-4, *Oxygen*.

11.6.2.1 Oxygen cylinders, containers, and associated equipment shall be protected from contact with oil or grease by means of the following specific precautions:

(1) Oil, grease, or readily flammable materials shall not be permitted to come in contact with oxygen cylinders, valves, pressure reducing regulators, gauges, or fittings.
(2) Pressure reducing regulators, fittings, or gauges shall not be lubricated with oil or any other flammable substance.
(3) Oxygen cylinders or apparatus shall not be handled with oily or greasy hands, gloves, or rags.

11.6.2.2 Equipment associated with oxygen shall be protected from contamination by means of the following specific precautions:

(1) Particles of dust and dirt shall be cleared from cylinder valve openings by slightly opening and closing the valve before applying any fitting to the cylinder valve.
(2) The high pressure valve on the oxygen cylinder shall be opened slowly before bringing the apparatus to the patient or the patient to the apparatus.
(3) An oxygen cylinder shall not be draped with any materials such as hospital gowns, masks, or caps.
(4) Cylinder-valve protection caps, where provided, shall be kept in place and be hand-tightened, except when cylinders are in use or connected for use.
(5) Valves shall be closed on all empty cylinders in storage.

11.6.2.3 Cylinders shall be protected from damage by means of the following specific procedures:

(1) Oxygen cylinders shall be protected from abnormal mechanical shock, which is liable to damage the cylinder, valve, or safety device.
(2) Oxygen cylinders shall not be stored near elevators or gangways or in locations where heavy moving objects will strike them or fall on them.
(3) Cylinders shall be protected from tampering by unauthorized individuals.
(4) Cylinders or cylinder valves shall not be repaired, painted, or altered.
(5) Safety relief devices in valves or cylinders shall not be tampered with.
(6) Valve outlets clogged with ice shall be thawed with warm — not boiling — water.
(7) A torch flame shall not be permitted, under any circumstances, to come in contact with a cylinder, cylinder valve, or safety device.
(8) Sparks and flame shall be kept away from cylinders.
(9) Even if they are considered to be empty, cylinders shall not be used as rollers, supports, or for any purpose other than that for which the supplier intended them.
(10) Large cylinders (exceeding size E) and containers larger than 45 kg (100 lb) weight shall be transported on a proper hand truck or cart complying with 11.4.3.1.
(11) Freestanding cylinders shall be properly chained or supported in a proper cylinder stand or cart.
(12) Cylinders shall not be supported by radiators, steam pipes, or heat ducts.

11.6.2.4 Cylinders and their contents shall be handled with care, which shall include the following specific procedures:

(1) Oxygen fittings, valves, pressure reducing regulators, or gauges shall not be used for any service other than that of oxygen.
(2) Gases of any type shall not be mixed in an oxygen cylinder or any other cylinder.
(3) Oxygen shall always be dispensed from a cylinder through a pressure reducing regulator.
(4) The cylinder valve shall be opened slowly, with the face of the indicator on the pressure reducing regulator pointed away from all persons.
(5) Oxygen shall be referred to by its proper name, *oxygen*, not air, and liquid oxygen shall be referred to by its proper name, not liquid air.
(6) Oxygen shall not be used as a substitute for compressed air.
(7) The markings stamped on cylinders shall not be tampered with, because it is against federal statutes to change these markings.
(8) Markings used for the identification of contents of cylinders shall not be defaced or removed, including decals, tags, and stenciled marks, except those labels/tags used for indicating cylinder status (e.g., full, in use, empty).
(9) The owner of the cylinder shall be notified if any condition has occurred that might allow any foreign substance to enter a cylinder or valve, giving details and the cylinder number.
(10) Neither cylinders nor containers shall be placed in the proximity of radiators, steam pipes, heat ducts, or other sources of heat.
(11) Very cold cylinders or containers shall be handled with care to avoid injury.

11.6.2.5 Oxygen equipment that is defective shall not be used until one of the following tasks has been performed:

(1) It has been repaired by competent in-house personnel.
(2) It has been repaired by the manufacturer or his or her authorized agent.
(3) It has been replaced.

11.6.2.6 Pressure reducing regulators that are in need of repair or cylinders having valves that do not operate properly shall not be used.

11.6.3 Special Precautions for Making Cylinder and Container Connections.

11.6.3.1 Cylinder valves shall be opened and connected in accordance with the following procedure:

(1) Make certain that apparatus and cylinder valve connections and cylinder wrenches are free of foreign materials.
(2) Turn the cylinder valve outlet away from personnel following these safety procedures:
   (a) Stand to the side — not in front and not in back.
   (b) Before connecting the apparatus to the cylinder valve, momentarily open the cylinder valve to eliminate dust.
(3) Make connection of the apparatus to the cylinder valve, and tighten the connection nut securely with a wrench.
(4) Release the low-pressure adjustment screw of the pressure-reducing regulator completely.
(5) Slowly open cylinder valve to the full-open position.
(6) Slowly turn in the low-pressure adjustment screw on the pressure reducing regulator until the proper operating pressure is obtained.
(7) Open the valve to the utilization apparatus.

11.6.3.2 Connections for containers shall be made in accordance with the container manufacturer's operating instructions.

11.6.4 Special Precautions for the Care of Safety Mechanisms.

11.6.4.1 Personnel using cylinders and containers and other equipment covered in this chapter shall be familiar with the CGA Pin-Index Safety System and the CGA Diameter-Index Safety System, which are both designed to prevent utilization of the wrong gas.

11.6.4.2 Safety relief mechanisms, noninterchangeable connectors, and other safety features shall not be removed, altered, or replaced.

11.6.5 Special Precautions — Storage of Cylinders and Containers.

11.6.5.1 Storage shall be planned so that cylinders can be used in the order in which they are received from the supplier.

11.6.5.2 If empty and full cylinders are stored within the same enclosure, empty cylinders shall be segregated from full cylinders.

11.6.5.2.1 When the facility employs cylinders with integral pressure gauge, it shall establish the threshold pressure at which a cylinder is considered empty.

11.6.5.3 Empty cylinders shall be marked to avoid confusion and delay if a full cylinder is needed in a rapid manner.

11.6.5.4 Cylinders stored in the open shall be protected as follows:

(1) Against extremes of weather and from the ground beneath to prevent rusting
(2) During winter, against accumulations of ice or snow
(3) During summer, screened against continuous exposure to direct rays of the sun in those localities where extreme temperatures prevail

11.7 Liquid Oxygen Equipment.

11.7.1 General. The storage and use of liquid oxygen in liquid oxygen base reservoir containers and liquid oxygen portable containers shall comply with the following, or storage and use shall be in accordance with the adopted fire prevention code.

11.7.2 Information and Instructions. The liquid oxygen seller shall provide the user with documentation that includes, but is not limited to, the following:

(1) Manufacturer's instructions, including labeling for storage and use of the containers
(2) Requirements for storage and use of containers away from ignition sources, exits, electrical hazards, and high-temperature devices
(3) Methods for container restraint to prevent falling
(4) Requirements for container handling
(5) Safeguards for refilling of containers

11.7.3 Container Storage, Use, and Operation.

11.7.3.1* Containers shall be stored, used, and operated in accordance with the manufacturer's instructions and labeling.

11.7.3.2 Containers shall not be placed in the following areas:

(1) Where they can be tipped over by the movement of a door
(2) Where they interfere with foot traffic
(3) Where they are subject to damage from falling objects
(4) Where exposed to open flames and high-temperature devices

11.7.3.3* Liquid oxygen base reservoir containers shall be secured by one of the following methods while in storage or use to prevent tipping over caused by contact, vibration, or seismic activity:

(1) Securing to a fixed object with one or more restraints
(2) Securing within a framework, stand, or assembly designed to resist container movement
(3) Restraining by placing the container against two points of contact

11.7.3.4 Liquid oxygen base reservoir containers shall be transported by a cart or hand truck designed for such use, unless a container is equipped with a roller base.

11.7.3.5* Liquid Oxygen Portable Containers.

11.7.3.5.1 Liquid oxygen portable containers shall be kept in an upright position.

11.7.3.5.2 Liquid oxygen portable containers shall not be carried under clothing or other covering.

11.7.3.5.3 Liquid oxygen portable containers shall be kept away from ignition sources, electrical hazards, and high temperature devices during filling and use.

11.7.3.6 The transfilling of containers shall be in accordance with the manufacturer's instructions and the requirements of 11.7.3.6.1 through 11.7.3.6.2.

11.7.3.6.1 Liquid oxygen containers shall be filled outdoors or in compliance with 11.5.2.3.1.

11.7.3.6.1.1* A drip pan compatible with liquid oxygen shall be provided under the liquid oxygen base reservoir container's filling and vent connections used during the filling process, unless the filling is performed on a noncombustible surface such as concrete.

11.7.3.6.2 Liquid oxygen portable containers shall be permitted to be filled indoors when the liquid oxygen base reservoir container is designed for filling such containers and the written instructions provided by the container manufacturer are followed.

11.7.4 Maximum Quantity. The maximum total quantity of liquid oxygen permitted in storage and in use in a patient bed location or patient care vicinity shall be 120 L (31.6 gal), provided that the patient bed location or patient care vicinity, or both, are separated from the remainder of the facility by fire barriers and horizontal assemblies having a minimum fire resistance rating of 1 hour in accordance with the adopted building code.

Chapter 12   Emergency Management

12.1* Applicability. This chapter shall apply to new and existing health care facilities.

12.1.1* This chapter shall provide those with the responsibility for emergency management in health care facilities with the criteria to assess, mitigate, prepare for, respond to, and recover from emergencies of any origin.

12.1.2 This chapter shall be the source for emergency management in health care facilities and is based on the foundations of *NFPA 1600, Standard on Disaster/Emergency Management and Business Continuity Programs.*

12.1.3 This chapter shall aid in developing, maintaining, and evaluating effective emergency management programs in new and existing facilities.

12.2 Responsibilities.

12.2.1* Authority Having Jurisdiction.

12.2.1.1 The authority having jurisdiction shall be cognizant of the requirements of a health care facility with respect to its uniqueness for continued operation of the facility in an emergency.

12.2.1.2 The authority having jurisdiction shall ensure health care facility emergency management programs meet the requirements of this chapter.

12.2.2 Senior Management.

12.2.2.1 The senior management shall actively participate in and support emergency management planning.

12.2.2.2 Senior management shall provide the required resources to develop and support the emergency management program.

12.2.2.3 Senior management shall appoint a program coordinator.

12.2.3* Emergency Management Committee. The emergency management committee shall include representatives of senior management and clinical and support services.

12.2.3.1 The membership of the emergency management committee shall include a chairperson, the emergency program coordinator, a member of senior management, nursing, and representatives from key areas within the organization, such as physicians, infection control, facilities engineering, safety/industrial hygiene, security, and other key individuals.

12.2.3.2 The emergency management committee shall have the responsibility for the emergency management program within the facility.

12.2.3.3* The emergency management committee shall model the emergency operations plan on an incident command system (ICS) in coordination with federal, state, and local emergency response agencies, as applicable.

12.3 Emergency Management Categories. The application of requirements in this chapter shall be based on the emergency management category of the health care facility as defined in Table 12.3.

12.4 General.

12.4.1 Health care facilities shall develop an emergency management program with a documented emergency operations plan based on the category of the health care facility as defined in Table 12.3.

12.4.1.1 The emergency management program shall include elements as required to manage an emergency during all four phases: mitigation, preparedness, response, and recovery.

12.4.1.2 The emergency management program shall comply with applicable regulations, directives, policies, and industry standards of practice.

12.4.2 When developing its emergency management program, the facility shall communicate its needs and vulnerabilities to community emergency response agencies and identify the capabilities of its community in supporting their mission.

Table 12.3 Emergency Management Categories

Emergency Management Category	Definition
1	Those inpatient facilities that remain operable to provide advanced life support services to injured responders and disaster victims. These facilities manage the existing inpatient load as well as plan for the influx of additional patients as a result of an emergency.
2	Those inpatient or outpatient facilities that augment the critical mission. These facilities manage the existing inpatient or outpatient loads but do not plan to receive additional patients as a result of an emergency or do not plan to remain operable should essential utilities or services be lost.

12.4.3 The medical facility, in combination with the local or federal authorities, or both, shall establish the required emergency management category as defined in Table 12.3.

12.5 Emergency Management Category 1 and Emergency Management Category 2 Requirements.

12.5.1 All emergency management Category 1 and emergency management Category 2 health care facilities shall be required to develop and maintain an emergency management program that addresses all program elements as prescribed in 12.5.2 and 12.5.3.

12.5.2 The elements and complexity of the subsequent code sections in this chapter shall apply, as appropriate to the hazard vulnerability analysis (HVA), the community's expectations, and the leadership's defined mission of the health care facility.

12.5.3 Program Elements.

12.5.3.1 Hazard Vulnerability Analysis (HVA).

12.5.3.1.1 A hazard vulnerability analysis (HVA) shall be conducted to identify and prioritize hazards that pose a threat to the facility and can affect the demand for its services.

12.5.3.1.2* The hazards to be considered shall include, but not be limited to, the following:

(1) Natural hazards (geological, meteorological, and biological
(2) Human-caused events (accidental or intentional)
(3) Technological events

12.5.3.1.3 The analysis shall include the potential impact of the hazards on conditions including, but not limited to, the following:

(1)*Continuity of operations
(2) Care for new and existing patients/residents/clients
(3) Health, safety, and security of persons in the affected area
(4) Support of staff
(5) Property, facilities, and infrastructure
(6) Environmental impact
(7) Economic and financial conditions
(8) Regulatory and contractual obligations
(9) Reputation of, or confidence in, the facility

12.5.3.1.4 The facility shall prioritize the hazards and threats identified in the HVA with input from the community.

12.5.3.2 Mitigation.

12.5.3.2.1 The facility shall develop and implement a strategy to eliminate hazards or mitigate the effects of hazards that cannot be eliminated.

12.5.3.2.2 A mitigation strategy shall be developed for priority hazards defined by the HVA.

12.5.3.2.3 The mitigation strategy shall consider, but not be limited to, the following:

(1) Use of applicable building construction standards
(2) Hazard avoidance through appropriate land-use practices
(3) Relocation, retrofitting, or removal of structures at risk
(4) Removal or elimination of the hazard
(5) Reduction or limitation of the amount or size of the hazard
(6) Segregation of the hazard from that which is to be protected
(7) Modification of the basic characteristics of the hazard
(8) Control of the rate of release of the hazard
(9) Provision of protective systems or equipment for both cyber or physical risks
(10) Establishment of hazard warning and communications procedures
(11) Redundancy or duplication of essential personnel, critical systems, equipment, information, operations, or materials

12.5.3.3 Preparedness.

12.5.3.3.1 The facility shall prepare for any emergency as determined by the HVA by organizing and mobilizing essential resources.

12.5.3.3.2 The facility shall maintain a current, documented inventory of the assets and resources it has on-site that would be needed during an emergency, such as medical, surgical, and pharmaceutical resources; water; fuel; staffing; food; and linen.

12.5.3.3.3 The facility shall identify the resource capability shortfalls from 96 hours of sustainability and determine if mitigation activities are necessary and feasible.

12.5.3.3.4 The facility shall establish a protocol for monitoring the quantity of assets and resources as they are utilized.

12.5.3.3.5 The facility shall write an emergency operations plan (EOP) that describes a command structure and the following critical functions within the facility during an emergency:

(1) Communications
(2) Resources and assets
(3) Safety and security
(4) Clinical support activities
(5) Essential utilities
(6) Exterior connections
(7) Staff roles

12.5.3.3.6 Critical Function Strategies. During the development of the EOP, the facility shall consider the strategies required in 12.5.3.3.6.1 through 12.5.3.3.6.8 in order to manage critical functions during an emergency within the facility.

12.5.3.3.6.1 Communications. The facility shall plan for the following during an emergency:

(1) Initial notification and ongoing communication of information and instructions to staff
(2) Initial notification and ongoing communication with the external authorities
(3) Communication with the following:
  (a) Patients and their families (responsible parties)
  (b) Responsible parties when patients are relocated to alternative care sites
  (c) Community and the media
  (d) Suppliers of essential materials, services, and equipment
  (e) Alternative care sites
(4) Definition of when and how to communicate patient information to third parties
(5)*Establishment of backup communications systems
(6) Cooperative planning with other local or regional health care facilities, including the following:
  (a) Exchange of information relating to command operations, including contact information
  (b) Staffing and supplies that could be shared
  (c) System to locate the victims of the event

12.5.3.3.6.2 Resources and Assets. The facility shall plan for the following during an emergency:

(1) Acquiring medical, pharmaceutical, and nonmedical supplies
(2) Replacing medical supplies and equipment that will be used throughout response and recovery
(3) Replacing pharmaceutical supplies that will be consumed throughout response and recovery
(4) Replacing nonmedical supplies that will be depleted throughout response and recovery
(5) Managing staff support activities, such as housing, transportation, incident stress debriefing, sanitation, hydration, nutrition, comfort, morale, and mental health
(6) Managing staff family support needs, such as child care, elder care, pet care, and communication to home
(7) Providing staff, equipment, and transportation vehicles needed for evacuation

12.5.3.3.6.3* Safety and Security. The facility shall plan for the following during an emergency:

(1) Internal security and safety operations
(2) Roles of agencies such as police, sheriff, and national guard
(3) Managing hazardous materials and waste
(4) Radioactive, biological, and chemical isolation and decontamination
(5) Patients susceptible to wandering
(6) Controlling entrance into the health care facility during emergencies
(7) Conducting a risk assessment with applicable authorities if it becomes necessary to control egress from the health care facility
(8) Controlling people movement within the health care facility
(9) Controlling traffic access to the facility

12.5.3.3.6.4 Clinical Support Activities. The facility shall plan for the following during an emergency:

(1) Clinical activities that could need modification or discontinuation during an emergency, such as patient scheduling, triage, assessment, treatment, admission, transfer, discharge, and evacuation
(2) Clinical services for special needs populations in the community, such as pediatric, geriatric, disabled, and chronically ill patients, and those with addictions (Emergency Management Category 1 only)

(3) Patient cleanliness and sanitation
(4) Behavioral needs of patients
(5) Mortuary services
(6) Evacuation both horizontally and, when required by circumstances, vertically, when the environment cannot support care, treatment, and services
(7) Transportation of patients, and their medications and equipment, and staff to an alternative care site(s) when the environment cannot support care, treatment, and services
(8) Transportation of pertinent patient information, including essential clinical and medication-related information, to an alternative care site(s) when the environment cannot support care, treatment, and services
(9) Documentation and tracking of patient location and patient clinical information

12.5.3.3.6.5* Essential Utilities. The facility shall plan for the following during an emergency:

(1) Electricity
(2) Potable water
(3) Nonpotable water
(4) HVAC
(5) Fire protection systems
(6) Fuel required for building operations
(7) Fuel for essential transportation
(8) Medical gas and vacuum systems (if applicable)

12.5.3.3.6.6 Exterior Connections. For essential utility systems in Emergency Management Category 1 facilities only, and based on the facility's HVA, consideration shall be given to the installation of exterior building connectors to allow for the attachment of portable emergency utility modules.

12.5.3.3.6.7 Staff Roles.

(A) Staff roles shall be defined for the areas of communications, resources and assets, safety and security, essential utilities, and clinical activities.

(B) Staff shall receive training for their assigned roles in the EOP.

(C) The facility shall communicate to licensed independent health care providers their roles in the EOP.

(D) The facility shall provide staff and other personnel with a form of identification, such as identification cards, wrist bands, vests, hats, badges, or computer printouts.

(E) The facility shall include in its plan the alerting and managing of all staff in an emergency.

12.5.3.3.6.8 The facility shall include the following in its EOP:

(1)*Standard command structure that is consistent with its community
(2) Reporting structure consistent with the command structure
(3) Activation and deactivation of the response and recovery phases, including the authority and process
(4) Facility capabilities and appropriate response efforts when the facility cannot be supported from the outside for extended periods in the six critical areas with an acceptable response, including examples such as the following:
  (a) Resource conservation
  (b) Service curtailment
  (c) Partial or total evacuation consistent with the staff's designated role in community response plan
(5) Alternative treatment sites to meet the needs of the patients

12.5.3.3.7 Staff Education.

12.5.3.3.7.1 Each facility shall implement an educational program in emergency management.

12.5.3.3.7.2 The educational program shall include an overview of the components of the emergency management program and concepts of the incident command system (ICS).

12.5.3.3.7.3 Individuals who are expected to perform as incident commanders or to be assigned to specific positions within the command structure shall be trained in and familiar with the ICS and the particular levels at which they are expected to perform.

12.5.3.3.7.4 Education concerning the staff's specific duties and responsibilities shall be conducted.

12.5.3.3.7.5 General overview education of the emergency management program and the ICS shall be conducted at the time of hire.

12.5.3.3.7.6 Department-/staff-specific education shall be conducted upon appointment to department/staff assignments or positions and annually thereafter.

12.5.3.3.8* Testing Emergency Plans and Operations.

12.5.3.3.8.1 The facility shall test its EOP at least twice annually, either through functional or full-scale exercises or actual events.

12.5.3.3.8.2 Exercises shall be based on the HVA priorities and be as realistic as feasible.

12.5.3.3.8.3 For Emergency Management Category 1 only, an influx of volunteer or simulated patients shall be tested annually, either through a functional or full-scale exercise or an actual event. *(See Table 12.3.)*

12.5.3.3.8.4 Annual table top, functional, or full-scale exercises shall include the following:

(1) Community integration
(2) Assessment of sustainability

12.5.3.3.8.5 For Emergency Management Category 1 only, if so required by the community designation to receive infectious patients, the facility shall conduct at least one exercise a year that includes a surge of infectious patients. *(See Table 12.3.)*

12.5.3.3.8.6 The identified exercises shall be conducted independently or in combination.

12.5.3.3.9 Scope of Exercises.

12.5.3.3.9.1 Exercises shall be monitored by at least one designated evaluator who has knowledge of the facility's plan and who is not involved in the exercise.

12.5.3.3.9.2 Exercises shall monitor the critical functions.

12.5.3.3.9.3 The facility shall conduct a debriefing session not more than 72 hours after the conclusion of the exercise or the event.

12.5.3.3.9.4 The debriefing shall include all key individuals, including observers; administration; clinical staff, including a physician(s); and appropriate support staff.

12.5.3.3.9.5 Exercises and actual events shall be critiqued to identify areas for improvement.

12.5.3.3.9.6 The critiques required by 12.5.3.3.9.5 shall identify deficiencies and opportunities for improvement based upon monitoring activities and observations during the exercise.

12.5.3.3.9.7 Opportunities for improvement identified in critiques shall be incorporated in the facility's improvement plan.

12.5.3.3.9.8* Improvements made to the emergency management program shall be evaluated in subsequent exercises.

12.5.3.4 Response.

12.5.3.4.1* The facility shall declare itself in an emergency mode based on current conditions that leadership considers extraordinary.

12.5.3.4.2 Once an emergency mode has been declared, the facility shall activate its EOP.

12.5.3.4.3 The decision to activate the EOP shall be made by the incident commander designated within the plan, in accordance with the facility's activation criteria.

12.5.3.4.4 The decision to deactivate the EOP shall be made by the incident commander in the health care organization in coordination with the applicable external command authority.

12.5.3.4.5* The organization shall make provisions for emergency credentialing of volunteer clinical staff.

12.5.3.4.5.1 At a minimum, a peer evaluation of skill shall be conducted to validate proficiency for volunteer clinical staff.

12.5.3.4.5.2 Prior to beginning work, the identity of other volunteers offering to assist during response activities shall be verified.

12.5.3.4.5.3 Personnel designated or involved in the EOP of the health care facility shall be supplied with a means of identification, which shall be worn at all times in a visible location.

12.5.3.4.6 The command staff shall actively monitor conditions present in the environment and remain in communication with community emergency response agencies during an emergency response.

12.5.3.4.7 When conditions approach untenable, the command staff, in combination with community emergency response agencies, shall determine when to activate the facility evacuation plan.

12.5.3.4.8 Evacuation to the alternative care site shall follow the planning conducted during the preparedness phase.

12.5.3.4.9* Crisis standards of care shall be developed through a community-wide approach.

12.5.3.4.10 The decision to implement crisis standards of care shall be coordinated with the community leadership.

12.5.3.4.11 Upon implementation of crisis standards of care in a community, the following shall be considered:

(1) The triage process
(2) The allocation of medical services across the population

12.5.3.4.12 Medical Surge Capacity and Capability. The requirements of 12.5.3.4.12.1 and 12.5.3.4.12.2 shall apply only to those facilities designated as Emergency Management Category 1 as defined by the HVA.

12.5.3.4.12.1* The facility shall plan for medical surge capacity and capability.

12.5.3.4.12.2 The triage process shall be implemented as follows:

(1) The arriving victim shall be assessed into the following cohorts:
   (a) Risk to others, as follows:
      i. Mentally unstable
      ii. Contaminated
      iii. Infectious
   (b) Risk to self, as follows:
      i. Emotionally impaired
      ii. Suicidal
   (c) Risk of death or permanent injury, as follows:
      i. Walking wounded
      ii. Severely injured but stable
      iii. Suffering from life-threatening injury
      iv. Beyond care
(2) Patients shall be admitted for treatment depending on facility capacity, the facility's specialty, and clinical need.
(3) Creation of ancillary clinical space shall have adequate utility support for the following:
   (a) HVAC
   (b) Sanitation
   (c) Lighting
   (d) Proximity to operating room (OR)

12.5.3.4.13 Recovery from controlled reduction in care standards shall be reversed at the earliest feasible time.

12.5.3.4.14 Health care facilities shall have a designated media spokesperson to facilitate news releases during the response process.

12.5.3.4.15 An area shall be designated for media representatives to assemble where they will not interfere with the operations of the health care facility.

12.5.3.5* Recovery.

12.5.3.5.1 Plans shall reflect measures needed to restore operational capability to pre-disaster levels.

12.5.3.5.2 Fiscal aspects shall be considered with respect to restoration costs and possible cash flow losses associated with the disruption.

12.5.3.5.3 Facility leadership shall accept and accommodate federal, state, and local assistance that will be beneficial for recovery of operations.

12.5.3.5.4 No party to recovery shall take action to unfairly limit lawful competition once recovery operations are completed.

12.5.3.5.5 Recovery shall not be deemed complete until infection control decontamination efforts are validated.

12.5.3.6 Administration.

12.5.3.6.1 The facility shall modify its HVA, EOP, supply chain (including the current emergency supplies inventory), and other components of the emergency management program, as a result of exercises, real event, and annual review.

12.5.3.6.2 The facility shall maintain written records of drills, exercises, and training as required by this chapter for a period of 3 years.

Chapter 13 Security Management

13.1* Applicability. This chapter shall apply to new and existing health care facilities.

13.2* Security Management Plan.

13.2.1 A health care facility shall have a security management plan.

13.2.2* The scope, objectives, performance, and effectiveness of the security plan shall be tested at a frequency shown to be necessary by review of the security vulnerability assessment (SVA) in accordance with Section 13.3.

13.3 Security Vulnerability Assessment (SVA).

13.3.1* At least annually, the health care facility shall conduct a security vulnerability assessment (SVA).

13.3.2 The SVA shall evaluate the potential security risks posed by the physical and operational environment of the health care facility to all individuals in the facility.

13.3.3 The facility shall implement procedures and controls in accordance with the risks identified by the SVA.

13.4 Responsible Person.

13.4.1 A person(s) shall be appointed by the leadership of the health care facility to be responsible for all security management activities.

13.4.2 The duties of the person assigned as required by 13.4.1 shall include, but not be limited to, the following, as identified in the SVA:

(1) Provide identification for patients, staff, and other people entering the facility
(2) Control access in and out of security-sensitive areas
(3) Define and implement procedures as follows:
  (a) Security incident
  (b) Hostage situation
  (c)*Bomb (explosive device or threat)
  (d) Criminal threat
  (e) Labor action
  (f) Disorderly conduct
  (g) Workplace violence
  (h) Restraining order
  (i) Prevention of, and response to, infant or pediatric abduction
  (j) Situations involving VIPs or the media
  (k) Maintenance of access to emergency areas
  (l) Civil disturbance
  (m) Forensic patients
  (n) Patient elopement
  (o) National Terrorism Advisory System (NTAS) or equivalent
  (p) Suspicious material or package
  (q) Suspicious powder or substance
  (r) Use of force policy
  (s) Security staffing augmentation
  (t) Active shooter
(4) Provide security at alternate care sites or vacated facilities
(5) Control vehicular traffic on the facility property
(6) Protect the facility assets, including property and equipment
(7) Provide policy for interaction with law enforcement agencies
(8) Comply with applicable laws, regulations, and standards regarding security management operations
(9) Educate and train the facility security force to address the following:
  (a) Customer service
  (b) Use of physical restraints
  (c) Use of force
  (d) Response criteria
  (e) Fire watch procedures
  (f) Lockdown procedures
  (g) Emergency notification procedures
  (h) Emergency communications procedures

13.5 Security-Sensitive Areas.

13.5.1 All security-sensitive areas, as identified by the SVA, shall be protected as appropriate.

13.5.2 Emergency department security shall include appropriate protection, including the following:

(1)*Control and limitation of access by the general public
(2) Private duress alarm at the nurses' station and reception for summoning immediate assistance
(3) Access-control of treatment area
(4) Lockdown procedure to secure the area when conditions threaten the viability of the department
(5) Bullet-resisting glazing material, as deemed necessary by review of the SVA

13.5.3 Pediatric and infant care areas shall have a security plan for the prevention of, and response to, pediatric and infant abduction that shall include appropriate protections, such as the following:

(1) Control and limitation of access by the general public
(2) Screening by nursing prior to allowing persons access to infant care areas
(3) Matching protocol with staff clearance to pair infants with parents
(4) System to monitor and track the location of pediatric and infant patients
(5)*Facility alert system, lockdown, and staff inspection of all packages leaving the premises
(6) Use of electronic monitoring, tracking, and access control equipment
(7) Use of an automated and standardized facility-wide alerting system to announce pediatric or infant abduction
(8) Remote exit locking or alarming
(9) Facility lockdown procedures and staff inspection of all persons and packages leaving the premises
(10) Prohibition on birth announcements by staff
(11) Detection of the presence of nonidentified individual constitutes security breach
(12) Movement of infants restricted to basinets only — no hand carries
(13) Health care staff wear unique identification or uniforms
(14) Secure storage of scrubs and uniforms, both clean and dirty
(15) Education in pediatric and infant abduction as follows:
  (a) Health care staff are familiar with infant abduction scenarios.
  (b) Parents know not to leave a child or an infant unattended or in the care of an unidentified person.
(16) Visiting family and friends not permitted to enter any nursery area with an infant or a newborn from the outside
(17) Infant abduction drills conducted periodically to test effectiveness of chosen measures

13.5.4* Medication storage and work areas shall be secured against admittance of unauthorized personnel through the use of the following:

(1) Physical access control
(2) Unique identification for the area
(3) Secure storage and controlled dispensing of drugs

13.5.5 Clinical and research laboratories shall be secured against admittance of unauthorized personnel through appropriate protections, such as the following:

(1) Physical access control
(2) Unique identification for the area
(3) Secure storage and controlled dispensing of regulated chemical, biological, and radiological materials

13.5.6 Dementia or behavioral health units shall be secured against the admittance or release of unauthorized personnel through appropriate protections, such as the following:

(1) Physical access control
(2) Unique identification for the area
(3)*Procedure to prevent entry of contraband prior to a person being admitted into the unit or department
(4) Elopement precautions
(5) Maintenance of color photos with the medical information of current patients to aid in identification

13.5.7 Forensic patient treatment areas shall provide appropriate protections, such as the following:

(1)*Law enforcement attending the patient at all times
(2) Treatment performed in an area separate from other patients
(3) Restraints applied or removed only under forensic staff control

13.5.8 Communications, data infrastructure, and medical records storage areas shall be secured against the admittance of unauthorized personnel or unauthorized release of confidential information through the use of appropriate protections, such as the following:

(1) Physical access control
(2) Unique identification for the area
(3) Surveillance equipment
(4) Data encryption and password protection

13.6 Access and Egress Security Measures.

13.6.1 Public visitation controls shall be enforced.

13.6.2 After-hours entrance by the public shall be restricted to designated areas, such as entrance lobbies and emergency departments.

13.6.3 Health care facility security controls and procedures shall comply with life safety requirements for egress.

13.6.3.1* Security plans for health care occupancies shall address access and egress control during periods of quarantine and other events in conjunction with emergency agencies.

13.7* Media Control.

13.7.1 The security management plan shall include procedures to accommodate media representatives.

13.7.1.1* A person shall be designated to serve as media contact and representative for the organization in regard to media interactions.

13.7.2* An area shall be designated for assembly of media representatives.

13.7.3 A security or facility staff member shall remain with the media representative(s) at all times.

13.7.4 Media representatives shall be escorted when granted access to the health care facility outside of the area designated in 13.7.2.

13.8* Crowd Control.

13.8.1 The security management plan shall provide procedures for crowd control for management of those demanding access to a health care facility.

13.8.2 The procedures for crowd control shall provide for the coordination and collaboration of security and law enforcement.

13.9 Security Equipment.

13.9.1 The security management plan shall provide procedures for crowd control demanding access to a health care facility.

13.9.2 The security management plan shall include processes and procedures for controlling access to the health care facility.

13.9.2.1 Exterior entrances shall be provided with locking devices.

13.9.2.2 Locking devices shall comply with applicable federal, state, and local requirements.

13.9.2.3 Locking devices shall be properly installed and be in good working order.

13.9.3* The facility shall operate a key control program.

13.10* Employment Practices. Employers shall ensure a high level of integrity in the workplace by using the following practices:

(1) Background checks of employees with access to critical assets
(2) Background checks of outside contractors' employees
(3) Drug testing program for employees

13.11* Security Operations.

13.11.1* Post orders shall be written for security personnel.

13.11.2 Security personnel training shall include, but not be limited to, the following:

(1) Customer service
(2) Emergency procedures
(3) Patrol methods
(4) De-escalation training
(5) Use of physical restraints
(6) Use of force

13.12 Program Evaluation.

13.12.1* Periodic drills shall be conducted at various times and locations.

13.12.2 The drills shall be critiqued for plan effectiveness and to identify opportunities for improvement.

13.12.3 Identified opportunities for improvement shall be incorporated into the security plan.

13.12.4 The security plan shall be evaluated at least annually.

13.12.5 The evaluation of the security management plan shall include a review of laws, regulations, and standards applicable to the security program.

Chapter 14 Hyperbaric Facilities

14.1* Scope. The scope of this chapter shall be as specified in 1.1.12.

14.1.1 Applicability.

14.1.1.1 This chapter shall apply to new facilities.

14.1.1.2 The following sections of this chapter shall apply to both new and existing facilities:

(1) 14.2.4.1.1 (excluding subsections)
(2) 14.2.4.1.1.1
(3) 14.2.4.1.2
(4) 14.2.4.1.3 (excluding subsections)
(5) 14.2.4.1.3.3
(6) 14.2.4.3.3 (and subsections)
(7) 14.2.4.4 (and subsections)
(8) 14.2.4.5.3
(9) 14.2.4.5.4 (and subsection)
(10) 14.2.5.1.4 (excluding subsection)
(11) 14.2.5.1.5
(12) 14.2.5.1.7
(13) 14.2.5.5 (and subsection)
(14) 14.2.7.1
(15) 14.2.7.2 (and subsection)
(16) 14.2.8.3 through 14.2.8.3.5
(17) 14.2.8.3.9 (and subsection)
(18) 14.2.8.3.15.4
(19) 14.2.8.3.16.5
(20) 14.2.8.3.17 (and subsections)
(21) 14.2.8.4.1.3
(22) 14.2.8.6 (and subsections)
(23) 14.2.9.3 through 14.2.9.8 (and subsections)
(24) 14.2.10.2.5
(25) 14.3.1 (and subsections)
(26) 14.3.2.1.1 through 14.3.2.1.8
(27) 14.3.2.4 through 14.3.2.6 (and subsection)
(28) 14.3.3 through 14.3.6 (and subsections)

14.1.1.3 This chapter shall also apply to the altered, renovated, or modernized portion of an existing system or individual component.

14.1.1.4 Existing construction or equipment shall be permitted to be continued in use when such use does not constitute a distinct hazard to life.

14.1.2 Classification of Chambers.

14.1.2.1 General. Chambers shall be classified according to occupancy in order to establish appropriate minimum essentials in construction and operation.

14.1.2.2* Occupancy. Hyperbaric chambers shall be classified according to the following criteria:

(1) Class A — Human, multiple occupancy
(2) Class B — Human, single occupancy
(3) Class C — Animal, no human occupancy

14.1.3 Category of Care.

14.1.3.1 Category 1 Care. Where interruption or failure of medical gas supply is likely to cause major injury or death of patients, staff, or visitors, the level of care shall be considered Category 1 in the requirements for medical gas systems in hyperbaric facilities.

14.1.3.2 Category 2 Care. Where interruption or failure of medical gas supply is likely to cause minor injury of patients, staff, or visitors, the level of care shall be considered Category 2 in the requirements for medical gas systems in hyperbaric facilities.

14.1.3.3 Category 3 Care. Where interruption or failure of medical gas supply is not likely to cause injury to patients, staff, or visitors, the level of care shall be considered Category 3 in the requirements for medical gas systems in hyperbaric facilities.

14.1.3.4 Category 4 Care. (Reserved)

14.2 Construction and Equipment.

14.2.1 Housing for Hyperbaric Facilities.

14.2.1.1 For Class A chambers located inside a building, the chamber(s) and all ancillary service equipment shall be protected by 2-hour fire-resistant-rated construction.

14.2.1.1.1* Freestanding, dedicated buildings containing only a Class A chamber(s) and ancillary service equipment shall not be required to be protected by 2-hour fire-resistant-rated construction.

14.2.1.1.2 Class B and C chambers located inside a building shall not be required to be protected by 2-hour fire-resistant-rated construction.

14.2.1.1.3 Trailer or vehicle-mounted facilities shall be permitted without a 2-hour fire-resistant-rated perimeter.

14.2.1.1.4 When trailer or vehicle-mounted facilities are located contiguous to a health care facility or another structure, a 2-hour fire-resistant-rated barrier shall be placed between the facility and the contiguous structure.

14.2.1.1.5 Where building exterior walls form part of the facility boundary, that portion of the facility boundary shall not require 2-hour fire-resistant-rated construction.

14.2.1.1.6* If there are connecting doors through such common walls of contiguity, they shall be at least B-label, 1½-hour fire doors.

14.2.1.1.7 When used for hyperbaric procedures, the room or rooms housing the Class A or Class B chambers shall be for the exclusive use of the hyperbaric operation.

14.2.1.1.8 Service equipment (e.g., compressors) shall be permitted to be located in multi-use spaces meeting the requirements of 14.2.1.1.

14.2.1.1.9 The supporting foundation for any chamber shall be designed to support the chamber.

14.2.1.1.9.1 If on-site hydrostatic testing will be performed, the chamber supporting foundation shall be designed to support an additional water weight.

14.2.1.2* A hydraulically calculated automatic wet pipe sprinkler system meeting the requirements of NFPA 13, *Standard for the Installation of Sprinkler Systems*, or an automatic water mist fire protection system installed in accordance with NFPA 750, *Standard on Water Mist Fire Protection Systems*, shall be installed in the room housing a Class A, Class B, or Class C chamber and in any ancillary equipment rooms.

14.2.1.2.1 Class A, Class B, or Class C chambers not contiguous to a health care facility and located in a mobile vehicle-mounted facility shall not be required to be protected as specified in 14.2.1.2.

14.2.1.3 Hyperbaric Piping Requirements.

14.2.1.3.1* Except where otherwise required by this chapter, piping systems dedicated to the hyperbaric chamber shall meet the requirements of ANSI/ASME PVHO-1, *Safety Standard for Pressure Vessels for Human Occupancy*, for hyperbaric facility piping systems.

14.2.1.3.2 Shutoff valves accessible to facility personnel shall be provided for piping specified in 14.2.1.3.1 at the point of entry to the room housing the chamber(s).

14.2.1.4 Hyperbaric Medical Oxygen System Requirements.

14.2.1.4.1 Where medical oxygen systems are installed for hyperbaric use, the hyperbaric area(s) or facility shall be treated as a separate zone.

14.2.1.4.2 The requirements of Chapter 5 shall apply to the medical oxygen system for hyperbaric use, from the source of supply to the first in-line valve located downstream of the zone valve(s).

14.2.1.4.3 The requirements of ANSI/ASME PVHO-1, *Safety Standard for Pressure Vessels for Human Occupancy*, shall apply to the medical oxygen system for hyperbaric use, starting immediately downstream of the first in-line valve located after the zone valve(s).

14.2.1.4.4 General. Where an oxygen system is installed for hyperbaric treatments, it shall comply with the requirements for the appropriate level as determined in 14.2.1.4.4.2 through 14.2.1.4.4.7.

14.2.1.4.4.1 Hyperbaric oxygen systems for Category 1, Category 2, and Category 3 care connected directly to a hospital's oxygen system shall comply with Section 5.1, as applicable, except as noted in 14.2.1.4.4.2.

14.2.1.4.4.2 Central Supply Systems. Oxygen systems shall comply with 5.1.3.5, as applicable, except as follows:

(1) An emergency oxygen supply connection (EOSC) is not required for the hyperbaric oxygen system.
(2) An in-building emergency reserve (IBER) is not required for the hyperbaric oxygen system.

14.2.1.4.4.3 Hyperbaric stand-alone oxygen systems for Category 1 and Category 2 care shall comply with Section 5.1, as applicable, except as noted in 14.2.1.4.4.4.

14.2.1.4.4.4 Central Supply Systems. Oxygen systems shall comply with 5.1.3.5, as applicable, except as follows:

(1) An EOSC is not required for the hyperbaric oxygen system.
(2) An IBER is not required for the hyperbaric oxygen system.

14.2.1.4.4.5 Warning Systems.

(A) Oxygen systems shall comply with 5.1.9, as applicable, except that warning systems shall be permitted to be a single master/area alarm panel.

(B) The alarm panel shall be located in the room housing the chamber(s) to allow for easy audio and visual monitoring by the chamber operator

14.2.1.4.4.6 Hyperbaric stand-alone oxygen systems for Category 3 care shall comply with Section 5.2, as applicable, except as noted in 14.2.1.4.4.7.

14.2.1.4.4.7 Central Supply Systems. Oxygen systems shall comply with 5.1.3.5, as applicable, except as follows:

(1) If the operating oxygen supply consists of high pressure cylinders designed with a primary and secondary source, no reserve supply is required.
(2) If the operating oxygen supply consists of liquid containers designed with a primary and secondary source, a reserve with a minimum supply of 15 minutes is required.
(3) If the operating oxygen supply consists of a bulk primary, a reserve with a minimum supply of 15 minutes is required.
(4) An EOSC is not required for the hyperbaric oxygen system.
(5) An IBER is not required for the hyperbaric oxygen system.

14.2.1.5 Storage and Handling of Medical Gases. Storage and handling of medical gases shall meet the applicable requirements of Chapter 5.

14.2.1.6 Hyperbaric Medical Air System Requirements.

14.2.1.6.1 Where medical air systems are installed for hyperbaric use, the hyperbaric area(s) or facility shall be treated as a separate zone.

14.2.1.6.2 Chapter 5 requirements shall apply to the medical air system for hyperbaric use, from the source of supply to the first in-line valve located downstream of the zone valve(s).

14.2.1.6.3 ANSI/ASME PVHO-1, *Safety Standard for Pressure Vessels for Human Occupancy*, requirements shall apply to the medical air system for hyperbaric use, starting immediately downstream of the first in-line valve located after the zone valve(s).

14.2.1.6.4 Where a medical air system is installed for hyperbaric treatments, it shall comply with the requirements for the appropriate level as determined in 14.2.1.6.4.1 through 14.2.1.6.4.7.

14.2.1.6.4.1 Hyperbaric medical air systems for Category 1, Category 2, and Category 3 care connected directly to a hospital's medical air system shall comply with Section 5.2, as applicable.

14.2.1.6.4.2 Reserved.

14.2.1.6.4.3 Hyperbaric stand-alone medical air systems for Category 1 and Category 2 care shall comply with Section 5.2, as applicable.

14.2.1.6.4.4 Reserved.

14.2.1.6.4.5 Medical air systems for Category 1 and Category 2 care shall comply with Section 5.2, as applicable, except that warning systems shall be permitted to be a single master/area alarm panel.

14.2.1.6.4.6 Hyperbaric stand-alone medical systems for Category 3 care shall comply with Section 5.2, as applicable, except as noted in 14.2.1.6.4.7.

14.2.1.6.4.7 Medical air systems shall comply with Section 5.2 as applicable, except as follows:

(1) Area and master alarms are not required for Category 3 care.
(2) A gas cylinder header per Section 5.2 with sufficient cylinder connections to provide for at least an average day's supply with the appropriate number of connections being determined after consideration of delivery schedule, proximity of the facility to alternate supplies, and the facility's emergency plan is permitted.

14.2.2 Fabrication of the Hyperbaric Chamber.

14.2.2.1* Chambers for human occupancy and their supporting systems shall be designed and fabricated to meet ANSI/ASME PVHO-1, *Safety Standard for Pressure Vessels for Human Occupancy*, by personnel qualified to fabricate vessels under such codes.

14.2.2.1.1 Piping systems for hyperbaric facilities shall be required to meet only the requirements of this chapter and section "Piping" of ANSI/ASME PVHO-1, *Safety Standard for Pressure Vessels for Human Occupancy.*

14.2.2.1.2 Piping that is installed in concealed locations in the building housing the hyperbaric facility, such as inside building walls or above false ceilings, shall use only those joining procedures permitted by Chapter 5.

14.2.2.2 The chamber shall be stamped in accordance with ANSI/ASME PVHO-1, *Safety Standard for Pressure Vessels for Human Occupancy.*

14.2.2.3 As a minimum, animal chambers shall be designed, fabricated, and stamped to meet ASME *Boiler and Pressure Vessel Code* Section VIII, Division 1 code requirements.

14.2.2.4 The floor of a Class A chamber shall be designed to support equipment and personnel necessary for the operation of the chamber according to its expected purpose.

14.2.2.4.1 The floor of Class A chambers shall be noncombustible.

14.2.2.4.2 If a bilge is installed, access to the bilge shall be provided for cleaning purposes.

14.2.2.4.3 If the interior floor of a Class A chamber consists of removable floor (deck) plates, the plates shall be mechanically secured and electrically bonded to the chamber to ensure a positive electrical ground and to prevent movement of the plate, which could cause injury to personnel.

14.2.2.5* The interior surface of Class A chambers shall be unfinished or treated with a paint/coating in accordance with 14.2.2.5.1.

14.2.2.5.1* Interior paint/coating shall meet the performance criteria of NFPA *101*, Class A interior finish, when tested in accordance with ASTM E 84, *Standard Test Method for Surface Burning Characteristics of Building Materials*, or ANSI/UL 723, *Standard for Test for Surface Burning Characteristics of Building Materials.*

14.2.2.5.2 One additional application of paint shall be permitted, provided total paint thickness does not exceed 1/28 in. (0.9 mm).

14.2.2.5.3 If the interior of a Class A chamber is treated (painted) with a finish described in 14.2.2.5, the cure procedure and minimum duration for each layer of paint/coating to off-gas shall be in accordance with the manufacturer's application instructions.

14.2.2.5.4* If sound-deadening materials are employed within a hyperbaric chamber, they shall be limited-combustible materials.

14.2.2.6* Viewing ports, access ports for piping and wiring or monitoring, and related leads shall be installed during initial fabrication of the chamber.

14.2.2.6.1 Access ports in Class A chambers, access ports for monitoring, and other electrical circuits shall be housed in enclosures that are weatherproof, both inside and outside the chamber, for protection in the event of sprinkler activation.

14.2.2.6.2 Viewports and penetrator plates shall be designed and fabricated according to ANSI/ASME PVHO-1, *Safety Standard for Pressure Vessels for Human Occupancy.*

14.2.3 Illumination.

14.2.3.1 Unless designed for chamber use, sources of illumination shall be mounted outside the pressure chamber and arranged to shine through chamber ports or through chamber penetrators designed for fiber-optic or similar lighting.

14.2.3.1.1 Lighting fixtures used in conjunction with viewports shall be designed so that temperature ratings for the viewport material given in ANSI/ASME PVHO-1 are not exceeded.

14.2.3.1.2 Gasket material shall be of a type that allows the movement of thermal expansion and shall be selected for the temperatures, pressures, and composition of gases involved.

14.2.3.1.2.1 Gaskets or O-rings shall be confined to grooves or enclosures, which will prevent their being blown out or squeezed from the enclosures or compression flanges.

14.2.3.2 Lighting permanently installed inside the chamber and portable lighting for temporary use inside the chamber shall meet the requirements of 14.2.8.3.15.

14.2.3.3 Emergency lighting for the interior of the chamber shall be provided.

14.2.4 Chamber Ventilation.

14.2.4.1 Ventilation of Class A Chambers.

14.2.4.1.1 The minimum ventilation rate for a Class A chamber shall be 0.085 m^3/min (3 ft^3/min) of air per chamber occupant who is not using a breathing-mask overboard dump system that exhausts exhaled gases.

14.2.4.1.1.1 The minimum threshold rate shall be 0.085 m^3/min (3 ft^3/min).

14.2.4.1.1.2 Provision shall be made for ventilation during nonpressurization of Class A chambers as well as during pressurization.

14.2.4.1.2* Ventilation shall not be required when saturation operations are conducted in the chamber, provided that carbon dioxide removal and odor control are accomplished and that the monitoring requirements of 14.2.9.4.1 and 14.2.9.5 are met.

14.2.4.1.3 Individual breathing apparatus shall be available inside a Class A chamber for each occupant for use in the event that the chamber atmosphere is fouled by combustion or otherwise.

14.2.4.1.3.1 The breathing mixture supplied to breathing apparatus shall be independent of chamber atmosphere.

14.2.4.1.3.2 The breathing gas supply shall be designed for simultaneous use of all breathing apparatus.

14.2.4.1.3.3 Breathing apparatus shall function at all pressures that can be encountered in the chamber.

14.2.4.1.3.4 In the event of a fire within a chamber, provision shall be made to simultaneously switch all breathing apparatus to an air supply that is independent of the chamber atmosphere.

14.2.4.2 Sources of Air for Chamber Atmospheres.

14.2.4.2.1* Sources of air for chamber atmospheres shall be such that toxic or flammable gases are not introduced.

14.2.4.2.2 Compressor intakes shall be located away from air contaminated by exhaust from activities of vehicles, internal combustion engines, stationary engines, or building exhaust outlets.

14.2.4.2.3 Air supply for chamber atmosphere shall be monitored as required in 14.2.9.6.

14.2.4.2.4 The use of conventional oil-lubricated compressors shall be permitted, provided that they are fitted with air treatment packages designed to meet the requirements of 14.2.9.6.

14.2.4.2.4.1 The air treatment packages shall include automatic safeguards.

14.2.4.2.5 Air compressor installations shall consist of two or more individual compressors with capacities such that required system flow rates can be maintained on a continuous basis with any single compressor out of operation, unless 14.2.8.2.5 is satisfied.

14.2.4.2.5.1 Each compressor shall be supplied from separate electrical branch circuits.

14.2.4.2.6 Air compressor installations that supply medical air to piped gas systems as well as to hyperbaric facilities shall meet the requirements of 5.1.3.6.3 and this chapter.

14.2.4.2.7 Air compressor installations that are used exclusively for hyperbaric facilities shall meet the requirements of this chapter only.

14.2.4.3 Temperature and Humidity Control.

14.2.4.3.1 Warming or cooling of the atmosphere within a Class A chamber shall be permitted by circulating the ambient air within the chamber over or past coils through which a constant flow of warm or cool water or water/glycol mixture is circulated.

14.2.4.3.2* Class A chambers that are not used in the capacity of an operating room shall maintain a temperature that is comfortable for the occupants [usually 22°C ±2°C (75°F ±5°F)].

14.2.4.3.3 Whenever the Class A chamber is used as an operating room, it shall be ventilated, and the atmosphere shall be conditioned according to the minimum requirements for temperature in hospital operating rooms.

14.2.4.3.3.1 If inhalation anesthetic agents are being utilized (e.g., halothane, isoflurane, sevoflurane, desflurane), a closed anesthetic system with exhaled gas scavenging and overboard dumping shall be employed.

14.2.4.3.3.2 Flammable inhalation anesthetics (e.g., cyclopropane, ethyl ether, ethylene, and ethyl chloride) shall not be employed.

14.2.4.3.4 Dehumidification shall be permitted through the use of cold coils.

14.2.4.3.5 Humidification by the use of an air-powered water nebulizer shall be permitted.

14.2.4.3.6 Noncombustible packing and nonflammable lubricant shall be employed on the fan shaft.

14.2.4.4 Ventilation of Class B Chambers.

14.2.4.4.1* The minimum ventilation rate for a Class B chamber shall be 0.0283 m^3/min (1 ft^3/min).

14.2.4.4.2 Class B chambers not designed for 100 percent oxygen environment shall comply with the monitoring requirements of 14.2.9.4.

14.2.4.4.3 For Class B chambers equipped with a breathing apparatus, the breathing apparatus shall function at all pressures that can be encountered in the chamber.

14.2.4.5 Emergency Depressurization and Facility Evacuation Capability.

14.2.4.5.1 Class A chambers shall be capable of depressurizing from 3 ATA (304.0 kPa) to ambient pressure in not more than 6 minutes.

14.2.4.5.2 Class B chambers shall be capable of depressurizing from 3 ATA (304.0 kPa) to ambient pressure in not more than 2 minutes.

14.2.4.5.3* A means for respiratory and eye protection from combustion products allowing unrestricted mobility shall be available outside a Class A or Class B chamber for use by personnel in the event the air in the vicinity of the chamber is fouled by smoke or other combustion products.

14.2.4.5.4 The time required to evacuate all persons from a hyperbaric area with a full complement of chamber occupants all at treatment pressure shall be measured annually during the fire training drill required by 14.3.1.4.5.

14.2.4.5.4.1 The occupants for this training drill shall be permitted to be simulated.

14.2.5 Fire Protection in Class A Chambers.

14.2.5.1 General.

14.2.5.1.1 A fire suppression system consisting of independently supplied and operating handline- and deluge-type water spray systems shall be installed in all Class A chambers.

14.2.5.1.2 Design of the fire suppression system shall be such that failure of components in either the handline or deluge system will not render the other system inoperative.

14.2.5.1.3 System design shall be such that activation of either the handline or the deluge system shall automatically cause the following:

(1) Visual and aural indication of activation shall occur at the chamber operator's console.
(2) All ungrounded electrical leads for power and lighting circuits contained inside the chamber shall be disconnected.
(3) Emergency lighting (see 14.2.3.3) and communication, where used, shall be activated.

14.2.5.1.3.1 Intrinsically safe circuits, including sound-powered communications, shall be permitted to remain connected when either the handline or the deluge system is activated.

14.2.5.1.4* A fire alarm signaling device shall be provided at the chamber operator's control console for signaling the emergency fire/rescue network of the institution containing the hyperbaric facility.

14.2.5.1.4.1 Trailer or vehicle-mounted facilities not contiguous to a health care facility shall conform to one of the following:

(1) They shall comply with 14.2.5.1.4.
(2) They shall have a means for immediately contacting the local fire department.

14.2.5.1.5* Fire blankets and portable carbon dioxide extinguishers shall not be installed in or carried into the chamber.

14.2.5.1.6 Booster pumps, control circuitry, and other electrical equipment involved in fire suppression system operation shall be powered from a critical branch of the essential electrical system as specified in 14.2.8.2.2.2.

14.2.5.1.7 Signs prohibiting the introduction of flammable liquids, gases, and other articles not permitted by this chapter into the chamber shall be posted at the chamber entrance(s).

14.2.5.1.8 The fire suppression system shall be permitted to be supplied from the local potable water service.

14.2.5.2 Deluge System. A fixed water deluge extinguishing system shall be installed in all chamber compartments that are designed for manned operations.

14.2.5.2.1 In chambers that consist of more than one chamber compartment (lock), the design of the deluge system shall meet the requirements of 14.2.5.2 when the chamber compartments are at different depths (pressures).

14.2.5.2.2 The deluge system in different compartments (locks) shall operate independently or simultaneously.

14.2.5.2.3 Fixed deluge systems shall not be required in chamber compartments that are used strictly as personnel transfer compartments (locks) and for no other purposes.

14.2.5.2.4* Manual activation and deactivation deluge controls shall be located at the operator's console and in each chamber compartment (lock) containing a deluge system.

14.2.5.2.4.1 Controls shall be designed to prevent unintended activation.

14.2.5.2.5 Water shall be delivered from the fixed discharge nozzles as specified in 14.2.5.2.7 within 3 seconds of activation of any affiliated deluge control.

14.2.5.2.6* Average spray density at floor level shall be not less than 81.5 L/min/m^2 (2 gpm/ft^2), with no floor area larger than 1 m^2 (10.76 ft^2) receiving less than 40.75 L/min/m^2 (1 gpm/ft^2).

14.2.5.2.7 Water shall be available in the deluge system to maintain the flow specified in 14.2.5.2.6 simultaneously in each chamber compartment (lock) containing the deluge system for 1 minute.

14.2.5.2.7.1 The limit on maximum extinguishment duration shall be governed by the chamber capacity (bilge capacity also, if so equipped) or its drainage system, or both.

14.2.5.2.8 The deluge system shall have stored pressure to operate for at least 15 seconds without electrical branch power.

14.2.5.3 Handline System. A handline extinguishing system shall be installed in all chamber compartments (locks).

14.2.5.3.1 At least two handlines shall be strategically located in treatment compartments (locks).

14.2.5.3.2 At least one handline shall be located in each personnel transfer compartment (lock).

14.2.5.3.3 If any chamber compartment (lock) is equipped with a bilge access panel, at least one handline shall reach the bilge area.

14.2.5.3.4 Handlines shall have a 12.7 mm (0.5 in.) minimum internal diameter and shall have a rated operating pressure greater than the highest supply pressure of the supply system.

14.2.5.3.5 Each handline shall be activated by a manual, quick-opening, quarter-turn valve located within the compartment (lock).

14.2.5.3.5.1 A hand-operated spring-return to close valves at the discharge end of handlines shall be permitted.

14.2.5.3.6 Handlines shall be equipped with override valves that are accessible to personnel outside the chamber.

14.2.5.3.7 The water supply for the handline system shall be designed to ensure a 345 kPa (50 psi) minimum water pressure above the maximum chamber pressure.

14.2.5.3.7.1 The system shall be capable of supplying a minimum of 18.9 L/min (5 gpm) simultaneously to each of any two of the handlines at the maximum chamber pressure for a period of not less than 4 minutes.

14.2.5.4 Automatic Detection System. Automatic fire detection systems shall not be required.

14.2.5.4.1 Surveillance fire detectors responsive to the radiation from flame shall be employed.

14.2.5.4.1.1 The type and arrangement of detectors shall be such as to respond within 1 second of flame origination.

14.2.5.4.2* The number of detectors employed and their location shall be selected to cover the chamber interior.

14.2.5.4.3 The system shall be powered from the critical branch of the essential electrical system or shall have automatic battery backup.

14.2.5.4.4 If used to automatically activate the deluge system, the requirements for manual activation/deactivation in 14.2.5.2.4 and deluge system response time in 14.2.5.2.5 shall still apply.

14.2.5.4.5 The system shall include self-monitoring functions for fault detection and fault alarms and indications.

14.2.5.4.6 Automatic fire detection equipment, when used, shall meet the applicable requirements in 14.2.8.3.

14.2.5.5* Testing. The deluge and handline systems shall be functionally tested at least semiannually per 14.2.5.2.7 for deluge systems and 14.2.5.3.7 for handline systems.

14.2.5.5.1 Following the test, all valves shall be placed in their baseline position.

14.2.5.5.2 If a bypass system is used, it shall not remain in the test mode after completion of the test.

14.2.5.5.3 During initial construction, or whenever changes are made to the installed deluge system that will affect the spray pattern, testing of spray coverage to demonstrate conformance to the requirements of 14.2.5.2.6 shall be performed at surface pressure and at maximum operating pressure.

14.2.5.5.3.1 The requirements of 14.2.5.2.6 shall be satisfied under both surface pressure and maximum operating pressure.

14.2.5.5.4 A detailed record of the test results shall be maintained and a copy sent to the hyperbaric facility safety director.

14.2.5.5.5 Inspection, testing, and maintenance of hyperbaric fire suppression systems shall be performed by a qualified person.

14.2.6 Pneumatic Controls for Class A Chambers. Class A chambers that utilize pneumatically operated controls that are related to fire suppression system operation, breathing gases, or rapid exhaust valves shall be equipped with a means to operate such controls or intended function in the event that the pneumatic supply fails.

14.2.7 Fire Protection in Class B and Class C Chambers. Class B and Class C chambers shall not be required to comply with 14.2.5.

14.2.7.1 Signs prohibiting the introduction of flammable liquids, gases, and other articles not permitted by this chapter into the chamber shall be posted at the chamber entrance(s).

14.2.7.2 A fire alarm signaling device shall be provided within the room housing the chamber(s) for signaling the emergency fire/rescue network of the institution containing the hyperbaric facility.

14.2.7.2.1 Trailer or vehicle-mounted facilities not contiguous to a health care facility shall conform to one of the following:

(1) They shall comply with 14.2.7.2.
(2) They shall have a means for immediately contacting the local fire department.

14.2.8 Electrical Systems.

14.2.8.1 General.

14.2.8.1.1 The requirements of *NFPA 70, National Electrical Code*, or local electrical codes shall apply to electrical wiring and equipment in hyperbaric facilities within the scope of this chapter, except as such rules are modified in 14.2.8.

14.2.8.1.2 All hyperbaric chamber service equipment, switchboards, panels, or control consoles shall be located outside of, and in the vicinity of, the chamber.

14.2.8.1.3 Console or module spaces containing both oxygen piping and electrical equipment shall be either one of the following:

(1) Mechanically or naturally ventilated
(2) Continuously monitored for excessive oxygen concentrations whenever the electrical equipment is energized

14.2.8.1.4 For the fixed electrical installation, none of the following shall be permitted inside the chamber:

(1) Circuit breakers
(2) Line fuses
(3) Motor controllers
(4) Relays
(5) Transformers
(6) Ballasts
(7) Lighting panels
(8) Power panels

14.2.8.1.4.1* If motors are to be located in the chamber, they shall meet the requirements of 14.2.8.3.14.

14.2.8.1.5 All electrical equipment connected to, or used in conjunction with, hyperbaric patients shall comply with the requirements of Chapter 10 and with the applicable subparagraphs of 14.2.8.3.

14.2.8.1.6 In the event of activation of the room sprinkler system, electrical equipment shall be protected from sprinkler water but shall not be required to remain functional if manual means to control and decompress the chamber are provided.

14.2.8.2 Electrical Service.

14.2.8.2.1 All hyperbaric facilities shall contain an electrical service that is supplied from two independent sources of electric power.

14.2.8.2.1.1 All hyperbaric facilities for human occupancies shall contain an electrical service that is supplied from two independent sources of electric power.

14.2.8.2.1.2 For hyperbaric facilities using a prime-mover-driven generator set, it shall be designated as the life safety and critical branches and shall meet the requirements of Chapter 6 for hyperbaric systems based in health care facilities.

14.2.8.2.1.3 Article 700 of *NFPA 70, National Electrical Code*, shall apply to hyperbaric systems located in facilities other than health care facilities.

14.2.8.2.2 Electrical equipment associated with life-support functions of hyperbaric facilities shall be connected to the critical branch of the life safety and critical branches, which requires that such equipment shall have electrical power restored within 10 seconds of interruption of normal power.

14.2.8.2.2.1 The equipment specified in 14.2.8.2.2 shall include, but is not limited to, the following:

(1) Electrical power outlets located within the chamber
(2) Chamber emergency lighting, whether internally or externally mounted
(3) Chamber intercommunications
(4) Alarm systems, including fire detectors
(5) Chamber fire suppression system equipment and controls
(6) Other electrical controls used for chamber pressurization and ventilation control
(7) A sufficient number of chamber room lights (either overhead or local) to ensure continued safe operation of the facility during a normal power outage

14.2.8.2.2.2 Booster pumps in the chamber fire suppression system shall be on separate branch circuits serving no other loads.

14.2.8.2.3 Electric motor–driven compressors and auxiliary electrical equipment normally located outside the chamber and used for chamber atmospheric control shall be connected to the equipment system *(see Chapter 6)* or the life safety and critical branches *(see NFPA 70, National Electrical Code, Article 700)*, as applicable.

14.2.8.2.4 Electric motor–driven compressors and auxiliary electrical equipment shall be arranged for delayed-automatic or manual connection to the alternate power source so as to prevent excessive current draw on the system during restarting.

14.2.8.2.5 When reserve air tanks or a nonelectric compressor(s) is provided to maintain ventilation airflow within the chamber and supply air for chamber pressurization, the compressor(s) and auxiliary equipment shall not be required to have an alternate source of power.

14.2.8.2.6 Electrical control and alarm system design shall be such that hazardous conditions (e.g., loss of chamber pressure control, deluge activation, spurious alarms) do not occur during power interruption or during power restoration.

14.2.8.3* Wiring and Equipment Inside Class A Chambers. The general rules of 14.2.8.3.1 through 14.2.8.3.17.6 shall be satisfied in the use of electrical devices and equipment. These requirements are intended to protect against the elevated fire risks known to exist in a pressurized air environment and shall not be construed as classifying the chamber interior as a Class I (as defined in *NFPA 70, National Electrical Code*, Article 500) hazardous location.

14.2.8.3.1 Equipment or equipment components installed in, or used in, the chamber shall not present an explosion or implosion hazard under the conditions of hyperbaric use.

14.2.8.3.2 All equipment shall be rated, or tested and documented, for intended hyperbaric conditions prior to use.

14.2.8.3.3 Only the electrical equipment necessary for the safe operation of the chamber and for required patient care shall be permitted in the chamber.

14.2.8.3.4 Only portable equipment necessary for the logistical and operational support shall be permitted in the chamber during manned pressurization.

14.2.8.3.5 Where conformance with Class I, Division 1 requirements is specified in 14.2.8.3.7, conformance with Class I, Division 2 requirements shall be permitted to be substituted.

14.2.8.3.6 Wires and Cables. Wires and cables used inside the chamber shall be resistant to the spread of fire by complying with 14.2.8.3.6.1 or shall be contained within equipment described in 14.2.8.3.6.2.

14.2.8.3.6.1 Wires and cables shall comply with the spread of fire requirements of "UL Flame Exposure, Vertical Tray Flame Test" in UL 1685, *Standard for Vertical-Tray Fire-Propagation and Smoke-Release Test for Electrical and Optical-Fiber Cables*, or shall exhibit damage (char length) not to exceed 1.5 m (4 ft 11 in.) when performing the CSA "Vertical Flame Test — Cables in Cable Trays," as described in CSA C22.2 No. 0.3-M, *Test Methods for Electrical Wires and Cables*.

14.2.8.3.6.2 Wires and cables that form an integral part of electrical equipment approved or listed specifically for use inside hyperbaric chambers, including patient leads, shall not be required to comply with the requirements of 14.2.8.3.6.1.

14.2.8.3.7 Wiring Methods.

14.2.8.3.7.1 Fixed wiring shall be installed in threaded RMC or IMC conduit utilizing the following waterproof components:

(1) Threaded metal joints
(2) Fittings
(3) Boxes
(4) Enclosures

14.2.8.3.7.2 A continuous ground shall be maintained between all conductive surfaces enclosing electrical circuits and the chamber hull using approved grounding means.

14.2.8.3.7.3 All threaded conduit shall be threaded with an NPT standard conduit cutting die that provides a 19 mm taper per 0.3 m (0.75 in. taper per 1 ft).

14.2.8.3.7.4 All threaded conduit shall be made wrench-tight to prevent sparking when fault current flows through the conduit system.

14.2.8.3.7.5 Wiring classified as intrinsically safe for any group location and installed in accordance with Article 504 of *NFPA 70, National Electrical Code*, shall be permitted.

14.2.8.3.7.6 Threaded, liquidtight flexible metal conduit installed in accordance with Article 350 of *NFPA 70, National Electrical Code*, shall be permitted when protected from damage by physical barriers such as equipment panels.

14.2.8.3.8 Drainage. Means of draining fixed conduit and fixed equipment enclosures shall be provided.

14.2.8.3.9 Flexible Electrical Cords. Flexible cords used to connect portable utilization equipment to the fixed electrical supply circuit shall meet all of the following requirements:

(1) They shall be of a type approved for extra-hard utilization in accordance with Table 400.4 of *NFPA 70, National Electrical Code*.
(2) They shall include a ground conductor.
(3) They shall meet the requirements of 501.140 of *NFPA 70, National Electrical Code*.

14.2.8.3.9.1 The normal cord supplied with the portable utilization equipment shall be permitted when the portable device is rated at less than 2 A and the cord is positioned out of traffic and protected from physical abuse.

14.2.8.3.10* Receptacles Installed Inside the Chamber.

14.2.8.3.10.1 Receptacles shall be waterproof.

14.2.8.3.10.2 Receptacles shall be of the type providing for connection to the grounding conductor of the flexible cord.

14.2.8.3.10.3 Receptacles shall be supplied from isolated power circuits meeting the requirements of 14.2.8.4.2.

14.2.8.3.10.4 The design of the receptacle shall be such that sparks cannot be discharged into the chamber environment when the plug is inserted or withdrawn under electrical load.

14.2.8.3.10.5 One of the following shall be satisfied to protect against inadvertent withdrawal of the plug under electrical load:

(1) The receptacle–plug combination shall be of a locking type.
(2) The receptacle shall carry a label warning against unplugging under load, and the power cord shall not present a trip hazard for personnel moving in the chamber.

14.2.8.3.11 Switches. Switches in the fixed wiring installation shall be waterproof.

14.2.8.3.11.1* Switch make and break contacts shall be housed in the electrical enclosure so that no sparks from arcing contacts can reach the chamber environment.

14.2.8.3.12* Temperature. No electrical equipment installed or used in the chamber shall have an operating surface temperature in excess of 85°C (185°F).

14.2.8.3.13 Exposed Live Electrical Parts. No exposed live electrical parts shall be permitted, except as specified in 14.2.8.3.13.1 and 14.2.8.3.13.2.

14.2.8.3.13.1 Exposed live electrical parts that are intrinsically safe shall be permitted.

14.2.8.3.13.2 Exposed live electrical parts that constitute patient monitoring leads, which are part of electromedical equipment, shall be permitted, provided that they meet the requirements of 14.2.8.3.17.

14.2.8.3.14 Motors. Motors shall meet one of the following requirements:

(1) They shall comply with 501.125(A)(1) of *NFPA 70, National Electrical Code*, for the chamber pressure and oxygen concentration.
(2) They shall be of the totally enclosed types meeting 501.125(A)(2) or 501.125(A)(3) of *NFPA 70, National Electrical Code*.

14.2.8.3.15* Lighting.

14.2.8.3.15.1 Lighting installed or used inside the chamber shall be rated for a pressure of 1½ times the chamber operating pressure.

14.2.8.3.15.2 Permanently installed fixtures shall meet the following requirements:

(1) They shall be rated and approved for Class I (Division 1 or 2) classified areas.
(2) They shall have lens guards installed.
(3) They shall be located away from areas where they would experience physical damage from the normal movement of people and equipment.

14.2.8.3.15.3 Ballasts and other energy storage components that are part of the lighting circuit shall be installed outside the chamber in accordance with 14.2.8.1.4.

14.2.8.3.15.4 Portable fixtures intended for spot illumination shall be shatterproof or protected from physical damage.

14.2.8.3.16 Low-Voltage, Low-Power Equipment. The requirements of 14.2.8.3.16.1 through 14.2.8.3.16.5 shall apply to sensors and signaling, alarm, communications, and remote-control equipment installed or used in the chamber for operation of the chamber.

14.2.8.3.16.1* Equipment shall be isolated from main power by one of the following means:

(1) Design of the power supply circuit
(2) Opto-isolation
(3) Other electronic isolation means

14.2.8.3.16.2 Circuits such as headset cables, sensor leads, and so forth, not enclosed as required in 14.2.8.3.7, shall meet one of the following requirements:

(1) They shall be part of approved intrinsically safe equipment.
(2) They shall be limited by circuit design to not more than 28 V and 0.5 A under normal or circuit-fault conditions.

14.2.8.3.16.3 Chamber speakers shall be of a design in which the electrical circuitry and wiring is completely enclosed.

14.2.8.3.16.4 The electrical rating of chamber speakers shall not exceed 28 V rms and 25 W.

14.2.8.3.16.5 Battery-operated, portable intercom headset units shall meet the requirements of 14.2.8.3.17.5 for battery-operated devices.

14.2.8.3.17* Portable Patient Care–Related Electrical Appliances.

14.2.8.3.17.1 The appliance shall be designed and constructed in accordance with Chapter 10.

14.2.8.3.17.2 The electrical and mechanical integrity of the appliance shall be verified and documented through an ongoing maintenance program as required in Chapter 10.

14.2.8.3.17.3 The appliance shall conform to the requirements of 14.2.8.3.1 and 14.2.8.3.12.

14.2.8.3.17.4 Appliances that utilize oxygen shall not allow oxygen accumulation in the electrical portions of the equipment under normal and abnormal conditions.

14.2.8.3.17.5 Battery-Operated Devices. Battery-operated devices shall meet the following requirements:

(1) Batteries shall be fully enclosed and secured within the equipment enclosure.
(2) Batteries shall not be damaged by the maximum chamber pressure to which they are exposed.
(3) Batteries shall be of a sealed type that does not off-gas during normal use.
(4) Batteries or battery-operated equipment shall not undergo charging while located in the chamber.
(5) Batteries shall not be changed on in-chamber equipment while the chamber is in use.
(6) The equipment electrical rating shall not exceed 12 V and 48 W.
(7) Lithium and lithium ion batteries shall be prohibited in the chamber during chamber operations, unless the product has been accepted or listed for use in hyperbaric conditions by the manufacturer or a nationally recognized testing agency.

14.2.8.3.17.6 Cord-Connected Devices. Cord-connected devices shall meet the following requirements:

(1) All portable, cord-connected equipment shall have an on/off power switch.
(2) The equipment electrical rating shall not exceed 120 V and 2 A, unless the electrical portions of the equipment are inert-gas purged.
(3) The plug of cord-connected devices shall not be used to interrupt power to the device.

14.2.8.4 Grounding and Ground-Fault Protection.

14.2.8.4.1 All chamber hulls shall be grounded to an electrical ground or grounding system that meets the requirements of Article 250, Grounding and Bonding, Section III, Grounding Electrode System and Grounding Electrode Conductor, of *NFPA 70, National Electrical Code*.

14.2.8.4.1.1 Grounding conductors shall be secured as required by Article 250, Grounding and Bonding, Section III, Grounding Electrode System and Grounding Electrode Conductor, of *NFPA 70, National Electrical Code*.

14.2.8.4.1.2 The material, size, and installation of the grounding conductor shall meet the requirements of Article 250, Grounding and Bonding, Section VI, Equipment Grounding and Equipment Grounding Conductors, of *NFPA 70, National Electrical Code*, for equipment grounding conductors.

14.2.8.4.1.3 The resistance between the grounded chamber hull and the electrical ground shall not exceed 1 ohm.

14.2.8.4.2 In health care facilities, electrical power circuits located within the chamber shall be supplied from an ungrounded electrical system equipped with a line isolation monitor with signal lamps and audible alarms.

14.2.8.4.2.1 The circuits specified in 14.2.8.4.2 shall meet the requirements of 517.160(A) and 517.160(B) of *NFPA 70, National Electrical Code*.

14.2.8.4.2.2 Branch circuits shall not exceed 125 V or 15 A.

14.2.8.4.3 Wiring located both inside and outside the chamber, that serves line level circuits and equipment located inside the chamber, shall meet the grounding and bonding requirements of 501.30 of *NFPA 70, National Electrical Code.*

14.2.8.5 Wiring Outside the Chamber. Those electrical components that must remain functional for the safe termination of a dive following activation of the room sprinkler system shall be enclosed in waterproof housing.

14.2.8.5.1 All associated conduits shall meet the following requirements:

(1) They shall be waterproof.
(2) They shall meet the requirements of *NFPA 70, National Electrical Code.*
(3) They shall be equipped with approved drains.

14.2.8.5.2* All other electrical devices outside the chamber shall meet the requirements of *NFPA 70.*

14.2.8.6 Additional Wiring and Equipment Requirements Inside Class B Chambers. The requirements in 14.2.8.6 shall apply to Class B chambers whether they are pressurized with oxygen or with air.

14.2.8.6.1 Electrical equipment inside Class B chambers shall be restricted to communications functions and patient physiological monitoring leads.

14.2.8.6.1.1* Each circuit shall be designed to limit the electrical energy to wire leads into the chamber under normal or fault conditions to not more than 28 V and 4.0 W. This requirement shall not exclude more stringent requirements imposed by other codes governing electromedical apparatus.

14.2.8.6.1.2 Communications wires shall be protected from physical damage and from coming into contact with flammable materials in the chamber by barriers or conduit.

14.2.8.6.1.3 Patient monitoring leads shall be part of approved electromedical apparatus meeting the requirements in 14.2.8.3.17.

14.2.8.6.2 Lighting inside the chamber shall be supplied from external sources.

14.2.8.6.3 No materials shall be permitted in a Class B chamber whose temperature exceeds 50° C (122° F), nor shall any electrical circuit inside a Class B chamber operate at a temperature exceeding 50°C (122°F).

14.2.9 Communications and Monitoring.

14.2.9.1 General.

14.2.9.1.1 Detectors, sensors, transducers, and communications equipment located inside the chamber shall meet the requirements of 14.2.8.3.16.

14.2.9.1.2 Wiring methods in the chamber shall meet the applicable requirements in 14.2.8.3.

14.2.9.1.3 The following equipment shall be installed outside the chamber or shall meet the requirements of 14.2.8.3.16:

(1) Control equipment
(2) Power amplifiers
(3) Output transformers
(4) Monitors associated with communications and monitoring equipment

14.2.9.2* Intercommunications.

14.2.9.2.1* An intercommunications system shall connect all personnel compartments (locks) and the chamber operator's control console.

14.2.9.2.2 Oxygen mask microphones shall be intrinsically safe at the maximum proposed pressure and 95 ± 5 percent oxygen.

14.2.9.3 Combustible Gas Detection.

14.2.9.3.1 The chamber atmosphere shall be continuously monitored for combustible gas concentrations whenever any volatile agents are used in the chamber. *(See 14.2.4.3.3.1.)*

14.2.9.3.1.1 The monitor shall be set to provide audible and visual alarms at 10 percent lower explosive limit (LEL) for the particular gas used.

14.2.9.4 Oxygen Monitoring.

14.2.9.4.1 Oxygen levels shall be continuously monitored in any chamber in which nitrogen or other diluent gas is added to the chamber to reduce the volumetric concentration of oxygen in the atmosphere.

14.2.9.4.1.1 Oxygen monitors shall be equipped with audible and visual alarms.

14.2.9.4.2 Oxygen levels shall be continuously monitored in Class A chambers when breathing mixtures containing in excess of 21 percent oxygen by volume are being breathed by patients or attendants or any flammable agents are present in the chamber, or when either of these conditions exists.

14.2.9.4.2.1 Audible and visual alarms shall indicate volumetric oxygen concentrations in excess of 23.5 percent.

14.2.9.5 Carbon Dioxide Monitoring. The chamber atmosphere shall be monitored for carbon dioxide levels during saturation operations whenever ventilation is not used.

14.2.9.6* Chamber Gas Supply Monitoring.

14.2.9.6.1* Air from compressors shall be sampled at least every 6 months and after major repair or modification of the compressor(s).

14.2.9.6.2* As a minimum, the air supplied from compressors to Class A chambers shall meet the requirements for CGA Grade E.

14.2.9.6.3 As a minimum, the air supplied from compressors to Class B chambers shall meet the requirements for CGA Grade E with the additional limit of no condensable hydrocarbons.

14.2.9.6.4 When air cylinders are used to provide breathing air in Class A or Class B chambers, the breathing air shall be medical air USP.

14.2.9.6.5 When cylinders are used to provide oxygen in Class A or Class B chambers, the gas shall be oxygen USP.

14.2.9.7 Electrical monitoring equipment used inside the chamber shall comply with the applicable requirements of 14.2.8.

14.2.9.8* Closed-circuit television monitoring of the chamber interior shall be employed for chamber operators who do not have direct visual contact with the chamber interior from their normal operating location.

14.2.10 Other Equipment and Fixtures.

14.2.10.1 All furniture permanently installed in the hyperbaric chamber shall be grounded.

14.2.10.2* Exhaust from all classes of chambers shall be piped outside of the building.

14.2.10.2.1 Each Class B chamber shall have an independent exhaust line.

14.2.10.2.2 The point of exhaust shall not create a hazard.

14.2.10.2.3 The point of exhaust shall not allow reentry of gases into the building.

14.2.10.2.4 The point of exhaust shall be protected by the provision of a minimum of 0.3 cm (0.12 in.) mesh screen and situated to prevent the intrusion of rain, snow, or airborne debris.

14.2.10.2.5 The point of exhaust shall be identified as an oxygen exhaust by a sign prohibiting smoking or open flame.

14.2.10.3 The supply piping for all air, oxygen, or other breathing mixtures from certified commercially supplied cylinders and portable containers shall be provided with a particulate filter of 66 microns or finer.

14.2.10.3.1 The particulate filter shall meet the construction requirements of ANSI/ASME PVHO-1, *Safety Standard for Pressure Vessels for Human Occupancy*, and be located as close as practical to the source.

14.3 Administration and Maintenance.

14.3.1 General.

14.3.1.1 Purpose. Section 14.3 contains requirements for administration and maintenance that shall be followed as an adjunct to physical precautions specified in Section 14.2.

14.3.1.2* Recognition of Hazards. The nature and recognition of hyperbaric hazards are outlined in Annex B of this document and shall be reviewed by the safety director.

14.3.1.3 Responsibility.

14.3.1.3.1 Personnel having responsibility for the hyperbaric facility, and those responsible for licensing, accrediting, or approving institutions or other facilities in which hyperbaric installations are employed, shall establish and enforce programs to fulfill the provisions of this chapter.

14.3.1.3.2* Each hyperbaric facility shall designate an on-site hyperbaric safety director to be in charge of all hyperbaric equipment and the operational safety requirements of this chapter.

14.3.1.3.2.1 The safety director shall participate with facility management personnel and the hyperbaric physician(s) in developing procedures for operation and maintenance of the hyperbaric facility.

14.3.1.3.2.2 The safety director shall make recommendations for departmental safety policies and procedures.

14.3.1.3.2.3 The safety director shall have the authority to restrict or remove any potentially hazardous supply or equipment items from the chamber.

14.3.1.3.3* The governing board shall be responsible for the care and safety of patients and personnel.

14.3.1.3.4* By virtue of its responsibility for the professional conduct of members of the medical staff of the health care facility, the organized medical staff shall adopt and enforce regulations with respect to the use of hyperbaric facilities located in health care facilities.

14.3.1.3.4.1 The safety director shall participate in the development of these regulations.

14.3.1.3.5* The safety director shall ensure that electrical, monitoring, life-support, protection, and ventilating arrangements in the hyperbaric chamber are inspected and tested as part of the routine maintenance program of the facility.

14.3.1.4 Rules and Regulations.

14.3.1.4.1* General. The administrative, technical, and professional staffs shall jointly develop policies for management of the hyperbaric facility.

14.3.1.4.1.1 Upon adoption, the management policies shall be available in the facility.

14.3.1.4.2 The medical director of hyperbaric medicine and the safety director shall jointly develop the minimum staff qualifications, experience, and complement based on the following:

(1) Number and type of hyperbaric chambers in use
(2) Maximum treatment capacity
(3) Type of hyperbaric therapy normally provided

14.3.1.4.3 All personnel, including those involved in maintenance and repair of the hyperbaric facility, shall be trained on the purpose, application, operation, and limitations of emergency equipment.

14.3.1.4.4 Emergency procedures specific to the hyperbaric facility shall be established.

14.3.1.4.4.1* All personnel shall be trained in emergency procedures.

14.3.1.4.4.2 Personnel shall be trained to control the chamber and decompress occupants when all powered equipment has been rendered inoperative.

14.3.1.4.5* Emergency procedures and fire training drills shall be conducted at least annually and documented by the safety director.

14.3.1.4.6 When an inspection, test, or maintenance procedure of the fire suppression system results in the system being placed out of service, a protocol shall be followed that notifies appropriate personnel and agencies of the planned or emergency impairment.

14.3.1.4.7 A sign indicating the fire suppression system is out of service shall be conspicuously placed on the operating console until the fire suppression system is restored to service.

14.3.1.4.8 During chamber operations with an occupant(s) in a chamber, the operator shall be physically present and shall maintain visual or audible contact with the control panel or the chamber occupant(s).

14.3.1.5 General.

14.3.1.5.1 Potential Ignition Sources.

14.3.1.5.1.1* The following shall be prohibited from inside the chamber and the immediate vicinity outside the chamber:

(1) Smoking
(2) Open flames
(3) Hot objects

14.3.1.5.1.2 The following shall be prohibited from inside the chamber:

(1) Personal warming devices (e.g., therapeutic chemical heating pads, hand warmers, pocket warmers)
(2) Cell phones and pagers
(3) Sparking toys
(4) Personal entertainment devices

14.3.1.5.2 Flammable Gases and Liquids.

14.3.1.5.2.1 Flammable agents, including devices such as laboratory burners employing bottled or natural gas and cigarette lighters, shall be prohibited inside the chamber and from the proximity of the compressor intake.

14.3.1.5.2.2 For Class A chambers, flammable agents used for patient care, such as alcohol swabs, parenteral alcohol-based pharmaceuticals, and topical creams, shall be permitted in the chamber if the following conditions are met:

(1) Such use is approved by the safety director or other authority having jurisdiction.
(2)*The quantities of such agents are limited so that they are incapable of releasing sufficient flammable vapor into the chamber atmosphere to exceed the LEL for the material.
(3) A safety factor is included to account for the localized concentrations, stratification, and the absence of ventilation.
(4) The oxygen monitoring requirement of 14.2.9.4.2 is observed.

14.3.1.5.2.3 Flammable liquids, gases, or vapors shall not be permitted inside any Class B chamber.

14.3.1.5.3* Personnel.

14.3.1.5.3.1 Antistatic procedures, as directed by the safety director, shall be employed whenever atmospheres containing more than 23.5 percent oxygen by volume are used.

14.3.1.5.3.2 In Class A and Class B chambers with atmospheres containing more than 23.5 percent oxygen by volume, electrical grounding of the patient shall be ensured by the provision of a high-impedance conductive pathway in contact with the patient's skin.

14.3.1.5.3.3 Shoes having ferrous nails that make contact with the floor shall not be permitted to be worn in Class A chambers.

14.3.1.5.4* Textiles.

14.3.1.5.4.1 Except where permitted in 14.3.1.5.4.3, silk, wool, or synthetic textile materials, or any combination thereof, shall be prohibited in Class A or Class B chambers.

14.3.1.5.4.2* Garments permitted inside of chambers shall be as follows:

(1) Garments fabricated of 100 percent cotton or a blend of cotton and polyester fabric shall be permitted in Class A chambers.
(2) Garments fabricated of 100 percent cotton, or a blend of cotton and polyester fabric containing no more than 50 percent polyester, shall be permitted in Class B chambers.

14.3.1.5.4.3* The physician or surgeon in charge, with the concurrence of the safety director, shall be permitted to use one of the following prohibited items in the chamber:

(1) Suture material
(2) Alloplastic devices
(3) Bacterial barriers
(4) Surgical dressings
(5) Biological interfaces
(6) Synthetic textiles

14.3.1.5.4.4 Physician and safety director approval to use prohibited items shall be stated in writing for all prohibited materials employed. *(See A.14.3.1.3.2.)*

14.3.1.5.4.5 Upholstered Furniture.

(A) Upholstered furniture (fixed or portable), shall be resistant to a cigarette ignition (i.e., smoldering) in accordance with one of the following:

(1) The components of the upholstered furniture shall meet the requirements for Class 1 when tested in accordance with NFPA 260, *Standard Methods of Tests and Classification System for Cigarette Ignition Resistance of Components of Upholstered Furniture*, ASTM E 1353, *Standard Test Methods for Cigarette Ignition Resistance of Components of Upholstered Furniture*, or California Technical Bulletin 133, *Flammability Test Procedure for Seating Furniture for Use in Public Occupancies*.
(2) Mocked-up composites of the upholstered furniture shall have a char length not exceeding 1½ in. (38 mm) when tested in accordance with NFPA 261, *Standard Method of Test for Determining Resistance of Mock-Up Upholstered Furniture Material Assemblies to Ignition by Smoldering Cigarettes*, or ASTM E 1352, *Standard Test Method for Cigarette Ignition Resistance of Mock-Up Upholstered Furniture Assemblies*.

(B) Upholstered furniture shall have limited rates of heat release when tested in accordance with ASTM E 1537, *Standard Test Method for Fire Testing of Upholstered Furniture*, as follows:

(1) The peak rate of heat release for the single upholstered furniture item shall not exceed 80 kW.
(2) The total heat released by the single upholstered furniture item during the first 10 minutes of the test shall not exceed 25 MJ.

14.3.1.5.4.6 Mattresses. Mattresses shall have a char length not exceeding 2 in. (51 mm) when tested in accordance with 16 CFR 1632, *Standard for the Flammability of Mattresses and Mattress Pads* (FF 4-72); 16 CFR Part 1633, *Standard for the Flammability (Open Flame) of Mattress Sets*; or California Technical Bulletin 129, *Flammability Test Procedure for Mattresses for Use in Public Buildings*. Mattresses shall have limited rates of heat release when tested in accordance with ASTM E 1590, *Standard Test Method for Fire Testing of Mattresses*, as follows:

(1) The peak rate of heat release for the mattress shall not exceed 100 kW. The peak rate of heat release for the mattress shall not exceed 100 kW.
(2) The total heat released by the mattress during the first 10 minutes of the test shall not exceed 25 MJ.

14.3.1.5.4.7 Fill materials shall comply with California Technical Bulletin 117 Requirements, *Test Procedure and Apparatus for Testing the Flame Retardance of Resilient Filling Materials Used in Upholstered Furniture*.

14.3.1.5.4.8 For materials with fire-retardant coatings, the material shall be maintained in accordance with the manufacturer's instructions to retain the fire-retardant properties.

14.3.1.5.4.9 Exposed foamed plastic materials shall be prohibited.

14.3.1.5.5 The use of flammable hair sprays, hair oils, and skin oils shall be forbidden for all chamber occupants/patients as well as personnel.

14.3.1.5.5.1 Whenever possible, patients shall be stripped of all clothing, particularly if it is contaminated by dirt, grease, or solvents, and then reclothed. *(See A.14.3.1.5.4.)*

14.3.1.5.5.2 All cosmetics, lotions, and oils shall be removed from the patient's body and hair.

14.3.1.5.6 All other fabrics used in the chamber, such as sheets, pillow cases, and blankets, shall conform to 14.3.1.5.4.1 and 14.3.1.5.4.2.

14.3.1.5.7 Drapes used within the chamber shall meet the flame propagation performance criteria contained in NFPA 701, *Standard Methods of Fire Tests for Flame Propagation of Textiles and Films.*

14.3.1.5.8 Clothing worn by patients in Class A or Class B chambers and personnel in Class A chambers shall, prior to each treatment, conform to the following:

(1) They shall be issued by the hyperbaric facility or specifically approved by the safety director for hyperbaric use.
(2) They shall be uncontaminated.
(3) They shall be devoid of prohibited articles prior to chamber pressurization.

14.3.2 Equipment.

14.3.2.1 All equipment used in the hyperbaric chamber shall comply with Section 14.2, including the following:

(1) All electrical and mechanical equipment necessary for the operation and maintenance of the hyperbaric facility
(2) Any medical devices and instruments used in the facility

14.3.2.1.1 Use of unapproved equipment shall be prohibited. *(See 14.3.1.5.4.3.)*

14.3.2.1.2 The following devices shall not be operated in the hyperbaric chamber unless approved by the safety director for such use:

(1) Portable X-ray devices
(2) Electrocautery equipment
(3) High-energy devices

14.3.2.1.3 Photographic equipment employing the following shall not remain in the chamber when the chamber is pressurized:

(1) Photoflash
(2) Flood lamps

14.3.2.1.4 The use of Class 1 or Class 2 lasers as defined by ANSI Z136.3 *American National Standard for the Safe Use of Lasers in Health Care Facilities*, shall be permitted.

14.3.2.1.5 Equipment known to be, or suspected of being, defective shall not be introduced into any hyperbaric chamber or used in conjunction with the operation of such chamber until repaired, tested, and accepted by qualified personnel and approved by the safety director. *(See 14.3.1.3.2.)*

14.3.2.1.6* Paper brought into the chamber shall be stored in a closed metal container.

14.3.2.1.7 Containers used for paper storage shall be emptied after each chamber operation.

14.3.2.1.8 Equipment that does not meet the temperature requirements of 500.8(A), 500.8(B), and 500.8(C) of *NFPA 70, National Electrical Code*, shall not be permitted in the chamber.

14.3.2.2* The following shall be all-metal to the extent possible:

(1) Oxygen containers
(2) Valves
(3) Fittings
(4) Interconnecting equipment

14.3.2.3 The following shall be compatible with oxygen under service conditions:

(1) Valve seats
(2) Gaskets
(3) Hose
(4) Lubricants

14.3.2.4 Equipment used inside the chamber requiring lubrication shall be lubricated with oxygen-compatible material.

14.3.2.4.1 Factory-sealed antifriction bearings shall be permitted to be used with standard hydrocarbon lubricants in Class A chambers that do not employ atmospheres of increased oxygen concentration.

14.3.2.5* Equipment made of the following shall be prohibited from the chamber interior:

(1) Cerium
(2) Magnesium
(3) Magnesium alloys

14.3.2.6* In the event that radiation equipment is introduced into a hyperbaric chamber, hydrocarbon detectors shall be installed.

14.3.2.6.1 In the event that flammable gases are detected in excess of 1000 ppm, radiation equipment shall not be operated until the chamber atmosphere is cleared.

14.3.3 Handling of Gases.

14.3.3.1 The institution's administrative personnel shall develop policies for safe handling of gases in the hyperbaric facility. *(See 14.3.1.5.2.)*

14.3.3.2 Oxygen and other gases shall not be introduced into the chamber in the liquid state.

14.3.3.3 Flammable gases shall not be used or stored in the chamber or in the hyperbaric facility.

14.3.3.4* Pressurized containers of gas shall be permitted to be introduced into the hyperbaric chamber, provided that the container and its contents are approved for such use by the safety director.

14.3.4 Maintenance.

14.3.4.1 General.

14.3.4.1.1 The hyperbaric safety director shall ensure that all valves, regulators, meters, and similar equipment used in the hyperbaric chamber are compensated for use under hyperbaric conditions and tested as part of the routine maintenance program of the facility.

14.3.4.1.1.1 Pressure relief valves shall be tested and calibrated as part of the routine maintenance program of the facility.

14.3.4.1.2 The hyperbaric safety director shall ensure that all gas outlets in the chambers are labeled or stenciled in accordance with CGA C-4, *Standard Method of Marking Portable Compressed Gas Containers to Identify the Material Contained.*

14.3.4.1.3 The requirements set forth in Section 5.1 and NFPA 55, *Compressed Gases and Cryogenic Fluids Code*, concerning the storage, location, and special precautions required for medical gases shall be followed.

14.3.4.1.4 Storage areas for hazardous materials shall not be located in the room housing the hyperbaric chamber. *(See 14.2.1.)*

14.3.4.1.4.1 Flammable gases, except as provided in 14.3.1.5.2.2(1), shall not be used or stored in the hyperbaric room.

14.3.4.1.5 All replacement parts and components shall conform to original design specification.

14.3.4.2 Maintenance Logs.

14.3.4.2.1 Installation, repairs, and modifications of equipment related to a chamber shall be evaluated by engineering personnel, tested under pressure, and approved by the safety director.

14.3.4.2.1.1 Logs of all tests shall be maintained.

14.3.4.2.2 Operating equipment logs shall be maintained by engineering personnel.

14.3.4.2.2.1 Operating equipment logs shall be signed before chamber operation by the person in charge. *(See A.14.3.1.3.2.)*

14.3.4.2.3 Operating equipment logs shall not be taken inside the chamber.

14.3.5 Electrical Safeguards.

14.3.5.1 Electrical equipment shall be installed and operated in accordance with 14.2.8.

14.3.5.1.1 All electrical circuits shall be tested in accordance with the routine maintenance program of the facility.

14.3.5.1.1.1 Electrical circuit tests shall include the following:

(1) Ground-fault check to verify that no conductors are grounded to the chamber
(2) Test of normal functioning *(see 14.2.8.2.2)*

14.3.5.1.2 In the event of fire, all nonessential electrical equipment within the chamber shall be de-energized before extinguishing the fire.

14.3.5.1.2.1 Smoldering, burning electrical equipment shall be de-energized before extinguishing a localized fire involving only the equipment. *(See 14.2.5.)*

14.3.6* Electrostatic Safeguards.

14.3.6.1 Administration. (Reserved)

14.3.6.2 Maintenance.

14.3.6.2.1 Furniture Used in the Chamber.

14.3.6.2.1.1 Conductive devices on furniture and equipment shall be inspected to ensure that they are free of wax, lint, or other extraneous material that could insulate them and defeat the conductive properties.

14.3.6.2.1.2* Casters or furniture leg tips shall not be capable of impact sparking.

14.3.6.2.1.3 Casters shall not be lubricated with oils or other flammable materials.

14.3.6.2.1.4 Lubricants shall be oxygen compatible.

14.3.6.2.1.5 Wheelchairs and gurneys with bearings lubricated and sealed by the manufacturer shall be permitted in Class A chambers where conditions prescribed in 14.2.9.4 are met.

14.3.6.2.2 Conductive Accessories. Conductive accessories shall meet conductivity and antistatic requirements.

14.3.6.2.3* Materials containing rubber shall be inspected as part of the routine maintenance program of the facility, especially at points of kinking.

14.3.6.3 Fire Protection Equipment Inside Hyperbaric Chambers.

14.3.6.3.1 Electrical switches, valves, and electrical monitoring equipment associated with fire detection and extinguishment shall be visually inspected before each chamber pressurization.

14.3.6.3.2 Fire detection equipment shall be tested each week, and full testing, including discharge of extinguishing media, shall be conducted annually.

14.3.6.3.3 Testing shall include activation of trouble circuits and signals.

14.3.6.4* Housekeeping. A housekeeping program shall be implemented, whether or not the facility is in regular use.

14.3.6.4.1 The persons assigned to the task of housekeeping shall be trained in the following:

(1) Potential damage to the equipment from cleaning procedures
(2) Potential personal injury
(3) Specific cleaning procedures
(4) Equipment not to be cleaned

Chapter 15 Features of Fire Protection

15.1 Applicability.

15.1.1 This chapter shall apply to all new and existing health care facilities.

15.1.2 An existing system that is not in strict compliance with the provisions of this code shall be permitted to be continued in use, unless the authority having jurisdiction has determined that such use constitutes a distinct hazard to life.

15.2 Construction and Compartmentation. Buildings or structures housing a health care facility shall meet the minimum construction and compartmentation requirements of the applicable code.

15.3 Special Hazard Protection for Flammable Liquids and Gases.

15.3.1 The storage and handling of flammable liquids or gases shall be in accordance with the following applicable standards:

(1) NFPA 30, *Flammable and Combustible Liquids Code*
(2) NFPA 54, *National Fuel Gas Code*
(3) NFPA 58, *Liquefied Petroleum Gas Code* [*101*:8.7.3.1]
(4) NFPA 55, *Compressed Gases and Cryogenic Fluids Code*

15.3.2* No storage or handling of flammable liquids or gases shall be permitted in any location where such storage would jeopardize egress from the structure, unless otherwise permitted by 15.3.1. [*101*:8.7.3.2]

15.4 Laboratories. Laboratories that use chemicals shall comply with NFPA 45, *Standard on Fire Protection for Laboratories Using Chemicals*, unless otherwise modified by other provisions of this code. [*101*:8.7.4.1]

15.5 Utilities.

15.5.1 General. Utilities shall comply with the requirements of 15.5.1.1 through 15.5.1.4. [*101*:12.5.1]

15.5.1.1 Gas. Equipment using gas and related gas piping shall be in accordance with NFPA 54, *National Fuel Gas Code*, or NFPA 58, *Liquefied Petroleum Gas Code*, unless such installations are approved existing installations, which shall be permitted to be continued in service. [*101*:9.1.1]

15.5.1.2 Electrical Systems. Electrical wiring and equipment shall be in accordance with *NFPA 70, National Electrical Code*, unless such installations are approved existing installations, which shall be permitted to be continued in service. [*101*:9.1.2]

15.5.1.3 Emergency Generators and Standby Power Systems. Emergency generators and standby power systems, where required for compliance with this code, shall be installed, tested, and maintained in accordance with NFPA 110, *Standard for Emergency and Standby Power Systems*.

15.5.1.4 Stored Electrical Energy Systems. Stored electrical energy systems shall be installed, tested, and maintained in accordance with NFPA 111, *Standard on Stored Electrical Energy Emergency and Standby Power Systems*. [*101*:9.1.4]

15.5.2 Heating, Ventilating, and Air-Conditioning. [*101*:9.2]

15.5.2.1* Heating, Ventilating, and Air Conditioning. Air-conditioning, heating, ventilating ductwork, and related equipment shall be in accordance with NFPA 90A, *Standard for the Installation of Air-Conditioning and Ventilating Systems*, unless such installations are approved existing installations, which shall be permitted to be continued in service.

15.5.2.2 Ventilating or Heat-Producing Equipment. Ventilating or heat-producing equipment shall be in accordance with NFPA 91, *Standard for Exhaust Systems for Air Conveying of Vapors, Gases, Mists, and Noncombustible Particulate Solids*; NFPA 211, *Standard for Chimneys, Fireplaces, Vents, and Solid Fuel–Burning Appliances*; NFPA 31, *Standard for the Installation of Oil-Burning Equipment*; NFPA 54, *National Fuel Gas Code*; or *NFPA 70, National Electrical Code*, as applicable, unless such installations are approved existing installations, which shall be permitted to be continued in service. [*101*:9.2.2]

15.5.2.3 Commercial Cooking Operations. Commercial cooking operations shall be protected in accordance with NFPA 96, *Standard for Ventilation Control and Fire Protection of Commercial Cooking Operations*, unless such installations are approved existing installations, which shall be permitted to be continued in service. [*101*:9.2.3]

15.5.2.4 Ventilating Systems in Laboratories Using Chemicals. Ventilating systems in laboratories using chemicals shall be in accordance with NFPA 45, *Standard on Fire Protection for Laboratories Using Chemicals*. [*101*:9.2.4]

15.5.3 Elevators, Escalators, and Conveyors. [*101*:9.4]

15.5.3.1 Code Compliance. [*101*:9.4.2]

15.5.3.1.1 Except as modified herein, new elevators, escalators, dumbwaiters, and moving walks shall be in accordance with the requirements of ASME A17.1/CSA B44, *Safety Code for Elevators and Escalators*. [*101*:9.4.2.1]

15.5.3.1.2 Except as modified herein, existing elevators, escalators, dumbwaiters, and moving walks shall be in accordance with the requirements of ASME A17.3, *Safety Code for Existing Elevators and Escalators*. [*101*:9.4.2.2]

15.5.3.2 Fire Fighters' Emergency Operations. [*101*:9.4.3]

15.5.3.2.1 All new elevators shall conform to the fire fighters' emergency operations requirements of ASME A17.1/CSA B44, *Safety Code for Elevators and Escalators*. [*101*:9.4.3.1]

15.5.3.2.2 All existing elevators having a travel distance of 25 ft (7620 mm) or more above or below the level that best serves the needs of emergency personnel for fire-fighting or rescue purposes shall conform to the fire fighters' emergency operations requirements of ASME A17.3, *Safety Code for Existing Elevators and Escalators*. [*101*:9.4.3.2]

15.5.3.3* Elevator Machine Rooms. Elevator machine rooms that contain solid-state equipment for elevators, other than existing elevators, having a travel distance exceeding 50 ft (15 m) above the level of exit discharge or exceeding 30 ft (9150 mm) below the level of exit discharge shall be provided with independent ventilation or air-conditioning systems to maintain temperature during fire fighters' emergency operations for elevator operation (see 9.4.3 of NFPA *101*). The operating temperature shall be established by the elevator equipment manufacturer's specifications. When standby power is connected to the elevator, the machine room ventilation or air-conditioning shall be connected to standby power. [*101*:9.4.5]

15.5.3.4 Elevator Testing. Elevators shall be subject to periodic inspections and tests as specified in ASME A17.1, *Safety Code for Elevators and Escalators*. All elevators equipped with fire fighters' emergency operations in accordance with 9.4.3 of NFPA *101* shall be subject to a monthly operation with a written record of the findings made and kept on the premises as required by ASME A17.1. [*101*:9.4.6]

15.6 Waste Chutes, Incinerators, and Linen Chutes. Waste chutes, linen chutes, and incinerators shall be installed and maintained in accordance with NFPA 82, *Standard on Incinerators and Waste and Linen Handling Systems and Equipment*, unless such installations are approved existing installations, which shall be permitted to be continued in service.

15.6.1 Any waste chute, including pneumatic rubbish and linen systems, shall be provided with automatic extinguishing protection in accordance with Section 9.7. [*101*:19.5.4.3]

15.7 Fire Detection, Alarm, and Communications Systems. [*101*:9.6]

15.7.1* General. [*101*:9.6.1]

15.7.1.1 Buildings or structures housing a health care facility shall meet the fire detection, alarm, and communications systems requirements of the applicable code.

15.7.1.2 A fire alarm system required for life safety shall be installed, tested, and maintained in accordance with the applicable requirements of *NFPA 70, National Electrical Code*, and *NFPA 72, National Fire Alarm and Signaling Code*, unless it is an approved existing installation, which shall be permitted to be continued in use. [*101*:9.6.1.3]

15.7.1.3 For the purposes of this code, a complete fire alarm system shall provide functions for initiation, notification, and control, and shall perform as follows:

(1) The initiation function provides the input signal to the system.
(2) The notification function advises that human action is required in response to a particular condition.
(3) The control function provides outputs to control building equipment to enhance protection of life.

15.7.2 Signal Initiation.

15.7.2.1 Buildings or structures housing a health care facility shall meet the minimum signaling and alarm initiation requirements of the applicable code.

15.7.2.2 Manual fire alarm boxes shall be used only for fire-protective signaling purposes.

15.7.2.2.1 Combination fire alarm and guard's tour stations shall be acceptable.

15.7.2.3 A manual fire alarm box shall be provided in the natural exit access path near each required exit from an area, unless modified by another section of this code.

15.7.2.4* Additional manual fire alarm boxes shall be located so that, on any given floor in any part of the building, no horizontal distance on that floor exceeding 60 m (200 ft) shall need to be traversed to reach a manual fire alarm box. [*101*:9.6.2.5]

15.7.2.5 For fire alarm systems using automatic fire detection or waterflow detection devices, not less than one manual fire alarm box shall be provided to initiate a fire alarm signal.

15.7.2.5.1 The manual fire alarm box shall be located where required by the authority having jurisdiction.

15.7.2.6 Manual fire alarm boxes shall be accessible, unobstructed, and visible. [*101*:9.6.2.7]

15.7.2.7 Where a sprinkler system provides automatic detection and alarm system initiation, it shall be provided with an approved alarm initiation device that operates when the flow of water is equal to or greater than that from a single automatic sprinkler.

15.7.3 Smoke Alarms.

15.7.3.1 Where required by the applicable code, single-station and multiple-station smoke alarms shall be in accordance with *NFPA 72, National Fire Alarm and Signaling Code.*

15.7.3.2 System smoke detectors in accordance with *NFPA 72, National Fire Alarm and Signaling Code*, and arranged to function in the same manner as single-station or multiple-station smoke alarms shall be permitted in lieu of smoke alarms. [*101*:9.6.2.10.6]

15.7.3.3 The alarms shall sound only within an individual dwelling unit, suite of rooms, or similar area and shall not actuate the building fire alarm system, unless otherwise permitted by the authority having jurisdiction. Remote annunciation shall be permitted.

15.7.4 Occupant Notification. [*101*:9.6.3]

15.7.4.1 Where required by the applicable code, occupant notification shall be provided to alert occupants of a fire or other emergency.

15.7.4.2 Occupant notification shall be in accordance with 15.7.4.3 unless otherwise provided in 15.7.4.2.1 and 15.7.4.2.2. [*101*:9.6.3.2]

15.7.4.2.1* Elevator lobby, hoistway, and associated machine room smoke detectors used solely for elevator recall, and heat detectors used solely for elevator power shutdown, shall not be required to activate the building evacuation alarm if the power supply and installation wiring to such detectors are monitored by the building fire alarm system, and if the activation of such detectors initiates a supervisory signal at a constantly attended location. [*101*:9.6.3.2.1]

15.7.4.2.2* Smoke detectors used solely for closing dampers or HVAC system shutdown shall not be required to activate the building evacuation alarm, provided that the power supply and installation wiring to the detectors are monitored by the building fire alarm system, and the activation of the detectors initiates a supervisory signal at a constantly attended location. [*101*:9.6.3.2.2]

15.7.4.3 Defend in Place. For new and existing facilities, where the response to a fire is to defend in place within a safe place in the building, occupant notification shall be in accordance with the facility fire plan.

15.7.4.3.1* Where buildings are required to be subdivided into smoke compartments, fire alarm notification zones shall coincide with one or more smoke compartment boundaries or shall be in accordance with the facility fire plan.

15.7.4.3.2* The private operating mode, as defined in *NFPA 72*, shall be permitted to be used for the placement of notification appliances within the health care and ambulatory health care occupancies of the building.

15.7.4.3.3 The notification signal shall readily identify the smoke zone or the floor area, floor, and building in need of staff response.

15.7.4.3.4 The notification signal shall be heard in all locations in accordance with the facility fire plan.

15.7.4.3.5 In critical care areas, visible alarm notification appliances shall be permitted to be used in lieu of audible alarm signals.

15.7.4.3.6 Visible signals shall not be required inside surgical operating rooms, patient sleeping rooms, or psychiatric care areas where their operation would interfere with patient treatment.

15.7.4.3.7 Visible signals shall not be required inside exam rooms, special procedure rooms, dressing rooms, and non-public toilet rooms where staff is required to respond to those areas in accordance with the facility fire plan.

15.8 Automatic Sprinklers and Other Extinguishing Equipment.

15.8.1 Automatic Sprinklers.

15.8.1.1 Buildings or structures housing a health care facility shall meet the automatic sprinkler system requirements of the applicable code.

15.8.1.2 Where provided, automatic sprinkler systems shall be installed in accordance with NFPA 13, *Standard for the Installation of Sprinkler Systems.*

15.8.1.3* Defend in Place. For new and existing facilities, where the response to a fire is to defend in place within a safe place in the building and not to automatically evacuate the building, sprinkler system zones shall coincide with smoke compartment boundaries or shall be in accordance with the facility fire plan.

15.8.1.4* Closets. Sprinklers shall not be required in clothes closets of patient sleeping rooms in hospitals where the area of the closet does not exceed 6 ft^2 (0.55 m^2) provided the distance from the sprinkler in the patient sleeping room to the back wall of the closet does not exceed the maximum distance permitted by NFPA 13, *Standard for the Installation of Sprinkler Systems.* [*101*:18.3.5.10]

15.9 Manual Extinguishing Equipment.

15.9.1* Portable fire extinguishers shall be selected, installed, inspected, and maintained in accordance with NFPA 10, *Standard for Portable Fire Extinguishers.*

15.9.2 Where provided, standpipe and hose systems shall be in accordance with NFPA 14, *Standard for the Installation of Standpipe and Hose Systems.*

15.9.2.1 Where standpipe and hose systems are installed in combination with automatic sprinkler systems, installation shall be in accordance with the appropriate provisions established by NFPA 13, *Standard for the Installation of Sprinkler Systems,* and NFPA 14, *Standard for the Installation of Standpipe and Hose Systems.*

15.9.2.2* Hose or hose outlets shall be permitted to be removed from existing standpipe and hose systems that are not required by the applicable code.

15.10* Compact Storage. Compact storage shall be protected by sprinklers in accordance with NFPA 13, *Standard for the Installation of Sprinkler Systems.*

15.11 Compact Mobile Storage.

15.11.1 Rooms with compact mobile storage units greater than 50 ft^2 (4.65 m^2) shall be protected as a hazardous area in accordance with the applicable code.

15.11.2 Smoke detection shall be installed above compact mobile storage units greater than 50 ft^2 (4.65 m^2) in accordance with *NFPA 72.*

15.11.3* Compact mobile storage units greater than 50 ft^2 (465 m^2) shall be protected by automatic sprinklers in accordance with NFPA 13.

15.12 Maintenance and Testing.

15.12.1 All water-based fire protection systems shall be inspected, tested, and maintained in accordance with NFPA 25, *Standard for the Inspection, Testing, and Maintenance of Water-Based Fire Protection Systems.*

15.12.2 All non-water-based fire protection systems shall be inspected, tested, and maintained in accordance with the applicable NFPA standards.

15.13* Fire Loss Prevention in Operating Rooms.

15.13.1 Hazard Assessment.

15.13.1.1 An evaluation shall be made of hazards that could be encountered during surgical procedures.

15.13.1.2 The evaluation shall include hazards associated with the properties of electricity, hazards associated with the operation of surgical equipment, and hazards associated with the nature of the environment.

15.13.1.3 Periodic reviews of surgical operations and procedures shall be conducted with special attention given to any change in materials, operations, or personnel.

15.13.2 Fire Prevention Procedures. Fire prevention procedures shall be established.

15.13.3 Germicides and Antiseptics.

15.13.3.1 Medicaments and alcohol-based hand sanitizers, including those dispersed as aerosols, shall be permitted to be used in anesthetizing locations

15.13.3.2* Flammable liquid germicides or antiseptics used in anesthetizing locations, whenever the use of electrosurgery, cautery, or a laser is contemplated, shall be packaged as follows:

(1) In a nonflammable package
(2) To ensure controlled delivery to the patient in unit dose applicators, swabs, and other similar applicators

15.13.3.3 Whenever the application of flammable liquid germicides or antiseptics is employed in surgeries where the use of electrosurgery, cautery, or a laser is contemplated, time shall be allowed to elapse between application of the germicide or antiseptic and the following:

(1) Application of drapes, to allow complete evaporation and dissipation of any flammable vehicle remaining
(2) Use of electrosurgery, cautery, or a laser, to ensure the solution is completely dry and to allow thorough evaporation and dissipation of any flammable vehicle remaining

15.13.3.4 Any solution-soaked materials shall be removed from the operating room prior to draping or use of electrosurgery, cautery, or a laser.

15.13.3.5 Pooling of flammable liquid germicides or antiseptics shall be avoided; if pooling occurs, excess solution shall be wicked, and the germicide or antiseptic shall be allowed to completely dry.

15.13.3.6 A preoperative "time out" period shall be conducted prior to the initiation of any surgical procedure using flammable liquid germicides or antiseptics to verify the following:

(1) Application site of flammable germicide or antiseptic is dry prior to draping and use of electrosurgery, cautery, or a laser.
(2) Pooling of solution has not occurred or has been corrected.
(3) Any solution-soaked materials have been removed from the operating room prior to draping and use of electrosurgery, cautery, or a laser.

15.13.3.7 Whenever flammable aerosols or antiseptics are employed, sufficient time shall be allowed to elapse between deposition and application of drapes to allow complete evaporation and dissipation of any flammable vehicle remaining.

15.13.3.8 Health care organizations shall establish policies and procedures outlining safety precautions related to the use of flammable liquid or aerosol germicides or antiseptics used in anesthetizing locations, as required in 15.13.3.9, whenever the use of electrosurgery, cautery, or a laser is contemplated.

15.13.3.9 Emergency Procedures.

15.13.3.9.1 Procedures for operating room/surgical suite emergencies shall be developed.

15.13.3.9.2 Procedures shall include alarm actuation, evacuation, and equipment shutdown procedures and provisions for control of emergencies that could occur in the operating room, including specific detailed plans for control operations by an emergency control group within the organization or a public fire department.

15.13.3.9.3 Emergency procedures shall be established for controlling chemical spills.

15.13.3.9.4 Emergency procedures shall be established for extinguishing drapery, clothing, or equipment fires.

15.13.3.10 Orientation and Training.

15.13.3.10.1 New operating room/surgical suite personnel, including physicians and surgeons, shall be taught general safety practices for the area and specific safety practices for the equipment and procedures they will use.

15.13.3.10.2 Continuing safety education and supervision shall be provided, incidents shall be reviewed monthly, and procedures shall be reviewed annually.

15.13.3.10.3 Fire exit drills shall be conducted annually or more frequently as determined by the applicable code.

Annex A Explanatory Material

Annex A is not a part of the requirements of this NFPA document but is included for informational purposes only. This annex contains explanatory material, numbered to correspond with the applicable text paragraphs.

A.1.1.10 Because no single model of an emergency management plan is feasible for every health care facility, this chapter is intended to provide criteria for the preparation and implementation of an individual plan. The principles involved are universally applicable; the implementation needs to be tailored to the specific facility.

A.1.1.12 During the past 20 years, there has been a widespread interest in the use of oxygen at elevated environmental pressure to increase the partial pressure of oxygen in a patient's tissues in order to treat certain medical conditions or to prepare a patient for surgery. These techniques are also employed widely for the treatment of decompression sickness (e.g., bends, caisson worker's disease) and carbon monoxide poisoning.

Recently, however, the level of knowledge and expertise has increased so dramatically that the codes are in need of updating. By the end of 1988, there were 218 hyperbaric facilities in operation in the United States and Canada. These facilities supported hyperbaric medical treatments for 62,548 patients between 1971 and 1987. As these facilities provide therapy for disorders indicated for treatment, these numbers will continue to increase. As the number of facilities increases, the number of patients treated will also increase.

Such treatment involves placement of the patient, with or without attendants, in a hyperbaric chamber or pressure vessel, the pressure of which is raised above ambient pressure. In the course of the treatment, the patient breathes up to 100 percent oxygen.

In addition to being used for patient care, these chambers also are being employed for research purposes using experimental animals and, in some instances, humans.

The partial pressure of oxygen present in a gaseous mixture is the determinate factor in the amount of available oxygen. This pressure will rise if the volume percentage of oxygen present increases, if the total pressure of a given gas mixture containing oxygen increases, or if both these factors increase. Because the sole purpose of the hyperbaric technique of treatment is to raise the total pressure within the treatment chamber, an increased partial pressure of oxygen always is available during treatment, unless positive means are taken to limit the oxygen content. In addition, the patient is often given an oxygen-enriched atmosphere to breathe.

The need for human diligence in the establishment, operation, and maintenance of hyperbaric facilities is continual. The chief administrator of the facility possessing the hyperbaric chamber is responsible to adopt and enforce appropriate regulations for hyperbaric facilities. In formulating and administering the program, full use should be made of technical personnel highly qualified in hyperbaric chamber operations and safety.

It is essential that personnel having responsibility for the hyperbaric facility establish and enforce appropriate programs to fulfill the provisions of Chapter 14.

Potential hazards can be controlled only when continually recognized and understood by all pertinent personnel.

The purpose of Chapter 14 is to set forth minimum safeguards for the protection of patients or others subject to, and personnel who administer, hyperbaric therapy and experimental procedures. Its purpose is also to offer some guidance for rescue personnel who are not ordinarily involved in hyperbaric chamber operation, but who could become so involved in an emergency.

Requirements cited in 1.1.12 are minimum requirements. Discretion on the part of chamber operators and others might dictate the establishment of more stringent regulations.

A.1.5 Although it is common practice for medical appliances to use metric units on their dials, gauges, and controls, many components of systems within the scope of this document are manufactured and used in the United States and employ nonmetric dimensions. Since these dimensions (such as nominal pipe sizes) are not established by the National Fire Protection Association, the Technical Correlating Committee on Health Care Facilities cannot independently change them. Accordingly, this document uses dimensions that are presently in common use by the building trades in the United States. Trade units vary from SI to U.S. customary units, depending on the equipment devices or material.

A.2.1 The documents referenced in this chapter or portions of such documents are referenced within this code and are considered part of the requirements of this document.

Documents referenced in this chapter or portions of such documents are only applicable to the extent called for within other chapters of this code.

Where the requirements of a referenced code or standard differ from the requirements of this code, the requirements of this code govern.

A.3.2.1 Approved. The National Fire Protection Association does not approve, inspect, or certify any installations, procedures, equipment, or materials; nor does it approve or evaluate testing laboratories. In determining the acceptability of installations, procedures, equipment, or materials, the authority having jurisdiction may base acceptance on compliance with NFPA or other appropriate standards. In the absence of such standards, said authority may require evidence of proper installation, procedure, or use. The authority having jurisdiction may also refer to the listings or labeling practices of an organization that is concerned with product evaluations and is thus in a position to determine compliance with appropriate standards for the current production of listed items.

A.3.2.2 Authority Having Jurisdiction (AHJ). The phrase "authority having jurisdiction," or its acronym AHJ, is used in NFPA documents in a broad manner, since jurisdictions and approval agencies vary, as do their responsibilities. Where public safety is primary, the authority having jurisdiction may be a federal, state, local, or other regional department or individual such as a fire chief; fire marshal; chief of a fire prevention bureau, labor department, or health department; building official; electrical inspector; or others having statutory authority. For insurance purposes, an insurance inspection department, rating bureau, or other insurance company representative may be the authority having jurisdiction. In many circumstances, the property owner or his or her designated agent assumes the role of the authority having jurisdiction; at government installations, the commanding officer or departmental official may be the authority having jurisdiction.

A.3.2.3 Code. The decision to designate a standard as a "code" is based on such factors as the size and scope of the document, its intended use and form of adoption, and whether it contains substantial enforcement and administrative provisions.

A.3.2.6 Listed. The means for identifying listed equipment may vary for each organization concerned with product evaluation; some organizations do not recognize equipment as listed unless it is also labeled. The authority having jurisdiction should utilize the system employed by the listing organization to identify a listed product.

A.3.3.10 Applicator. In the given sense, an applicator is not an electrode, because it does not use a conductive connection to the patient in order to function. A radio frequency "horn" of a diathermy machine is a typical applicator.

A.3.3.12 Atmosphere. As employed in this code, the term *atmosphere* can refer to the environment within or outside of a hyperbaric facility. When used as a measure of pressure, atmosphere is expressed as a fraction of standard air pressure [101.4 kPa (14.7 psi)]. *(See the first column of Table D.1 in NFPA 99B.)*

A.3.3.12.2 Atmosphere of Increased Burning Rate. The degree of fire hazard of an oxygen-enriched atmosphere varies with the concentration of oxygen and diluent gas and the total pressure. The definition contained in the current edition of NFPA 53, *Recommended Practice on Materials, Equipment, and Systems Used in Oxygen-Enriched Atmospheres*, and in editions of NFPA 56D, *Standard for Hyperbaric Facilities*, prior to 1982 did not necessarily reflect the increased fire hazard of hyperbaric and hypobaric atmospheres.

The definition of *atmosphere of increased burning rate* used in Chapter 14 and in NFPA 99B, *Standard for Hypobaric Facilities*, defines an oxygen-enriched atmosphere with an increased fire hazard as it relates to the increased burning rate of material in the atmosphere. It is based on a 1.2 cm/sec (0.47 in./sec) burning rate (at 23.5 percent oxygen at 1 atmosphere absolute) as described in Figure A.3.3.12.2.

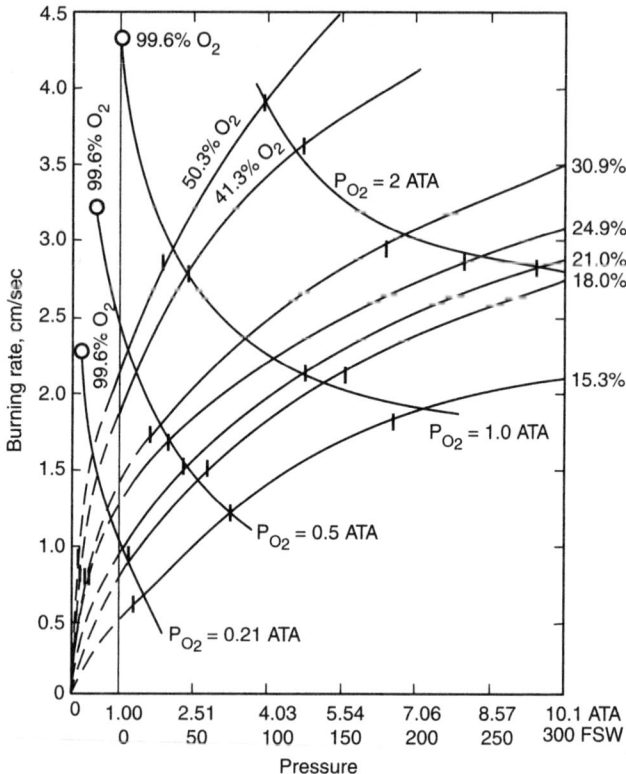

ATA = Atmospheres absolute
FSW = Feet of sea water

FIGURE A.3.3.12.2 Burning Rates of Filter Paper Strips at an Angle of 45 Degrees in N_2–O_2 Mixtures. (Adapted from Figure 4 of "Technical Memorandum UCRI-721, Chamber Fire Safety.")

This rate can be determined as follows:

$$\frac{23.45}{\sqrt{TP_{atmos}}} \quad [A.3.3.12.2]$$

where:
TP_{atmos} = total pressure in atmospheres

A.3.3.19.3 Bulk Oxygen System. The oxygen containers can be stationary or movable, and the oxygen can be stored as gas or liquid. The bulk oxygen system terminates at the point where oxygen at service pressure first enters the supply line.

A.3.3.23 Combustible Liquid. See NFPA 30, *Flammable and Combustible Liquids Code*, for further information on flash point test procedures.

A.3.3.24 Combustion. Combustion is not limited to a chemical reaction always involving oxygen. Certain metals, such as calcium and aluminum, will burn in nitrogen; nitrous oxide will support the combustion of phosphorus and carbon; and so on. However, this document deals with the more common process of fuels burning in air.

A.3.3.32 Defend in Place. The concept of the term *defend in place* includes, but is not limited to, elements related to moving building occupants from an area of immediate danger to a safe location in the building and containment of the emergency or dangerous condition.

A.3.3.35 Direct Electrical Pathway to the Heart. Electrodes, such as those used for pacing the heart, and catheters filled with conductive fluids, are examples of direct electrical pathways to the heart.

A.3.3.36 Disaster. A disaster can be either an event that causes, or threatens to cause, physical damage and injury to facility personnel or patients within the facility, or an event that requires expansion of facilities to receive and care for a large number of casualties resulting from a disaster that produces no damage or injury to the health care facility and staff, or a combination thereof.

Such a situation creates the need for emergency expansion of facilities, as well as operation of this expanded facility in an unfamiliar environment. Under this definition, the recognition of a disaster situation will vary greatly from one facility to another and from time to time in any given facility. Such recognition and concomitant activation of the Health Care Emergency Preparedness Plan is dependent on mutual aid agreements, facility type, geographic location, bed capacity, bed occupancy at a given time, staff size, staff experience with disaster situations, and other factors. For example, the routine workload of the emergency department of a large metropolitan general hospital would constitute a disaster, requiring activation of the Health Care Emergency Preparedness Plan, were this same workload to be suddenly applied to a small community hospital.

Disasters have a variety of causes, all of which should be considered for effective emergency preparedness planning. Among the most common are natural disasters such as earthquakes, hurricanes, tornadoes, and floods; mass food poisoning; industrial accidents involving explosion or environmental release of toxic chemicals; transportation accidents involving crashes of trains, planes, or automobiles with resulting mass casualties; civil disturbances; building fires; extensive or prolonged utility failure; collapse of buildings or other occupied structures; and toxic smogs in urban areas. Arson attempts and bomb threats have been made on health care facilities and should, therefore, be considered. Potential admission to the facility of high profile persons should be addressed. Although a high profile admission does not involve mass casualties or the potential for mass casualties, the degree of disruption of normal routine will be sufficient to qualify it as a disaster like situation.

Disaster plans should reflect a facility's location from potential internal and external disasters. As an example, areas subject to frequent wildland fires should invoke countermeasures for smoke management and air quality maintenance.

A.3.3.38 Double-Insulated Appliances. Double-insulated appliances can be identified by a symbol consisting of a square within a square or wording such as "double-insulated" marked on the appliance. Appliance packaging and documents can also provide identification. Although double-insulated appliances do not require a third wire or pin, some double-insulated appliances have a third conductor or pin solely for purposes of electromagnetic compatibility (EMC).

A.3.3.45 Essential Electrical System. The essential electrical system can be comprised of three branches: life safety branch, critical branch, and equipment branch.

A.3.3.48 Facility Fire Plan. This plan can be either a standalone plan or be a part of the emergency operations plan.

A.3.3.49 Failure. Failure includes failure of a component; loss of normal protective paths, such as grounding; and short circuits or faults between energized conductors and the chassis.

A.3.3.52 Flammable. Flammables can be solids, liquids, or gases exhibiting these qualities. Many substances that are nonflammable in air become flammable if the oxygen content of the gaseous medium is increased above 0.235 ATA.

A.3.3.55 Flash Point. Note that the flash point temperature is heavily dependent on the test used to determine it.

A.3.3.57 Flowmeter. A pressure compensated flowmeter should be used to indicate an accurate flow of gas whether the gas is discharged into ambient pressure or into a system at nonambient pressure.

A.3.3.58 Frequency. Formerly, the unit of frequency was cycles per second, a terminology no longer preferred. The waveform can consist of components having many different frequencies, in which case it is called a complex or nonsinusoidal waveform.

A.3.3.59 Fume Hood. Laboratory fume hoods prevent toxic, flammable, or noxious vapors from entering the laboratory; present a physical barrier from chemical reactions; and serve to contain accidental spills.

This definition does not include canopy hoods or recirculation laminar-flow biological-safety cabinets that are not designed for use with flammable materials.

A.3.3.61 General Anesthesia and Levels of Sedation/Analgesia. It should be noted that these are not static conditions. Minimal sedation can easily become moderate sedation, and moderate sedation can progress to deep sedation or general anesthesia.

A.3.3.63 Ground-Fault Circuit Interrupter (GFCI). Class A ground-fault circuit interrupters trip when the current to ground is 6 mA or higher and do not trip when the current to ground is less than 4 mA. For further information, see UL 943, *Standard for Ground-Fault Circuit Interrupters*. [**70**, 2014]

A.3.3.65 Grounding System. It coordinates with, but can be locally more extensive than, the grounding system described in Article 250 of *NFPA 70, National Electrical Code*.

A.3.3.67 Health Care Facilities. Health care facilities include, but are not limited to, hospitals, nursing homes, limited care facilities, clinics, medical and dental offices, and ambulatory health care centers, whether permanent or movable. This definition applies to normal, regular operations and does not pertain to facilities during declared local or national disasters. A health care facility is not a type of occupancy classification as defined by NFPA *101, Life Safety Code*. Therefore, the term *health care facility* should not be confused with the term *health care occupancy*. All health care occupancies (and ambulatory health care occupancies) are considered health care facilities; however, not all health care facilities are considered health care occupancies, as health care facilities also include ambulatory health care occupancies and business occupancies.

A.3.3.76 Impedance. The circuit element can consist of any combination of resistance, capacitance, or inductance.

A.3.3.78 Instrument Air. Instrument air is intended for the powering of medical devices unrelated to human respiration (e.g., to remove excess moisture from instruments before further processing, or to operate medical–surgical tools, air-driven booms, pendants, or similar applications).

A.3.3.80 Intrinsically Safe. "Abnormal conditions" can include accidental damage to any part of the equipment or wiring, damage to insulation or other failure of electrical components, application of overvoltage, adjustment and maintenance operations, and other similar conditions.

A.3.3.83 Isolated Power System. See *NFPA 70, National Electrical Code.*

A.3.3.85 Laboratory. These laboratories are not intended to include isolated frozen section laboratories; areas in which oxygen is administered; blood donor rooms in which flammable, combustible, or otherwise hazardous materials normally used in laboratory procedures are not present; and clinical service areas not using hazardous materials.

A.3.3.88 Limited-Combustible (Material). Material subject to increase in combustibility or flame spread index beyond the limits herein established through the effects of age, moisture, or other atmospheric condition is considered combustible.

See NFPA 259, *Standard Test Method for Potential Heat of Building Materials*, and NFPA 220, *Standard on Types of Building Construction.*

A.3.3.90 Liquid. When not otherwise identified, the term *liquid* includes both flammable and combustible liquids. *(See also B.11.1.1.)*

A.3.3.91 Local Signal. Examples would include a gauge, a flag, a light, or some other possible manifestation that allows a maintenance person to stand at the equipment and know what conditions are present (e.g., which header of cylinders is in service). The elements to be displayed are typically those that will also be monitored at the master alarm, but the local signal is visible at the equipment rather than remotely.

A.3.3.94 Manufactured Assembly. Examples are headwalls, columns, ceiling columns, ceiling-hung pendants, movable track systems, and so on.

A.3.3.96 Medical Air. Air supplied from on-site compressors and associated air treatment systems (as opposed to medical air USP supplied in cylinders) that complies with the specified limits is considered medical air. Hydrocarbon carryover from the compressor into the pipeline distribution system could be detrimental to the safety of the end user and to the integrity of the piping system. Mixing of air and oxygen is a common clinical practice, and the hazards of fire are increased if the air is thus contaminated. Compliance with these limits is thus considered important to fire and patient safety. The quality of local ambient air should be determined prior to its selection for compressors and air treatment equipment.

A.3.3.98 Medical/Dental Office. Examples include a dental office/clinic, a medical office/clinic, an immediate care facility, and a podiatry office.

A.3.3.115 Nonflammable Medical Gas System. See Chapter 5, Gas and Vacuum Systems.

A.3.3.118 Oxidizing Gas. Oxygen and nitrous oxide are examples of oxidizing gases. There are many others, including halogens.

A.3.3.119 Oxygen. Oxygen's outstanding property is its ability to sustain life and to support combustion. Although oxygen is nonflammable, materials that burn in air will burn much more vigorously and create higher temperatures in oxygen or in oxygen-enriched atmospheres.

A.3.3.119.2 Liquid Oxygen. If spilled, the liquid can cause frostbite on contact with skin.

A.3.3.120 Oxygen Delivery Equipment. If an enclosure such as a mask, hood, incubator, canopy, or tent is used to contain the oxygen-enriched atmosphere, that enclosure is considered to be oxygen-delivery equipment.

A.3.3.122 Oxygen Hood. For additional information, see A.3.3.12.2 and Figure A.3.3.12.2.

A.3.3.123 Oxygen Toxicity (Hyperbaric). Under the pressures and times of exposure normally encountered in hyperbaric treatments, toxicity is a direct function of concentration and time of exposure.

A.3.3.127 Patient Care Space. Business offices, corridors, lounges, day rooms, dining rooms, or similar areas typically are not classified as patient care spaces.

A.3.3.127.1 Category 1 Space. These spaces, formerly known as critical care rooms, are typically where patients are intended to be subjected to invasive procedures and connected to line-operated, patient care–related appliances. Examples include, but are not limited to, special care patient rooms used for critical care, intensive care, and special care treatment rooms such as angiography laboratories, cardiac catheterization laboratories, delivery rooms, operating rooms, post-anesthesia care units, trauma rooms, and other similar rooms.

A.3.3.127.2 Category 2 Space. These spaces were formerly known as general care rooms. Examples include, but are not limited to, inpatient bedrooms, dialysis rooms, in vitro fertilization rooms, procedural rooms, and similar rooms.

A.3.3.127.3 Category 3 Space. These spaces, formerly known as basic care rooms, are typically where basic medical or dental care, treatment, or examinations are performed. Examples include, but are not limited to, examination or treatment rooms in clinics, medical and dental offices, nursing homes, and limited care facilities.

A.3.3.127.4 Category 4 Space. These spaces were formerly known as support rooms. Examples of support spaces include, but are not limited to, anesthesia work rooms, sterile supply, laboratories, morgues, waiting rooms, utility rooms, and lounges.

A.3.3.130 Patient Lead. A patient lead can be a surface contact (e.g., an ECG electrode), an invasive connection (e.g., implanted wire or catheter), or an incidental long-term connection (e.g., conductive tubing).

It is not intended to include adventitious or casual contacts, such as a push button, bed surface, lamp, hand-held appliance, and so forth.

(Also see 3.3.79, Isolated Patient Lead.)

A.3.3.135.5 Operating Pressure. The operating pressure for patient medical gases is typically 345 kPA to 380 kPa (50 psig to 55 psig). The operating pressure for medical support gases is typically 1100 kPa to 1275 kPA (160 psig to 185 psig).

A.3.3.135.6 Partial Pressure. The pressure contributed by other gases in the mixture is ignored. For example, oxygen is one of the constituents of air; the partial pressure of oxygen in standard air, at a standard air pressure of 14.7 psia, is 3.06 psia or 0.208 ATA or 158 mm Hg.

A.3.3.136 Pressure-Reducing Regulator. In hospitals, the term *regulator* is frequently used to describe a regulator that incorporates a flow-measuring device.

A.3.3.139 psig. Under standard conditions, 0 psig is equivalent to 14.7 psia.

A.3.3.144 Remote. A gas storage supply system can be remote from the single treatment facility, but all use points must be contiguous within the facility.

A.3.3.153 Single Treatment Facility. The definition of single treatment facility was established to take into consideration principally single-level installations or those of a practice that could be two-level, but are reached by open stairs within the confines of the single treatment facility. *(See Figure A.3.3.153.)*

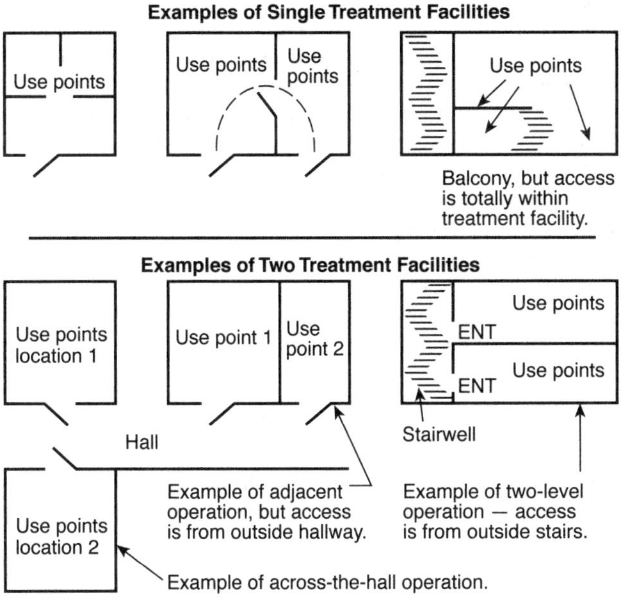

FIGURE A.3.3.153 Examples of Treatment Facilities.

A.3.3.154 Site of Intentional Expulsion. This definition addresses the site of intended expulsion. Actual expulsion can occur at other sites remote from the intended site due to disconnections, leaks, or rupture of gas conduits and connections. Vigilance on the part of the patient care team is essential to ensure system integrity.

For example, for a patient receiving oxygen via a nasal cannula or face mask, the site of expulsion normally surrounds the mask or cannula; for a patient receiving oxygen while enclosed in a canopy or incubator, the site of intentional expulsion normally surrounds the openings to the canopy or incubator; for a patient receiving oxygen while on a ventilator, the site of intentional expulsion normally surrounds the venting port on the ventilator.

A.3.3.160 Surface-Mounted Medical Gas Rail Systems. It is the intent that surface-mounted medical gas rail systems would be permitted in individual patient rooms but would not be permitted to go directly through room walls to adjacent patient rooms. However, it is the intent to permit surface-mounted medical gas rails to be used in a given critical care area where there can be a partition separating certain patient care functions, essentially leaving the system within the given critical care area. As an example, two adjacent patient rooms outside of a critical care unit would not be permitted to have a surface-mounted medical gas rail interconnect between the two rooms through the wall. However, in a nursery where there might be one or two segregated areas for isolation, a medical gas rail system supplying more than one isolation room, but within the nursery area, would be permitted to be interconnected with the nursery system.

A.3.3.165.1 Endotracheal Tube. An endotracheal tube can be equipped with an inflatable cuff.

A.3.3.171 Wet Procedure Locations. Routine housekeeping procedures and incidental spillage of liquids do not define a wet procedure location.

A.4.1 Four levels of systems categories are defined in this code, based on the risks to patients and caregivers in the facilities. The categories are as follows:

(1) Category 1: Systems are expected to work or be available at all times to support patient needs.
(2) Category 2: Systems are expected to provide a high level of reliability; however, limited short durations of equipment downtime can be tolerated without significant impact on patient care. Category 2 systems support patient needs but are not critical for life support.
(3) Category 3: Normal building system reliabilities are expected. Such systems support patient needs, but failure of such equipment would not immediately affect patient care. Such equipment is not critical for life support.
(4) Category 4: Such systems have no impact on patient care and would not be noticeable to patients in the event of failure.

The category definitions apply to equipment operations and are not intended to consider intervention by caregivers or others. Potential examples of areas/systems and their categories of risk follow. A risk assessment should be conducted to evaluate the risk to the patients, staff, and visitors.

(1) Ambulatory surgical center, two patients with full OR services, Category 1
(2) Reconstructive surgeon's office with general anesthesia, Category 1
(3) Procedural sedation site for outpatient services, Category 2
(4) Cooling Towers in Houston, TX, Category 2
(5) Cooling Towers in Seattle, WA, Category 3
(6) Dental office, no general anesthesia, Category 3
(7) Typical doctor's office/exam room, Category 4
(8) Lawn sprinkler system, Category 4

A.4.1.1 Major injury can include the following:

(1) Any amputation
(2) Loss of the sight of an eye (whether temporary or permanent)
(3) Chemical or hot metal burn to the eye or any penetrating injury to the eye
(4) Any injury that results in electric shock and electric burns leading to unconsciousness and that requires resuscitation or admittance to a hospital for 24 hours or more
(5) Any other injury leading to hypothermia, heat induced illness, or unconsciousness requiring resuscitation or admittance to a hospital for 24 hours or more
(6) Loss of consciousness caused by asphyxia or lack of oxygen or exposure to a biological agent or harmful substance
(7) Absorption of any substance by inhalation, skin, or ingestion causing loss of consciousness or acute illness requiring medical treatment
(8) Acute illness requiring medical treatment where there is reason to believe the exposure was to biological agents, its toxins, or infected materials

A.4.1.2 A minor injury means *not serious* or *involving risk of life*.

A.4.2 Risk assessment should follow procedures such as those outlined in ISO/IEC 31010, *Risk Management — Risk Assessment Techniques*; NFPA 551, *Guide for the Evaluation of Fire Risk Assessments*; SEMI S10-0307E, *Safety Guideline for Risk Assessment and Risk Evaluation Process*; or other formal process. The results of the assessment procedure should be documented and records retained. Figure A.4.2 is a sample risk assessment model that can be used to evaluate the categories.

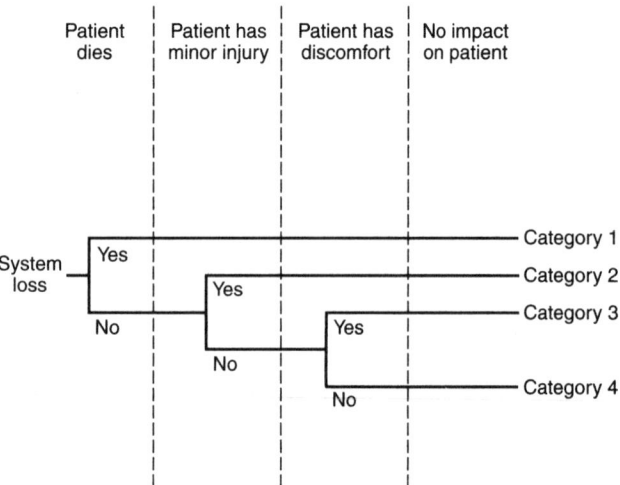

FIGURE A.4.2 Sample Risk Assessment.

A.4.4.1 The provisions of 4.4.1 do not require inherently noncombustible materials to be tested in order to be classified as noncombustible materials.

A.4.4.2 Materials subject to increase in combustibility or flame spread index beyond the limits herein established through the effects of age, moisture, or other atmospheric condition are considered combustible. (See NFPA 259, *Standard Test Method for Potential Heat of Building Materials*, and NFPA 220, *Standard on Types of Building Construction*.) [*101*: A.4.6.14]

A.5.1.1 Section 5.1 covers requirements for Category 1 piped gas and vacuum systems; Section 5.2 covers Category 2 piped gas and vacuum systems; and Section 5.3 covers Category 3 piped gas and vacuum systems. Laboratory systems are no longer covered by Chapter 5 (2002 edition).

A.5.1.1.3 These requirements do not restrict the distribution of other inert gases through piping systems.

A.5.1.3 See Figure A.5.1.3. Category 1 source drawings in this annex are representational, demonstrating a possible arrangement of components required by the text. The diagrams are not intended to imply method, materials of construction, or more than one of many possible and equally compliant arrangements. Alternative arrangements are permitted if they meet the intent of the text. Listed paragraphs might not be the only paragraphs that apply.

A.5.1.3.1.1 Regulations of the U.S. Department of Transportation (formerly U.S. Interstate Commerce Commission) outline specifications for transportation of explosives and dangerous articles (49 CFR 171–190). In Canada, the regulations of the Canadian Transport Commission, Union Station, Ottawa, Ontario, apply.

A.5.1.3.1.2 CGA documents contain both mandatory and nonmandatory language. Enforceable language uses the word "shall"; nonmandatory language uses the word "should." This section indicates that NFPA 99 is making reference only to the mandatory requirements in the CGA document.

A.5.1.3.3 The bulk supply system should be installed on a site that has been prepared to meet the requirements of NFPA 55, *Compressed Gases and Cryogenic Fluids Code*, or CGA G-8.1, *Standard for Nitrous Oxide Systems at Consumer Sites*. A storage unit(s), reserve, pressure regulation, and a signal actuating switch(es) are components of the supply system. Shutoff valves, piping from the site, and electric wiring from a signal switch(es) to the master signal panels are components of the piping system.

The bulk supply system is normally installed on the site by the owner of this equipment. The owner or the organization responsible for the operation and maintenance of the bulk supply system is responsible for ensuring that all components of the supply system — main supply, reserve supply, supply system signal-actuating switch(es), and delivery pressure regulation equipment — function properly before the system is put in service.

A.5.1.3.3.1.10 Examples of inert gases include but are not limited to helium and nitrogen.

A.5.1.3.3.2 Electric wiring and equipment in storage rooms for oxygen and nitrous oxide are not required to be explosion-proof.

A.5.1.3.3.2(6) The test for walls and floors is ASTM E119, *Standard Test Methods for Fire Tests of Building Construction and Materials*; or UL 263, *Fire Resistance Ratings*. The test for doors is NFPA 252, *Standard Methods of Fire Tests of Door Assemblies*.

A.5.1.3.3.2(7) Electrical devices should be physically protected, such as by use of a protective barrier around the electrical devices, or by location of the electrical device such that it will avoid causing physical damage to the cylinders or containers. For example, the device could be located at or above 1.5 m (5 ft) above finished floor or other location that will not allow the possibility of the cylinders or containers to come into contact with the electrical device as required by this section.

A.5.1.3.3.2(10) Chapter 6 specifies medical gas equipment that should be powered by the essential electrical systems. Electrical equipment that is not essential for the operation of the supply system can be powered by nonessential power (e.g., telemetry, site lighting).

A.5.1.3.5 See Figure A.5.1.3.5. A four-valve bypass arrangement is illustrated. Three-way valves are permitted in lieu of the four valves shown.

A.5.1.3.5.4 Components include, but are not limited to, containers, valves, valve seats, lubricants, fittings, gaskets, and interconnecting equipment, including hose. Easily ignitable materials should be avoided.

Compatibility involves both combustibility and ease of ignition. Materials that burn in air will burn violently in pure oxygen at normal pressure and explosively in pressurized oxygen.

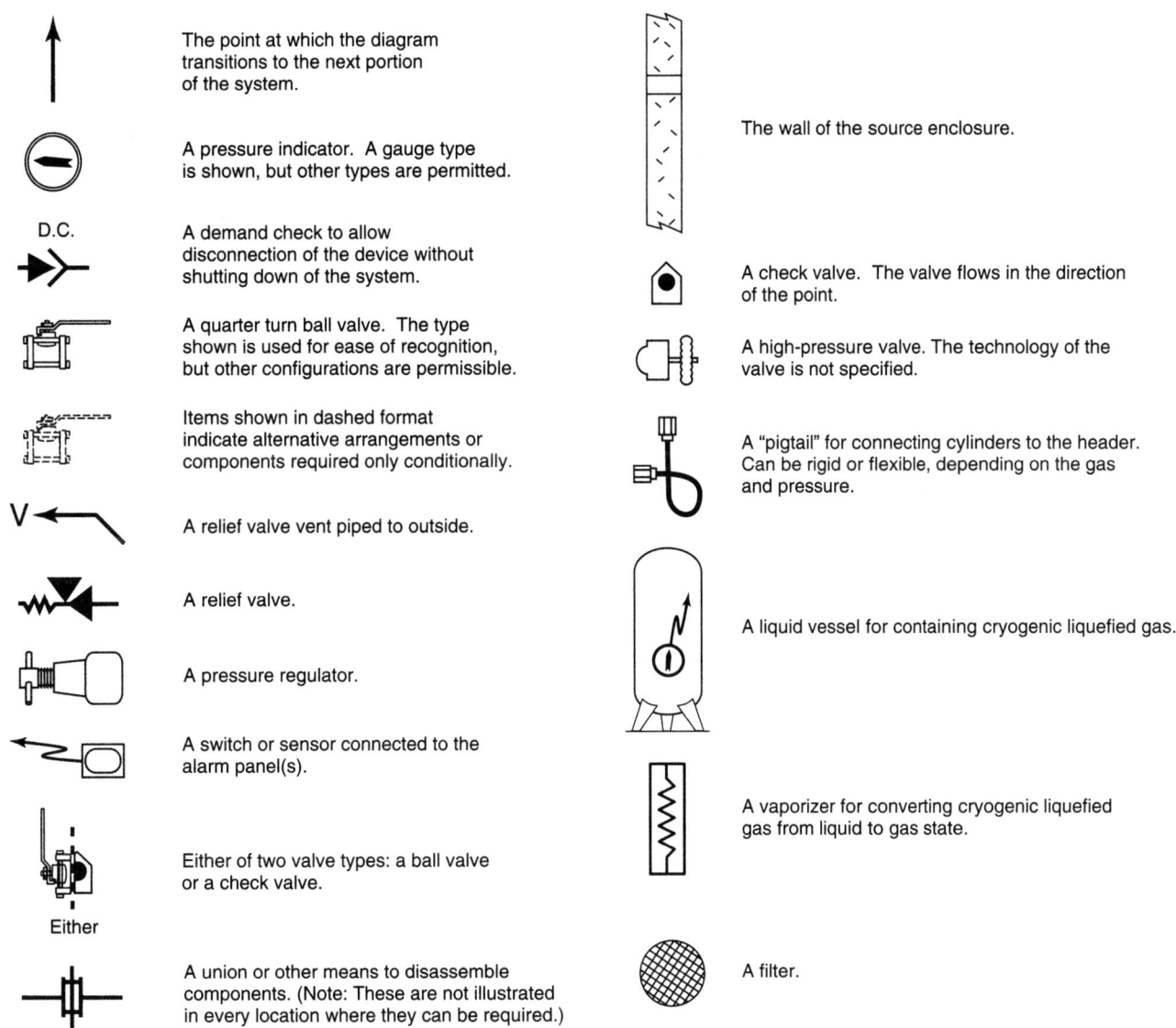

FIGURE A.5.1.3 Legend for Typical Category 1 Source Drawings.

Also, many materials that do not burn in air will do so in pure oxygen, particularly under pressure. Metals for containers and piping have to be carefully selected, depending on service conditions. The various steels are acceptable for many applications, but some service conditions can call for other materials (usually copper or its alloys) because of their greater resistance to ignition and lower rate of combustion.

Similarly, materials that can be ignited in air have lower ignition energies in oxygen. Many such materials can be ignited by friction at a valve seat or stem packing or by adiabatic compression produced when oxygen at high pressure is rapidly introduced into a system initially at low pressure.

A.5.1.3.5.10 See Figure A.5.1.3.5.10(a) and Figure A.5.1.3.5.10(b). Connection to the gas outlet connection is illustrated. If the liquid outlet connection were used, an external vaporizer could be required.

A.5.1.3.5.10(1) The appropriate number of cylinders should be determined after consideration of delivery schedules, proximity of the facility to alternate supplies, and the emergency plan.

A.5.1.3.5.11 See Figure A.5.1.3.5.11.

A.5.1.3.5.12 See Figure A.5.1.3.5.12.

A.5.1.3.5.13.3 The provisions of Section 15.7 cover the basic functions of a complete fire alarm system, including fire detection, alarm, and communications. These systems are primarily intended to provide the indication and warning of abnormal conditions, the summoning of appropriate aid, and the control of occupancy facilities to enhance protection of life.

Some of the provisions of Section 15.7 originated with *NFPA 72, National Fire Alarm and Signaling Code.* For purposes of this *Code*, some provisions of Section 15.7 are more stringent than those of *NFPA 72*, which should be consulted for additional details. [*101*: A.9.6.1]

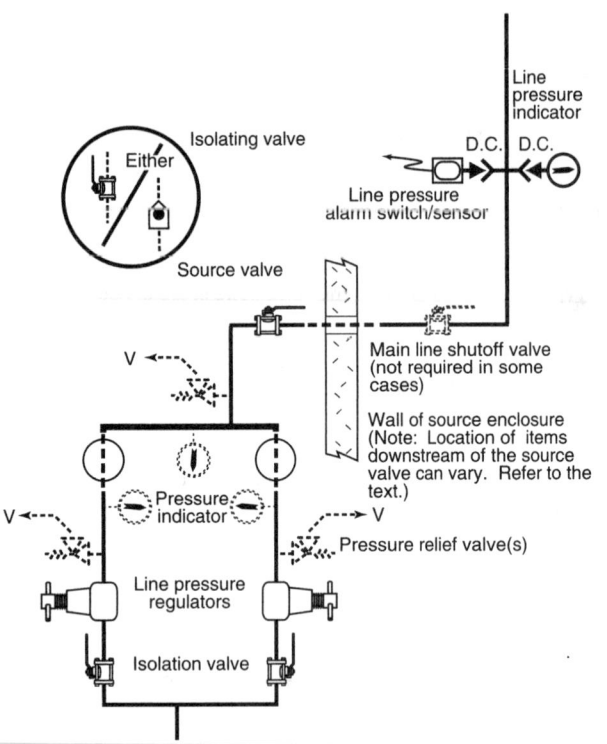

FIGURE A.5.1.3.5 Typical Arrangement for Line Controls at Pressure Sources.

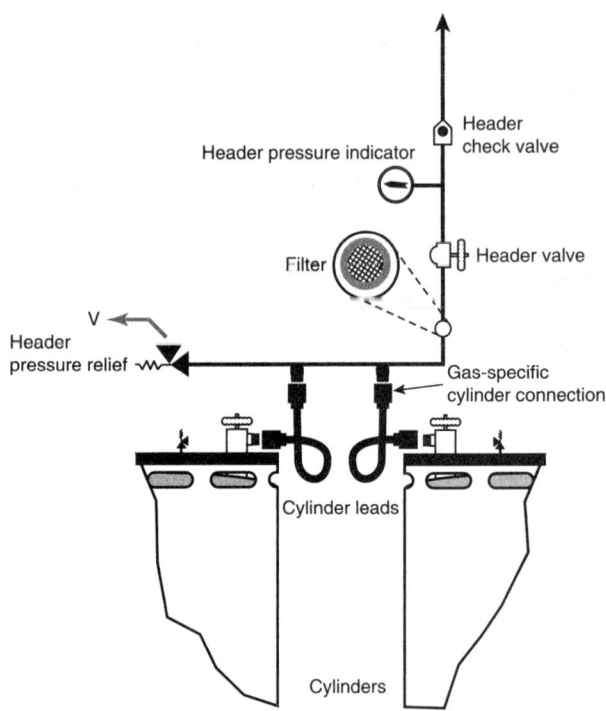

FIGURE A.5.1.3.5.10(b) Header for Cryogenic Gas in Containers.

A.5.1.3.5.14 For bulk oxygen systems, see NFPA 55, *Compressed Gases and Cryogenic Fluids Code.* See Figure A.5.1.3.5.14(a) and Figure A.5.1.3.5.14(b). Two possible choices of reserves are illustrated. Both are not required.

A.5.1.3.5.14.4 The local signal arose from the simple need of a maintenance person to know what is going on with any given piece of source equipment. Note that it is not an alarm in the sense of a local or master alarm. It is simply an indicator, which might be a gauge, a flag, a light, or some other possible manifestation that allows a maintenance person to stand at the equipment and know what conditions are present (e.g., which header of cylinders is in service). The elements to be displayed are typically those that will also be monitored at the master alarm, but the local signal is visible at the equipment rather than remotely.

A.5.1.3.5.15 See Figure A.5.1.3.5.15.

If the relief valve on the emergency oxygen supply connection is moved downstream from the check valve in the emergency oxygen line, it should be connected to the system with a demand check fitting.

The emergency oxygen supply connection (EOSC) can be used as a part of the emergency operation plan (EOP) for an unplanned loss of oxygen supply. However, a risk assessment should be conducted by the facility to determine the contingency plan for vital life support and critical care areas. There might need to be interim measures for dealing with the loss of oxygen (e.g., high-pressure oxygen cylinders available for back feeding critical care areas).

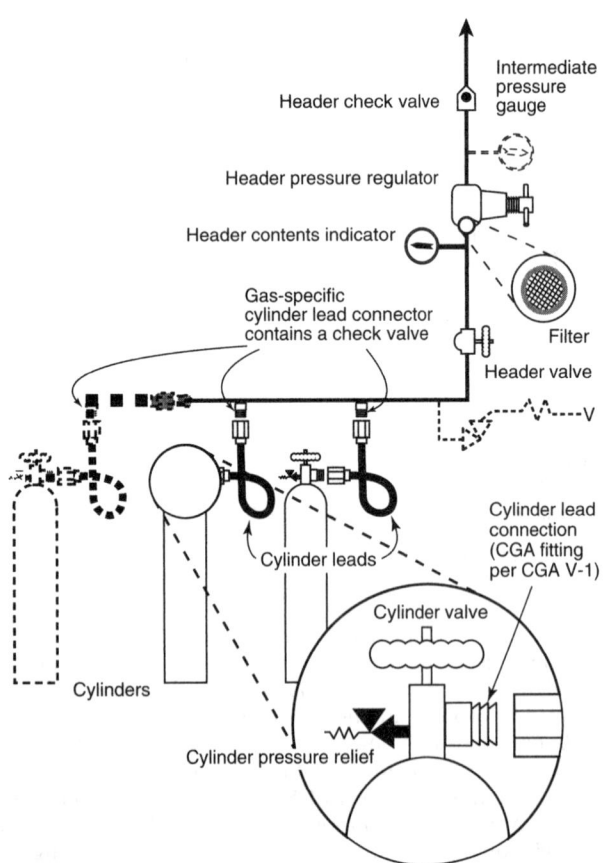

FIGURE A.5.1.3.5.10(a) Header for Gas in Cylinders.

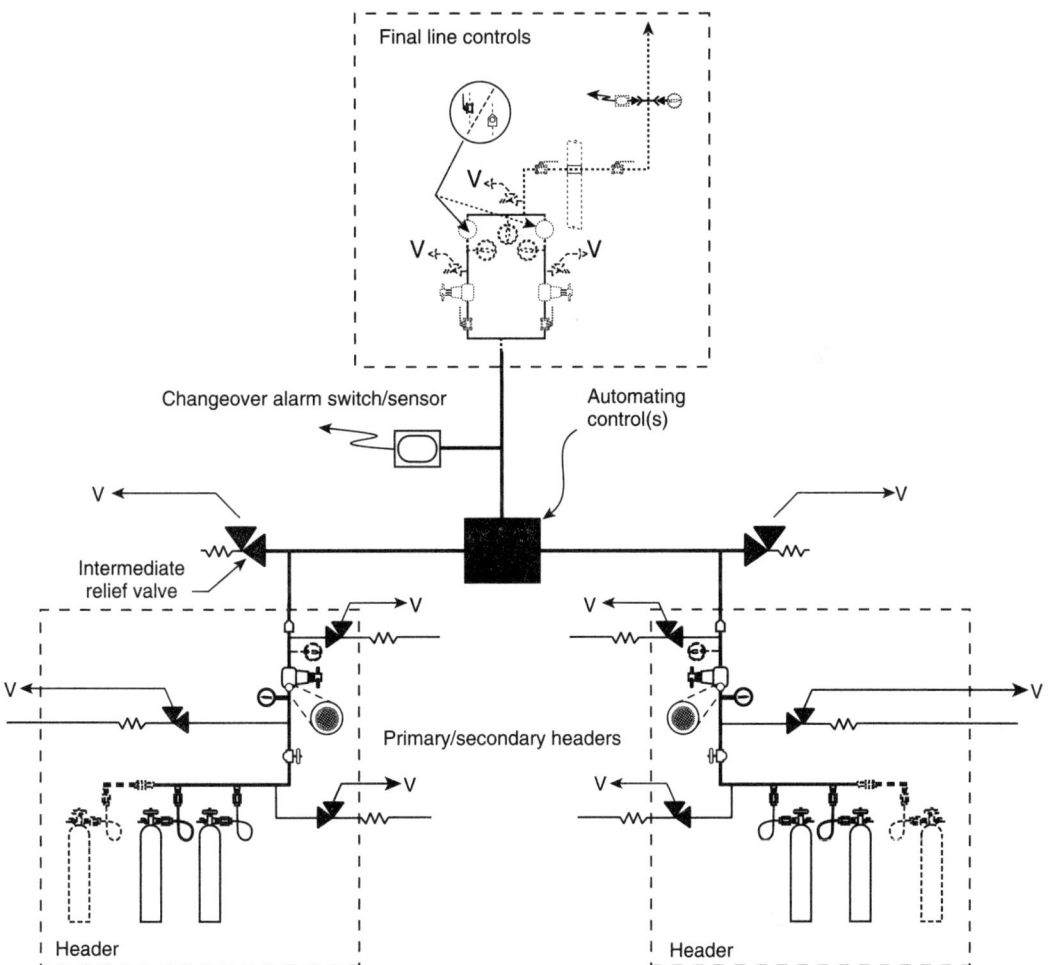

FIGURE A.5.1.3.5.11 Manifold for Gas Cylinders.

A.5.1.3.6 Air supplied from on-site compressor and associated air treatment systems (as opposed to medical air USP supplied in cylinders) that complies with the specified limits is considered medical air. Hydrocarbon carryover from the compressor into the pipeline distribution system could be detrimental to the safety of the end user and to the integrity of the piping system. Mixing of air and oxygen is a common clinical practice, and the hazards of fire are increased if the air is contaminated. Compliance with these limits is thus considered important to fire and patient safety. The quality of local ambient air should be determined prior to its selection for compressors and air treatment equipment. See Figure A.5.1.3.6.

A.5.1.3.6.1 Supply systems for medical air using compressors draw air of the best available quality from a source of clean local ambient air; add no contaminants in the form of particulate matter, odor, or other gases; and dry, filter, regulate, and supply that air only via the medical air piping distribution system for use exclusively in the application of human respiration.

The utilization of an air treatment system is the joint responsibility of the system designer, the hospital clinical and engineering staffs, and the authority having jurisdiction. Different types of compressors have characteristics that affect the selection of the type of air treatment system. Some air treatment systems impose an additional load upon the compressors that has to be accounted for in the sizing of the system (usable capacity). The compressor duty cycle has to be chosen in accordance with the manufacturer's recommendation.

The type of air compressor and air condition at the intake will govern the type of filter provided for the air compressor supply system. All filters should be examined quarterly for the presence of liquids or excessive particulates and replaced according to the manufacturer's instructions.

One procedure for reaching a decision on the quality of the medical air is the following:

(1) Test at the intake and at the sample connection valve.
(2) If the two purities agree within the limits of accuracy of the test, the compressor system can be accepted.

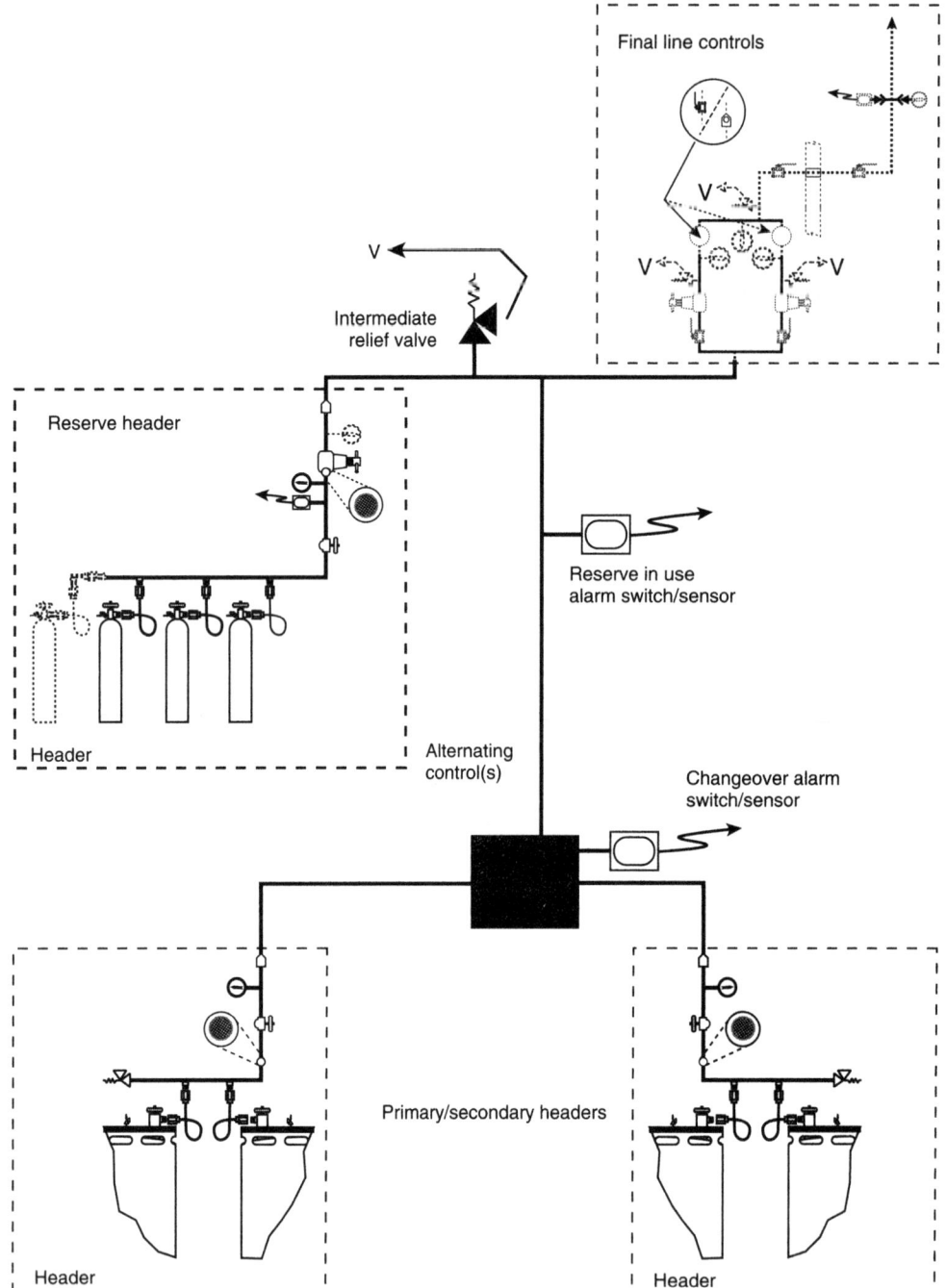

FIGURE A.5.1.3.5.12 Typical Source of Supply for Cryogenic Gas in Containers.

(3) If the air is found to exceed the values for medical compressed air as defined in 5.1.3.6.1, the facility can elect to install purification apparatus for the contaminants in question.

A.5.1.3.6.2 It is the intent that the medical air piping distribution system support only the intended need for breathable air for such items as intermittent positive pressure breathing (IPPB) and long-term respiratory assistance needs, anesthesia machines, and so forth. The system is not intended to be used to provide engineering, maintenance, and equipment needs for general hospital support use. It is the intent that the life safety nature of the medical air be protected by a system dedicated solely for its specific use.

As a compressed air supply source, a medical air compressor should not be used to supply air for other purposes, because such use could increase service interruptions, reduce service life, and introduce additional opportunities for contamination.

A.5.1.3.6.3 See Figure A.5.1.3.6.

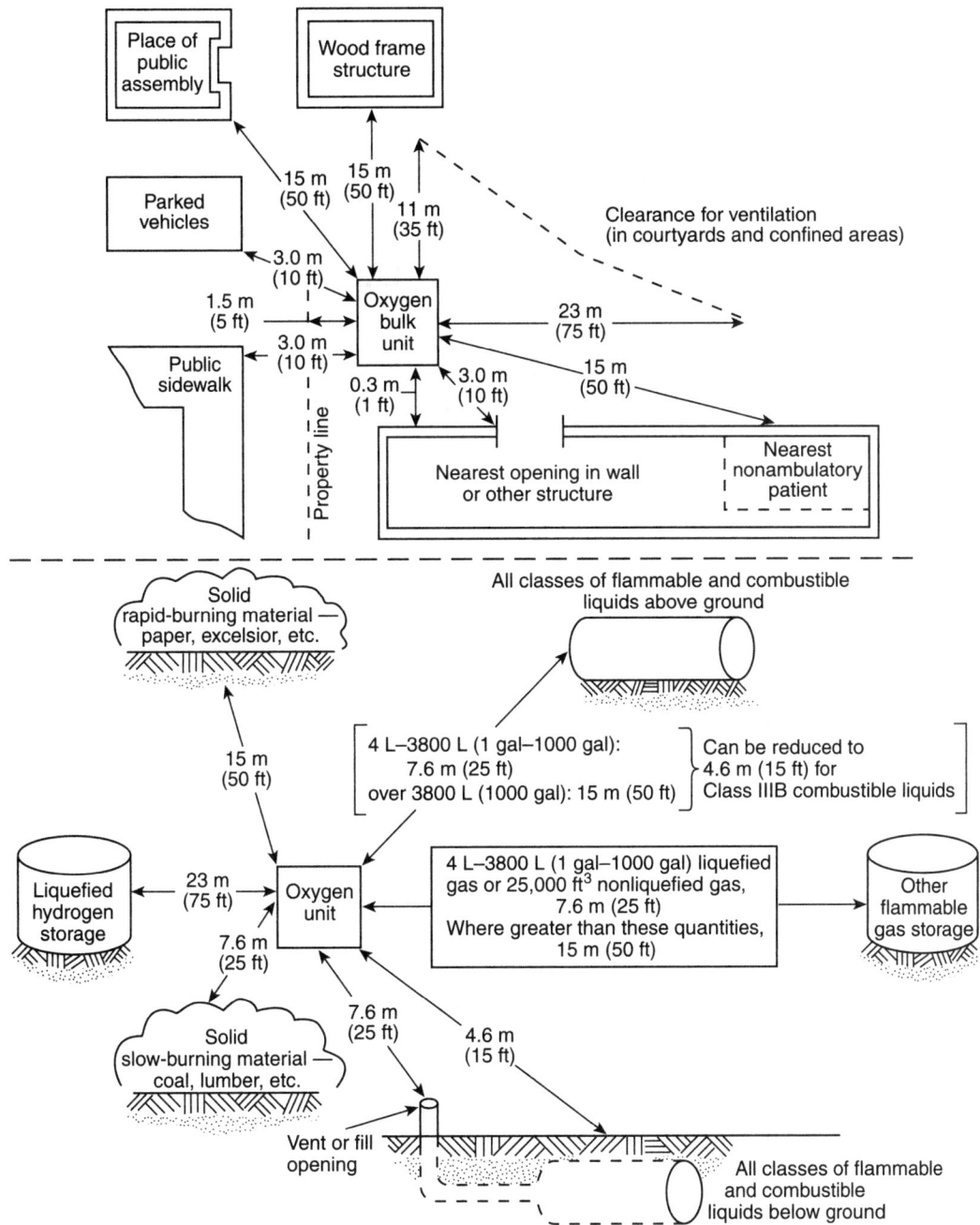

FIGURE A.5.1.3.5.14(a) Distance Between Bulk Oxygen Systems and Exposures.

A.5.1.3.6.3.4(A) Examples of 5.1.3.6.3.4(A)(1) are liquid ring and permanently sealed bearing compressors.

An example of 5.1.3.6.3.4(A)(2) is an extended head reciprocating compressor with an atmospheric vent between the compression chamber and the crankcase.

An example of 5.1.3.6.3.4(A)(3) is a rotating element compressor with the compression chamber being nonlubricated and separated from the lubricated gears by at least one shaft seal with an atmospheric vent on both sides. The vent on the lubricated side is provided with a gravity drain to atmosphere.

A.5.1.3.6.3.9(D) A typical example of valving the receiver is shown in Figure A.5.1.3.6.3.9(D).

A.5.1.3.6.3.9(F) The two configurations are equally acceptable. The components can be arranged in either of the arrangements shown in Figure A.5.1.3.6.3.9(F).

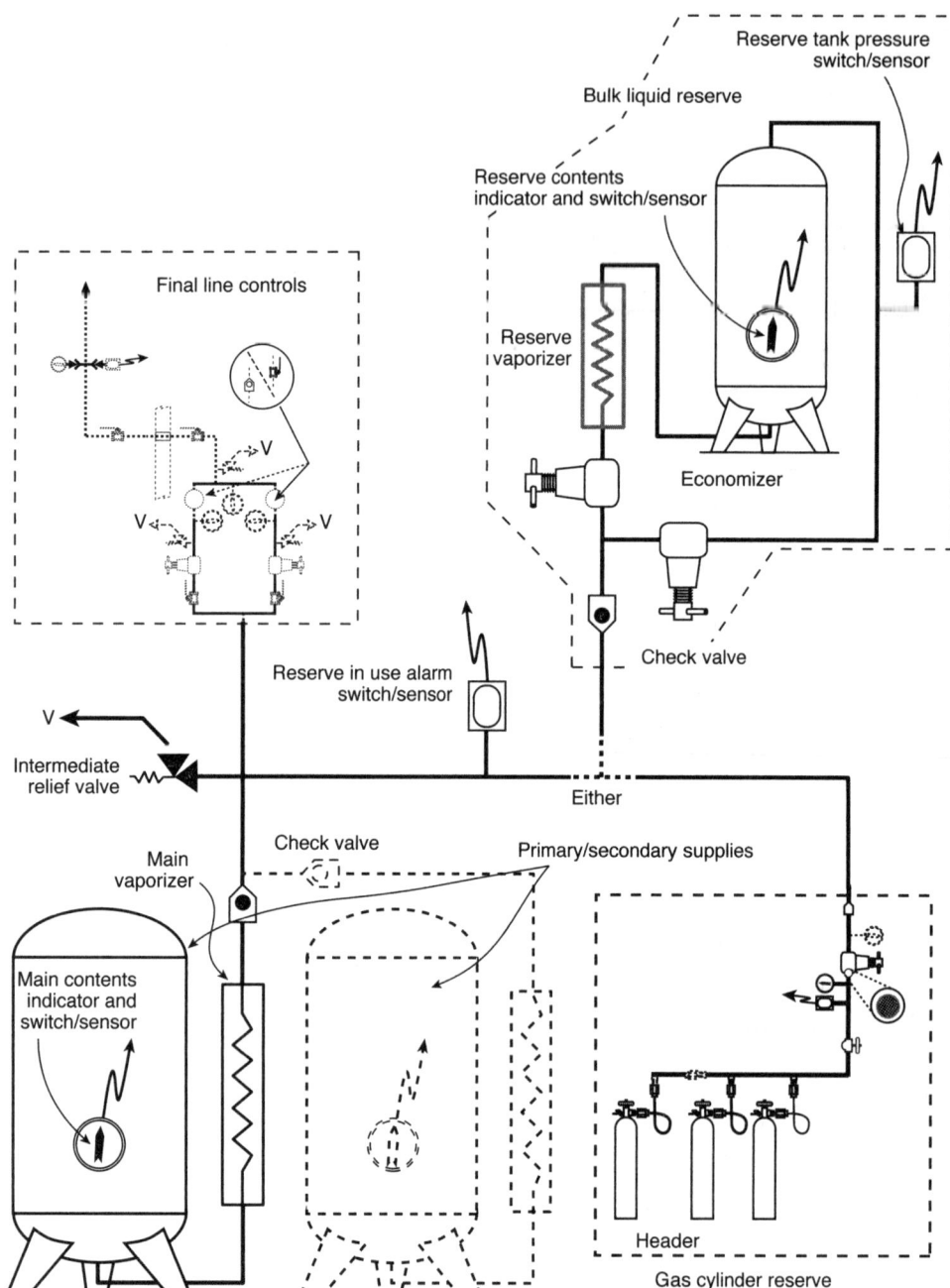

FIGURE A.5.1.3.5.14(b) Typical Source of Supply for Cryogenic Gas in Bulk.

A.5.1.3.6.3.14(A)(9) The proportioning system should be monitored for conditions that can affect air quality during use or in the event of failure, based on the type of proportioning system design used in the system, including monitoring for the following systems and conditions:

(1) Where proportioning systems used are configured with a primary proportioning system and a reserve medical air manifold per 5.1.3.5.11
(2) Where proportioning systems used are configured with a primary proportioning system and a reserve proportioning system
(3) Where proportioning systems used are configured with a primary proportioning system and a reserve medical air compressor per 5.1.3.5.3
(4) Alarm at a predetermined set point, before the reserve supply begins to supply the system, indicating reserve supply in use
(5) Alarm at a predetermined set point, before the reserve supply contents fall to one day's average supply, indicating reserve low

Water-in-receiver alarms are not required for a proportioning system.

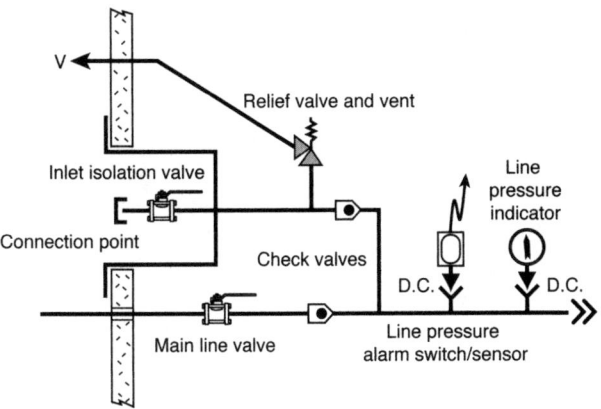

FIGURE A.5.1.3.5.15 Emergency Oxygen Supply Connection.

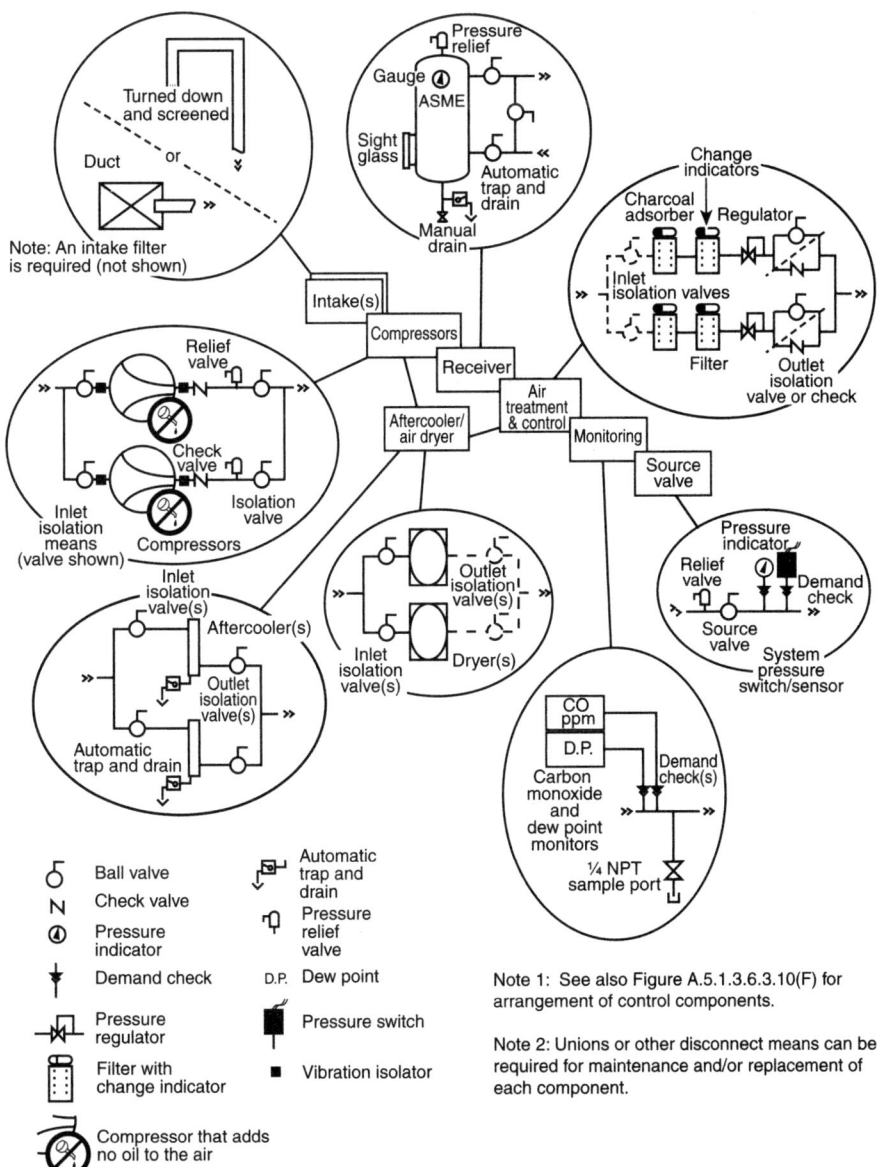

FIGURE A.5.1.3.6 Elements of a Typical Duplex Medical Air Compressor Source System (Category 1 Gas Systems).

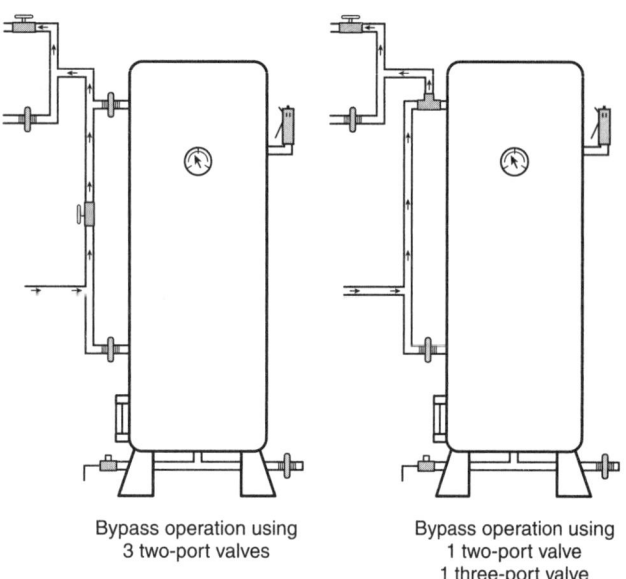

FIGURE A.5.1.3.6.3.9(D) Receiver Valving Arrangement.

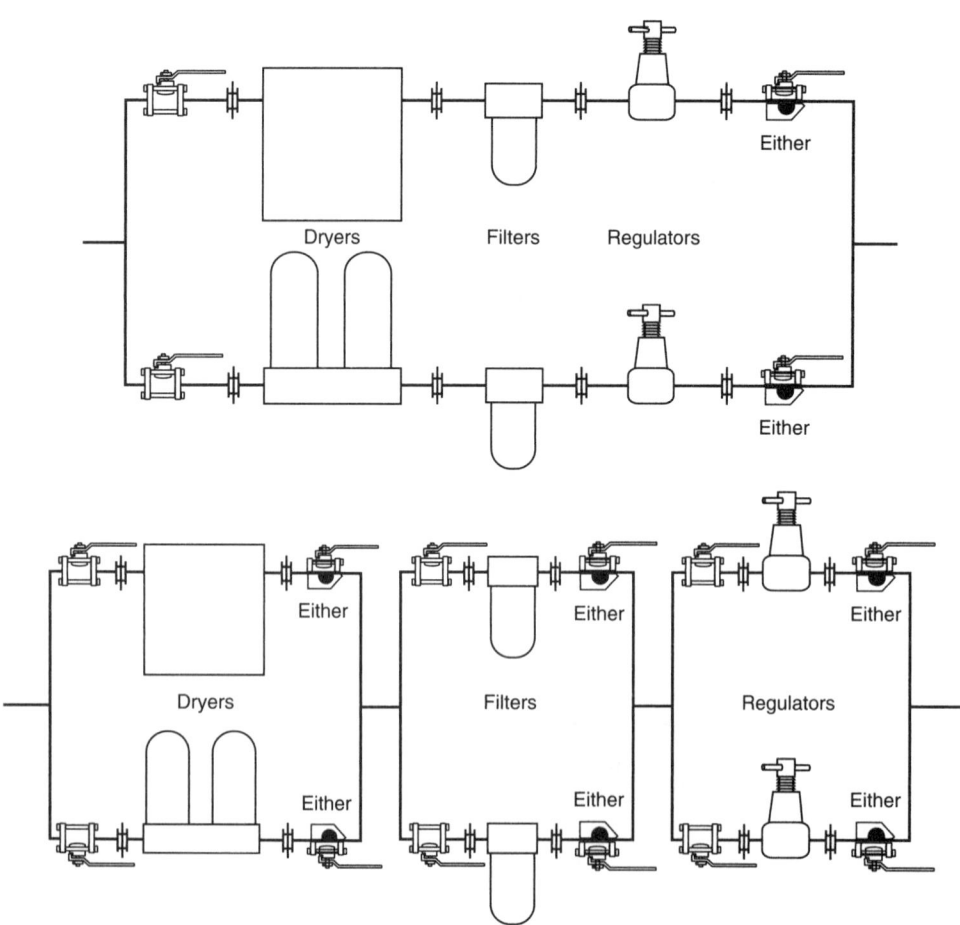

FIGURE A.5.1.3.6.3.9(F) Alternate Valving Sequences for Line Controls in Medical Air.

The engineering controls should include, as a minimum, the following:

(1) Engineering controls should be in place when the sources of oxygen USP or nitrogen NF, or both, are the same sources as those supplying the oxygen USP or nitrogen NF pipelines, or both, for other uses within the health care facility, and the following should apply:

(a) In cases where a new supply system is installed, or in cases where one or more bulk supplies are used to supply the mixer, bulk systems and vaporizers should be sized for total peak demand flow, including peak demand flow to the mixer and any other areas of utilization.

(b) Operating limits should be established, at a minimum, for the oxygen and nitrogen source pressures, both high and low, and for the medical air oxygen concentration, both high and low, based upon USP specifications. A process upset can be defined as an excursion in the process windows established for oxygen and nitrogen source pressures or medical air oxygen concentration, or both. A means to detect excursion from these process limits and power failure should be provided.

(c) At least one dedicated valve or other control should be installed in the proportioning system and/or the line(s) between the oxygen and/or nitrogen supply system(s) and proportioning system. The purpose of the dedicated control(s) is to prevent the cross contamination of the oxygen and nitrogen lines due to product backflow as follows:

   i. The control(s) should be separate from the valve(s) or other device(s) used to control oxygen flow and nitrogen flow in normal operation.
   ii. The control(s) should not cycle in normal operation.
   iii. If installed in the line(s) between the oxygen or nitrogen supply system(s), or both, and proportioning system, upon activation of the control(s), an alarm should be sent to the facility. The control(s) cannot exist exclusively via the use of check valves.

(d) In the event of a process upset, the dedicated control(s) should either positively isolate the supply of oxygen or nitrogen, or both, from the mixer, or the dedicated control(s) should reduce the mixer pressure to less than half of the minimum final line pressure values, each, for the oxygen and nitrogen lines. In the event of a process upset, the control(s) should operate. Manual reset should be required to restart the proportioning system.

A.5.1.3.6.3.14(C)(9) The proportioning system should be monitored for conditions that can affect air quality during use or in the event of failure, based on the type of proportioning system design used in the system, including the following situations:

(1) Where proportioning systems that are configured with a primary proportioning system and a reserve medical air manifold per 5.1.3.5.11 are used

(2) Where proportioning systems that are configured with a primary proportioning system and a reserve proportioning system are used

(3) Where proportioning systems that are configured with a primary proportioning system and a reserve medical air compressor per 5.1.3.5.11 are used

(4) When proportioning systems are configured to alarm at a predetermined set point before the reserve supply begins to supply the system, indicating reserve supply in use

(5) When proportioning systems are configured to alarm at a predetermined set point before the reserve supply contents fall to one day's average supply, indicating reserve low

Water-in-receiver alarms are not required for proportioning systems.

A.5.1.3.7 See Figure A.5.1.3.7.

A.5.1.3.8 A functioning WAGD system allows the facility to comply with occupational safety requirements by preventing the accumulation of waste anesthetic gases in the work environment.

WAGD using an HVAC (heating, ventilation, and air-conditioning) system are not within the scope of Chapter 5.

Flammable and nonflammable gases are known to be incompatible with some seals and piping used in medical–surgical vacuum systems. If WAGD is to be included as part of the medical–surgical vacuum system, it should be recognized that this activity might cause deterioration of the vacuum system. The station inlet performance tests outlined in 5.1.12.3.10 are extremely important in maintaining the integrity of the medical–surgical vacuum system, and they should be made at more frequent intervals if WAGD is included in the vacuum system.

A.5.1.3.8.1 Interfaces are provided with overpressure, underpressure, overflow, and underflow compensation to ensure the breathing circuit is isolated from the WAGD system.

A.5.1.4 See Figure A.5.1.4.

Area alarms are required in critical care locations (e.g., intensive care units, coronary care units, angiography laboratories, cardiac catheterization laboratories, post-anesthesia recovery rooms, and emergency rooms) and anesthetizing locations (e.g., operating rooms and delivery rooms). Refer to definitions for these areas.

A.5.1.4.3 The presence of a main line shutoff valve is optional where the source valve can equally or more effectively perform the same function. An example is a case where the source is within the building or just on the outside of the building and, therefore, there would be no great distance separating the two valves. A source that was physically separate from the building would require both valves to ensure the intervening piping could be controlled.

A.5.1.5 Station outlets/inlets should be located at an appropriate height above the floor to prevent physical damage to equipment attached to the outlet.

A.5.1.6 Manufactured assembly examples include headwalls, columns, ceiling columns, ceiling-hung pendants, movable track systems, and so forth. See Figure A.5.1.6.

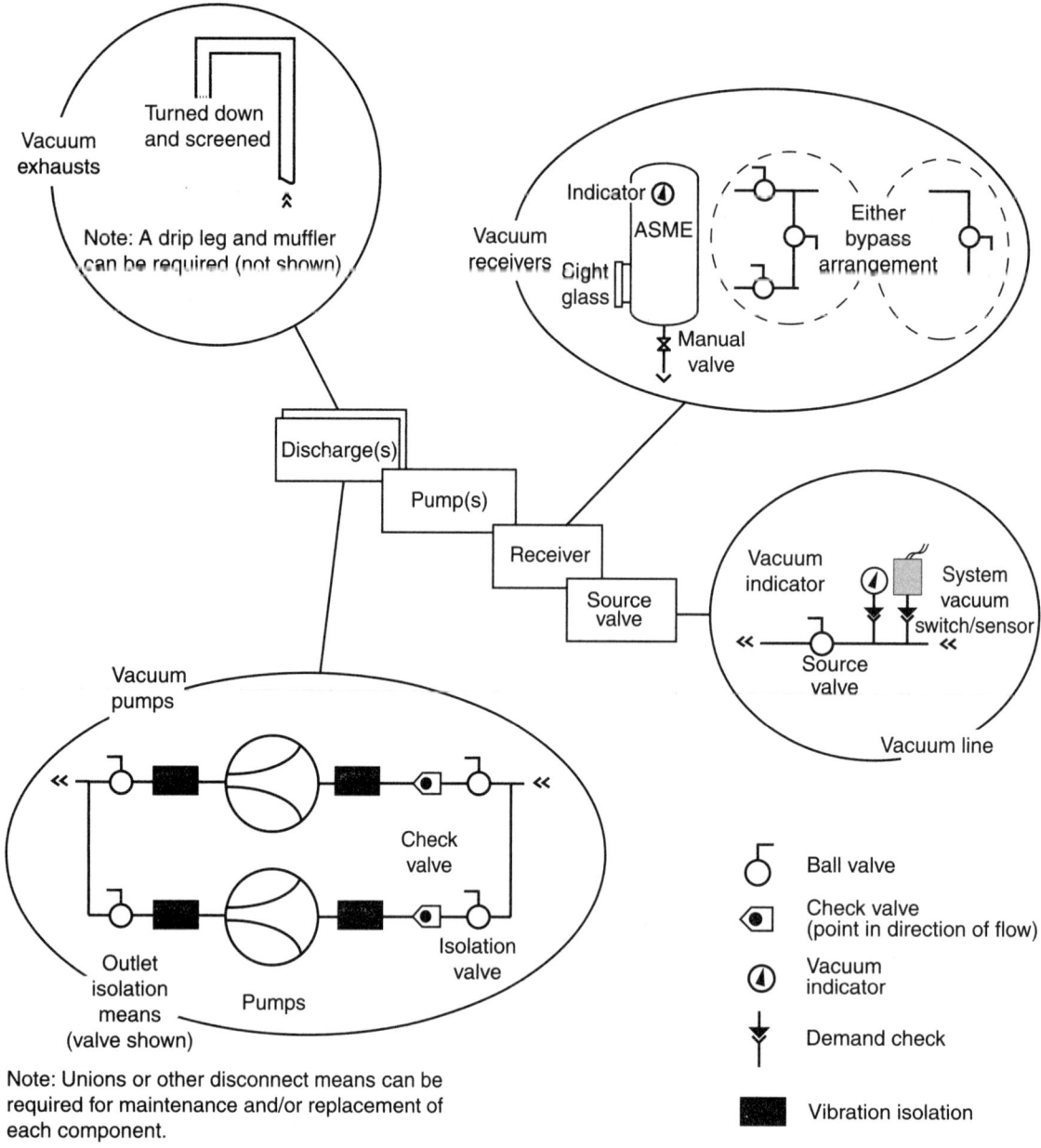

FIGURE A.5.1.3.7 Elements of Typical Duplex Vacuum Source System (Category 1 Vacuum Systems).

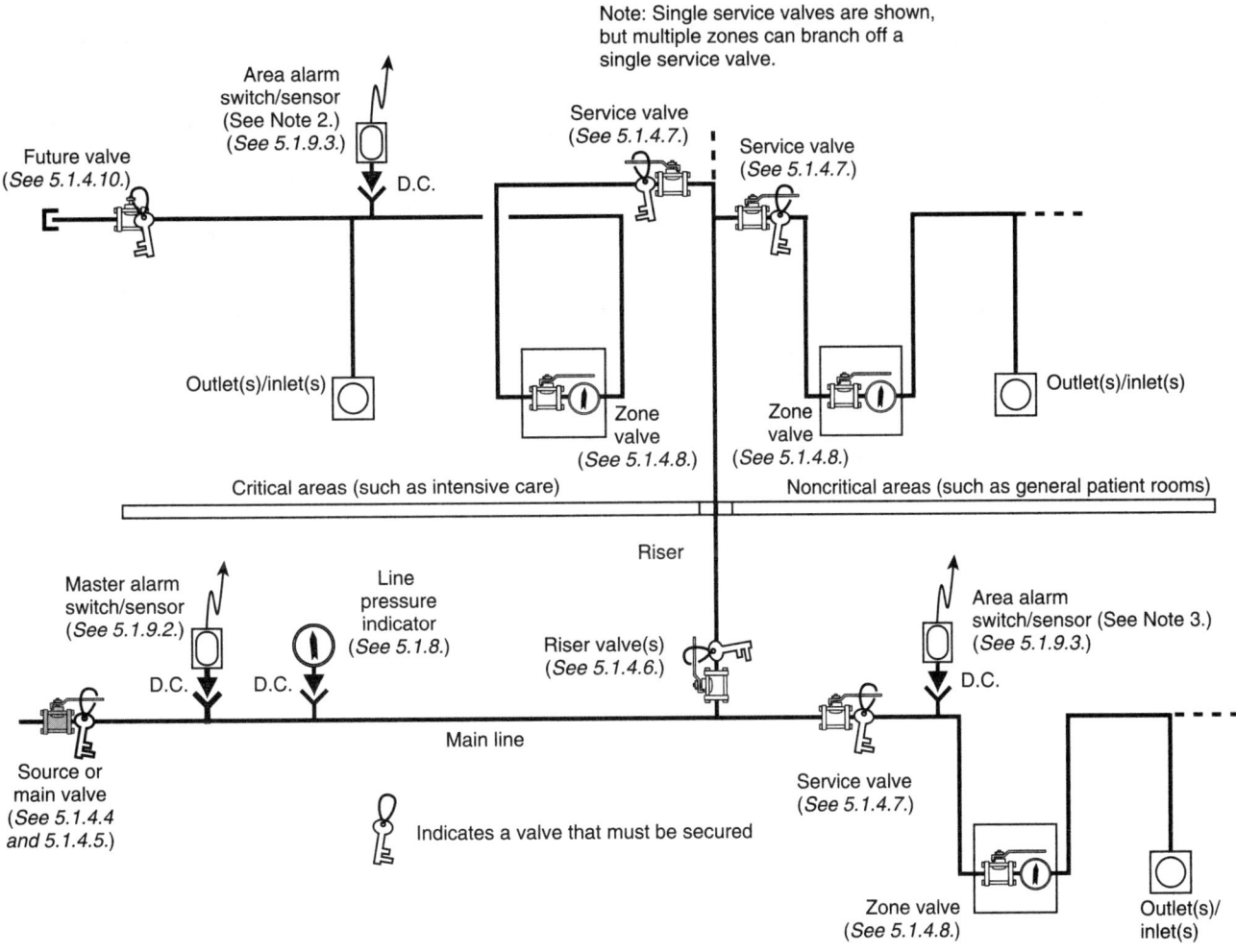

FIGURE A.5.1.4 Arrangement of Pipeline Components.

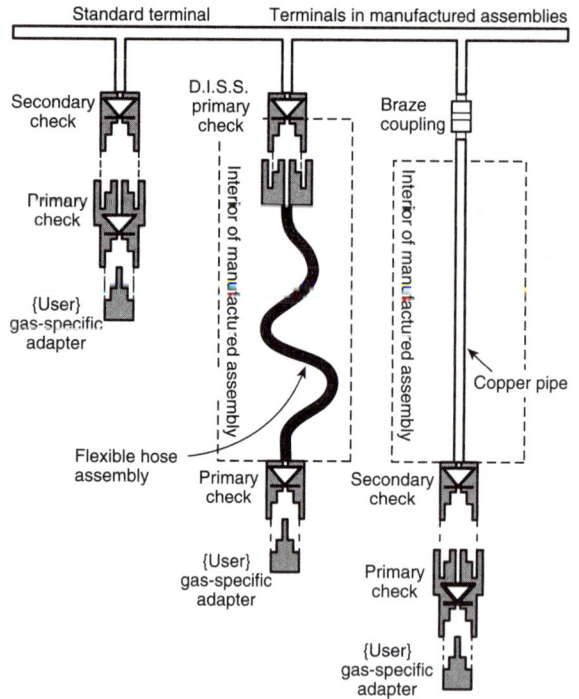

FIGURE A.5.1.6 Terminals in Manufactured Assemblies.

A.5.1.7 It is the intent that surface-mounted medical gas rail systems would be permitted in individual patient rooms but would not be permitted to go directly through room walls to adjacent patient rooms. However, it is the intent to permit surface-mounted medical gas rails to be used in a given critical care area where there can be a partition separating certain patient care functions, essentially leaving the system within the given critical care area. As an example, two adjacent patient ooms outside of a critical care unit would not be permitted to have a surface-mounted medical gas rail interconnect between the two rooms through the wall. However, in a nursery where there might be one or two segregated areas for isolation, a medical gas rail system supplying more than one isolation room, but within the nursery area, would be permitted to be interconnected with the nursery system.

A.5.1.7.9 Typical plating would be nickel plating over copper or brass per Federal Specification QQ-N290, Class I, Type 7.

A.5.1.8.1.3 This gauge would therefore be suitable for any operating pressure of 690 kPa to 1380 kPa (100 psig to 200 psig).

A.5.1.9 See Figure A.5.1.4.

A.5.1.9.2 See Table A.5.1.9.2.

Table A.5.1.9.2 Requirements for Category 1 Master Alarms for Gas and Vacuum Systems

Alarm Condition	Manifold for Gas Cylinders (5.1.3.5.11)	Manifold for Cryogenic Liquid Cylinders with Reserve (5.1.3.5.12)	Cryogenic Bulk with Cryogenic Reserve (5.1.3.5.13)	Cryogenic Bulk with Cylinder Reserve (5.1.3.5.14)	Medical Air Proportioning System (5.1.3.6.3.13)	Medical Air Compressors (5.1.3.6)	Instrument Air Compressors (5.1.13.3.5)	Medical–Surgical Vacuum Pumps (5.1.3.7)	WAGD Producers (5.1.3.8)
Nitrogen main line pressure high	5.1.9.2.4(7)	5.1.9.2.4(7)	5.1.9.2.4(7)	5.1.9.2.4(7)					
Nitrogen main line pressure low	5.1.9.2.4(7)	5.1.9.2.4(7)	5.1.9.2.4(7)	5.1.9.2.4(7)					
Nitrogen changeover to secondary supply	5.1.3.5.11.6 5.1.9.2.4(1)	5.1.3.5.12.9(1) 5.1.9.2.4(1)							
Nitrogen main supply less than 1 day (low contents)			5.1.9.2.4(2) 5.1.3.5.13.4(1)	5.1.9.2.4(2) 1.3.5.13.4(1)					
Nitrogen reserve in use		5.1.3.5.12.9(3) 5.1.9.2.4(3)	5.1.9.2.4(3) 5.1.3.5.13.4(2)	5.1.9.2.4(3) 5.1.3.5.13.4(2)					
Nitrogen reserve supply less than 1 day (low contents)		5.1.3.5.12.9(4)	5.1.9.2.4(5) 5.1.3.5.13.4(3)	5.1.9.2.4(5) 5.1.3.5.13.4(3)					
Nitrogen reserve pressure low (not functional)			5.1.9.2.4(6) 5.1.3.5.13.4(4)						
Carbon dioxide main line pressure high	5.1.9.2.4(7)	5.1.9.2.4(7)	5.1.9.2.4(7)	5.1.9.2.4(7)					
Carbon dioxide main line pressure low	5.1.9.2.4(7)	5.1.9.2.4(7)	5.1.9.2.4(7)	5.1.9.2.4(7)					
Carbon dioxide changeover to secondary supply	5.1.3.5.11.6 5.1.9.2.4(1)	5.1.3.5.12.9(1) 5.1.9.2.4(1)							
Carbon dioxide main supply less than 1 day (low contents)			5.1.9.2.4(2) 5.1.3.5.13.4(1)	5.1.9.2.4(2) 5.1.3.5.13.4(1)					
Carbon dioxide reserve in use		5.1.3.5.12.9(3) 5.1.9.2.4(3)	5.1.9.2.4(3) 5.1.3.5.13.4(2)	5.1.9.2.4(3) 5.1.3.5.13.4(2)					
Carbon dioxide reserve supply less than 1 day (low contents)		5.1.3.5.12.9(4)	5.1.9.2.4(5) 5.1.3.5.13.4(3)	5.1.9.2.4(5) 5.1.3.5.13.4(3)					

Table A.5.1.9.2 *Continued*

Alarm Condition	Manifold for Gas Cylinders (5.1.3.5.11)	Manifold for Cryogenic Liquid Cylinders with Reserve (5.1.3.5.12)	Cryogenic Bulk with Cryogenic Reserve (5.1.3.5.13)	Cryogenic Bulk with Cylinder Reserve (5.1.3.5.14)	Medical Air Proportioning System (5.1.3.6.3.13)	Medical Air Compressors (5.1.3.6)	Instrument Air Compressors (5.1.13.3.5)	Medical–Surgical Vacuum Pumps (5.1.3.7)	WAGD Producers (5.1.3.8)
Carbon dioxide reserve pressure low (not functional)			5.1.9.2.4(6) 5.1.3.5.13.4(3)						
Medical air main line pressure high	5.1.9.2.4(7)					5.1.9.2.4(7)			
Medical air main line pressure low	5.1.9.2.4(7)					5.1.9.2.4(7)			
Medical air changeover to secondary supply	5.1.3.5.11.6 5.1.9.2.4(1)								
Medical air dew point high						5.1.3.6.3.13(1) 5.1.9.2.4(10)			
Medical air production stop					5.1.9.2.4(13)				
Oxygen main line pressure high	5.1.9.2.4(7)	5.1.9.2.4(7)	5.1.9.2.4(7)	5.1.9.2.4(7)					
Oxygen main line pressure low	5.1.9.2.4(7)	5.1.9.2.4(7)	5.1.9.2.4(7)	5.1.9.2.4(7)					
Oxygen changeover to secondary supply	5.1.3.5.11.6 5.1.9.2.4(1)	5.1.3.5.12.9(1) 5.1.9.2.4(1)							
Oxygen main supply less than 1 day (low contents)			5.1.9.2.4(2) 5.1.3.5.13.4(1)	5.1.9.2.4(2) 5.1.3.5.13.4(1)					
Oxygen reserve in use		5.1.3.5.12.9(3) 5.1.9.2.4(3) 5.1.3.5.12.9(4) 5.1.9.2.4(5)	5.1.9.2.4(3) 5.1.3.5.13.4(2)	5.1.9.2.4(3) 5.1.3.5.13.4(2)					
Oxygen reserve supply less than 1 day (low contents)			5.1.3.5.13.4(3)	5.1.3.5.13.4(3)					
Oxygen reserve pressure low (not functional)			5.1.9.2.4(6) 5.1.3.5.13.4(3)						
Nitrous oxide main line pressure high	5.1.9.2.4(7)	5.1.9.2.4(7)	5.1.9.2.4(7)	5.1.9.2.4(7)					
Nitrous oxide main line pressure low	5.1.9.2.4(7)	5.1.9.2.4(7)	5.1.9.2.4(7)	5.1.9.2.4(7)					
Nitrous oxide changeover to secondary supply	5.1.3.5.11.6 5.1.9.2.4(1)	5.1.3.5.12.9(1) 5.1.9.2.4(1)							
Nitrous oxide main supply less than 1 day (low contents)			5.1.9.2.4(2) 5.1.3.5.13.4(1)	5.1.9.2.4(2) 5.1.3.5.13.4(1)					
Nitrous oxide reserve in use		5.1.9.2.4(3) 5.1.3.5.12.9(3) 5.1.3.5.12.9(4)	5.1.9.2.4(3) 5.1.3.5.13.4(2)	5.1.9.2.4(3) 5.1.3.5.13.4(2)					
Nitrous oxide reserve supply less than 1 day (low contents)			5.1.9.2.4(5) 5.1.3.5.13.4(3)	5.1.9.2.4(5) 5.1.3.5.13.4(2)					
Nitrous oxide reserve pressure low (not functional)			5.1.9.2.4(6) 5.1.3.5.13.4(4)						
Medical–surgical main line vacuum low								5.1.9.2.4(8)	
WAGD main line vacuum low									5.1.9.2.4(11)
Local alarm					5.1.9.2.4(9) 5.1.9.5.2 5.1.3.6.3.13(C)(9)	5.1.3.9.10 5.1.3.6.3.11 5.1.9.2.4(9) 5.1.9.5.2	5.1.13.3.5.12 5.1.9.2.4(9) 5.1.9.5.2	5.1.3.7.7 5.1.9.2.4(9) 5.1.9.5.2	5.1.3.8.4 5.1.9.2.4(9) 5.1.9.5.2
Instrument air main line pressure high							5.1.9.2.4(7)		
Instrument air main line pressure low							5.1.9.2.4(7)		
Instrument air dew point high							5.1.13.3.5.12(A)(2) 5.1.9.2.4(12)		
Instrument air cylinder reserve in use (if provided)							5.1.13.3.5.12(B)(1)		
Instrument air cylinder reserve less than 1 hour supply							5.1.13.3.5.12(B)(2)		

A.5.1.9.4 See Table A.5.1.9.4.

A.5.1.9.4(2) Examples of critical care areas include post-anesthesia recovery, intensive care units, and emergency departments.

Table A.5.1.9.4 Requirements for Category 1 Area Alarms

Alarm Condition	Requirement Location
High line pressure (for each gas piped to the area)	5.1.9.3 5.1.9.3.1 5.1.9.3.2 5.1.9.3.4
Low line pressure (for each gas piped to the area)	5.1.9.3 5.1.9.3.1 5.1.9.3.2 5.1.9.3.4
Low medical–surgical vacuum (if piped to the area)	5.1.9.3 5.1.9.3.1 5.1.9.3.3 5.1.9.3.4
Low WAGD vacuum (if piped to the area)	5.1.9.3 5.1.9.3.1 5.1.9.3.3 5.1.9.3.4

A.5.1.9.4.1 Area alarm panels should be placed in a location that will most closely fulfill the following criteria (recognizing that no existing location might fulfill all criteria):

(1) Near or within the location where the staff will most often be present (e.g., a staff base, a nurses' station)
(2) Where the audible alert will best carry throughout the unit being surveilled
(3) Where the panel is visible from the largest number of rooms, beds, or stations within the zone
(4) Where visualization of the panel will not be blocked (e.g., by cabinet doors, carts, room doors, curtains, supplies)
(5) At a height above the floor at which the panel can be comfortably viewed and at which the mute button can be conveniently accessed

A.5.1.9.4.4(1) This signal is intended to provide immediate warning for loss of, or increase in, system pressure for each individual vital life support and critical care area.

A.5.1.9.4.4(2) This signal is intended to provide immediate warning for loss of, or increase in, system pressure for all anesthetizing locations supplied from a single branch line — not for each individual operating or delivery room.

A.5.1.9.5 Activation of any of the warning signals should immediately be reported to the department of the facility responsible for the medical gas piping system involved. If the medical gas is supplied from a bulk supply system, the owner or the organization responsible for the operation and maintenance of that system, usually the supplier, should also be notified. As much detail as possible should be provided. See Table A.5.1.9.5.

Table A.5.1.9.5 Requirements for Category 1 Local Alarms

Alarm Condition	Medical Air Compressors			Instrument Air Compressors	Medical–Surgical Vacuum Pumps	WAGD Producers
	Oil-less (Sealed Bearing) 5.1.3.6.3.4(A)(1)	Oil-Free (Separated) 5.1.3.6.3.4(A)(2)	Liquid Ring (Water-Sealed) 5.1.3.6.3.4(A)1			
Backup (lag) compressor in operation	5.1.3.6.3.12(F) 5.1.9.5.4(1)	5.1.3.6.3.12(F) 5.1.9.5.4(1)	5.1.3.6.3.12(F) 5.1.9.5.4(1)			
Backup (lag) medical–surgical vacuum pump in operation					5.1.3.7.7 5.1.9.5.4(4)	
Backup (lag) WAGD producer in operation						5.1.3.8.3.2 5.1.9.5.4(5)
Backup (lag) instrument air compressor in operation				5.1.13.3.5.12(1) 5.1.9.5.4(1)		
Carbon monoxide high	5.1.3.6.3.13(2)	5.1.3.6.3.13(2)	5.1.3.6.3.13(2)			
High discharge air temperature	5.1.9.5.1(2) 5.1.3.6.3.12(D) 5.1.9.5.4(9)	5.1.9.5.1(2) 5.1.3.6.3.12(E)(1) 5.1.9.5.4(9)	5.1.9.5.1(2)			
High water in receiver	5.1.3.6.3.12(B) 5.1.9.5.4(7)	5.1.3.6.3.12(B) 5.1.9.5.4(7)	5.1.3.6.3.12(B) 5.1.9.5.4(7)			
High water in separator			5.1.3.6.3.12(C)			
Medical air dew point high	5.1.3.6.3.13(1)	5.1.3.6.3.13(1)	5.1.9.5.4(8) 5.1.3.6.3.13(1)			
Instrument air dew point high	5.1.9.5.4(3)	5.1.9.5.4(3)	5.1.9.5.4(3)	5.1.13.3.5.12(2) 5.1.9.5.4(6)		

A.5.1.10.1.4 Operation of piped medical gas systems at gauge pressures in excess of 1275 kPa (185 psi) involves certain restrictions because of the limitations in materials.

A.5.1.10.3.1 A distinction is made between deep-socket solder-joint fittings (ASME B16.22, *Wrought Copper and Copper Alloy Solder Joint Pressure Fitting*) and those having shallow sockets for brazing (ANSI/ASME B16.50, *Wrought Copper and Copper Alloy Braze Joint Pressure Fitting*). The use of shallow-socket brazing fittings improves the quality of the brazement without decreasing its strength, particularly in larger sizes, which are difficult to heat. See Table A.5.1.10.3.1 for socket depths conforming to ANSI/ASME B16.50. The installer can use ANSI/ASME B16.50 fittings (if available) or have the sockets on ASME B16.22 fittings cut down to ASME B16.50 depths. Where shallow-socket fittings are used for the medical gas piping, care should be taken to avoid their use in other piping systems where joints could be soldered instead of brazed.

Table A.5.1.10.3.1 Socket Depths for ASME B16.50 Brazing Fittings

Tube Size (in.)	Socket Depth (in.)
¼ (⅜ O.D.)	0.17
⅜ (½ O.D.)	0.2
½ (⅝ O.D.)	0.22
¾ (⅞ O.D.)	0.25
1 (1⅛ O.D.)	0.28
1¼ (1⅜ O.D.)	0.31
1½ (1⅝ O.D.)	0.34
2 (2⅛ O.D.)	0.40
2½ (2⅝ O.D.)	0.47
3 (3⅛ O.D.)	0.53
4 (4⅛ O.D.)	0.64
5 (5⅛ O.D.)	0.73
6 (6⅛ O.D.)	0.83

A.5.1.10.4.5 The intent is to provide an oxygen-free atmosphere within the tubing and to prevent the formation of copper oxide scale during brazing. This is accomplished by filling the piping with a low-volume flow of low pressure inert gas.

A.5.1.10.4.5.12 This is to ensure a quality joint and to prevent the formation of copper oxide on the inside and outside surfaces of the joint.

A.5.1.10.5.1.5 Gas mixtures are commonly used in GTAW autogenous fusion welding. The identification of a gas mixture as "75He 25Ar" is a common industry term to define a commercially available grade from gas suppliers. If test welding results lead to questions about the mixture percentage or gas quality, another bottle should be substituted and test welds performed.

A.5.1.10.8(3) It is intended that the "recommended for oxygen service" apply to both polytetrafluoroethylene tape as well as the "other thread sealant."

A.5.1.11 It is recommended that the facility's normal operating pressure of nitrous oxide be initially set and continually maintained at least 34.5 kPag (5 psig) below the normal operating pressures of the oxygen and medical air.

Piping systems that are connected though blending devices are in effect cross-connected through the device. In the rare event of a failure of the safeties inside the equipment, the possibility of having the gases flow across the device exists. When the device is an anesthesia machine, and one of the gases is nitrous oxide, a pressure in the nitrous oxide pipeline greater than the pressure in the medical air or oxygen system opens the possibility of nitrous oxide flowing into the other pipelines. A patient could then receive a lethal quantity of nitrous oxide from a labeled and indexed medical air or oxygen outlet. Adjusting the pressure as recommended can reduce the likelihood of the causative equipment failure and also reduce the severity of the problem in the event it does occur.

A.5.1.11.2.7 It is not intended that every room be listed on the label, but an area that is easily identifiable by staff needs to be indicated. This can be accomplished with text or by graphical means such as a map or color coding. The label should be permanently affixed outside and near valve box. The label should not be affixed to a removable cover.

A.5.1.11.4.2 It is not intended that every room be listed on the label, but an area that is easily identifiable by staff needs to be indicated. This can be accomplished with text or by graphical means such as a map or color coding.

A.5.1.12 All testing should be completed before putting a new piping system, or an addition to an existing system, into service. Test procedures and the results of all tests should be made part of the permanent records of the facility of which the piping system forms a part. They should show the room and area designations, dates of the tests, and name(s) of the person(s) conducting the tests.

A.5.1.12.2.3.5 Ammonia is known to cause stress cracking in copper and its alloys.

A.5.1.12.2.6.5 The effect of temperature changes on the pressure of a confined gas is based on the Ideal Gas Law. The final absolute pressure (P2a) equals the initial absolute pressure (P1a) times the final absolute temperature (T2a), divided by the initial absolute temperature (T1a). The relationship is the same for nitrogen, nitrous oxide, oxygen, and compressed air.

Absolute pressure is the gauge pressure reading plus the absolute atmospheric pressure. See Table A.5.1.12.2.6.5 for the absolute atmospheric pressures for elevations at and above sea level.

Absolute temperature K (°R) is the temperature gauge reading °C (°F) plus the absolute zero temperature 273°C (460°F).

Examples of pressure test data at sea level in SI and IP units follow.

The initial test pressure is 415 kPag (60 psig) at 27°C (80°F). A temperature decrease to 18°C (65°F) will cause the test pressure to drop to 400 kPag (57.9 psig).

P1g = 415 kPag, T1g = 27°C, T2g = 18°C, P1g = 60 psig, T1g = 80°F, T2g = 65°F

P1a = 415 + 101 = 516 kPa   P1a = 60 + 14.7 = 74.7 psia
T1a = 27 + 273 = 300K   T1a = 80 + 460 = 540°R
T2a = 18 + 273 = 291K   T2a = 65 + 460 = 525°R
P2a = 516 × 291/300 = 501 kPa   P2a = 74.7 × 525/540 = 72.6 psia
P2g = 501 − 101 = 400 kPag   P2g = 72.6 − 14.7 = 57.9 psig

Table A.5.1.12.2.6.5 Pressure Corrections for Elevation

Elevation (ft)	Absolute Atmospheric Pressure			
	kPa	psia	mmHg	inHg
0	101.33	14.70	760.0	29.92
500	99.49	14.43	746.3	29.38
1000	97.63	14.16	733.0	28.86
1500	95.91	13.91	719.6	28.33
2000	94.19	13.66	706.6	27.82
2500	92.46	13.41	693.9	27.32
3000	90.81	13.17	681.2	26.82
3500	89.15	12.93	668.8	26.33
4000	87.49	12.69	656.3	25.84
4500	85.91	12.46	644.4	25.37
5000	84.33	12.23	632.5	24.90

A.5.1.12.2.7.5 The effect of temperature changes on the vacuum of a confined gas is based on the Ideal Gas Law. The final absolute vacuum (V2a) equals the initial absolute vacuum (V1a) times the final absolute temperature (T2a), divided by the initial absolute temperature (T1a).

Absolute vacuum is the absolute zero pressure 101 kPa (30 inHg) less the vacuum reading below atmospheric. See Table A.5.1.12.2.6.5 for the absolute atmospheric pressures for elevations at and above sea level.

Absolute temperature K (°R) is the temperature gauge reading °C (°F) plus the absolute zero temperature 273°C (460°F).

Examples of vacuum test data at sea level in SI and IP units follow.

The initial test vacuum is 54 kPa or 16 inHg at 18°C (65°F). A temperature increase to 27°C (80°F) will cause the test vacuum to decrease to 52.5 kPa (15.6 inHg).

$V1g = 54$ kPa, $T1g = 18°C$, $T2g = 27°C$, $V1g = 16$ inHg, $T1g = 65°F$, $T2g = 80°F$

$V1a = 101 - 54 = +47$ kPaV
$T1a = 18 + 273 = 291$K
$T2a = 27 + 273 = 300$K
$V2a = 47 \times 300/291 = +48.5$ kPaV
$V2g = 101 - 48.5 = 52.5$ kPa

$V1a = 30 - 16 = +14$ inHgV
$T1a = 65 + 460 = 525°R$
$T2a = 80 + 460 = 540°R$
$V2a = 14 \times 540/525 = +14.4$ inHgV
$V2g = 30 - 14.4 = 15.6$ inHg

A.5.1.12.3.2 This is the final pressure test of the completely installed system and is intended to locate any leaks that would be more likely to occur at lower pressure (e.g., leaks in station outlet valve seals).

A.5.1.12.3.6.4 Odor is checked by sniffing a moderate flow of gas from the outlet being tested. Specific measure of odor in gas is impractical. Gas might have a slight odor, but the presence of a pronounced odor should render the piping unsatisfactory.

A.5.1.12.3.8 The detector used for total hydrocarbons is calibrated with a gas that has a known quantity of methane. When a sample is run with this calibrated detector, the result will be total hydrocarbons as methane. Since methane is the one hydrocarbon that does not interact with the body and is present in all air and most oxygen, the actual amount of methane in the sample is subtracted from the total hydrocarbon result to give total non-methane hydrocarbons.

A.5.1.12.3.10.6 Sleep laboratories are an example of where gas flow and concentration are frequently modified.

A.5.1.12.3.11(3) The committee recognizes that current clinical practice is to use analyzers that might not be able to analyze oxygen to current USP requirements of 99 percent and that these analyzers frequently have an error of up to 3 percent.

A.5.1.13.1 Support gas systems are subject to the same hazards as are present in any piped medical gas system, with the additional hazard of operating at higher pressures.

A.5.1.13.3.5 See Figure A.5.1.13.3.5.

A.5.1.13.3.5.8 Drawing intake air from outside in compliance with 5.1.3.6.3.11 is recommended.

A.5.1.14 All cylinders containing compressed gases, such as anesthetic gases, oxygen, or other gases used for medicinal purposes, whether these gases are flammable or not, should comply with the specifications and be maintained in accordance with regulations of the U.S. Department of Transportation.

Cylinder and container temperatures greater than 52°C (125°F) can result in excessive pressure increase. Pressure relief devices are sensitive to temperature and pressure. When relief devices actuate, contents are discharged.

A.5.1.14.1.1 Piping systems for the distribution of flammable gases (e.g., hydrogen, acetylene, natural gas) are outside the scope of this chapter.

A.5.1.14.1.3 Vacuum systems from station inlets to the exhaust discharge should be considered contaminated unless proven otherwise. Methods exist to disinfect the system or portions thereof.

Clogging of regulators, for example, with lint, debris, or dried body fluids, reduces vacuum system performance.

A.5.1.14.1.4 Other examples of prohibited use of medical–surgical vacuum would be scope cleaning, decontamination, and laser plume.

A.5.1.14.2.1 The facility should retain a written or an electronic copy of all findings and any corrections performed.

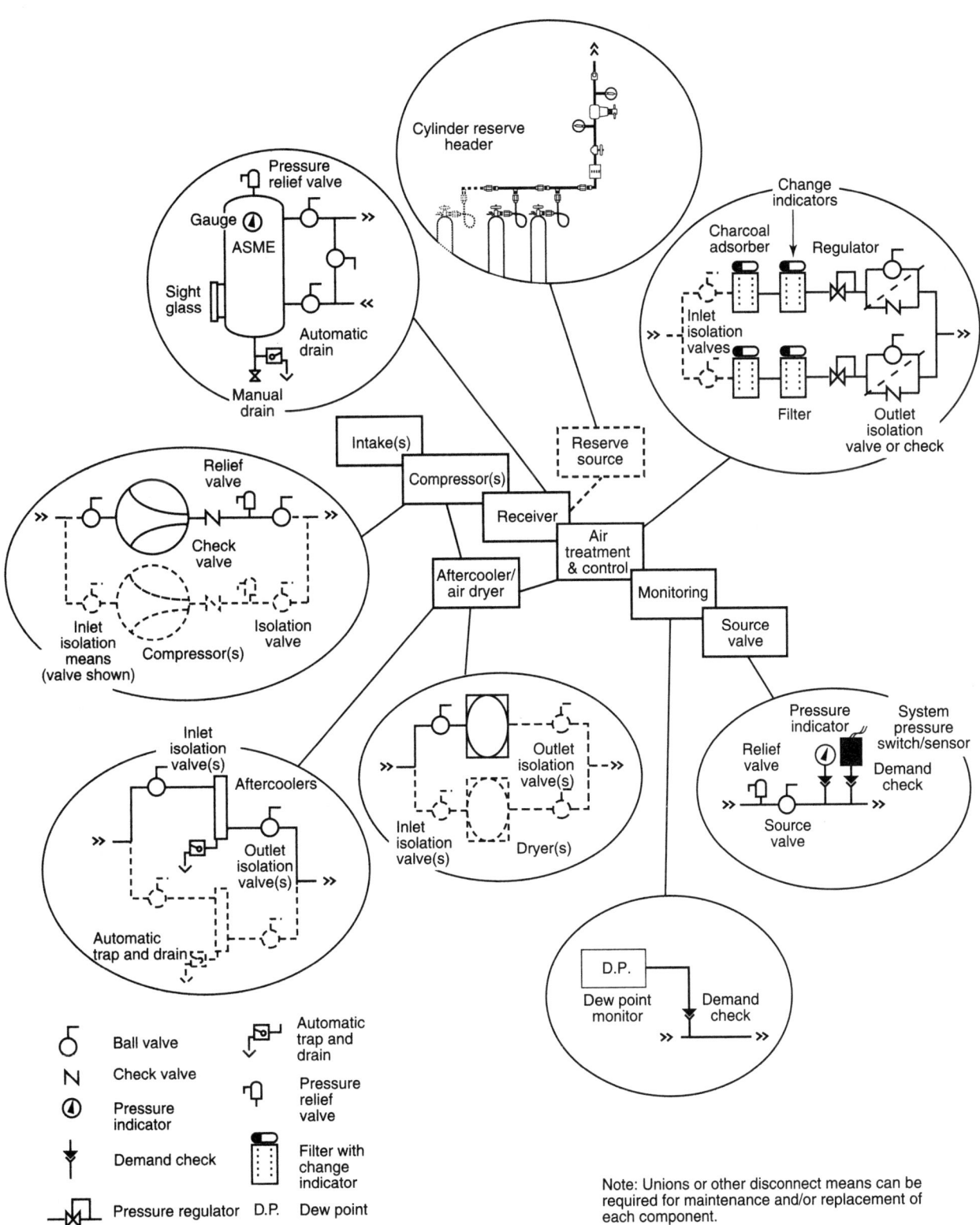

FIGURE A.5.1.13.3.5 Elements of Typical Instrument Air Source.

A.5.1.14.2.2.2 In addition to the minimum inspection and testing in 5.1.14, facilities should consider annually inspecting equipment and procedures and correcting any deficiencies.

A.5.1.14.2.3 The following should be considered in a routine testing, maintenance, and inspection program:

(1) Medical air source, as follows:
 (a) Room temperature
 (b) Shaft seal condition
 (c) Filter condition
 (d) Presence of hydrocarbons
 (e) Room ventilation
 (f) Water quality, if so equipped
 (g) Intake location
 (h) Carbon monoxide monitor calibration
 (i) Air purity
 (j) Dew point
(2) Medical vacuum source — exhaust location
(3) WAGD source — exhaust location
(4) Instrument air source — filter condition
(5) Manifold sources (including systems complying with 5.1.3.5.10, 5.1.3.5.11, 5.1.3.5.12, and 5.1.3.5.13), as follows:
 (a) Ventilation
 (b) Enclosure labeling
(6) Bulk cryogenic liquid source inspected in accordance with NFPA 55, *Compressed Gases and Cryogenic Fluids Code*
(7) Final line regulation for all positive pressure systems — delivery pressure
(8) Valves — labeling
(9) Alarms and warning systems — lamp and audio operation
(10) Alarms and warning systems, as follows:
 (a) Master alarm signal operation
 (b) Area alarm signal operation
 (c) Local alarm signal operation
(11) Station outlets/inlets, as follows:
 (a) Flow
 (b) Labeling
 (c) Latching/detaching
 (d) Leaks
(12) Medical gas quality
 (a) Purity — Percent Concentration
 (b) Permanent particulates and contaminants
 (c) Odor and moisture

A.5.2.1 Section 5.1 covers requirements for Category 1 piped gas and vacuum systems; Section 5.2 covers Category 2 piped gas and vacuum systems; and Section 5.3 covers Category 3 piped gas and vacuum systems. Laboratory systems are no longer covered by Chapter 5 (2002 edition).

A.5.2.14 Medical gas and vacuum systems should be surveyed at least annually for the items that follow and deficient items corrected. Survey of medical air and instrument air sources should include, but not be limited to, the following:

(1) Dew point monitor (operation and calibration)
(2) Carbon monoxide monitor (medical air only) (operation and calibration)
(3) Aftercoolers (condition, operation of drains)
(4) Operating pressures (cut-in, cut-out, and control pressures)
(5) All local alarms (verify presence of required alarms, perform electrical test, test lag alarm)
(6) Receiver elements (auto drain, manual drain, sight glass, pressure gauge)
(7) Filters (condition)
(8) Pressure regulators (condition, output pressure)
(9) Source valve (labeling)
(10) Intake (location and condition)
(11) Housekeeping around compressors

Survey of the medical vacuum and the WAGD source(s) should include, but not be limited to, the following:

(1) Operating vacuum (cut-in, cut-out, and control pressures)
(2) All local alarms (verify presence of required alarms, perform electrical test, test lag alarm)
(3) Receiver elements (manual drain, sight glass, vacuum gauge)
(4) Source valve (labeling)
(5) Exhaust (location and condition)
(6) Housekeeping around pump

Survey of the medical gas manifold source(s) should include, but not be limited to, the following:

(1) Number of cylinders (damaged connectors)
(2) Cylinder leads (condition)
(3) Cascade (switching from one header to another)
(4) All local alarms (verify presence of required alarms, perform electrical test, test all alarms)
(5) Source valve (labeling)
(6) Relief valves (discharge location and condition)
(7) Leaks
(8) Security (door or gate locks and signage)
(9) Ventilation (general operation, housekeeping)
(10) Housekeeping around manifolds

Survey of medical gas area alarms should include, but not be limited to, the following:

(1) Locations (visible to staff)
(2) Signals (audible and visual, use test function)
(3) Activation at low pressure
(4) Housekeeping around alarm

Survey of medical gas master alarms should include, but not be limited to, the following:

(1) Locations (visible to appropriate staff)
(2) Signals (audible and visual, use test function)
(3) Activation at low pressure
(4) Housekeeping around alarm

Survey of zone valves should include, but not be limited to, the following:

(1) Locations (relationship to terminals controlled)
(2) Leaks
(3) Labeling
(4) Housekeeping around alarm

Survey of medical gas outlet/inlets should include, but not be limited to, the following:

(1) Flow and function
(2) Latching/detaching
(3) Leaks
(4) General condition (noninterchangeable indexing)

The facility should retain a written or an electronic copy of all findings and any corrections performed.

A.5.3 A Category 3 vacuum system is not intended for Category 1 medical–surgical vacuum applications. A Category 3 wet piping system is designed to accommodate liquid, air–gas, and solids through the service inlet. A Category 3 dry piping system is designed to accommodate air–gas only through the service inlet, with liquids and solids being trapped before entering the system.

A.5.3.1 Section 5.1 covers requirements for Category 1 piped gas and vacuum systems; Section 5.2 covers Category 2 piped gas and vacuum systems; and Section 5.3 covers Category 3 piped gas and vacuum systems. Laboratory systems are no longer covered by Chapter 5 (2002 edition).

A.5.3.1.2(1) Category 3 medical gas systems are intended to be used where minimal or moderate sedation is administered. Deep sedation and general anesthesia are not allowed in Category 3; therefore, WAGD is not required. *(See Scavenging, 5.3.8.)*

A.5.3.3.6 Category 3 drive gas systems are supplied by one or more of the following:

(1) Compressed air from compressors
(2) Compressed air from cylinders
(3) Nitrogen from cylinders

These systems are used primarily to drive gas-powered power devices. See Figure A.5.3.3.6 for an illustration of this type of system. Similar applications are found in podiatry and plastic surgery. Examples of these are air used to drive turbine-powered drills and air used to dry teeth and gums. Some dental hand pieces have an internal self-contained air return system, while other hand pieces discharge air into the atmosphere. Some discharge a mixture of air and water. Nitrogen is often piped as an alternate or reserve supply to the compressor system.

Dental compressed air is not used for life-support purposes, such as respirators, intermittent positive pressure breathing (IPPB) machines, analgesia, anesthesia, and so forth. Air discharged into the oral cavity is incidental and not a primary source of air to sustain life.

A dental compressed air system should not be used to provide power for an air-powered evacuation system without specific attention paid to the discharge of the evacuated gases and liquids. An open discharge of evacuated gases into the general environment of an operatory could compromise the quality of breathing air in the treatment facility. Air discharge should be vented to the outside of the building through a dedicated vent.

An air-powered evacuation system might require significant quantities of air to operate.

Manufacturer's recommendations should be followed regarding proper sizing of the air compressor. Inadequate sizing can result in overheating, premature compressor failures, and inadequate operating pressures and flows.

A.5.3.3.6.1.3 A color dew point monitor downstream of the receiver indicating the quality of air coming into the receiver is desirable.

A color dew point monitor in the main treatment facility is appropriate to help the staff promptly identify when the system is being degraded with air of a dew point higher than is acceptable.

The design of the color monitor should be such that the normal tolerance of variations will limit the maximum moisture at 3.9°C at 690 kPag (39°F at 100 psig) at activation.

A.5.3.3.6.1.5 The environmental air source for the compressor inlet should take into consideration possible contamination by particulates, concentrations of biological waste contaminants, ozone from nearby brush-type electric motors, and exhaust fumes from engines.

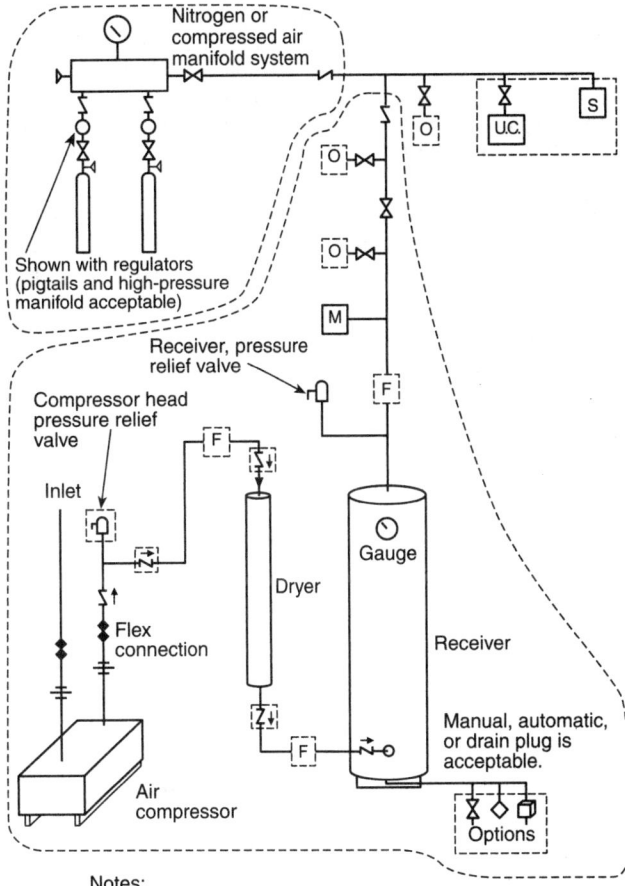

FIGURE A.5.3.3.6 Category 3 Drive Gas Supply System.

Air taken from an outside atmosphere could cause harmful condensation problems in the compressor. Long runs of inlet tube should also be avoided, as they will degrade compressor performance. The compressor manufacturer's recommendations should be followed regarding appropriate pipe size to prevent possible degradation of system performance.

A dental air compressor and dental vacuum system can be in the same equipment room as long as the inlet for the dental air compressor does not draw air from a room or space containing an open discharge for the dental vacuum system.

Atmospheric air from an operatory can have traces of mercury vapor, nitrous oxide, and other contaminants. A compressor inlet location that would draw its supply directly from an operatory should be avoided.

A.5.3.3.8 If nitrogen is used as a backup supply to a compressed gas system, the nitrogen operating pressure should be regulated so as not to exceed the operating pressure of the Category 3 compressed air system.

2015 Edition

A.5.3.3.10.1.3 A Category 3 vacuum system is not intended for Category 1 vacuum applications. A wet piping system is designed to accommodate liquid, air–gas, and solids through the service inlet. A dry piping system is designed to accommodate air–gas only through the service inlet, with liquids and solids being trapped before entering the system. *[See Figure A.5.3.3.10.1.3(a) through Figure A.5.3.3.10.1.3(d).]*

A.5.3.3.10.1.3(4) Improper design allows gas pressure to build up in the ventilation system, which might blow the trap on liquid seals. See Figure A.5.3.3.10.1.3(4)(a) and Figure A.5.3.3.10.1.3(4)(b).

A.5.3.3.10.1.4(8) Care should be taken to ensure the dual exhaust systems do not develop excessive back pressure when using a common exhaust line.

A.5.3.4.1(1) See Figure A.5.3.4.1(1) for an illustration of single treatment locations.

A.5.3.5 Service outlets can be recessed or otherwise protected from damage.

A.5.3.12.2.4(6) Ammonia is known to cause stress cracking in copper and its alloys.

A.6.1 Although complete compliance with this chapter is desirable, variations in existing health care facilities should be considered acceptable in instances where wiring arrangements are in accordance with prior editions of this document or afford an equivalent degree of performance and reliability. Such variations could occur, particularly with certain wiring in separate or common raceways, with certain functions connected to one or another system or branch, or with certain provisions for automatically or manually delayed restoration of power from the alternate (emergency) source of power.

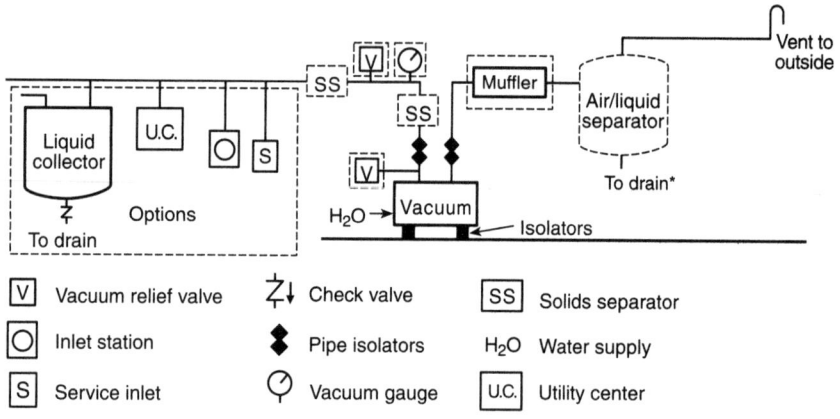

FIGURE A.5.3.3.10.1.3(a) Typical Category 3 Wet or Dry Piping System with Single Vacuum Pump Source.

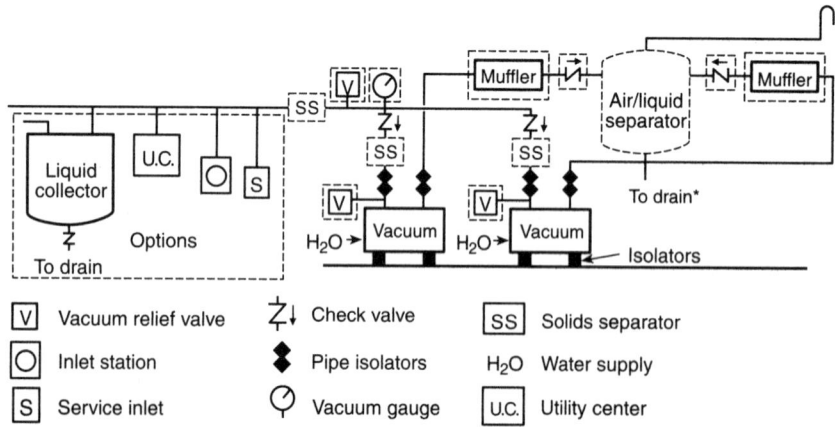

FIGURE A.5.3.3.10.1.3(b) Typical Category 3 Wet or Dry Piping System with Duplex Vacuum Source with Air/Liquid Separator.

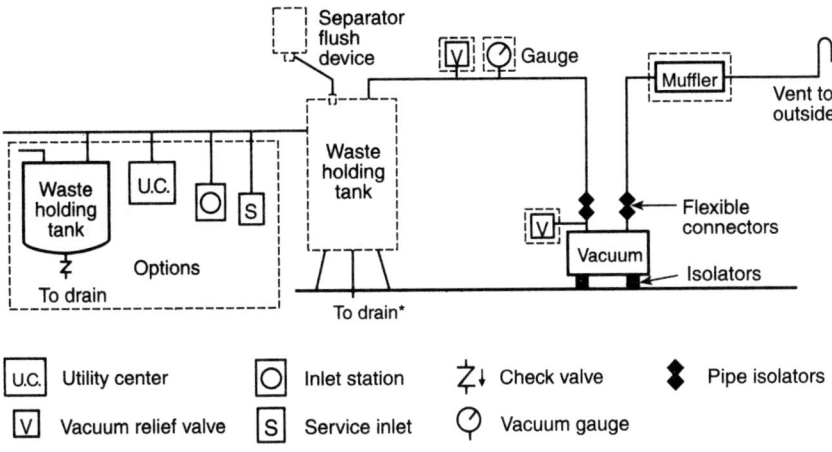

FIGURE A.5.3.3.10.1.3(c) Typical Category 3 Wet or Dry Piping System with Single Vacuum Source.

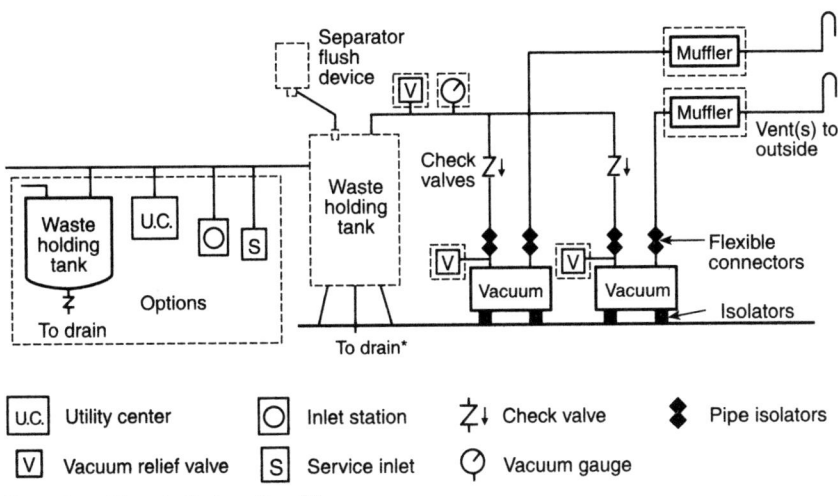

FIGURE A.5.3.3.10.1.3(d) Typical Category 3 Wet or Dry Piping System with Duplex Vacuum Source with Waste Holding Tank.

A.6.2.1 Electrical systems can be subject to the occurrence of electrical fires. Grounding systems, overcurrent protective devices, and other subjects discussed in this code could be intended for fire prevention as well as other purposes. This aspect of electrical systems is the primary focus of other NFPA standards and will not be emphasized herein.

A.6.3.2.1.1 Assignment of degree of reliability of electrical systems in health care facilities depends on the careful evaluation of the variables at each particular installation. For further information, see ANSI/IEEE 493-2007, *Recommended Practice for the Design of Reliable Industrial and Commercial Power Systems.*

A.6.3.2.2.1 At the time of installation of branch circuit wiring, 600 V or less, steps should be taken to ensure that the insulation on each conductor intended to be energized, and on the equipment grounding conductor in systems containing isolated ground receptacles, has not been damaged in the process of installation. When disconnected and unenergized, the resistance should be at least 20 megohms when measured with an ohmmeter having an open-circuit test voltage of at least 500 V dc.

Consideration should be given to providing reasonable accessibility to branch-circuit switching and overcurrent protection devices by the hospital staff in the patient care space. Consideration should also be given to providing labels at each receptacle and on installed equipment indicating the location and identity of the distribution panel serving that power outlet or equipment, especially where the location or identity might not be readily apparent.

A.6.3.2.2.1.1 The requirement that branch circuits be fed from not more than one distribution panel was introduced for several reasons. A general principle is to minimize possible

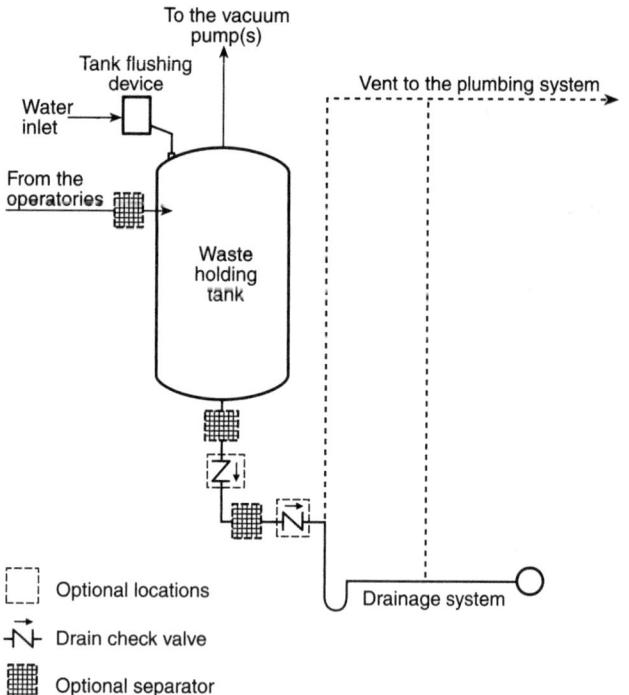

FIGURE A.5.3.3.10.1.3(4)(a) Drainage from Gravity Drained Liquid Collector Tank.

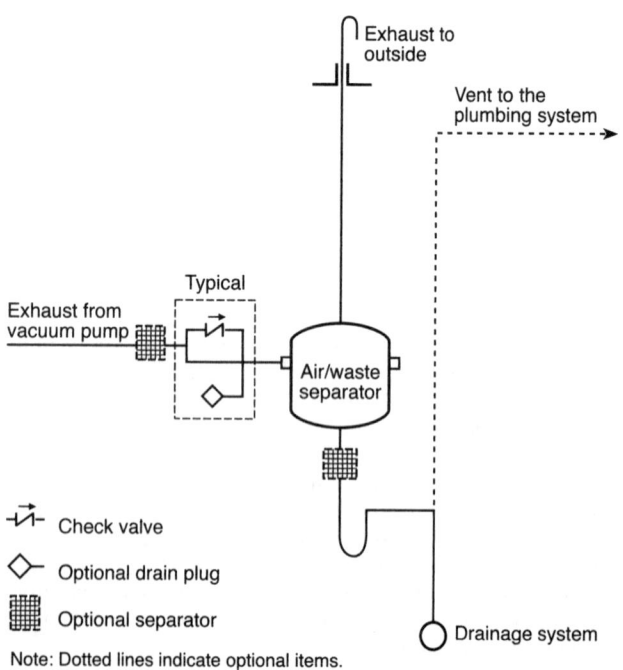

FIGURE A.5.3.3.10.1.3(4)(b) Drainage from Positive Discharge Vacuum Pump Through Air/Liquid Separator.

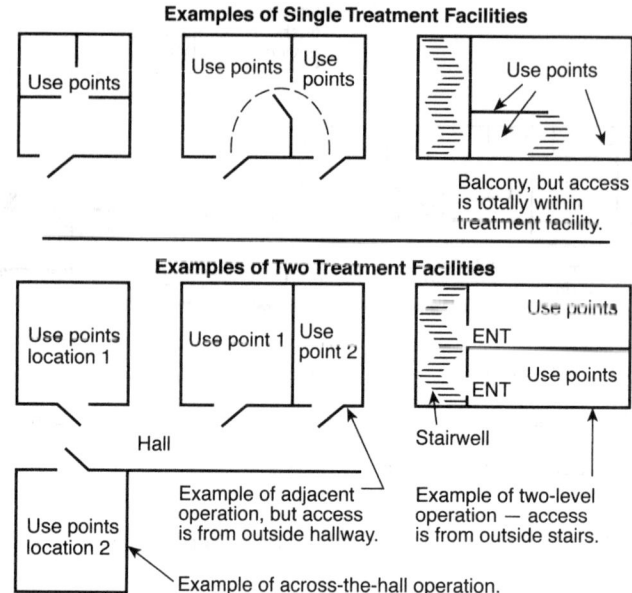

FIGURE A.5.3.4.1(1) Examples of Single Treatment Locations.

potential differences between the grounding pins of receptacles in one area by bringing the grounding conductors to a common point. A specific reason is to simplify maintenance by making it easier to find the source for the receptacles in a room. This is particularly a problem in hospitals where emergency conditions might require rapid restoration of power.

A.6.3.2.2.3 The requirement for grounding interconnection between the normal and essential power systems follows the principle of minimizing possible potential differences between the grounding pins of receptacles in one area by bringing the grounding conductors to a common point.

A.6.3.2.2.4.1 Within the constraints of the equipment provided, consideration should be given to coordinating circuit breakers, fuses, and other overcurrent protective devices so that power interruption in that part of the circuit that precedes the interrupting device closest to a fault is not likely to occur.

A.6.3.2.2.6.1 It is best, if possible, to employ only one type of receptacle (standard three-prong type) for as many receptacles being served by the same line voltage to avoid the inability to connect life-support equipment in emergencies. The straight-blade, three-prong receptacle is now permitted in all locations in a hospital. Previously, special receptacles were specified in operating room locations and have caused compatibility problems.

A.6.3.2.2.7.1 Care should be taken in specifying a system containing isolated ground receptacles, because the grounding impedance is controlled only by the grounding wires and does not benefit from any conduit or building structure in parallel with it.

A.6.3.2.2.7.3 Special grounding methods could be required in patient vicinities immediately adjacent to rooms containing high-power or high-frequency equipment that causes electrical interference with monitors or other electromedical devices. In extreme cases, electromagnetic induction can cause the voltage limits of 6.3.3.1 to be exceeded.

Electromagnetic interference problems can be due to a variety of causes, some simple, others complex. Such problems are best solved one at a time. In some locations, grounding of stretchers, examining tables, or bed frames will be helpful. Where necessary, a patient equipment grounding point should be installed. This can usually be accomplished even after completion of construction by installing a receptacle faceplate fitted with grounding posts. Special grounding wires should not be used unless they are found to be essential for a particular location, because they can interfere with patient care procedures or present trip hazards.

A.6.3.2.2.8.1 Moisture can reduce the contact resistance of the body, and electrical insulation is more subject to failure.

A.6.3.2.2.8.4 In conducting a risk assessment, the health care governing body should consult with all relevant parties, including, but not limited to, clinicians, biomedical engineering staff, and facility safety engineering staff.

A.6.3.2.2.8.7 The health care governing body and designer of record should evaluate the type of protection to be provided against electrical shock to patients and caregivers in wet procedure locations. The application considerations should include but not be limited to the reliability of power to critical equipment and systems.

A.6.3.2.6 Patient protection is provided primarily by an adequate grounding system. The ungrounded secondary of the isolation transformer reduces the cross-sectional area of grounding conductors necessary to protect the patient against voltage resulting from fault current by reducing the maximum current in case of a single probable fault in the grounding system. The line isolation monitor is used to provide warning when a single fault occurs. Excessive current in the grounding conductors will not result in a hazard to the patient unless a second fault occurs. If the current in the grounding system does not exceed 10 mA, even under fault conditions, the voltage across 3 m (9.84 ft) of No. 12 AWG wire will not exceed 0.2 mV, and the voltage across 3 m (9.84 ft) of No. 18 AWG grounding conductor in a flexible cord will not exceed 0.8 mV. Allowing 0.1 mV across each connector, the voltage between two pieces of patient-connected equipment will not exceed 2 mV.

The reference grounding point is intended to ensure that all electrically conductive surfaces of the building structure, which could receive heavy fault currents from ordinary (grounded) circuits, are grounded in a manner to bypass these heavy currents from the operating room.

A.6.3.2.6.2.1 It is desirable to limit the size of the isolation transformer to 10 kVA or less and to use conductor insulation with low leakage to meet the impedance requirements. Keeping branch circuits short and using insulation with a dielectric constant less than 3.5 and insulation resistance constant greater than 6100 megohmmeters at 16°C (20,000 megohm-ft at 60°F) reduces leakage from line to ground.

To correct milliammeter reading to line impedance, use the following equation:

$$\text{Line impedance (in ohms)} = \frac{V \cdot 1000}{I} \quad [\text{A.6.3.2.6.2.1}]$$

where:
V = isolated power system voltage
I = milliammeter reading made during impedance test

A.6.3.2.6.3.1 Protection for the patient is provided primarily by a grounding system. The ungrounded secondary of the isolation transformer reduces the maximum current in the grounding system in case of a single fault between either isolated power conductor and ground. The line isolation monitor provides warning when a single fault occurs, or when excessively low impedance to ground develops, which might expose the patient to an unsafe condition if an additional fault occurs. Excessive current in the grounding conductors will not result from a first fault. A hazard exists if a second fault occurs before the first fault is cleared.

A.6.3.2.6.3.3 It is desirable to reduce this monitor hazard current, provided that this reduction results in an increased "not alarm" threshold value for the fault hazard current.

A.6.3.2.6.3.4 It is desirable to locate the ammeter such that it is conspicuously visible to persons in the anesthetizing location.

The line isolation monitor can be a composite unit, with a sensing section cabled to a separate display panel section on which the alarm and test functions are located, if the two sections are within the same electric enclosure.

A.6.3.3.1.1 In a conventional grounded power distribution system, one of the line conductors is deliberately grounded, usually at some distribution panel or the service entrance. This grounded conductor is identified as the neutral conductor. The other line conductor (s) is the high side of the line. The loads to be served by this distribution system are fed by the high and neutral conductors.

In addition to the high and neutral conductors, a grounding conductor is provided. One end is connected to the neutral at the point where the neutral is grounded, and the other end leads out to the connected loads. For purposes here, the load connection point will be considered to be a convenience receptacle, with the grounding conductor terminating at the grounding terminal of that receptacle.

This grounding conductor can be a separate wire running from the receptacle back to the remote grounding connection (where it joins the neutral conductor). If that separate conductor does not make any intermediate ground contacts between the receptacle and the remote ground, then the impedance of the connection between the receptacle and the remote ground is primarily the resistance of the grounding conductor itself and is, therefore, predictable.

If, however, the receptacle is also interconnected with the remote ground point by metallic conduit or other metallic building structures, the impedance of the circuit between the receptacle and remote ground is not easily predictable; nor is it easy to measure accurately, although one can be sure that the impedance will be less than that of the grounding wire itself because of the additional parallel paths.

Fortunately, as will become apparent in the paragraphs that follow, the absolute value of the apparent impedance between the grounding contact of an outlet and the remote ground point need not be known or measured with great accuracy.

Ideally, and under no-fault conditions, the grounding system described earlier is supposed to be carrying no current at all. If this were true, then no voltage differences would be found between exposed conductive surfaces of any electrical appliances that were grounded to the grounding contacts of the receptacles that powered them. Similarly, there would be no voltage differences between these appliances and any other exposed metal surface that was also interconnected with the grounding system, provided that no currents were flowing in that interconnection.

Ideal conditions, however, do not prevail, and even when there are no "faults" within an appliance, residual "leakage" current does flow in the grounding conductor of each of the appliances, producing a voltage difference between the chassis of that appliance and the grounding contact of the receptacle that feeds it. Furthermore, this current can produce voltage differences among other appliances plugged into various receptacles on the system.

Fortunately, these leakage currents are small, and for reasonably low grounding-circuit impedances, the resulting voltage differences are entirely negligible.

If, however, a breakdown of insulation between the high side of the line and the chassis of an appliance occurs, the leakage condition becomes a fault condition, the magnitude of which is limited by the nature of the breakdown, or, in the case of a dead short circuit in the appliance, the magnitude of the fault current is limited only by the residual resistance of the appliance power cord conductors and that of the power distribution system.

In the event of such a short circuit, the impedance of the grounding circuit, as measured between the grounding contact of the receptacle that feeds the defective appliance and the remote ground point where the neutral and grounding conductors are joined, should be so small that a large enough fault current will flow to ensure a rapid breaking of the circuit by the overcurrent protective device that serves that receptacle.

For a 20-A branch circuit, a fault current of 40 A or more would be required to ensure a rapid opening of the branch-circuit overcurrent protective device. This corresponds to a circuit impedance of 3 ohms or less, of which the grounding system should contribute 1 ohm or less.

During the time this large fault current flows in the grounding system, the chassis of the defective appliance is raised many volts above other grounded surfaces in the same vicinity. The hazard represented by this condition is minimized by the fact that it exists for only a short time, and, unless a patient simultaneously contacts both the defective appliance and some other grounded surface during this short time interval, there is no hazard. Furthermore, the magnitude of an applied voltage required to produce a serious shock hazard increases as its duration decreases, so the rapidity with which the circuit is interrupted helps reduce shock hazard even if such a patient contact occurs.

If, however, the defect in the appliance is not such as to cause an immediate circuit interruption, then the effect of this intermediate level of fault current on the voltages appearing on various exposed conductive surfaces in the patient care vicinity should be considered.

Because all of this fault current flows in the grounding conductor of the defective appliance's power cord, the first effect is to raise the potential of this appliance above that of the receptacle that feeds it by an amount proportional to the power cord grounding conductor resistance. This resistance is required to be less than 0.15 ohm, so fault currents of 20 A or less, which will not trip the branch-circuit overcurrent protective device, will raise the potential of the defective appliance above the grounding contact of its supply receptacle by only 3 V or less. This value is not hazardous for casual contacts.

The fault current that enters the grounding system at the grounding contact of any receptacle in the patient care vicinity could affect the potential at the grounding contacts of all the other receptacles, and, more importantly, it could produce significant voltage differences between them and other grounded surfaces, such as exposed piping and building structures.

If one grounded point is picked as a reference (a plumbing fixture in or near the patient care vicinity, for example), and the voltage difference is then measured between that reference and the grounding contact of a receptacle, produced by driving some known current into that contact, a direct measure of the effectiveness of the grounding system within the patient care vicinity is obtained. The "figure of merit" can be stated as so many volts per ampere of fault current. The ratio volts per ampere is, of course, impedance; but because the exact path taken by the fault current is not known, and because the way in which the reference point is interconnected with the grounding system is not known, it cannot be stated that this value is the impedance between the receptacle and some specific point, such as the joining of the neutral and grounding conductors. However, it can be stated that this measured value of "effective impedance" is indicative of the effectiveness with which the grounding system minimizes voltage differences between supposedly grounded objects in the patient care vicinity that are produced by ground faults in appliances used in that vicinity. This impedance, which characterizes the ability of the grounding system to maintain nearly equipotential conditions within the patient care vicinity, is of prime importance in assessing shock hazard; but this impedance is not necessarily the same as the impedance between receptacle and remote ground point, which controls the magnitude of the short-circuit current involved in tripping the branch-circuit overcurrent protective device.

Fault currents on the grounding system can also come from neutral-to-ground faults, which allow some current to flow in the neutral and some in the ground. This type of fault is often the cause of interference on EEG and ECG equipment. It is often not recognized easily because, except for 60 Hz interference, the equipment works perfectly properly. It is most easily found by causing a substantial change in the line-to-line load and noting changes in the ground-to-reference voltage.

A.6.3.3.1.1.4 The grounding system (reference ground and conduit) is to be tested as an integral system. Lifting of grounds from receptacles and fixed equipment is not required or recommended for the performance of this test.

A.6.3.3.1.3 Effective grounding to safely handle both fault and leakage currents requires following the requirements of both Chapter 6 of NFPA 99 and Article 250 of *NFPA 70, National Electrical Code*; having good workmanship; and using some techniques that are not found in these documents.

The performance of the grounding system is made effective through the existence of the green grounding wire, the metal raceway, and all of the other building metal. Measurements have shown that it is the metal raceway and building steel that provide most of the effective grounding path of less than 10 milliohms at the receptacle, including plug-to-receptacle impedance. The green grounding wire becomes a backup, not a primary grounding path performer.

Good practice calls for each receptacle to have a good jumper grounding connection to the metal raceway at the receptacle location in addition to having the green grounding wire connecting these points to the grounding bus in the distribution panel. Good workmanship includes seeing that these grounding connections are tight at each receptacle and that all metal raceway joints are secure and tight.

The voltage difference measurements listed in 6.3.3.1.3 in connection with power distribution grounding systems should ideally be made with an oscilloscope or spectrum analyzer in order to observe and measure components of leakage current and voltage differences at all frequencies.

For routine testing, such instruments could be inconvenient. An alternative is to use a metering system that weighs the contribution to the meter reading of the various components of the signal being measured in accordance with their probable physiological effect.

A meter specifically designed for this purpose would have an impedance of approximately 1000 ohms, and a frequency characteristic that was flat to 1 kHz, dropped at the rate of 20 decibels per decade to 100 kHz, and then remaining flat to 1 MHz or higher. This frequency response characteristic could be achieved by proper design of the internal circuits of the amplifier that probably precedes the indicating instrument or by appropriate choice of a feedback network around the amplifier. These details are, of course, left to the instrument designer.

If a meter specifically designed for these measurements is not available, a general-purpose laboratory millivoltmeter can be adapted for the purpose by adding a frequency response–shaping network ahead of the meter. One such suggested network is shown in Figure A.6.3.3.1.3(a).

The circuit shown in Figure A.6.3.3.1.3(a) is especially applicable to measurements of leakage current, where the current being measured is derived from a circuit whose source impedance is high compared to 1000 ohms. Under these conditions, the voltage developed across the millivoltmeter will be proportional to the impedance of the network. The network impedance will be 1000 ohms at low frequencies and 10 ohms at high frequencies, and the transition between these two values will occur in the frequency range between 1 kHz and 100 kHz.

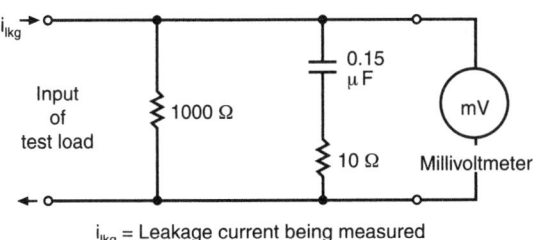

FIGURE A.6.3.3.1.3(a) Circuit Used to Measure Leakage Current with High Source Impedance.

The basic low-frequency sensitivity will be 1 mV of meter reading for each 1 mA of leakage current.

The millivoltmeter's own input impedance needs to be very large compared to 1000 ohms (100 kilohms), and the meter should have a flat frequency response to well beyond 100 kHz. (If the meter impedance is lower than 100 kilohms, then the 1000 ohm resistor can be raised to a higher value, such that the impedance of that resistor in parallel with the meter will still be 1000 ohms.)

The circuit in Figure A.6.3.3.1.3(a) can be used for the voltage difference measurements required in Section 6.5, but, because the source impedance will be very low compared to 1000 ohms, the frequency response of the measurement system will remain flat. If any high-frequency components produced, for example, by pickup from nearby radio frequency transmitters appear on the circuit being measured, then they will not be attenuated, and the meter reading will be higher than it should be.

For meter readings below any prescribed limits, this possible error is of no consequence. For borderline cases, it could be significant. To avoid this uncertainty when making voltage-difference measurements, a slightly more elaborate version of a frequency response–shaping network is given in Figure A.6.3.3.1.3(b).

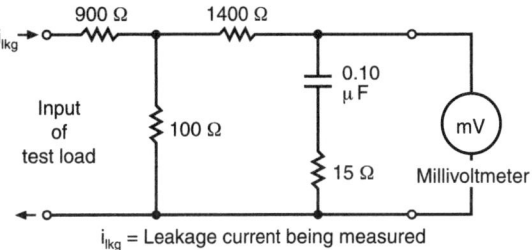

FIGURE A.6.3.3.1.3(b) Circuit Used to Measure Leakage Current with Low Source Impedance.

Here the source being measured is separated from the frequency response–shaping network by the combination of the 900 ohm and 100 ohm resistors. The frequency response characteristic is now independent of the circuit being tested.

This independence is achieved, however, at a loss in signal delivered to the millivoltmeter. The basic low-frequency sensitivity of this metering circuit is 1 mV of meter reading for 10 A of leakage current or, on a voltage basis, 1 mV of meter reading for 10 mV at the input terminals of the network.

The millivoltmeter should have an input impedance of 150 kilohms and a frequency response that is flat to well beyond 100 kHz.

For either of the suggested networks, the resistors and capacitors should be mounted in a metal container close to the millivoltmeter to avoid stray pickup by the leads going to the meter.

A.6.3.3.1.4 It is not the intent that each receptacle be tested. It is intended that compliance be demonstrated through random testing. The 10 percent random testing should include a mixture of both normal and emergency receptacles.

A.6.3.4 Administration is in conjunction with 6.3.4.1.

A.6.3.4.2.1 Although several approaches to documentation exist in hospitals, the minimum acceptable documentation should identify what was tested, when it was tested, and whether it performed successfully. Adopting a system of exception reporting can be the most efficient form of record keeping for routine rechecks of equipment or systems, thereby minimizing technicians' time in recording the value of each measurement taken. For example, once a test protocol is established, which simply means testing the equipment or system consistent with Chapter 6, the only item (value) that needs to be recorded is the failure or the deviation from the requirements of the chapter that was detected when a corrective action (repair) was undertaken. This approach can serve to eliminate, for example, the need to keep individual room sheets to record measured results on each receptacle or to record measurement values of all types of leakage current tests.

A.6.4.1.1.1 *Connection to Dual Source of Normal Power.* For the greatest assurance of continuity of electrical service, the normal source should consist of two separate full-capacity services, each independent of the other. Such services should be selected and installed with full recognition of local hazards of interruption, such as icing and flooding.

Where more than one full-capacity service is installed, they should be connected in such a manner that one will pick up the load automatically upon loss of the other, and should be so arranged that the load of the emergency and equipment systems will be transferred to the alternate source (generator set) only when both utility services are de-energized, unless this arrangement is impractical and waived by the authority having jurisdiction. Such services should be interlocked in such a manner as to prevent paralleling of utility services on either primary or secondary voltage levels.

Note that, in any installation where it is possible to parallel utility supply circuits (e.g., to prevent interruption of service when switching from one utility source to another) it is imperative to consult the power companies affected as to problems of synchronization.

Facilities whose normal source of power is supplied by two or more separate central station–fed services (dual sources of normal power) experience greater reliability than those with only a single feed.

Installation of Generator Sets. For additional material on diesel engines, see National Research Council Publication 1132, *Diesel Engines for Use with Generators to Supply Emergency and Short Term Electric Power (see Annex D).*

A.6.4.1.1.1.2(5) Careful consideration should be given to the location of the spaces housing the components of the essential electrical system to minimize interruptions caused by natural forces common to the area (e.g., storms, floods, or earthquakes; or hazards created by adjoining structures or activities). Consideration should also be given to the possible interruption of normal electrical services resulting from similar causes as well as possible disruption of normal electrical service due to internal wiring and equipment failures.

Consideration should be given to the physical separation of the main feeders of the essential electrical system from the normal wiring of the facility to prevent possible simultaneous destruction as a result of a local catastrophe.

In selecting electrical distribution arrangements and components for the essential electrical system, high priority should be given to achieving maximum continuity of the electrical supply to the load. Higher consideration should be given to achieving maximum reliability of the alternate power source and its feeders rather than protection of such equipment, provided that the protection is not required to prevent a greater threat to human life such as fire, explosion, electrocution, and so forth, than would be caused by the lack of an essential electrical supply.

A.6.4.1.1.7.3 This can be accomplished by operating the system continuously.

A.6.4.1.1.8.3 The intent of 6.4.1.1.8.3 is as follows:

(1) Contiguous or same-site nonhospital buildings can be served by the generating equipment. However, such loads should not compromise the integrity of the system serving the hospital. Thus, any such contiguous or same-site nonhospital buildings can be served by the generating equipment only if the transfer means operates in accordance with 6.4.1.1.8.3.

(2) Within a hospital building, 6.4.2.2.4.2(9) permits "additional" loads on the critical branch and 6.4.2.2.5.4(9) permits "other equipment" on the equipment system in order to provide limited flexibility to a facility to add one or two loads not otherwise listed in 6.4.2.2.4.2(1) through (8), 6.4.2.2.5.3, or 6.4.2.2.5.4(1) through (9) to a critical branch panel or an equipment system panel. This is permitted to prevent the need for an additional panel to serve a small number of selected circuits in a particular area. These sections are not intended to permit large blocks of loads not listed in these sections to be on the critical branch or equipment system. The intent of the division of the essential system loads into systems and branches is to ensure maximum reliability of service to loads considered essential. Every additional load placed onto a system somewhat increases the probability of a failure on the system that threatens the integrity of service to the balance of loads served by the system. Therefore, while "additional" loads and "other equipment" are permitted to be placed onto the critical branch and equipment system in very limited situations, where a facility wants to put large blocks of loads not listed in 6.4.2.2.4.2(1) through (8), 6.4.2.2.5.3, or 6.4.2.2.5.4(1) through (9) onto the generating equipment, the facility is permitted to do so, but only by designating these large blocks of loads as "optional loads" and by complying with 6.4.1.1.8.3.

A.6.4.1.1.10 It is the intent of 6.4.1.1.10 to mandate generator sizing based upon actual demand likely to be produced by the connected load of the essential electrical system(s) at any one time. It is not the intent that generator sizing be based upon connected load or feeder calculation procedures described in *NFPA 70, National Electrical Code.* Demand calculations should be based upon prudent demand factors and historical data.

A.6.4.1.1.13 During operation, EPS and related equipment reject considerable heat that needs to be removed by proper ventilation or air-cooling. In some cases, outdoor installations rely on natural air circulation, but enclosed installations need properly sized, properly positioned ventilation facilities, to prevent recirculation of cooling air. The optimum position of air-supply louvers and radiator air discharge is on opposite walls, both to the outdoors. [110: A.7.7.1]

A.6.4.1.1.18.1 As a supplement to hard-wired alarm annunciations, it is permissible to have Level 1 and Level 2 EPS and ATS functions monitored off-site. Monitoring stations can include pagers, cell phones, and Internet-connected devices.

A.6.4.2 It should be emphasized that the type of system selected and its area and type of coverage should be appropriate to the medical procedures being performed in the facility. For example, a battery-operated emergency light that switches "on" when normal power is interrupted and an alternate source of power for suction equipment, along with the immediate availability of some portable hand-held lighting, would be advisable where oral and maxillofacial surgery (e.g., extraction of impacted teeth) is performed. On the other hand, in dental offices where simple extraction, restorative, prosthetic, or hygienic procedures are performed, only remote corridor lighting for purposes of egress would be sufficient. Emergency power for equipment would not be necessary. As with oral surgery locations, a surgical clinic requiring use of life-support or emergency devices, such as suction machines, ventilators, cauterizers, or defibrillators, would require both emergency light and power.

A.6.4.2.1.2 It is important that the various overcurrent devices be coordinated, as far as practicable, to isolate faulted circuits and to protect against cascading operation on short-circuit faults. In many systems, however, full coordination could compromise safety and system reliability. Primary consideration also should be given to prevent overloading of equipment by limiting the possibilities of large current inrushes due to instantaneous reestablishment of connections to heavy loads. The terms *coordination* and *coordinated* as used in 6.4.2.1.2 do not cover the full range of overcurrent conditions.

A.6.4.2.1.5.1(A) Where special loads require more rapid detection of power loss, underfrequency monitoring also might be provided. Upon frequency decay below the lower limit necessary for proper operation of the loads, the transfer switch should automatically initiate transfer to the alternate source. *(See A.6.2.15 of NFPA 110.)* [110: A.6.2.2.1]

A.6.4.2.1.5.1(A)(2) See 6.2.5 and 6.2.7 of NFPA 110.

A.6.4.2.1.5.3 Authorized personnel should be available and familiar with manual operation of the transfer switch and should be capable of determining the adequacy of the alternate source of power prior to manual transfer. [110: A.6.2.4]

A.6.4.2.1.5.4 For most applications, a nominal delay of 1 second is adequate. The time delay should be short enough so that the generator can start and be on the line within the time specified for the type classification. [110: A.6.2.5]

A.6.4.2.1.5.7 It is recommended that the timer for delay on retransfer to the primary source be set for 30 minutes. The 30-minute recommendation is to establish a "normalized" engine temperature, when it is beneficial for the engine. *NFPA 70, National Electrical Code*, establishes a minimum time requirement of 15 minutes. [110: A.6.2.8]

A.6.4.2.1.5.12 For maintenance purposes, consideration should be given to a transfer switch counter. [110: A.6.2.13]

A.6.4.2.1.5.14 Automatic transfer switches (ATS) can be provided with accessory controls that provide a signal to operate remote motor controls that disconnect motors prior to transfer, and to reconnect them after transfer when the residual voltage has been substantially reduced. Another method is to provide inphase monitors within the ATS in order to prevent retransfer to the primary source until both sources are nearly synchronized. A third method is to use a programmed neutral position transfer switch. See Section 230.95(B) of *NFPA 70, National Electrical Code*. [110: A.6.2.15]

A.6.4.2.1.5.15 Standards for nonautomatic transfer switches are similar to those for automatic transfer switches, as defined in 3.3.7.1 and 3.3.7.3 of NFPA 110, *Standard for Emergency and Standby Power Systems*, with the omission of automatic controls. [110: A.6.2.16]

A.6.4.2.1.8.3 Consideration should be given to the effect that load interruption could have on the load during maintenance and service of the transfer switch.

A.6.4.2.2.1 Type 1 essential electrical systems are comprised of three separate branches capable of supplying a limited amount of lighting and power service that is considered essential for life safety and effective facility operation during the time the normal electrical service is interrupted for any reason. These three separate branches are the life safety, critical, and equipment branches.

A.6.4.2.2.3.2(3) Departmental installations such as digital dialing systems used for intradepartmental communications could have impaired use during a failure of electrical service to the area. In the event of such failure, those systems that have lighted selector buttons in the base of the telephone instrument or in the desk units known as "director sets" will be out of service to the extent that the lights will not function and that the buzzer used to indicate incoming calls will be silenced. The lack of electrical energy will not prevent the use of telephones for outgoing calls, but incoming calls will not be signaled, nor will intercommunicating calls be signaled. This communication failure should be taken into consideration in planning essential electrical systems.

A.6.4.2.2.4 It is recommended that facility authorities give consideration to providing and properly maintaining automatic battery-powered lighting units or systems to provide minimal task illumination in operating rooms, delivery rooms, and certain special-procedure radiology rooms, where the loss of lighting due to failure of the essential electrical system could cause severe and immediate danger to a patient undergoing surgery or an invasive radiographic procedure.

A.6.4.2.2.4.2(7) Departmental installations such as digital dialing systems used for intradepartmental communications could have impaired use during a failure of electrical service to the area. In the event of such failure, those systems that have lighted selector buttons in the base of the telephone instrument or in the desk units known as "director sets" will be out of service to the extent that the lights will not function and that the buzzer used to indicate incoming calls will be silenced. The lack of electrical energy will not prevent the use of telephones for outgoing calls, but incoming calls will not be signaled, nor will intercommunicating calls be signaled. This communication failure should be taken into consideration in planning essential electrical systems.

A.6.4.2.2.5.3 The equipment in 6.4.2.2.5.3(A)(1) through (3) can be arranged for sequential delayed-automatic connection to the alternate power source to prevent overloading the generator where engineering studies indicate that it is necessary.

A.6.4.2.2.5.4 For elevator cab lighting control and signal system requirements, see 6.4.2.2.3.1(6).

In instances where interruption of normal power would result in other elevators stopping between floors, throw-over facilities should be provided to allow the temporary operation of any elevator for the release of patients or other persons who are confined between floors.

A.6.4.2.2.5.4(2) The outside design temperature is based on the 97½ percent design value, as shown in Chapter 24 of the ASHRAE *Handbook of Fundamentals*.

A.6.4.2.2.5.4(9) Consideration should be given to selected equipment in kitchens, laundries, and radiology rooms and to selected central refrigeration.

It is desirable that, where heavy interruption currents can be anticipated, the transfer load be reduced by the use of multiple transfer devices. Elevator feeders, for instance, might be less hazardous to electrical continuity if they are fed through an individual transfer device.

A.6.4.2.2.6.1 See *NFPA 70, National Electrical Code*, for installation requirements.

A.6.4.2.2.6.2(C) If color is used to identify these receptacles, the same color should be used throughout the facility.

A.6.4.4.1.1.4(A) When events, such as the issuance of storm warnings, indicate that power outages might be likely, good practice recommends the warming up of generator sets by a regular exercise period. Operation of generator sets for short intervals should be avoided, particularly with compression ignition engines, since it is harmful to the engines.

Records of changes to the essential electrical system should be maintained so that the actual demand likely to be produced by the connected load will be within the available capacity.

A.6.4.4.1.2.1 Main and feeder circuit breakers should be periodically tested under simulated overload trip conditions to ensure reliability.

A.6.5.2.1.1 It is important that the various overcurrent devices be coordinated, as far as practicable, to isolate faulted circuits and to protect against cascading operation on short-circuit faults. In many systems, however, full coordination could compromise safety and system reliability. Primary consideration also should be given to prevent overloading of equipment by limiting the possibilities of large current inrushes due to instantaneous re-establishment of connections to heavy loads. The terms *coordination* and *coordinated* as used in 6.5.2.1.1 do not cover the full range of overcurrent conditions.

A.6.5.2.2.1 Type 2 essential electrical systems are comprised of two separate branches capable of supplying a limited amount of lighting and power service that is considered essential for the protection of life and safety and effective operation of the institution during the time normal electrical service is interrupted for any reason. These two separate branches are the life safety and equipment branches.

The number of transfer switches to be used should be based upon reliability, design, and load considerations. Each branch of the essential electrical system should have one or more transfer switches. One transfer switch should be permitted to serve one or more branches in a facility with a maximum demand on the essential electrical system of 150 kVA (120 kW).

A.6.5.2.2.2.1(4) Departmental installations such as digital dialing systems used for intradepartmental communications could have impaired use during a failure of electrical service to the area. In the event of such failure, those systems that have lighted selector buttons in the base of the telephone instrument or in the desk units known as "director sets" will be out of service to the extent that the lights will not function and that the buzzer used to indicate incoming calls will be silenced. The lack of electrical energy will not prevent the use of telephones for outgoing calls, but incoming calls will not be signaled, nor will intercommunicating calls be signaled. This communication failure should be taken into consideration in planning essential electrical systems.

A.6.5.2.2.3.4 Other selected equipment can be served by the equipment branch.

Note that consideration should be given to selected equipment in kitchens and laundries and to selected central refrigeration.

It is desirable that, where heavy interruption currents can be anticipated, the transfer load be reduced by the use of multiple transfer devices. Elevator feeders, for instance, might be less hazardous to electrical continuity if they are fed through an individual transfer device.

A.6.5.2.2.3.4(A)(1) The outside design temperature is based on the 97½ percent design value, as shown in Chapter 24 of the ASHRAE *Handbook of Fundamentals*.

A.6.5.2.2.3.4(B) For elevator cab lighting, control, and signal system requirements, see 6.5.2.2.2(6).

A.6.5.2.2.4.1 See *NFPA 70, National Electrical Code*, for installation requirements.

A.6.5.2.2.4.2 If color is used to identify these receptacles, the same color should be used throughout the facility.

A.7.1 Additional information on these systems can be found in IEEE 602, *Recommended Practice for Electric Systems in Health Care Facilities*, and the FGI Guidelines.

A.7.3.1.2 Additional information can be found in TIA/EIA 569-B, *Commercial Building Standard for Telecommunications Pathways and Spaces*.

A.7.3.1.2.1.3(B) Off-site electronic storage of patient records should also be considered.

A.7.3.1.2.1.4(C) Such systems can include security, nurse call, cable television, patient education, voice, data, head end equipment for clinical systems, and similar low voltage systems.

A.7.3.1.2.1.4(F) Such sources of electromagnetic interference include, but are not limited to, medical imaging equipment, transformers, motors, variable frequency drives, induction heaters, arc welders, radios, and radar systems.

A.7.3.1.2.1.8(B) Supplying the circuits serving equipment in the telecommunications entrance facility through an uninterrupted power system (UPS) provides a desirable level of redundancy.

A.7.3.1.2.1.8(C) Consideration should be given to the reliability of power supply to the HVAC equipment because of its important function within the telecommunications entrance facility.

A.7.3.1.2.2.2 In combined spaces, care should be taken to provide separation of, and adequate service access for, service provider equipment.

A.7.3.1.2.2.4(E) Such sources of electromagnetic interference include, but are not limited to, medical imaging equipment, transformers, motors, variable frequency drives, induction heaters, arc welders, radios, and radar systems.

A.7.3.1.2.3.4(D) Such sources can include medical imaging equipment, transformers, motors, variable frequency drives, induction heaters, arc welders, radio transmission systems, or other sources of electromagnetic interference.

A.7.3.3.1.1 Depending on the size and scope of the Category 1 space, an audiovisual type system or tone visual type system may be used for the nurse call system. While both system types provide audible tones and illuminated light sources to annunciate call events, the audiovisual system provides voice communication capabilities to enable staff to speak with patients or other staff at locations remote from the patient's room. Depending on the requirements for the different care areas, combinations of audiovisual and tone visual nurse call system equipment can be used.

A.7.3.3.1.1.3 The fundamental operation of a listed nurse call system provides alerts and notifications of call system events. In addition to the call notifications activated by call stations, a nurse call system should also provide alert notifications for system self-monitoring events to annunciate trouble conditions that can occur within the system itself.

A.7.3.3.1.1.5 A nurse call system can be integrated with a wireless communication system for the purposes of providing supplemental call notifications to staff-carried wireless devices. Such notifications are considered supplemental unless the wireless communication system is listed to ANSI/UL 1069. Supplemental communication systems should be provided with appropriate NRTL safety certifications and listings that are consistent with the intended use as a stand-alone wireless communication system.

The requirements in ANSI/IEC/ISO 80001-1-1 *Risk Management of Medical IT-Networks* and ANSI/IEC/ISO 80001-2-5 *Guidance for the Risk Management of Distributed Alarm Systems Utilizing Medical IT-Networks* should be followed whenever a nurse call system is integrated with a supplemental communication system.

A.7.3.3.1.2.1 Patient care areas and nursing unit support areas may contain many types of call stations with varying combinations of call initiation functions (e.g., code call, staff emergency, medical device alarm, help, assistance). A single call station can be equipped and configured to activate a single call type or a number of different call types, and may have bidirectional voice communication capability.

A.7.3.3.1.2.2 Patient stations provide a means for patients to summon assistance from the nursing staff. Calling devices such as listed wired or wireless pillow speakers, pendant controls call cords, and patient- or staff-worn personal pendants are permitted to initiate patient or staff calls. A call station that serves two beds is permitted when beds are located adjacent to each other.

A.7.3.3.1.2.3 Primary signaling of a medical device alarm is a requirement of the medical device itself, per ANSI/IEC/ISO 60601-1 or other governing regulatory standard, and is beyond the scope of this code.

A.7.3.3.1.3.1 When two or more call stations are located in the same area and all are visible from any call location, the alarm should be capable of being canceled at any of these locations. This method of call cancelation can be applied to all call station types.

A.7.3.3.1.4 A code call can also be referred to as an emergency resuscitation alarm.

A.7.3.3.1.4.1 When two or more call stations are located in the same area and all are visible from any call location, the alarm should be capable of being canceled at any of these locations. This method of call cancelation can be applied to all call station types.

A.8.2.1 There are no interdependencies for each type of system (e.g., medical gas, electrical, potable water, nonpotable water, nonmedical compressed air, heating). A risk assessment of each system should be conducted to evaluate the risk to the patient, staff, and visitors.

A.8.3.3 Another source of maximum hot water temperatures would be the FGI Guidelines.

A.9.2.1 There are no interdependencies for each type of system (e.g., medical gas, electrical, potable water, nonpotable water, nonmedical compressed air, plumbing). A risk assessment of each system should be conducted to evaluate the risk to the patient, staff, and visitors. It is possible when applying this section to identify multiple categories of systems serving a single patient. For example see A.4.1.

A.9.3.1.3 Previous editions had required smoke purge systems in these locations. The elimination of flammable anesthetics and limited use of combustible material in modern ORs makes this requirement obsolete.

A.9.3.3.2 ASHRAE Guideline 0, *The Commissioning Process*, and ASHRAE Guideline 1.1, *HVAC&R Technical Requirements for the Commissioning Process*, while not mandatory, can provide guidance for this requirement.

A.9.3.6.3 Paragraph 9.3.6.3 only covers fluids that are stored in enclosed spaces.

A.9.3.6.5.1 See Table A.11.3.1.

A.10.1 An appliance that yields erroneous data or functions poorly is potentially harmful. Quality and assurance of full appliance performance is not covered, except as it relates to direct electrical or fire injury to patients or personnel.

The material in this annex, as it relates to electrical safety, interprets some of the basic criteria by presenting different methodologies and alternative procedures to achieve the level of safety defined by the criteria.

A.10.1.3 Risk categorization is not appropriate for this chapter. Much of the chapter deals with electrical safety issues. Any line-powered device that does not meet these requirements poses the risk of electric shock to patients or personnel, regardless of where it is used or the clinical application of the device. Shock and fire risk is not dependent on the required reliability of the device or the extent to which patient diagnosis and treatment depends on the device. Even categorization regarding direct cardiac connection or patient contact are not relevant as death or injury can occur in any case.

A.10.2.3.2.4 IEC 60601-1, *Medical Electrical Equipment — Part 1: General Requirements for Basic Safety and Essential Performance*, defines the terms *protective earth conductor* and *functional earth conductor*. A protective earth conductor is relied upon for safety and provides one means of protection from electric shock. A functional earth conductor has no safety function and does not provide a means of protection against electric shock. A double-insulated medical product is permitted (but not required) to have a functional earth conductor that is referred to as a *functional ground conductor*.

A.10.2.3.6(2) Whole-body hyperthermia/hypothermia units should be powered from a separate branch circuit.

A.10.2.3.6(4) See Chapter 6 for criteria of receptacles.

A.10.2.6 Where existing equipment exceeds 500 µA, methods to reduce leakage current, such as the addition of small isolation transformers to that equipment, or methods that provide equivalent safety by adding redundant equipment ground, are permissible.

A.10.3.1 Visual inspections do not have to be formal or documented by any particular staff member. All staff are expected to be observant of the condition of the equipment they use, including power cord assemblies.

A.10.3.2 There are several methods for measuring ground-wire resistance accurately. Three examples are described as follows and shown in Figure A.10.3.2(a) through Figure A.10.3.2(c):

(1) *Two-Wire Resistance Technique.* A known current is fed through the unknown resistance. A high-input-impedance voltmeter measures the voltage drop across the resistance, R, and R is calculated as voltage divided by impedance, V/I. This technique measures the lead resistance in series with the unknown resistance. When the unknown resistance is a ground wire (less than 0.15 ohm), the lead resistance is appreciable. This is accounted for by shorting the lead wires together and "zeroing" the voltmeter. The actual resistance, in effect, subtracts out the lead wire resistance. In order for this technique to be reasonably accurate for measuring ground wires, an active high-impedance millivoltmeter has to be used.

(2) *Four-Wire Resistance Technique.* This technique is very similar to the two-wire resistance technique. The difference is that the known current is fed to the resistance to be measured through a pair of leads separate from the pair of leads to the voltmeter. The voltmeter is measuring the true voltage across the resistance to be measured, regardless of the resistance of the measuring leads. This method eliminates the need for zeroing out the measuring lead resistance.

(3) *AC Current Method.* This technique utilizes a step-down transformer of known voltage output to feed current through the ground wire and measure the current that flows. The impedance of the ground wire is then calculated by Ohm's law.

Note that the internal impedance of the measuring circuit has to be established with the test leads shorted. This value needs to be subtracted from the test measurement.

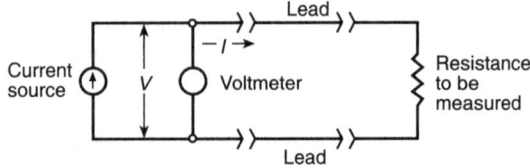

FIGURE A.10.3.2(a) Two-Wire Resistance Technique.

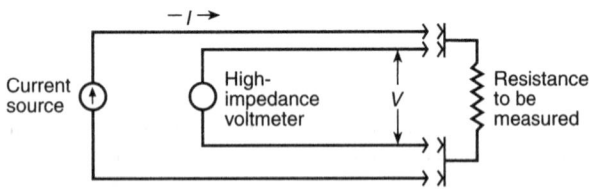

FIGURE A.10.3.2(b) Four-Wire Resistance Technique.

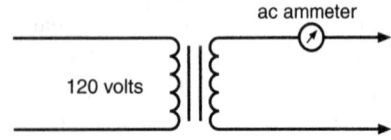

FIGURE A.10.3.2(c) AC Current Method.

A.10.3.3 For complex leakage current waveforms, a single reading from an appropriate metering system can represent the physiologically effective value of the composite waveform, provided that the contribution of each component to the total reading is weighted in accordance with 10.3.3.

This weighting can be achieved by a frequency response–shaping network that precedes a flat-response meter, or by a meter whose own frequency response characteristic matches that of 10.3.3.

If the required performance is obtained by a meter with integral response-shaping properties, then that meter should have a constant input resistance of 1000 ohms. (A high-input-impedance meter can be used by shunting a 1000 ohm resistor across the meter's input terminals.)

If, however, the required frequency response is obtained by a network that precedes an otherwise flat-response meter, then the input impedance of the network should be 1000 ohms ± 10 percent, over the frequency range from 0 to 1 MHz, and the frequency response of the network–meter combination should be substantially independent of the impedance of the signal source.

For maximum chassis leakage current permitted (i.e., 300 µA) below 1 kHz, this network will yield the limiting current of 10 mA above 30 kHz.

A suggested input network is shown in Figure A.10.3.3.

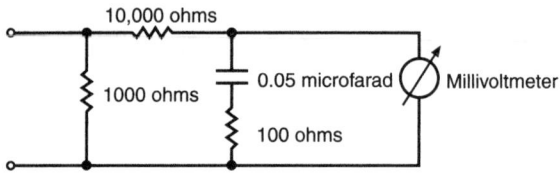

FIGURE A.10.3.3 Leakage Current Measurements (1.0 Millivoltmeter Reading Corresponds to Input Current of 1.0 Microampere).

A.10.3.3.3 This test is not valid when performed on the load side of an isolation transformer or an isolated power system, because the values obtained will be falsely low.

A.10.3.6 Although the touch current value is 500 µA, the patient lead leakage current limit for nonisolated input has been intentionally limited to 100 µA. This decision is in recognition of the need for a greater level of electrical safety for those portions of devices that make direct electrical patient connection.

A.10.5.2.5 Systems should comply with the appropriate medical device or system standards such as ANSI/AAMI ES60601-1, *Medical Electrical Equipment,* and IEC 60601-1-1, *Medical Electrical Equipment — Part 1: General Requirements for Basic Safety and Essential Performance.* Consideration should also be given to the guidance provided as part of the documentation supplied with the individual medical equipment.

A.10.5.4.5 The use of electrical equipment in spaces where there is a high oxygen content is a matter of concern because of the fire hazard. It is particularly a problem where the oxygen is "pure," that is, 80 percent to 90 percent, because materials that are not very flammable in ordinary air become extremely flammable in pure oxygen.

In medical practice, particularly in surgery, patients are often given supplemental oxygen via respirator, anesthesia machines, and so forth. Such supplemental oxygen can range from room air to 100 percent oxygen. Clearly, different levels of protection are needed.

This code addresses the problem by defining the following three elements of the situation:

(1) *Type of Air.* An oxygen-enriched atmosphere (OEA) is air that ranges from slightly enriched (23.5 percent, rather than 21 percent) to total oxygen (100 percent).
(2) *Type of Apparatus.* Oxygen delivery equipment (ODE) is a device that delivers an OEA to a patient.
(3) *Type of Space.* A site of intentional expulsion (SIE) is a small-volume space where oxygen that has been delivered to the patient is discharged to the ambient air.

When an OEA is within ODE, it is much more likely to have a high concentration of oxygen. Paragraph 10.5.4 therefore advises manufacturers, and A.10.5.4.5 advises users, of precautions to take to reduce the fire hazard. Paragraph 10.5.4 lists four ways of attacking the problem. Note that an OEA can be created not only in a ventilator or oxygen tubing, but also in an oxygen tent or incubator. Special precautions should be taken.

At the other extreme of hazard is a space in the open air, the site of intentional expulsion (SIE). This space is defined as located within 30.5 cm (12 in.) of the exhaust port, because, in most instances, dilution to ambient levels occurs within a few inches of the port; 30.5 cm (12 in.) provides an adequate safety factor. Paragraph 10.5.4.1 provides guidance to minimize this hazard by requiring that only those parts of the apparatus that are intended to be within the SIE are to be of concern. Even these components, such as nurse call buttons, leads, and so forth, do not necessarily need to be listed for use in OEA, because they usually conform to provisions of 10.5.4.1(4); that is, they do not have hot surfaces, and they meet the requirements of Figure 10.5.4.1(a) through Figure 10.5.4.1(f).

The intent of A.10.5.4.5 is to advise users to specify appliances that meet higher requirements where the hazard is higher but not to overspecify where the hazard is minimal. Thus, as they are ordinarily used, nurse call buttons, pillow speakers, and so forth, do not need to be listed for use in oxygen-enriched atmospheres.

Note, however, that these requirements apply only to the intended use. The user should exercise vigilance to guard against an unintended use or an accidental failure, which can vastly increase the hazard.

A.10.5.4.6 Where possible, combustible materials such as hair, fabric, and paper should be removed from the vicinity where the energy is delivered. Water-soluble surgical jelly has been shown to dramatically reduce the combustibility of such materials.

A.10.5.5.1 Most laboratory fires involve biomedical or other electronic equipment failures. The most common ignition factors are short circuits or ground faults. Electrical wire or cable insulation is the material most likely to first ignite in a clinical laboratory fire. *(See Hoeltge, G.A., Miller, A., Klein, B.R., Hamlin, W.B., "Accidental fires in clinical laboratories.")*

A.10.5.6.2 Although several approaches to documentation exist in hospitals, the minimum acceptable documentation should identify what was tested, when it was tested, and whether it performed successfully. Adopting a system of exception reporting can be the most efficient form of record keeping for routine rechecks of equipment or systems, thereby minimizing technicians' time in recording the value of each measurement taken. For example, once a test protocol is established, which simply means testing the equipment or system consistent with Chapter 10, the only item (value) that needs to be recorded is the failure or the deviation from the requirements of the chapter that was detected when a corrective action (repair) was undertaken. This approach can serve to eliminate, for example, the need to keep individual room sheets to record measured results on each receptacle or to record measurement values of all types of leakage current tests.

A.10.5.8.1 "Personnel" includes physicians, nurses, nursing assistants, engineers, and technicians.

A.10.5.8.3 Qualification for equipment servicing does not always include manufacturer training, as required knowledge and skills can be obtained by other means.

A.11.1.2 Respiratory therapy is an allied health specialty employed with medical direction in the treatment, management, control, diagnostic evaluation, and care of patients with deficiencies and abnormalities of the cardiopulmonary system. (Courtesy of the American Association for Respiratory Therapy, 1720 Regal Row, Dallas, TX 75235.)

Respiratory therapy includes the therapeutic use of the following: medical gases and administration apparatus, environmental control systems, humidification, aerosols, medications, ventilatory support, bronchopulmonary drainage, pulmonary rehabilitation, cardiopulmonary resuscitation, and airway management. (Courtesy of the American Association for Respiratory Therapy, 1720 Regal Row, Dallas, TX 75235.)

There is a continual need for human diligence in the establishment and maintenance of safe practices for respiratory therapy. It is essential for personnel having responsibility for respiratory therapy to establish and enforce appropriate programs to fulfill the provisions of this chapter.

It is the responsibility of the administrative and professional staff of a hospital, or safety director, if one is appointed, to adopt and enforce appropriate regulations for a hospital. In other health care facilities, responsibility could be assigned to a safety director, or other responsible person, who is, in turn, responsible to the administration.

In institutions having a respiratory therapy service, it is recommended that this service be directly responsible for the administration of Chapter 11. Hazards can be mitigated only when there is continual recognition and understanding.

A.11.1.4 See Chapter 14.

A.11.1.5 Risk categorization is not appropriate for this chapter. Much of the chapter deals with gas equipment safety issues. Any storage or use of gas equipment in a health care facility presents risks that the requirements of this chapter reduce. Any gas-related device that does not meet these requirements poses the same risk to patients or personnel, regardless of where it is used or its clinical application.

A.11.2.8 It is particularly important that the intermixing of oxidizing and flammable gases under pressure be scrupulously avoided. Such mixing can result in a violent explosion.

A.11.3.1 See Table A.11.3.1.

A.11.3.2 When determining the volume of storage, do not consider cylinders and containers that are in use. Only the volume of stored gas that is in excess of 300 ft^3 is required to be

Table A.11.3.1 Typical Medical Gas Cylinders' Volume and Weight of Available Contents [All Volumes at 21.1°C (70°F) and 101.325 kPa (14.696 psi)]

Cylinder Style and Dimensions	Nominal Volume [L (in.³)]	Contents	Air	Carbon Dioxide	Helium	Nitrogen	Nitrous Oxide	Oxygen	Mixtures of Oxygen	
									Helium	CO₂
B 8.89 × 33 cm (3½ in. O.D. × 13 in.)	1.43 (87)	kPa (psig) L (ft³)		5778 (838) 370 (13)				13,100 (1900) 200 (7)		
		kg (lb-oz)		0.68 (1–8)				—		
D 10.8 × 43 cm (4¼ in. O.D. × 17 in.)	2.88 (176)	kPa (psig) L (ft³)	13,100 (1900) 375 (13)	5778 (838) 940 (33)	11,032 (1600) 300 (11)	13,100 (1900) 370 (13)	5137 (745) 940 (33)	13,100 (1900) 400 (14)	* 300 (11)	* 400 (14)
		kg (lb-oz)	—	1.73 (3–13)	—	—	1.73 (3–13)	—	*	*
E 10.8 × 66 cm (4¼ in. O.D. × 26 in.)	4.80 (293)	kPa (psig) L (ft³)	13100 (1900) 625 (22)	5778 (838) 1590 (56)	11,032 (1600) 500 (18)	13,100 (1900) 610 (22)	5137 (745) 1590 (56)	13,100 (1900) 660 (23)	* 500 (18)	* 660 (23)
		kg (lb-oz)	—	2.92 (6–7)	—	—	2.92 (6–7)	—	*	*
M 17.8 × 109 cm (7 in. O.D. × 43 in.)	21.9 (1337)	kPa (psig) L (ft³)	13,100 (1900) 2850 (101)	5778 (838) 7570 (267)	11,032 (1600) 2260 (80)	15,169 (2200) 3200 (113)	5137 (745) 7570 (267)	15,169 (2200) 3450 (122)	* 2260 (80)	* 3000 (106)
		kg (lb-oz)	—	13.9 (30–10)	—	—	13.9 (30–10)	—	*	*
G 21.6 × 130 cm (8½ in. O.D. × 51 in.)	38.8 (2370)	kPa (psig) L (ft³)	13,100 (1900) 5050 (178)	5778 (838) 12,300 (434)	11,032 (1600) 4000 (141)	15,169 (2200) 5000 (176)	5137 (745) 13,800 (487)	15,169 (2200) 6000 (211)	* 4000 (141)	* 5330 (188)
		kg (lb-oz)	—	22.7 (50–0)	—	—	25.4 (56–0)	—	*	*
H or K 23.5 × 130 cm (9¼ in. O.D. × 51 in.)	43.6 (2660)	kPa (psig) L (ft³)	15,169 (2200) 6550 (231)	5778 (838) 15,840 (559)	15,169 (2200) 6000 (212)	15,169 (2200) 6400 (226)	5137 (745) 15,800 (558)	15,169† (2200†) 6900 (244)	* 6000 (212)	* 15,840 (559)
		kg (lb-oz)	—	29.1 (64)	—	—	29.1 (64)	—	*	*

Notes:
These are computed contents based on nominal cylinder volumes and rounded to no greater variance than ±1 percent.
* The pressure and weight of mixed gases will vary according to the composition of the mixture.
†275 ft³/7800 L cylinders at 2490 psig are available on request.
Source: Compressed Gas Association, Inc.

located in an enclosure, since 11.3.3 already permits up to 300 ft³ without any special storage requirements.

A.11.4.1.1 If the sole source of supply of nonflammable medical gases, such as nitrous oxide and oxygen, is a system of cylinders attached directly to, and supported by, the device (such as a gas anesthesia apparatus) used to administer these gases, it is recommended that two cylinders of each gas be attached to the administering device.

A.11.4.1.2 The Pin-Index Safety System consists of a combination of two pins projecting from the yoke assembly of the apparatus and so positioned as to fit into matching holes drilled into the cylinder valves. It is intended to protect against the possibility of error in attaching the flush-type valves, with which gas cylinders and other sources of gas supply are equipped, to gas apparatus having yoke connections.

A.11.4.1.4 Fabrication specifications are contained in CGA V-1 (ANSI B57.1), *Standard for Compressed Gas Cylinder Valve Outlet and Inlet Connections*. Connection No. 860, shown in that document, illustrates the system. Connection No. 870 (Oxygen, Medical), Connection No. 880 (Oxygen–Carbon Dioxide Mixture), Connection No. 890 (Oxygen–Helium Mixture), Connection No. 900 (Ethylene), Connection No. 910 (Nitrous Oxide), Connection No. 920 (Cyclopropane), Connection No. 930 (Helium), and Connection No. 940 (Carbon Dioxide) are for specific medical gases and gas mixtures and utilize the basic dimensions of Connection 860.

A.11.4.3.2 For example, ozone sterilizers using medical grade oxygen from the piped distribution system should meet the following requirements:

(1) Not be permanently attached to the piped distribution system
(2) Be a medical device that has been listed for the intended purpose with the United States Food & Drug Administration
(3) Operate at or below 5 psig (34.5 kPa)

A.11.5.1.1.2 Outside of a patient care room, 11.5.1.1.2 prohibits sources of open flames within the site of intentional expulsion [1 ft (0.3 m)] of a nasal cannula. No sources of open flame are permitted within the area of administration [15 ft (4.3 m)] for other types of oxygen delivery equipment or in patient care rooms *(see 11.5.1.1.3)*.

The amount of oxygen delivered by a nasal cannula is limited. One foot (0.3 m) is sufficient separation from an oxygen-enriched atmosphere produced by a nasal cannula, which is oxygen delivery equipment used outside of patient care areas. In the open air, dilution goes to ambient levels (not oxygen-enriched atmosphere) within a few inches of the cannula openings, but 12 in. (300 mm) provides an adequate safety factor. Other oxygen delivery equipment, such as masks, are not included since masks would not typically be associated with mobile patients in health care facilities and can deliver greater quantities of oxygen than nasal cannula.

The household-style nursing homes that include kitchens intended for residents' use and enclosed gas fireplaces present a source of flame ignition to which residents will be exposed. Residents utilizing a nasal cannula would potentially not be allowed to participate in the cooking because it would place the cooking flame within the site of intentional expulsion. However, they would be allowed in the kitchen area to assist in preparing the food and to socialize with other residents and staff in the kitchen similar to what happens in the kitchens of residential environments.

The primary concern is that flame-producing equipment exists in many places in a nursing home and that it would be impractical to maintain a resident with a nasal cannula a minimum of 15 ft (4.3 m) (Area of Administration) away from the flame-producing equipment. Typical flame-producing equipment found in a nursing home includes the following:

(1) Candles in chapels
(2) Open kitchens using gas cooking equipment
(3) Fireplaces
(4) Fuel-fired heating equipment
(5) Private family dining rooms using fuel-fired equipment
(6) Canned cooking fuel (e.g., used under chafing dishes)

A.11.5.1.1.3 Patients and hospital personnel in the area of administration should be advised of respiratory therapy hazards and regulations.

Visitors should be cautioned of these hazards through the prominent posting of signs. *(See 11.3.4.)*

A.11.5.1.1.4 Solid fuel–burning appliances include wood-burning fireplaces, wood stoves, and similar appliances. These pose a greater risk in locations where oxygen is being provided than gas-fueled appliances, in part due to their ability to emit embers into the environment.

A.11.5.1.1.5 Such toys have been associated with fire incidents in health care facilities.

A suggested text for precautionary signs for oxygen tent canopies and oxygen hoods used in pediatric nursing units is the following:

CAUTION: OXYGEN IN USE

ONLY TOYS APPROVED BY

NURSES MAY BE GIVEN TO CHILD

A.11.5.1.3.3 Service manuals, instructions, and procedures provided by the manufacturer should be considered in the development of a program for maintenance of equipment. The experience accumulated by the facility and others (evidence-based maintenance) should be used to adjust manufacturer's recommendations whenever and wherever appropriate, even if they indicate no maintenance program is required.

A.11.5.2.1.1 "Personnel" typically includes physicians, nurses, nursing assistants, respiratory therapists, engineers, technicians, and others.

A.11.5.2.2.2 CGA P.2-5 contains both mandatory and nonmandatory language. Enforceable language uses the term shall; nonmandatory language uses the term should. This section indicates that NFPA 99 is making reference only to the mandatory requirements in that document.

A.11.5.2.3.2 CGA P-2.6, *Transfilling of Liquid Oxygen to Be Used for Respiration*, describes the recommended precautions and safety procedures when transfilling liquid oxygen used for respiration. Mishandling of oxygen presents potential hazards to both trained and untrained persons. Organizations engaged in the transfilling of liquid oxygen should ensure that personnel are familiar with the hazards of cryogenic liquid oxygen and that they comply with applicable regulations and safety practices.

A.11.5.2.4 Oxygen concentrator filling systems are FDA approved and have been in use for many years with excellent safety records. The inherent risks associated with typical transfilling do not apply to oxygen concentrator filling compressors.

Limitations were placed on these systems in order to prevent the risks associated with larger, higher flow, or higher pressure systems being introduced into the patient environment. The cylinder size was limited to cylinders normally used for patient ambulation. The filling rate was limited to prevent excessive heating of the cylinder contents. The filling pressure was limited based on the existing industry practice.

A.11.5.3.2 Precautionary signs should be at least 21 cm × 28 cm (8 in. × 11 in.).

Any material that can burn in air will burn more rapidly in the presence of oxygen.

Special signs and additional precautionary measures should be employed whenever foreign languages present a communication problem. *(See Figure A.11.5.3.2.)*

FIGURE A.11.5.3.2 Suggested Minimum Text for Precautionary Signs.

No electrical equipment is permitted within an oxygen enclosure or within 1.5 m (5 ft) of the enclosure.

A.11.7.3.1 The seller has a responsibility to provide written instructions to the user in accordance with 11.7.2. In fulfilling this responsibility, the seller should explain to the user the use of the equipment being delivered and precautions that are to be taken. The seller's written instructions are intended to make the user aware of the hazards of the material and to provide recommendations that will address the location, restraint, movement, and refill of ambulatory containers when these containers are to be refilled by the user. However, the user has the responsibility to receive, read, and understand the written material regarding storage and use of liquid oxygen and the containers and equipment that are furnished by the seller. In addition to specific information or instructions provided by the seller or equipment manufacturer regarding the storage or use of the equipment and of the liquid oxygen or the containers used, the user remains responsible to see that the containers are used or maintained in accordance with the seller's instructions to ensure that they are as follows:

(1) Located and maintained in accordance with the requirements of 11.7.3.2
(2) Restrained in accordance with the requirements of 11.7.3.3
(3) Handled or transported in accordance with the requirements of 11.7.3.4
(4) Refilled in accordance with the requirements of 11.7.3.6 and the manufacturer's instructions when liquid oxygen ambulatory containers are to be refilled by the user

CGA P-2.7 *Guide for the Safe Storage, Handling, and Use of Small Portable Liquid Oxygen Systems in Health Care Facilities*, describes the recommended precautions and safety procedures to be followed when liquid oxygen systems are used within health care facilities. Mishandling of oxygen presents potential hazards to both trained and untrained persons. It is, therefore, important that personnel who assume responsibility for oxygen equipment and its use be familiar with the hazards of oxygen, the operational characteristic of the equipment, and the precautions to be observed while using it.

A.11.7.3.3 Two points of contact can be provided by using elements of a room or furnishings in the room, such as the walls of a corner of a room, or a wall and a furnishing or object, such as a table or a desk.

A.11.7.3.5 CGA P-2.7, *Guide for the Safe Storage, Handling, and Use of Small Portable Liquid Oxygen Systems in Health Care Facilities*, describes the recommended precautions and safety procedures to be followed when liquid oxygen systems are used within health care facilities, such as a 5 ft separation from electrical appliances during filling and use.

Mishandling of oxygen presents potential hazards to both trained and untrained persons. It is therefore important that personnel who assume responsibility for oxygen equipment and its use be familiar with the hazards of oxygen, the operational characteristic of the equipment, and the precautions to be observed while using it.

A.11.7.3.6.1.1 Drip pans or similar containment devices are used in order to protect against liquid oxygen spillage coming into contact with combustible surfaces, including asphalt, which would elevate the potential for ignition.

A.12.1 Such facilities include, but are not limited to, hospitals, convalescent or nursing homes, and emergency receiving stations. A government authority could formally designate such facilities as disaster treatment centers. Such facilities would not normally include doctors' or dentists' offices, medical laboratories, or school nurseries, unless such facilities are used for treatment of disaster victims. National bioterrorism preparedness efforts call for the use of schools and other large public facilities to provide facilities for mass immunization.

A.12.1.1 Throughout this chapter, wherever the term *hospital* is used, the term should also apply to other types of health care facilities. Applicable facilities include, but are not limited to, hospitals, convalescent or nursing homes, and emergency receiving stations. A government authority could formally designate such facilities as disaster treatment centers. Such facilities would not normally include doctors' or dentists' offices, medical laboratories, or school nurseries, unless such facilities are used for treatment of disaster victims. National bioterrorism preparedness efforts call for the use of schools and other large public facilities to provide facilities for mass immunization. An emergency management program (formerly known as a disaster plan or internal/external plan) encompasses activities across four phases: mitigation, preparedness, response, and recovery. Mitigation activities are those designed to reduce or eliminate the impact of hazards. Preparedness activities include those that build organizational and individual capabilities to deal with disasters. Response activities include all necessary actions to stop ongoing negative effects of a disaster. Recovery activities are those that restore the organization, its employees, and the community back to normal.

NFPA 1600, Standard on Disaster/Emergency Management and Business Continuity Programs, is an internationally accepted framework for an emergency program. NFPA 99, Chapter 12, recognizes this overall structure and provides additional information useful to health care organizations. Table A.12.1.1 illustrates the relationship between the elements of NFPA 99, Chapter 12, and *NFPA 1600*.

A.12.2.1 In time of disaster, all persons are subject to certain constraints or authorities not present during normal circumstances. The emergency operations plans written by a health care facility should be reviewed and coordinated with such authorities so as to prevent confusion. Such authorities include, but are not limited to, civil authorities (such as a fire department, a police department, a public health department, or emergency medical service councils), Centers for Disease Control, Federal Bureau of

Table A.12.1.1 How NFPA 99, Chapter 12, Relates to *NFPA 1600*

NFPA 1600	NFPA 99, Chapter 12
Introduction	—
Scope	12.1 Applicability
Purpose	12.1.1.1 Framework
Program Management	—
Policy	12.2.1 Authority Having Jurisdiction
Program Coordinator	12.2.2 Senior Management
Program Committee	12.2.3 Emergency Management Committee
Program Assessment	12.2.3 Emergency Management Committee
Program Elements	—
General	—
Laws and Authority	—
Hazard Identification and Risk Management	A.12.5.3.1.2 Hazard Identification
Hazard Management (Mitigation)	—
Resource Management	12.5.3.3.6.2 Resource Assessment
Planning	12.5.3.3.5 Emergency Operations Plan
Direction, Control, and Coordination	A.12.2.3.3 Incident Command System
Communications and Warning	A.12.5.3.3.6.1(5) Communications
Operations and Procedures	A.12.5.3.3.6.3 Identification of Personnel
	12.5.3.3.6.5 Essential Utilities
	12.5.3.3.6.7 Staff Roles
	12.5.3.4.12 Medical Surge Capacity and Capability
	12.5.3.3.6.3 Safety and Security
	12.5.3.5 Recovery
Logistics and Facilities	12.5.3.4 Response
Training	12.5.3.3.8 Staff Education
Exercises, Evaluations, and Corrective Actions	12.5.3.3.8 Testing Emergency Plans and Operations
Public Education and Information	12.5.3.3.6.1(3) Communication
Finance and Administration	—

Investigation, and emergency management or military authorities. See Comprehensive Emergency Plan, Annex G for publications explaining how the out-of-hospital response is organized to multiple and mass casualty incidents. Further, an authority having jurisdiction can impose upon the senior management of the facility the responsibility for participating in a community emergency management program.

A.12.2.3 It is strongly recommended that medical leadership representatives play a key role in the emergency management committee and planning process. The following list is not intended to be all-inclusive, and additional representatives might be needed based on the level of care provided or the structure of the organization:

(1) Bioterrorism coordinator
(2) Communications/data management
(3) Finance
(4) Human resources
(5) Legal/risk management
(6) Public relations
(7) Purchasing/materials management
(8) Quality management
(9) Training and education

A.12.2.3.3 Federal, state, and local governments are required to use an incident command system (ICS) based on the National Incident Management System (NIMS). Although private sector hospitals are not required to be NIMS compliant, many are choosing to comply, not only to integrate with other emergency responders but also to remain eligible to receive certain federal grant monies. HICS, the Hospital Incident Command System, was specifically designed to be NIMS compliant, and, therefore, many hospitals use this model, either as developed or with some customization. HICS can be customized and adapted to other types of health care facilities.

HICS is led by an incident commander and assisted by command staff consisting of the public information officer, safety officer, and liaison officer, and those medical/technical specialists who are appropriate to the event. Section chiefs are responsible for each of the following sections:

(1) Operations Section: Conducts the tactical operations to carry out the incident action plan using defined objectives and directing all necessary resources
(2) Planning Section: Collects and evaluates information for decision support, maintains resource status information, prepares the Incident Action Plan, and maintains documentation
(3) Logistics Section: Provides support, resources, and other services to meet the operational objectives
(4) Finance/Administration: Monitors costs related to the incident and provides accounting services, time recording, and cost analyses

Each section is composed of subordinate positions that are divided into branches or units.

Features of HICS include the following:

(1) Clear chain of command
(2) Manageable span of control
(3) Common terminology
(4) Adaptability to unified command

HICS tools include the following:

(1) Job action sheets detailing position responsibilities
(2) Forms to document the event

HICS was intended to be used not only for emergencies but also for planned events. Complete HICS documentation is free and available for download at www.emsa.ca.gov.

A.12.5.3.1.2 By basing the planning of health care emergency management on realistic conceptual events, the program reflects those issues or events that are predictable for the environment in which the organization operates. Thus, such conceptual planning should focus on issues, such as severe weather typical in the locale, situations that can occur due to close proximity of industrial, government, or transportation complexes, or earthquake possibilities due to local seismic activity. Planning should also incorporate knowledge available in the emergency management research about how individuals, small groups, organizations, communities, and societies behave during emergencies.

A.12.5.3.1.3(1) Continuity of operations can include, but is not limited to, maintaining staffing levels, resources and assets, ability to obtain support from the outside environment, and leadership sustainability.

A.12.5.3.3.6.1(5) Emergency internal and external communications systems should be established to facilitate communication with security forces and other authorities having jurisdiction, as well as internal patient care and service units in the event normal communications methods are rendered inoperative. The basic form of communication in a disaster is the telephone system. As part of the contingency plan to maintain communication, a plan for restoring telephone systems or using alternate systems is necessary. Typically, the first line of internal defense for a system outage is strategically placed power-failure telephones that are designed to continue to function in the event of system failure (e.g., dedicated lines, fax lines). Plans for external outages and load control should include the use of pay phones, where available, that have first priority status in external system restoration. Facilities should preplan restoration activities and prioritization with their telephone service providers. A review with the state and other communications agencies (Government Emergency Telecommunications Service, Wireless Priority Service, Health and Homeland Alert Network) should be conducted.

Contingency plans should also contain strategies for the use of radio frequency communications to supplement landline usage. The plan should include a means to distribute and use two-way radio communication throughout the facility. A plan for the incorporation and use of amateur radio operators should also be considered.

It should be recognized that single-channel radio communication is less desirable than telephone system restoration due to the limited number of messages that can be managed. Cellular telephones, although useful in some disaster situations, should not be considered a contingency that has high reliability due to their vulnerability to the load control schemes of telephone companies. Text messaging has been proven to be more reliable than cellular phone calls. Social media can be an important tool for emergency communication, but it must be managed so that responses to inquiries can be provided. Portable e-mail devices, satellite telephones, and audio- and video-conferencing services are useful tools to link key staff and organizations.

A.12.5.3.3.6.3 Prior to a disaster, facilities should formally coordinate their security needs with local law enforcement agencies. The health care institution will find it necessary to share its emergency operations plans with local law enforcement agencies or, better still, involve them in the process of planning for security support during disasters. The information should at least include availability of parking for staff, patients, and visitors, and normal vehicular, emergency vehicular, and pedestrian traffic flow patterns in and around the facility. The extent of the security and traffic control problems for any given health care facility will depend upon its geographical location, physical arrangement, availability of visitor parking areas, number of entrances, and so forth.

Crowd Control. Visitors can be expected to increase in number with the severity of the disaster. They should not be allowed to disrupt the functioning of the facility disaster plan. Ideally, a visitor reception center should be established away from the main facility itself, particularly in the case of major disasters. Volunteer personnel such as community emergency response teams (CERT), Red Cross, Explorer Scouts, or other helpers can be utilized as liaisons between the visitors and the health care facility itself.

Vehicular Traffic Control. Arrangement for vehicular traffic control into and on the facility premises should be made in the disaster planning period. It will be necessary to direct ambulances and other emergency vehicles carrying casualties to triage areas or the emergency room entrance, and to direct incoming and outgoing vehicles carrying people, supplies, and equipment. Charts showing traffic flow and indicating entrances to be used, evacuation routes to be followed, and so forth, should be prepared and included in the emergency operations plan. Parking arrangements should not be overlooked.

Internal Security and Traffic Control. Internal security and traffic control are best conducted by facility-trained personnel (i.e., regular health care facility security forces) with reinforcements as necessary. Potential additional assistance from the local law enforcement agencies should be coordinated in the disaster planning phase. Upon activation of the emergency operations plan, security guards should be stationed at all unlocked entrances and exits to the extent possible. Entrance to the facility should be restricted to personnel bearing staff identification cards and to casualties. In the case of major access corridors between key areas of the facility, pedestrian traffic should be restricted to one side of the corridor, keeping the other side of the corridor free for movement of casualties. Traffic flow charts for internal traffic should also be prepared in the planning phase, as is the case with external traffic control.

A.12.5.3.3.6.5 Consideration should be given to preemptively installing parallel components such that maintenance can be performed on operating equipment. This will necessitate the installation of additional valves, circuits, or controls to isolate those parts to be removed and replaced, such as air or fuel filters. This work should not violate any other code, standard, or safety device. The desired outcome is system resiliency despite part failure.

A.12.5.3.3.6.8(1) The command structure should also follow the National Incident Management System (NIMS) as provided in *NFPA 1600, Standard on Disaster/Emergency Management and Business Continuity Programs.*

A.12.5.3.3.8 Experiences show the importance of drills to rehearse the implementation of all elements of a specific response, including the entity's role in the community, space management, staff management, and patient management activities. To document an exercise, the following aspects are typically incorporated. A general overview of the scenario, documented activation of the emergency operations plan, reports from an identified evaluator(s), evaluation of all involved participants (departments) and any observer(s), a written critique following the drill, and any identified follow-up training or improvement action(s) to correct or manage any deficiencies.

A.12.5.3.3.9.8 When improvements require substantive resources that cannot be accomplished by the next planned exercise, interim improvements should be put in place until final resolution.

A.12.5.3.4.1 In emergency situations that occur without warning and impact the facility, staff at the scene of the problem are expected to follow established protocols to protect life, notify others, and conserve property. Senior management can establish a hospital command center (HCC) or participate in unified command with other responding agencies at a designated emergency operations center (EOC). In emergency situations with warning or whose impacts require extended periods to resolve, designated leadership reports to the HCC. Not all incidents require an HCC.

The HCC provides centralized locations for information collection, display, coordination, documentation, and dissemination.

A.12.5.3.4.5 Note that care should be taken to ensure that identification cards are recalled whenever personnel terminate association with the health care facility. Members of the news media should be asked to wear some means of identification, such as a press card, on their outside garments so that they are readily identifiable by security guards controlling access to the facility or certain areas therein. Clergy also will frequently accompany casualties or arrive later for visitations and require some means of identification. Water storage systems should be inventoried and protected to the greatest extent possible.

A.12.5.3.4.9 For additional information see "Crisis Standards of Care: A Systems Framework for Catastrophic Disaster Response," 2012 Institute of Medicine (IOM) Report.

A.12.5.3.4.12.1 For additional information see *Medical Surge Capacity and Capability* Handbook (http://www.phe.gov/preparedness/planning/mscc/pages/default.aspx).

A.12.5.3.5 Recovery measures could involve a simple repositioning of staff, equipment, supplies, and information services; or recovery could demand extensive cleanup and repair. It can, under certain circumstances, be a means to identify opportunities for structural and nonstructural mitigation efforts. Filing of loss claims might require special approaches. Health care facilities should have access to cash or negotiable instruments to procure immediately needed supplies.

A.13.1 This chapter is the source for security management in health care facilities and is based on the foundations of NFPA 730, *Guide for Premises Security.*

A.13.2 A health care facility security plan can be formulated from security-sensitive areas that need the highest level of protection outward to the perimeter of the health care facility campus in concentric rings. Viewed from the outside, security is thus open and welcoming to patients and visitors. As an individual proceeds into the interior, public spaces might have minimal surveillance, but those sensitive areas that cannot be entered are layered with protections and countermeasures.

A.13.2.2 The security plan should be reviewed annually or more frequently if new challenges present themselves.

A.13.3.1 For general information regarding the SVA and premises security, see NFPA 730, *Guide for Premises Security.*

A.13.4.2(3)(c) The emergency potential inherent in the telephoned bomb threat warrants inclusion of this contingency in the health care emergency operations plan. Experience has shown that facility personnel have to accompany police or military bomb demolition personnel in searching for the suspected bomb, because speed is of the essence, and only individuals familiar with a given area can rapidly spot unfamiliar or suspicious objects or conditions in the area. This is particularly true in health care facilities. The facility switchboard operator should have a checklist, to be kept available at all times, in order to obtain as much information as possible from the caller concerning the location of the supposed bomb, time of detonation, and other essential data, which should be considered in deciding whether or not to evacuate all or part of the facility.

A.13.5.2(1) A visible presence is normally accomplished by the placement of a security officer at the ambulance entrance. This serves the dual purpose of monitoring the security cameras throughout the emergency department as well as the activity at the ambulance entrance.

A.13.5.3(5) The facility-wide alerting system should be activated for all reports of pediatric or infant abduction. The use of a standardized "code alert" system can facilitate the announcement; for example, "code pink" for an infant abduction or "code purple" for a pediatric abduction.

A.13.5.4 Video surveillance and motion detection can be used as additional protection for these areas. Some controlled drugs should be stored in safes.

A.13.5.6(3) Reasons for a contraband check procedure would be to control items such as tobacco, drugs, or tools that could cause harm to the patient or staff.

A.13.5.7(1) Law enforcement personnel should have orientation on the emergency procedures and layout of the facility. There should be good communication between law enforcement and health care facility security staff.

A.13.6.3.1 There can be times where full or partial facility access or egress is not desirable. Planning for these events should be conducted in coordination with local emergency agencies, such as police, fire, and public health agencies.

A.13.7 Patients that generate media interest should be subject to special security procedures. VIP or media representatives present the need for a unique set of security requirements. Protection of VIPs is normally accomplished by restricting the use of names on charts and rooms and by assigning a dedicated security watch.

Admission of a high-profile person to a health care facility creates two sets of problems that might require partial activation of the health care emergency management plan. These problems are security and the reception of news media.

Provision of security forces in this situation might be provided by a governmental agency or private security forces. However, activation of facility security forces might be required to prevent hordes of curious onlookers from entering facility work areas and interfering with routine facility functioning. Routine visiting privileges and routine visiting hours might need to be suspended in parts of the facility.

A.13.7.1.1 The marketing department of the hospital might be best suited to assist security personnel with media control.

A.13.7.2 Ideally, news media personnel should be provided with a media briefing area or a media staging area, or both, with access to telephone communication and, if possible, an expediter who, though not permitted to act as a spokesperson for news releases, could provide other assistance to such personnel. News media personnel should not be allowed into the health care facility without proper identification. Media representatives should be requested to wear some means of identification for security purposes. Members of the news media should be asked to wear some means of identification, such as a press card, on their outside garments so that they are readily identifiable by security guards controlling access to the facility or certain areas therein.

A.13.8 Crowd control of persons demanding access to care will create additional demands on security. Because of the intense public interest in disaster casualties, news media representatives should be given as much consideration as the situation will allow. To alert off-duty health care staff and to reassure the public, use of broadcast media should be planned.

Where feasible, photo identification or other means to ensure positive identification should be used. Visitor and crowd control create the problem of distinguishing staff from visitors. Such identification should be issued to all facility personnel, including volunteer personnel who might be utilized in disaster functions. Note that care should be taken to ensure that identification cards are recalled whenever personnel terminate association with the health care facility. Clergy also will frequently accompany casualties or arrive later for visitations and require some means of identification.

A.13.9.3 Key cards are preferable to traditional keys because they can be immediately deactivated if lost or not returned by a terminated employee.

Facility keys should not be identified in any manner such that a person finding a lost key could trace it back to the facility. A policy should be established to restrict duplication of keys without written permission. All keys should be marked "DO NOT DUPLICATE" to deter the unauthorized copying of keys.

There should be a log of keys issued to employees and vendors maintained at the facility. A responsible individual should be in charge of issuing keys and maintaining complete, up-to-date records of the disposition of keys, including copies. The records should show the issuance and return of keys, including the name of the person to whom the key was issued, as well as the date and time. Records of key issuance should be secured and kept separate from keys.

Keys should be restricted to those who need them, and extra copies of keys should be kept locked in a secure cabinet with access control.

Procedures should be established for collecting keys from terminated employees, employees on vacation, and vacated tenants. Lost keys should be reported immediately and procedures established for the rekeying or replacement of the affected locks.

A master key system should be designed so that the grandmaster key is the only key that will open every restricted area of the facility. A master key system is used to limit the number of keys carried by personnel requiring access to multiple areas of the building. It is important that such a system not be designed so that the loss of a single key could provide an unauthorized individual unrestricted access to all areas of the building. The sophistication of the master key system should depend upon an assessment of employees' or tenants' needs and the criticality, risk, and sensitivity of restricted areas.

The number of grandmaster keys should be limited to the least number necessary for operation of the health care facility. Master key distribution should be limited to the personnel requiring access to multiple restricted areas. A log should be maintained showing who is in possession of master keys.

A.13.10 Background checks should include criminal record checks, employment histories, and references. This function is typically managed by the human resources department.

A.13.11 The number of guards needed at any given time will depend on the size of the facility, the hours of operation, and the current risk factors. Many states have laws that require background checks and specific training for security personnel, especially armed personnel. It is essential that facilities using security personnel train them in the legal and practical applications of their employment. Training must reflect changes in regulations and the enactment of new laws.

A.13.11.1 Post orders should contain a list of the duties of the security officer and instructions to cover all foreseeable events the security officer can encounter. Post orders should list the name of the facility, the date issued, the effective date, and the purpose. Duties of security personnel should be listed, including job classification, uniforms, carrying of firearms, reporting times, watch tours, hours of coverage, and other duties to be assigned. Instructions should be lawful and protect the safety of the security officer and those they encounter. Reviews of post orders should be conducted regularly with facility management and security officers. Post orders should be updated regularly and at least annually. A procedure should be established to inform security officers of changes in post orders.

A.13.12.1 The effectiveness of the security plan is tested by performing drills. Drills should be conducted on all work schedules, so that all personnel are familiar with the plan. Practicing the plan helps personnel react as needed during a security incident.

A.14.1 Chapter 14 does not apply to respiratory therapy employing oxygen-enriched atmospheres at ambient pressures. *(See Chapter 11.)*

A.14.1.2.2 Chambers designed for animal experimentation but equipped for access of personnel to care for the animals are classified as Class A for the purpose of Chapter 14.

A.14.2.1.1.1 For guidance on minimum construction requirements, depending on occupancy classification, see NFPA *101, Life Safety Code.*

A.14.2.1.1.6 Characteristics of building construction housing hyperbaric chambers and ancillary facilities are no less important to safety from fire hazards than are the characteristics of the hyperbaric chambers themselves. It is conceivable that a fire emergency occurring immediately outside a chamber, given sufficient fuel, could seriously endanger the life or lives of those inside the chamber. Since the service facilities, such as compressors, cooling equipment, reserve air supply, oxygen, and so forth, will, in all probability, be within the same building, these facilities will also need protection while in themselves supplying life-maintaining service to those inside.

A.14.2.1.2 When the area to be covered is small (six sprinklers or less), 9.7.1.2 of NFPA *101, Life Safety Code*, permits fire sprinkler systems required to be installed in accordance with NFPA 13, *Standard for the Installation of Sprinkler Systems*, to be supplied from the local domestic water system, provided that the local domestic water system has sufficient pressure and flow capacity.

In addition to the functions of building protection, the chamber room sprinkler system should be designed to ensure a degree of protection to chamber operators who likely will not be able to immediately evacuate the premises in the event of a fire.

A.14.2.1.3.1 Hyperbaric chamber systems often require piping materials, pressure ratings, and joining techniques that are not permitted by Chapter 5 of this code.

A.14.2.2.1 Other chapters in NFPA 99 contain many requirements that could appear to relate to hyperbaric facilities but could be inappropriate. The requirements of other chapters in NFPA 99 should be applied to hyperbaric facilities only where specifically invoked by this chapter.

A.14.2.2.5 One common hazard of paint fires in ships is related to welding or burning operations on one side of a metal bulkhead that heats the metal to a point where the paint on the opposite side ignites. Most paints are not flammable when installed as thin layers over a substantial heat sink, such as the thick steel walls of a hyperbaric chamber, unless the walls are heated first. The same paints, when ground into a powder or installed over a very thin metal substrate, can burn readily. The paint selected for use in the interior walls of a hyperbaric chamber should be selected both for suitability to the requirements of the application and for its combustibility properties. The hazard of a fire increases as the amount of heat sink is reduced. Therefore, combustion is easier to achieve when paint is applied over thin materials and when there are multiple layers of paint. On thin section materials that are easily heated, care should be exercised in selecting the flammability characteristics of the paint and the amount of paint applied.

A.14.2.2.5.1 In past editions of this code, "high quality epoxy" was specified as interior finish in these chambers, without a specific fire performance. Although not the only option, this type of material offers suitable physical properties. The interior finish of a Class A chamber should be smooth, impermeable, durable, provide corrosion resistance, and be compatible with infection control procedures.

A.14.2.2.5.4 Many commercial sound-deadening materials that might be nonflammable are porous and will absorb water from activation of the fire-suppression system and retain odor. Metallic panels that contain a large quantity of small holes or are made of wire mesh and are installed about 2.5 cm (1 in.) away from the chamber wall can be used to form an acoustic baffle. These panels should be made from corrosive-resistant materials, such as stainless steel or aluminum, and are permitted to be painted in accordance with 14.2.2.5.3.

A.14.2.2.6 Prudent design considerations suggest that at least 50 percent excess pass-through capacity be provided, for future use, given the difficulty of adding pass-throughs to the chamber after it is constructed and tested.

A.14.2.4.1.2 Experience and practice can dictate the need for a threshold ventilation rate in excess of the minimum specified for sanitary reasons. It is recommended that consideration be given, if necessary, to the use of odor filters in the chamber circulation system as a means of keeping sanitary ventilation rate requirements to a minimum.

A.14.2.4.2.1 If intakes are located where it could be possible for maintenance to be conducted in the immediate vicinity, a warning sign should be posted.

A.14.2.4.3.2 Efforts should be made in the design and operation of thermal control systems to maintain the temperature as close to 22°C (75°F) as possible. The air-handling system of all Class A chambers should be capable of maintaining relative humidity in the range of 50 percent to 70 percent during stable depth operations.

The thermal control system should be designed to maintain the temperature below 29°C (85°F) during pressurization, if possible, and above 19°C (65°F) during depressurization, if possible.

A.14.2.4.4.1 Ventilation is permitted to be provided by closed- or open-circuit systems.

A.14.2.4.5.3 The intent of this requirement is to allow facility staff to evacuate the facility and avoid breathing contaminated air. This requirement is permitted to be met using either a self-contained breathing apparatus, smoke hood with integral filter/air supply, or similar technology.

The number of units available should be adequate to meet facility staffing.

The breathing duration of the personal protection devices should be predicated upon the time necessary for evacuation of the facility.

Facility evacuation time should be determined during fire drills conducted by the hyperbaric facility.

A.14.2.5.1.4 This requirement does not preclude the use of an alarm system affording direct fire department contact.

A.14.2.5.1.5 Experience has shown that fire blankets, portable carbon dioxide extinguishers, and other methodology intended to "snuff out" fires by excluding air are not effective in controlling fires in oxygen-enriched atmospheres. Valuable time can be lost in attempting to use such devices.

A.14.2.5.2.4 More than one control station could be required in a compartment (lock), depending on its size.

A.14.2.5.2.6 Experience has shown that, when water is discharged through conventional sprinkler heads into a hyperbaric atmosphere, the spray angle is reduced because of increased resistance to water droplet movement in the denser atmosphere. This is so, even though the water pressure differential is maintained above chamber pressure. Therefore, it is necessary to compensate by increasing the number of sprinkler heads. It is recommended that spray coverage tests be conducted at maximum chamber pressure.

Some chamber configurations, such as small-diameter horizontal cylinders, could have a very tiny floor, or even no floor at all. For horizontal cylinder chambers and spherical chambers, the term *floor level* should be taken to mean the level at ¼ diameter below the chamber centerline or actual floor level, whichever yields the larger floor area.

A.14.2.5.4.2 Additional detectors are recommended to avoid "blind" areas if the chamber contains compartmentation.

A.14.2.5.5 The primary focus for the semiannual test of a water-based extinguishing system is to ensure water flow through the system (i.e., inspector's test). Other vitally important benefits are the activation of water flow devices, alarm appliances, and notification and annunciator systems.

A.14.2.8.1.4.1 It is recommended that system design be such that electric motors not be located inside the chamber.

A.14.2.8.3 This subsection contains requirements for the safe use of electrical equipment in the hyperbaric, oxygen-enriched environment of the Class A chamber.

A.14.2.8.3.10 It should be recognized that interruption of any powered circuit, even of very low voltage, could produce a spark sufficient to ignite a flammable agent.

A.14.2.8.3.11.1 It is recommended that all control switching functions inside the chamber be accomplished using intrinsically safe circuits that control power and control circuits located outside of the chamber.

A.14.2.8.3.12 It is the intention of 14.2.8.3.12 that equipment used in the chamber be incapable of igniting, by heating, any material or fabric that could come into contact with the surface of the equipment.

A.14.2.8.3.15 It is strongly recommended that high-intensity local task lighting be accomplished using through-hull fiber-optic lights. Many high-intensity lights will not meet the temperature requirements specified in 14.2.8.3.15.

A.14.2.8.3.16.1 The requirement for isolation from mains supply in 14.2.8.3.16.1 is not the same as the requirement in 14.2.8.4.2 that circuits supplying power to portable utilization equipment inside the chamber be isolated, monitored, and alarmed.

It is recommended that intrinsically safe sensors and controls be used whenever possible.

A.14.2.8.3.17 These requirements are only the minimum requirements for electrical safety. There are many other safety concerns that should be addressed on a case-by-case basis. Meeting the requirements of 14.2.8.3.17 does not indicate that proper device performance will occur in the hyperbaric environment and that the device will be safe for use with patients.

A.14.2.8.5.2 It is necessary that these circuits be protected from exposure to water from the room sprinkler system protecting the chamber housing in the event of a fire in the vicinity of the chamber while it is in operation.

A.14.2.8.6.1.1 Limiting current using a suitable current sensing device (e.g., a rapid acting fuse or circuit breaker, located outside the chamber) would provide appropriate protection and prevent circuits from exceeding the 4.0 W power limit.

A.14.2.9.2 Intercommunications equipment is mandatory for safe operation of a hyperbaric facility.

A.14.2.9.2.1 It is recommended that multiple-compartment (lock) Class A chambers be equipped with multiple channel systems and that, in addition, a sound-powered telephone or surveillance microphone be furnished.

A.14.2.9.6 The purity of the various gas supplies should be ensured.

A purity statement for any cryogenic or high pressure cylinder gas should be supplied by the vendor.

Gas cylinder purity statements should be cross-referenced, where possible, with the delivered gas.

For additional verification, some facilities have installed sampling ports for monitoring oxygen and other gases.

A.14.2.9.6.1 The frequency of such monitoring should depend on the location of the air intake relative to potential sources of contamination.

A.14.2.9.6.2 CGA Grade E permits quantities of hydrocarbons and water in air. In piping systems where air and oxygen might be used interchangeably, hydrocarbon buildup can occur and increase the risk of fire when oxygen is used. There is also a concern about pneumatic components being fouled and functionally impaired by hydrocarbons or water from compressed air. Ideally, there should be no condensed hydrocarbons in an oxygen system and no liquid water in pneumatic control systems.

A.14.2.9.8 It is recommended that information about the status of an anesthetized or otherwise monitored patient be transmitted to the inside chamber attendants via the intercommunications system. As an alternative, the monitor indicators can be placed adjacent to a chamber viewport (or viewports) for direct observation by inside personnel.

A.14.2.10.2 Exhaust piping extending from the building can create a lightning risk. Lightning protection should be considered.

A.14.3.1.2 The hazards involved in the use of hyperbaric facilities can be mitigated successfully only when all of the areas of hazard are fully recognized by all personnel and when the physical protection provided is complete and is augmented by attention to detail by all personnel of administration and maintenance having any responsibility for the functioning of the hyperbaric facility. Since Section 14.3 is expected to be used as a text by those responsible for the mitigation of hazards of hyperbaric facilities, the requirements set forth are frequently accompanied by explanatory text.

A.14.3.1.3.2 The complexity of hyperbaric chambers is such that one person should be designated chamber operator, such as a person in a position of responsible authority. Before starting a hyperbaric run, this person should record, in an appropriate log, the purpose of the run or test, the duties of all personnel involved, and a statement that he or she is satisfied with the condition of all equipment. Exceptions should be itemized in the statement.

Safety, operational, and maintenance criteria of other organizations are published, for example, in the Undersea & Hyperbaric Medical Society Safety Committee documents and the Compressed Gas Association pamphlets and should be reviewed by the safety director. The safety director should serve on the health care facility safety committee.

Due to a conflict of responsibility, the same individual should not serve as both medical director and safety director.

The term *safety director* is used for convenience. It is the intent of 14.3.1.3.2 to establish a set of safety responsibilities for the responsible person, regardless of the job title.

A.14.3.1.3.3 It is incumbent upon the governing body to insist that rules and regulations with respect to practices and conduct in hyperbaric facilities, including qualifications and training of hyperbaric personnel, be adopted by the medical or administrative staff of the institution, and that regulations for inspection and maintenance are in use by the administrative, maintenance, and ancillary (and, in the case of a hospital, nursing and other professional) personnel.

In meeting its responsibilities for safe practices in hyperbaric facilities, the administration of the facility should adopt or correlate regulations and standard operating procedures to ensure that both the physical qualities and the operating maintenance methods pertaining to hyperbaric facilities meet the standards set in Chapter 14. The controls adopted should cover the conduct of personnel in and around hyperbaric facilities and the apparel and footwear allowed. They should cover periodic inspection of static-dissipating materials and of all electrical equipment, including testing of ground contact indicators.

A.14.3.1.3.4 It is recommended that training of hyperbaric chamber personnel be closely monitored, following the guidelines and publications of the Undersea & Hyperbaric Medical Society, the Baromedical Nurses Association, and the National Board of Diving and Hyperbaric Medical Technology.

A.14.3.1.3.5 In the case of a hyperbaric facility located in a hospital, hospital licensing and other approval bodies, in meeting their responsibilities to the public, should include in their inspections not only compliance with requirements for physical installations in hyperbaric facilities, but also compliance with the requirements set forth in Section 14.3.

A.14.3.1.4.1 It is recommended that all personnel, including trainees and those involved in the operation and maintenance of hyperbaric facilities, and including professional personnel and (in the case of hospitals) others involved in the direct care of patients undergoing hyperbaric therapy, be familiar with Chapter 14. Personnel concerned should maintain proficiency in the matters of life and fire safety by periodic review of Chapter 14, as well as any other pertinent material.

Positive measures are necessary to acquaint all personnel with the rules and regulations established and to ensure enforcement. Training and discipline are necessary.

A.14.3.1.4.4.1 All full- and part-time personnel should receive training in emergency management appropriate to their job descriptions.

A.14.3.1.4.5 A calm reaction (without panic) to an emergency situation can be expected only if the recommendations are familiar to and rehearsed by all concerned.

A suggested outline for emergency action in the case of fire is contained in B.14.2.

A.14.3.1.5.1.1 Oxygen-filled chambers dump oxygen into the room each time the door is opened at the end of a treatment. Oxygen could also be dumped into the room by the chamber pressure relief device. Air-filled chambers could leak oxygen into the room from the breathing gas piping. This oxygen enrichment lowers the ignition temperature of combustible materials. Therefore, extra caution should be used in the area around the chamber as well as inside the chamber.

A.14.3.1.5.2.2(2) Allowable quantities complying with 14.3.1.5.2.2(2) can be determined from the chamber volume, flammable agent vapor density, and lower explosive limit (LEL). Experience has shown that increased pressure has little effect on LEL for a given flammable gas and oxygen concentration. A safety factor of 10 is recommended. Flammable liquids should be confined to nonbreakable, nonspill containers.

Sample Determination. An example of limiting quantity of flammable agent substance:

Isopropyl alcohol (2-propanol)
LEL = 2%/vol. (irrespective of chamber pressure)
Vapor density = 2.1 relative to air
Liquid density = 786 g/L (49.1 lb/ft^3)
Air density = 0.075 lb/ft^3 (1.2 kg/m^3) at STP

The limiting case occurs at the lowest ambient pressure, that is, 1 atmosphere:

$$\begin{aligned}\text{Alcohol vapor density at LEL} &= 0.02 \times 2.1 \times 0.075 \\ &= 0.00315 \text{ lb/ft}^3 \ (0.05 \text{ kg/m}^3) \\ &= 1.43 \text{ g/ft}^3 \ (0.05 \text{ kg/m}^3)\end{aligned}$$

For a relatively small 500 ft^3 (14.2 m^3) chamber, this implies:

$1.43 \times 500 = 715$ g (1.58 lb) alcohol vapor at LEL

Using a safety factor of 10 to account for uneven vapor concentrations gives 71.5 g = 91 mL (3 oz) alcohol.

One could conclude that even 90 mL (3 oz) of alcohol is more than would be needed for almost any medical procedure. The preceding calculation also does not account for the mitigating effect of ventilation.

Many "inert" halogenated compounds have been found to act explosively in the presence of metals, even under normal atmospheric conditions, despite the fact that the halogen compound itself does not ignite in oxygen or, in the case of solids such as polytetrafluoroethylene, is self-extinguishing. Apparently these materials are strong oxidizers, whether gases, liquids (solvents, greases), or solids (electrical insulation, fabric, or coatings). Some halogenated hydrocarbons that will not burn in the presence of low pressure oxygen will ignite and continue to burn in high pressure oxygen. Customarily, Class A chambers maintain internal oxygen concentration that does not exceed 23.5 percent.

Parts of Chapter 14 deal with the elements required to be incorporated into the structure of the chamber to reduce the possibility of electrostatic spark discharges, which are a possible cause of ignition in hyperbaric atmospheres. The elimination of static charges is dependent on the vigilance of administrative activities in materials, purchase, maintenance supervision, and periodic inspection and testing. It cannot be emphasized too strongly that an incomplete chain of precautions generally will increase the electrostatic hazard. For example, conductive flooring can contribute to the hazard unless all personnel wear conductive shoes, all objects in the room are electrically continuous with the floor, and humidity is maintained.

The limitations in 14.3.1.5.2.2 on the use in the chamber of alcohol and other agents that emit flammable vapors should be strictly observed, and such restrictions should be prominently posted.

A.14.3.1.5.3 The number of occupants of the chamber should be kept to the minimum number necessary to carry out the procedure.

A.14.3.1.5.4 It is recommended that all chamber personnel should wear garments of the overall or jumpsuit type that completely cover all skin areas to the extent possible and that are as tight-fitting as possible. It can be impractical to clothe some patients (depending upon their disease or the site of any surgery) in such garments. Hospital gowns can be employed in such a case.

A.14.3.1.5.4.2 Selection of textiles for the hyperbaric chamber should be based on a variety of factors, including comfort, lint production, ignition temperature, static-producing properties, and fuel load of the material. The amount of polyester in a cotton/polyester blend will likely have an effect on all of these factors.

Historically, all synthetic fabrics were prohibited from the chamber. Previous editions of this code allowed an "antistatic blend of cotton and polyester" because of one specific fabric — a blend of cotton and polyester with steel fibers to make it conductive. This blended fabric was intended for surgical scrubs, but its conductive properties made it a good choice for hyperbaric garments. The polyester in the fabric was acceptable because the conductive properties of the fabric actually afforded some protection from static production that cotton fabric did not. This particular fabric is no longer made. Selection of textiles has always been about balancing various safety concerns; primarily fire-resistance and static production. For further guidance on selecting appropriate textiles, see A.14.3.1.5.4.3.

A.14.3.1.5.4.3 The textiles definitions and risk assessment process for hyperbaric wound dressings are as follows:

Combustion. A chemical process of oxidation that occurs at a rate fast enough to produce heat in the form of either a glow or a flame.

Flammable. Refers to a combustible (solid, liquid, or gas) that is capable of easily being ignited and rapidly consumed by fire.

Flash Point. The minimum temperature of a liquid or solid at which it gives off vapor sufficient to form an ignitible mixture with oxygen under specified environmental conditions.

Ignition Temperature. The minimum temperature required to initiate or cause self-sustaining combustion under specified environmental conditions.

Lower Explosive Limit (LEL) or Lower Flammable Limit (LFL). The minimum concentration of fuel vapor (percent by volume) over which combustion will occur on contact with an ignition source.

General Risk Assessment Information. This risk assessment process was designed to evaluate wound dressing products for use in a hyperbaric chamber. However, the same decision process can be applied to the evaluation of textiles for hyperbaric use. Wound dressings are commonly used inside hyperbaric chambers. They play an important role in infection control and patient outcome. Important safety concerns include production of heat, production of static electricity, production of flammable vapor, ignition temperature, and total fuel load. Many wound dressings employ fabrics and other materials that are gas-permeable. It is a common misconception that a gauze bandage will isolate an undesirable product from the chamber environment. Gauze is gas-permeable and will allow oxygen from the chamber to interact with the product and vapors from the product to interact with the chamber environment. Also, gas-permeable materials exposed to hyperbaric oxygen will hold additional oxygen for some period of time after the exposure. These materials should be kept away from open flames for at least 20 minutes after the hyperbaric treatment.

Risk Assessment Process (see Figure A.14.3.1.5.4.3).

(1) Is there a more suitable alternative to this dressing? The issue of need must first be addressed. There might be a substitute dressing that has already been deemed acceptable for the hyperbaric environment. The wound dressing orders can be changed to the more desirable substitute (if there is no negative impact on patient outcome). It might be viable to remove the dressing before the hyperbaric treatment, leave it off during the treatment, and replace it after the treatment. Before making this decision, it is important to remember that some dressings should not be disturbed (e.g., in the case of a new skin graft); some dressings are designed to stay in place for several days; some dressings are very expensive; and it can be detrimental for the wound to remain undressed during the treatment. If there is a suitable alternative to using this dressing, the rest of the decision process can be eliminated.

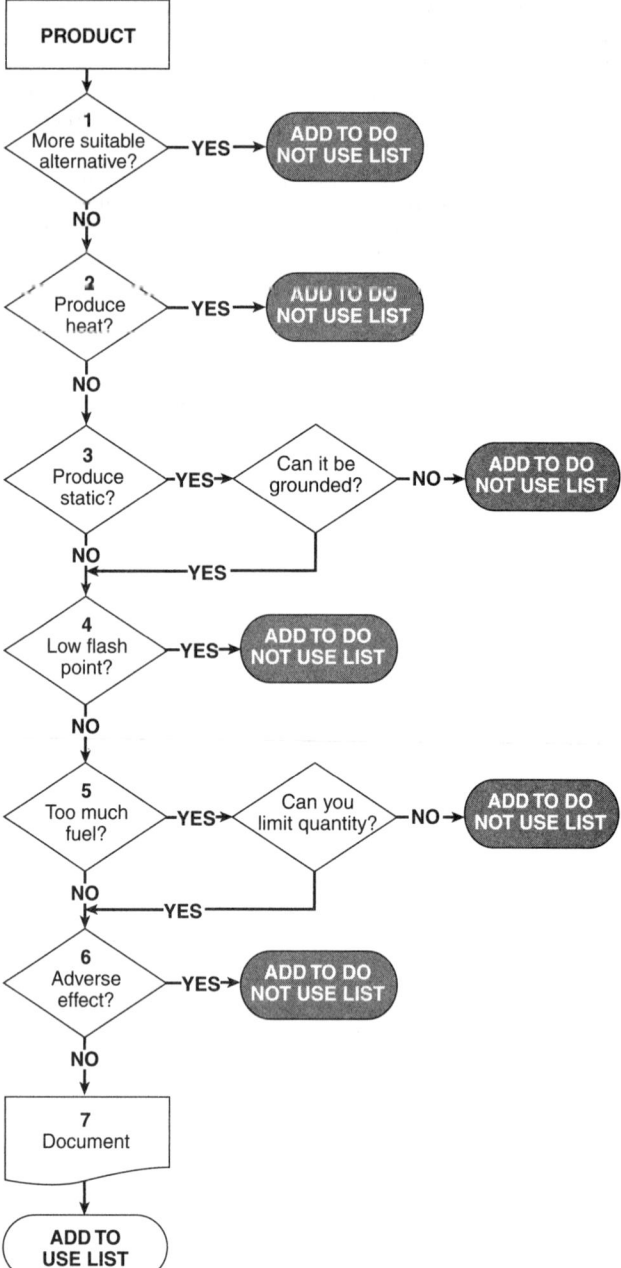

FIGURE A.14.3.1.5.4.3 Risk Assessment Process.

(2) Does this dressing produce heat in the chamber? Dressings are made from a large variety of materials. The concern is that materials in a dressing can rapidly oxidize and produce heat (exothermic reaction) when exposed to additional oxygen. For example, air-activated heat patches (commonly used for pain relief) have been tested in hyperbaric environments. The average operating temperature increased from 48.1°C (119°F) in normobaric air to 121.8°C (251°F) in hyperbaric oxygen. In this circumstance, the patient's skin would be burned, and the heat could ignite combustible material in the chamber. Information on oxygen compatibility can be found in a product material safety data sheet (MSDS).

(3) Does this dressing produce too much static electricity? All common textiles will contribute to static production. Wool and synthetic materials generally contribute more to static production than cotton. Although static charge is constantly accumulating, it will dissipate into the environment when humidity is present. At less than 30 percent relative humidity, static charge can accumulate faster than it can dissipate. At greater than 60 percent relative humidity, static charge is all but completely eliminated. Use of conductive surfaces and electrical grounding will allow static charge to dissipate. Paragraph 14.2.8.4.1 requires all hyperbaric chambers to be grounded. Paragraph 14.2.10.1 requires any furniture installed inside a chamber to be grounded. Paragraph 14.3.1.5.3.2 requires all occupants of the chamber to be grounded when the oxygen percentage in the chamber is above 23.5 percent. The continuity of electrical grounds should be verified periodically.

(4) Does this dressing have a low ignition temperature/flash point? In all hyperbaric environments, the partial pressure of oxygen is higher than at normal atmospheric conditions. Increasing the partial pressure of oxygen can change the classification of a material from non-flammable to flammable. Many materials are flammable in a 100 percent oxygen environment. Any material used in a hyperbaric chamber should have an ignition temperature higher than it can be exposed to. Paragraph 14.2.8.3.12 limits electrical equipment inside a Class A (multi-place) chamber to a maximum operating surface temperature of 85°C (185°F). Paragraph 14.2.8.6.3 limits electrical circuits inside a Class B (monoplace) chamber to a maximum operating temperature of 50°C (122°F). As the oxygen percentage increases, it takes less energy to ignite materials. This leads to more conservative decisions in a 100 percent oxygen environment. A greater margin of safety is achieved when there is a greater difference between the temperature limit of the equipment inside a Class A and B chamber and the ignition temperature of material in question. A material will release vapor into the chamber environment as it approaches its flash point temperature. Once a sufficient quantity of vapor is present in the chamber (LEL), it takes very little energy for ignition to occur. Paragraph 14.3.1.5.2.2 sets limits on flammable agents inside Class A (multi-place) chambers. Paragraph 14.3.1.5.2.3 specifically prohibits flammable liquids, gases, and vapors inside Class B (monoplace) chambers. Information on ignition temperature and flash point in air can be found in a product MSDS.

(5) Is the total fuel load too high? If a fire does occur, the energy produced is a function of the partial pressure of oxygen and the total fuel load. In a hyperbaric environment, the partial pressure of oxygen is higher and contributes to greater energy production. Any dressing product placed inside of a hyperbaric chamber is a combustible material and, therefore, adds to the fuel load. Therefore, total fuel load inside the chamber should be minimized to only what is necessary.

(6) Is there an adverse effect when this product is used inside the hyperbaric chamber? It has been reported that the antibacterial agent mafenide acetate (Sulfamylon®), in combination with hyperbaric oxygen, has a poorer clinical result than either one by itself. There can be other drug interactions with hyperbaric oxygen that are undesirable. The mechanical effects of pressure change can cause a dressing material to rupture. If the material is capable of venting/equalizing during pressure change, this should not occur.

(7) The hyperbaric facility should maintain a "use list" and a "do not use list" of items that have been evaluated for hyperbaric use. In addition to this list, it is important to keep documentation on file explaining the risk assessment for each item. This will prevent future duplication of effort. It also serves as evidence that due diligence was used.

A.14.3.2.1.6 The use of paper should be kept to an absolute minimum in hyperbaric chambers.

A.14.3.2.2 Users should be aware that many items, if ignited in pressurized oxygen-enriched atmospheres, are not self-extinguishing. Iron alloys, aluminum, and stainless steel are, to various degrees, in this category, as well as human skin, muscle, and fat, and plastic tubing such as polyvinyl chloride (Tygon®). Testing for oxygen compatibility is very complicated. Very little data exist, and many standards still have to be determined. Suppliers do not normally have facilities for testing their products in controlled atmospheres, especially high pressure oxygen. Both static conditions and impact conditions are applicable. Self-ignition temperatures normally are unknown in special atmospheres.

A.14.3.2.5 See A.14.3.2.2.

A.14.3.2.6 Radiation equipment, whether infrared or roentgen ray, can make hyperbaric chambers even more hazardous.

A.14.3.3.4 Quantities of oxygen stored in the chamber should be kept to a minimum.

A.14.3.6 The elimination of static charges is dependent on the vigilance of administrative supervision of materials purchased, maintenance, and periodic inspection and testing.

A.14.3.6.2.1.2 Ferrous metals can cause such sparking, as can magnesium or magnesium alloys, if contact is made with rusted steel.

A.14.3.6.2.3 Materials containing rubber deteriorate rapidly in oxygen-enriched atmospheres.

A.14.3.6.4 It is absolutely essential that all areas of, and components associated with, the hyperbaric chamber be kept meticulously free of grease, lint, dirt, and dust.

A.15.3.2 NFPA 58, *Liquefied Petroleum Gas Code*, permits portable butane-fueled appliances in restaurants and in attended commercial food catering operations where fueled by not in excess of two 0.28 kg (10 oz) LP-Gas capacity, nonrefillable butane containers having a water capacity not in excess of 0.4 kg (1.08 lb) per container. Containers are required to be directly connected to the appliance, and manifolding of containers is not permitted. Storage of cylinders is also limited to 24 containers, with an additional 24 permitted where protected by a 2-hour fire resistance–rated barrier. [*101*: A.8.7.3.2]

A.15.5.2.1 This section is different from 9.2.1 of NFPA *101, Life Safety Code*, because NFPA *90B, Standard for the Installation of Warm Air Heating and Air-Conditioning Systems*, referenced in NFPA *101*, is not applicable to health care facilities.

A.15.5.3.3 Continued operation of solid-state elevator equipment is contingent on maintaining the ambient temperature in the range specified by the elevator manufacturer. If the machine room ventilation/air-conditioning is connected to the general building system, and that system is shut down during a fire, the fire department might lose the use of elevators due to excessive heat in the elevator machine room. [*101*: A.9.4.5]

A.15.7.1 The provisions of Section 15.7 cover the basic functions of a complete fire alarm system, including fire detection, alarm, and communications. These systems are primarily intended to provide the indication and warning of abnormal conditions, the summoning of appropriate aid, and the control of occupancy facilities to enhance protection of life.

Some of the provisions of Section 15.7 originated with NFPA 72, *National Fire Alarm and Signaling Code*. For purposes of this *Code*, some provisions of Section 15.7 are more stringent than those of NFPA *72*, which should be consulted for additional details. [*101*: A.9.6.1]

A.15.7.2.4 It is not the intent of 15.7.2.4 to require manual fire alarm boxes to be attached to movable partitions or to equipment, nor is it the intent to require the installation of permanent structures for mounting purposes only. [*101*: A.9.6.2.5]

A.15.7.4.2.1 Elevator lobbies have been considered areas subject to unwanted alarms due to factors such as low ceilings and smoking. In the past several years, new features have become available to reduce this problem. These features are, however, not necessarily included in any specific installation. [*101*: A.9.6.3.2.1]

A.15.7.4.2.2 The concept addressed is that detectors used for releasing service, such as door or damper closing and fan shutdown, are not required to sound the building alarm. [*101*: A.9.6.3.2.2]

A.15.7.4.3.1 It is not the intent of this paragraph to require fire alarm system zones to coincide with smoke compartment boundaries, provided that the facility fire plan addresses the differences between fire alarm system zones and building smoke compartments.

A.15.7.4.3.2 In the private operating mode, audible and visible signaling is required only to those persons directly concerned with the implementation and direction of emergency action. Provided that those persons receive alarm notification, audible and visible signaling is not required to patients or other building occupants who are not responsible for the implementation and direction of emergency action.

A.15.8.1.3 It is not the intent of this paragraph to require sprinkler system zones to coincide with smoke compartment boundaries, provided that the facility fire plan addresses the differences between sprinkler systems zones and building smoke compartments.

A.15.8.1.4 This exception is limited to hospitals as nursing homes and many limited care facilities can have more combustibles within the closets. The limited amount of clothing found in the small clothes closets in hospital patient rooms is typically far less than the amount of combustibles in casework cabinets that do not require sprinkler protection such as nurse servers. In many hospitals, especially new hospitals, it is difficult to make a distinction between clothes closets and cabinet work. NFPA 13, *Standard for the Installation of Sprinkler Systems*, already permits the omission of sprinklers in wardrobes *[see 8.1.1(7) of NFPA 13]*. It is not the intent of this paragraph to affect the wardrobe provisions of NFPA 13. It is the intent that the sprinkler protection in the room covers the closet as if there was no door on the closet *(see 8.5.3.2.3 of NFPA 13)*. [*101*: A.18.3.5.8]

A.15.9.1 The selection of portable fire extinguishers for health care facilities is a vital step in preparing the facility to effectively deal with a fire in its incipient stage. There are special extinguisher requirements such as nonferrous fire extinguisher components of fire extinguishers in an MRI room or area, and Class K extinguishers in kitchens. There is a need to consider extinguishing agents for various areas of a facility that are nontoxic, noncorrosive, and/or nonconductive. Chapters 5 and 6 and their Annex A notes of NFPA 10, *Standard for Portable Fire Extinguishers*, need very careful review. Annex C, "Fire Extinguisher Selection" and Annex D, "Operation and Use" of NFPA 10 provide additional valuable guidance.

A.15.9.2.2 Hose and hose outlets can be required by a building code or by the authority having jurisdiction.

A.15.10 Compact storage is characterized by shelving units that are manually or electrically moved on fixed tracks to provide access aisles. Such systems are also known as mobile shelving, track files, compaction files, or movable files.

A floor loading calculation should be performed.

A.15.11.3 NFPA 13, *Standard for the Installation of Sprinkler Systems*, contains protection criteria for limited configurations of compact mobile storage units and materials stored. Storage arrangements not specifically addressed in NFPA 13 are outside the scope of the standard (i.e., protection for plastic commodities in compact mobile storage units does not simply follow high piled storage protection criteria for shelves or racks). Where compact mobile storage configurations outside the scope of NFPA 13 are to be utilized, they must be addressed on a case-by-case basis, with consideration given to the fact that no known sprinkler protection criteria is currently available. The storage of paper administrative and medical records with limited plastic labels or folders is typically a Class III commodity with less than 5 percent plastics. The storage of most office supplies, engineering parts, food products, garments, and other general supplies is typically a Class IV commodity with 5 percent to 25 percent plastics. The storage of X-rays, medicines, and plastic-based medical supplies is typically a Group A plastics commodity with greater than 25 percent plastics.

A.15.13 The following definitions were adapted from the ACS publication 04GR-0001, *Guidelines for Optimal Ambulatory Surgical Care and Office-Based Surgery*, which was developed by the Board of Governors Committee on Ambulatory Surgical Care. Class A, Class B, and Class C operating rooms are classified as follows:

Class A — Provides for minor surgical procedures performed under topical and local infiltration blocks with or without oral or intramuscular preoperative sedation. (Excluded are procedures that make use of spinal, epidural axillary, and stellate ganglion blocks; regional blocks (e.g., interscalene) and supraclavicular, infraclavicular, and intravenous regional anesthesia.) These procedures are also appropriately performed in Class B and C facilities.

Class B — Provides for minor or major surgical procedures performed in conjunction with oral, parenteral, or intravenous sedation or under analgesic or dissociative drugs. These procedures are also appropriately performed in Class C facilities.

Class C — Provides for major surgical procedures that require general or regional block anesthesia and support of vital bodily functions.

A.15.13.3.2 Some tinctures and solutions of disinfecting agents provide significant clinical benefits in reducing the risk of surgical infections. However, they can be flammable and can be used improperly during surgical procedures. Tipping containers, accidental spillage, and the pouring of excessive amounts of such flammable agents on patients expose them to injury in the event of accidental ignition of the flammable solvent. To control this risk, flammable germicides or antiseptics that are used when electrosurgery, cautery, or a laser is contemplated should be packaged to ensure controlled delivery to the patient (e.g., unit dose applicator, swab) in small volumes appropriate for single application.

Annex B Additional Explanatory Notes

This annex is not a part of the requirements of this NFPA document but is included for informational purposes only.

B.1 Reserved.

B.2 Reserved.

B.3 Reserved.

B.4 Reserved.

B.5 Additional Information on Chapter 5. Numbers in brackets refer to paragraphs in Chapter 5.

B.5.1 General. This section sets out a minimum recommended guide for testing. Testing requirements are listed in 5.1.12 and summarized in Table B.5.1. Tests specified in 5.1.12 should be carried out by an experienced person or persons designated by the administration of the health care facility. Such a person(s) should certify the results of tests to the administration. The designated person(s) should be experienced in medical gas testing and verification of piping systems with cross-connection testing. A member of the health care facility should be present to verify the testing.

B.5.2 Retesting and Maintenance of Nonflammable Medical Piped Gas Systems (Level 1 Systems).

B.5.2.1 [5.1.3.5.10] These systems should be checked daily to ensure that proper pressure is maintained and that the changeover signal has not malfunctioned. Periodic retesting of the routine changeover signal is not necessary, as it will normally be activated on a regular basis.

B.5.2.2 [5.1.3.5.12] These systems should be checked daily to ensure that proper pressure is maintained and that the changeover signal has not malfunctioned. Periodic retesting of the routine changeover signal is not required. Annual retesting of the operation of the reserve and activation of the reserve-in-use signal should be performed.

B.5.2.3 [5.1.3.5.12] If the system has an actuating switch and signal to monitor the contents of the reserve, it should be retested annually.

B.5.2.4 [5.1.3.5.14] Maintenance and periodic testing of the bulk system is the responsibility of the owner or the organization responsible for the operation and maintenance of that system.

The staff of the facility should check the supply system daily to ensure that medical gas is ordered when the content's gauge drops to the reorder level designated by the supplier. Piping system pressure gauges and other gauges designated by the supplier should be checked regularly, and gradual variation, either increases or decreases, from the normal range should be reported to the supplier. These variations might indicate the need for corrective action.

Periodic testing of the master signal panel system, other than the routine changeover signal, should be performed. Assistance should be requested from the supplier or detailed instruction if readjustment of bulk supply controls is necessary to complete these tests.

B.5.2.5 [5.1.8.2.3] The main line pressure gauge should be checked daily to ensure the continued presence of the desired pressure. Variation, either increases or decreases, should be investigated and corrected.

B.5.2.6 [5.1.3.6.3.14] Quarterly rechecking of the location of the air intake should be made to ensure that it continues to be a satisfactory source for medical compressed air.

B.5.2.7 [5.1.3.6.3.14] Proper functioning of the pressure gauge and high water level sensor should be checked at least annually. The receiver drain should be checked daily to determine if an excessive quantity of condensed water has accumulated in the receiver.

B.5.2.8 [5.1.3.6] An important item required for operation of any medical compressed air supply system is a comprehensive preventive maintenance program. Worn parts on reciprocating compressors can cause high discharge temperatures, resulting in an increase of contaminants in the discharge gas. Adsorber beds, if not changed at specified time intervals, can become saturated and lose their effectiveness. It is important that all components of the system be maintained in accordance with the manufacturers' recommendations. It is important that any instrumentation, including analytical equipment, be calibrated routinely and maintained in operating order. Proper functioning of the dew point sensor should be checked at least annually.

B.5.2.9 [5.1.9] When test buttons are provided with signal panels, activation of the audible and visual signals should be performed on a regular basis (monthly).

B.5.2.10 [5.1.9.2.4] Changeover Warning Signals. As these are routine signals that are activated and deactivated at frequent intervals, there is no need for retesting unless they fail. If the reserve-in-use signal is activated because both units of the operating supply are depleted without the prior activation of the changeover signal, it should be repaired and retested.

B.5.2.11 [5.1.9.2.4] Reserve-In-Use Warning Signal. All components of this warning signal system should be retested annually. Audible and visual signals should be tested periodically during the year (monthly).

B.5.2.12 [5.1.9.2.4] Reserve Supply Low (Down to an Average One-Day Supply) High Pressure Cylinder or Liquid Reserve. All components of these signal warning systems should be retested annually. If test buttons are provided, audible and visual signals should be periodically tested throughout the year (monthly).

B.5.2.13 [5.1.9.2.4] The medical compressed air system alarms in 5.1.3.6.3.12 should be checked at least annually.

Table B.5.1 Performance Criteria and Testing — Level 1 (Gases, Medical–Surgical Vacuum, and WAGD)

Responsibility	Test Reference	Test (as Applicable)	Purpose of Test
Installer	5.1.12.2.1	General	
Installer	5.1.12.2.2	Initial blow down	Distribution piping is blown down to remove particulates
Installer	5.1.12.2.3	Initial pressure test	Distribution piping is free from pressure loss
Installer	5.1.12.2.4	Cross-connection test	Distribution piping is free from cross-connections
Installer	5.1.12.2.5	Piping purge test	Distribution piping is purged to remove particulates
Installer	5.1.12.2.6	Standing pressure test for positive pressure medical gas piping	Distribution piping is free from excessive pressure loss
Installer	5.1.12.2.7	Standing vacuum test for vacuum system	Distribution piping is free from excessive vacuum loss
System verification	5.1.12.3.1	General	
System verification	5.1.12.3.2	Standing pressure test	Distribution piping is free from leaks
System verification	5.1.12.3.3	Cross-connection test	Distribution piping is free from cross-connections
	5.1.12.3.3.1	Individual pressurization	
	5.1.12.3.3.2	Pressure differential	
System verification	5.1.12.3.4	Valve test	Shutoff valves are functioning and labeled properly
System verification	5.1.12.3.5	Alarm test	Alarms are functioning and labeled properly
	5.1.12.3.5.1	General	
	5.1.12.3.5.2	Master alarm	
	5.1.12.3.5.3	Area alarm	
System verification	5.1.12.3.6	Piping purge test	Distribution piping is purged to remove particulates
System verification	5.1.12.3.7	Piping particulate test	Distribution piping is free from particulates
System verification	5.1.12.3.7.6	Piping purity test	Distribution piping is free from excessive water vapor, total hydrocarbons, and halogenated hydrocarbons
System verification	5.1.12.3.9	Final tie-in test	The new and existing distribution system is free from leaks at the point of connection, and no additional contamination was added to the existing system
System verification	5.1.12.3.10	Operational pressure test	Distribution piping is free from excessive pressure/vacuum loss
System verification	5.1.12.3.11	Medical gas concentration test	Proper concentration of system gas is present at each outlet
System verification	5.1.12.3.12	Medical air purity test (compressor system)	Proper quality of medical air is present
System verification	5.1.12.3.13	Labeling	Distribution piping, outlets/inlets, shutoff valves, alarms, and source equipment are labeled correctly
System verification	5.1.12.3.14	Source equipment verification	Source equipment properly functions
	5.1.12.3.14.1	General	
	5.1.12.3.14.2	Gas supply sources	
	5.1.12.3.14.3	Medical air compressor systems	
	5.1.12.3.14.5	Medical–surgical vacuum systems	

B.5.2.14 [5.1.8.2.2(1)] This pressure gauge should be checked on a daily basis to ensure proper piping system pressure. A change, increase or decrease, if noted, could be evidence that maintenance is required on the line pressure regulator and could thus avoid a problem.

B.5.2.15 [5.1.9] Annual retesting of all components of warning systems, if it can be done without changing piping system line pressure, should be performed.

B.5.2.16 [5.1.9] If test buttons are provided, the retesting of audible and visual alarm indicators should be performed monthly.

B.5.2.17 [5.1.4] Shutoff valves should be periodically checked for external leakage by means of a test solution or other equally effective means of leak detection that is safe for use with oxygen.

B.5.2.18 [5.1.5] Station outlets should be periodically checked for leakage and flow. Manufacturer instructions should be followed in making this examination.

B.5.3 Oxygen Service–Related Documents. The following publications can be used for technical reference:

(1) ASTM G 63, *Standard Guide for Evaluating Nonmetallic Materials for Oxygen Service*
(2) ASTM G 88, *Standard Guide for Designing Systems for Oxygen Service*
(3) ASTM G 93, *Practice for Cleaning Methods and Cleanliness Levels for Material and Equipment Used in Oxygen-Enriched Environments*
(4) ASTM G 94, *Standard Guide for Evaluating Metals for Oxygen Service*

B.6 Additional Information on Chapter 6.

B.6.1 Typical Hospital Wiring Arrangement. See Figure B.6.1. Separate transfer switches for each branch, as shown, are required only if dictated by load considerations. Smaller facilities can be served by a single transfer switch.

B.6.2 Maintenance Guide for an Essential Electrical System. This generalized maintenance guide is provided to assist administrative, supervisory, and operating personnel in establishing and evaluating maintenance programs for emergency electric generating systems. See Figure B.6.2.

B.6.3 Suggested Format for Listing Functions to Be Served by the Essential Electrical System in a Hospital. It might be advantageous in listing the specific functions for a given construction project or building review to list them, at the outset, by geographical location within the project in order to ensure comprehensive coverage. Every room or space should be reviewed for possible inclusion of the following:

(1) Lighting (partial or all)
(2) Receptacles (some or all)
(3) Permanently wired electrical apparatus

The format suggested herein is offered as a convenient tool, not only for identifying all functions to be served and their respective time intervals for being re-energized by the alternate electric source, but also for documenting other functions that were considered, discussed, and excluded as nonessential. The last column in Figure B.6.3 is considered worthy of attention *(see Figure B.6.3)*. It might be that the hospital engineer or the reviewing authority will wish to keep on file a final copy of the list, which would be the basis for the electrical engineer's detailed engineering design.

Although this suggested format is intended for use by a hospital, it might, with suitable changes, be useful for other health care facilities.

B.7 Reserved.

B.8 Reserved.

B.9 Reserved.

B.10 Reserved.

B.11 Additional Information on Chapter 11.

B.11.1 Medical Safeguards — Respiratory Therapy.

B.11.1.1 General.

B.11.1.1.1 Personnel setting up, operating, and maintaining respiratory therapy equipment, including suction apparatus, should familiarize themselves with the problems of the use of each individual unit.

B.11.1.1.2 Personnel must be aware of the exact location of equipment in storage to facilitate emergency use.

B.11.1.1.3 Suction tubing employed in a hazardous location is to be electrically conductive.

B.11.2 Glossary of Respiratory Therapy Terminology.

B.11.2.1 Arrhythmia. Irregularity of heartbeats.

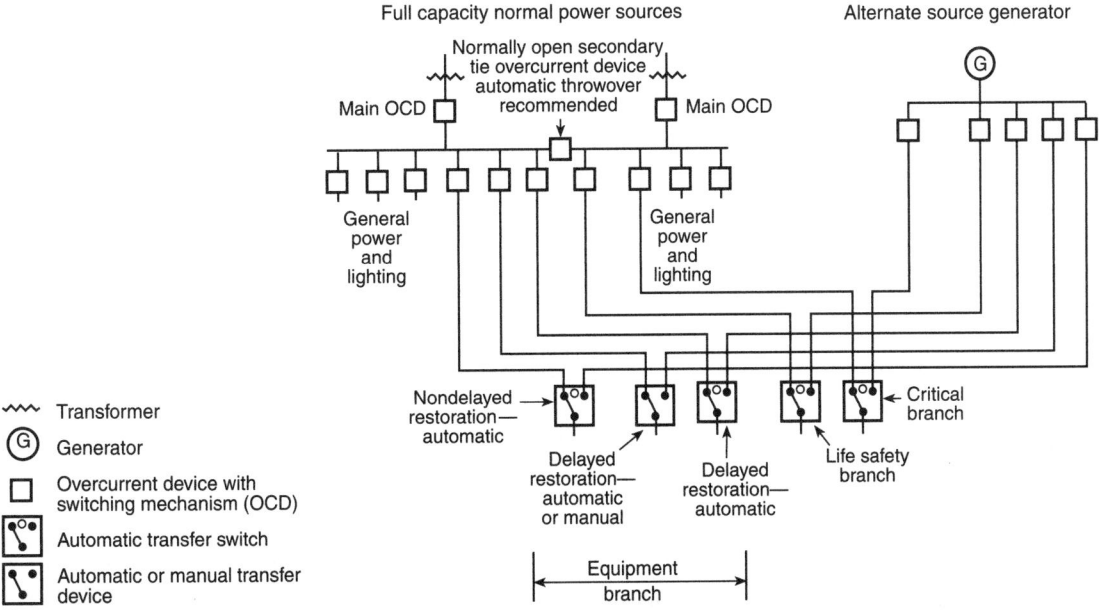

FIGURE B.6.1 Typical Hospital Wiring Arrangement.

MAINTENANCE GUIDE

Monthly

a. Testing of generator sets and transfer switches under load and operating temperature conditions at least every 30 days. A 30-minute exercise period is an absolute minimum, or the engine manufacturer's recommendations should be followed.

b. Permanently record all available instrument readings during the monthly test.

c. During the monthly test, check the following system or systems applicable to your installation:

Natural Gas or Liquid Petroleum Gas System:
- ❏ Operation of solenoids and regulators
- ❏ Condition of all hoses and pipes
- ❏ Fuel quantity

Gasoline Fuel System:
- ❏ Main tank fuel level
- ❏ Operation of system

Diesel Fuel System:
- ❏ Main tank fuel level
- ❏ Day tank fuel level
- ❏ Operation of fuel supply pump and controls

Turbine Prime Movers:
- ❏ Follow manufacturer's recommended maintenance procedure

Engine Cooling System:
- ❏ Coolant level
- ❏ Rust inhibitor in coolant
- ❏ Antifreeze in coolant (if applicable)
- ❏ Adequate cooling water to heat exchangers
- ❏ Adequate fresh air to engine and radiators
- ❏ Condition of fan and alternator belts
- ❏ Squeeze and check condition of hose and connections
- ❏ Functioning of coolant heater (if installed)

Engine Lubricating System:
- ❏ Lubricating oil level
- ❏ Crankcase breather not restricted
- ❏ Appearance of lubricating oil
- ❏ Correct lubricating oil available to replenish or change
- ❏ Operation of lubricating oil heater (if installed)
- ❏ Oil pressure correct

Engine Electrical Starting System:
- ❏ Battery terminals clean and tight
- ❏ Add distilled water to maintain proper electrolyte level
- ❏ Battery charging rate
- ❏ Battery trickle charging circuit operating properly
- ❏ Spare batteries charged (if provided)

Engine Compressed Air Starting System:
- ❏ Air compressor operating properly
- ❏ Air compressor lubricating oil level
- ❏ Spare compressed air tanks full
- ❏ Main compressed air tanks full
- ❏ Drain water from compressed air tanks

Engine Exhaust System:
- ❏ Condensate trap drained
- ❏ No exhaust leaks
- ❏ Exhaust not restricted
- ❏ All connections tight

Transfer Switch:
- ❏ Inside clean and free of foreign matter
- ❏ No unusual sounds
- ❏ Terminals and connectors normal color
- ❏ Condition of all wiring insulation
- ❏ All covers tight
- ❏ Doors securely closed

General:
- ❏ Any unusual condition of vibration, deterioration, leakage, or high surface temperatures or noise
- ❏ Maintenance manuals, service log, basic service tools, jumpers, and supplies readily available
- ❏ Check and record the time intervals of the various increments of the automatic start-up and shutdown sequences
- ❏ Overall cleanliness of room
- ❏ No unnecessary items in room

d. After the monthly test: Take prompt action to correct all improper conditions indicated during test. Check that the standby system is set for automatic start and load transfer.

Quarterly

a. On generator sets:

Engine Electrical Starting System:
- ❏ Check battery electrolyte specific gravity
- ❏ Check battery cap vents

Engine Lubricating System:
- ❏ Check lubricating oil (or have analyzed if part of an engineered lube oil program)

b. Fuel System:
- ❏ Drain water from fuel filters (if applicable)
- ❏ Drain water from day tank (if applicable)
- ❏ Check fuel gauges and drain water from main fuel tanks
- ❏ Inspect all main fuel tank vents

© 2014 National Fire Protection Association (NFPA 99, 1 of 2)

FIGURE B.6.2 Maintenance Guide for Essential Electrical System.

MAINTENANCE GUIDE (Continued)

Semiannually

On generator sets:

Engine Lubricating System:
- ❏ Change oil filter (if sufficient hours)
- ❏ Clean crankcase breather

Fuel System:
- ❏ General inspection of all components
- ❏ Change fuel filter
- ❏ Change or clean air filter

Governor:
- ❏ Check all linkages and ball joints
- ❏ Check oil level (if applicable)
- ❏ Observe for unusual oil leakage

Generator:
- ❏ Check brush length and pressure
- ❏ Check appearance of slip rings and clean if necessary
- ❏ Blow out with clean, dry compressed air

Engine Safety Controls:
- ❏ Check operation of all engine-operating alarms and safety shutdown devices (generator not under load during this check)

Annually

a. On generator sets:

Fuel System:

Diesel:
- ❏ Analyze fuel for condition (replace if required)

Gasoline:
- ❏ Replace fuel

Natural Gas or Liquefied Petroleum Gas:
- ❏ Examine all supply tanks, fittings, and lines

Lubricating Systems:
- ❏ Change oil
- ❏ Change oil filter
- ❏ Replace carburetor air filter

Cooling System:
- ❏ Check condition and rod-out heat exchangers if necessary
- ❏ Change coolant on closed systems
- ❏ Clean exterior of all radiators
- ❏ Check all engine water pumps and circulating pumps
- ❏ Examine all duct work for looseness
- ❏ Clean and check motor-operated louvers

Exhaust System:
- ❏ Check condition of mufflers, exhaust lines, supports, and connections

Ignition System:
- ❏ Spark ignition engines
- ❏ Replace points and plugs
- ❏ Check ignition timing
- ❏ Check condition of all ignition leads

Generator:
- ❏ Clean generator windings
- ❏ Check generator bearings
- ❏ Measure and record resistance readings of generator windings using insulation tester (megger)

Engine Control:
- ❏ General cleaning
- ❏ Check appearance of all components
- ❏ Check meters

b. Transfer Switch:
- ❏ Inspect transfer switch and make repairs or replacements if indicated

c. On main switchgear and generator switchgear:
- ❏ Operate every circuit breaker manually
- ❏ Visually check bus bars, bracing, and feeder connections for cleanliness and signs of overheating

Every 3 Years

a. System Controls:
- ❏ Reevaluate the settings of the voltage-sensing and time delay relays.

b. Main Switchgear and Generator Switchgear:
- ❏ Determine whether changes to the electrical supply system have been made that require a revision of the main circuit breaker, fuse, or current-limiting bus duct coordination.
- ❏ Calibrate and load test main circuit breakers. Spot-check bus bar bolts and supports for tightness. Obtain and record insulation tester readings on bus bars and circuit breakers. Obtain and record insulation tester readings on internal distribution feeders.

Periodically

a. Prime Mover Overhaul:
- ❏ Each prime mover should have a periodic overhaul in compliance with the manufacturer's recommendation or as conditions warrant.

b. Connected Load:
- ❏ Update the record of demand and connected load and check for potential overload.

© 2014 National Fire Protection Association

(NFPA 99, 2 of 2)

FIGURE B.6.2 *Continued*

Essential Electrical Systems

Hospital _____ Date _____

Room No.	Room Name	Function Served*	Emergency system		Equipment system		Nonessential
			Life Safety Branch	Critical Branch	Delayed Auto.†	Delayed Manual	

* Indicate precise lighting, receptacles, and/or equipment. Use a separate line for each function.
† Indicate time interval.

FIGURE B.6.3 Essential Electrical Systems.

B.11.2.2 Asphyxia. Suffocation from lack of oxygen and an accumulation of carbon dioxide.

B.11.2.3 Aspiration. Removal of accumulated mucus by suction.

B.11.2.4 Bronchi. The two primary divisions of the trachea.

B.11.2.5 CPAP. Continuous positive airway pressure.

B.11.2.6 CPR. Cardiopulmonary resuscitation.

B.11.2.7 Croup Tent. Equipment utilized to provide environmental control inside a canopy in relation to oxygen concentration, temperature, humidity, and filtered gas.

B.11.2.8 Cyanosis. A bluish discoloration of skin and mucus membranes due to excessive concentration of reduced hemoglobin in the blood.

B.11.2.9 Defibrillate. Use of electrical shock to synchronize heart activity.

B.11.2.10 Diffusion. Transfer of gases across the alveolar capillary membrane.

B.11.2.11 EKG, ECG. Electrocardiogram.

B.11.2.12 Hemoglobin. The chemical compound in red blood cells that carries oxygen.

B.11.2.13 Hypoxia. An abnormally decreased supply or concentration of oxygen.

B.11.2.14 IMV. Intermittent mandatory ventilation.

B.11.2.15 IPPB. Intermittent positive pressure breathing.

B.11.2.16 PEEP. Positive end expiratory pressure.

B.11.2.17 Respiration. The exchange by diffusion of gases between the alveoli, the blood, and the tissue.

B.11.2.18 Retrolental Fibroplasia. A disease entity of the premature infant causing blindness.

B.11.2.19 Thorax. The chest; the upper part of the trunk between the neck and the abdomen.

B.11.2.20 Trachea. The windpipe leading from the larynx to the bronchi.

B.11.2.21 Ultrasonic Nebulizer. A device that produces sound waves that are utilized to break up water into aerosol particles.

B.11.2.22 Ventilation. Movement of air into and out of the lungs.

B.11.2.23 Ventilator. Machine used to support or assist non-breathing or inadequately breathing patient.

B.11.3 Suggested Fire Response — Respiratory Therapy. The suggested procedure in the event of fire involving respiratory therapy apparatus is given in B.11.3.1 through B.11.3.6.

B.11.3.1 General. Fires in oxygen-enriched atmospheres spread rapidly, generate intense heat, and produce large volumes of heated and potentially toxic gases. Because of the immediate threat to patients and personnel, as well as the damage to equipment and possible spread to the structure of the building, it is important that all personnel be aware of the steps necessary to save life, to preserve limb, and, within reason, to extinguish or contain the fire.

B.11.3.2 Steps to Take in Event of Fire.

B.11.3.2.1 The following steps are recommended in the event of a fire, in the approximate order of importance:

(1) Remove the patient or patients immediately exposed from the site of the fire if their hair and clothing are not burning; if they are burning, extinguish the flames. *(See B.11.3.4 and B.11.3.5.)*
(2) Sound the fire alarm by whatever mode the hospital fire plan provides.
(3) Close off the supply of oxygen to the therapy apparatus involved if this step can be accomplished without injury to personnel. *(See B.11.3.3.)*
(4) Carry out any other steps specified in the fire plan of the hospital including, but not limited to, the following examples:
   (a) Remove patients threatened by the fire.
   (b) Close the doors leading to the site of the fire.
   (c) Attempt to extinguish or contain the fire *(see B.11.3.4)*.
   (d) Direct fire fighters to the site of the fire.
   (e) Take whatever steps necessary to protect or evacuate patients in adjacent areas.

B.11.3.3 Closing Off of Oxygen Supply.

B.11.3.3.1 In the event of a fire involving respiratory therapy equipment connected to an oxygen station outlet, the zone valve supplying that station is to be closed.

B.11.3.3.1.1 All personnel are cautioned to be aware of the hazard of the step specified in B.11.3.3.1 to other patients receiving oxygen supplied through the same zone valve. Steps should be taken to minimize such hazards, realizing that closing the valve is of foremost importance.

B.11.3.3.2 In the case of oxygen therapy apparatus supplied by a cylinder or container of oxygen, it is desirable to close the valve of the cylinder or container, provided that such closure can be accomplished without injury to personnel.

Note that metallic components of regulators and valves can become exceedingly hot if exposed to flame. Personnel are cautioned not to use their bare hands to effect closure.

B.11.3.4 Extinguishment or Containment of Fire.

B.11.3.4.1 Fire originating in or involving respiratory therapy apparatus generally involves combustibles such as rubber, plastic, linen, blankets, and the like. Water or water-based extinguishing agents are most effective in such fires.

B.11.3.4.1.1 Precautions should be observed if electrical equipment is adjacent to, or involved in, the fire because of the danger of electrocution of personnel if streams of water contact live 115-V circuits.

B.11.3.4.1.2 Before attempting to fight the fire with water or a water-based extinguishing agent, electrical apparatus should be disconnected from the supply outlet, or the supply circuit should be de-energized at the circuit panel.

B.11.3.4.1.3 If de-energization at the circuit panel cannot be accomplished, water should not be employed. *(See B.11.3.4.2.)*

B.11.3.4.2 Fires involving or adjacent to electrical equipment with live circuits can be fought with extinguishers suitable for Class C fires in accordance with NFPA 10, *Standard for Portable Fire Extinguishers.*

Note that chemical extinguishers are not effective against fires in oxygen-enriched atmospheres unless the source of oxygen is shut off. See B.11.3.3 for closing off the oxygen supply.

B.11.3.5 Protection of Patients and Personnel.

B.11.3.5.1 Because of the intense heat generated, serious and even fatal burns of the skin or of the lungs from inhaling heated gases are possible sequelae to a fire in an oxygen-enriched atmosphere. Thus, it is essential that patients be removed from the site of the fire whenever practical.

Note that, where a nonambulatory patient is connected to a burning piece of therapy equipment, it might be more practical, as the initial step, to remove the equipment or extinguish the fire, or both, than to remove the patient.

B.11.3.5.2 The large quantities of noxious gases produced constitute a threat to life from asphyxia, beyond the thermal burn problem.

B.11.3.5.2.1 Personnel are cautioned not to remain in the fire area after patients are evacuated if quantities of gaseous combustion products are present.

B.11.3.6 Indoctrination of Personnel. It is highly desirable that personnel involved in the care of patients, including nurses, aides, ward secretaries, and physicians, irrespective of whether or not they are involved in respiratory therapy practices, be thoroughly indoctrinated in all aspects of fire safety, including the following:

(1) Location of zone valves of nonflammable medical gas systems where employed, and the station outlets controlled by each valve
(2) Location of electrical service boxes and the areas served thereby
(3) Location of fire extinguishers, indications for their use, and techniques for their application
(4) Recommended methods of evacuating patients and routes by which such evacuation is accomplished most expeditiously, with reference made to the facility's fire plan
(5) Steps involved in carrying out the fire plan of the hospital
(6) Location of fire alarm boxes, or knowledge of other methods, for summoning the local fire department

B.11.4 Typical Gas Cylinders. See Table A.11.3.1.

B.12 Additional Information on Chapter 12.

B.12.1 Emergency Management Program Development. The program development process illustrated in Figure B.12.1 is an example of a tool that can be used to develop an emergency management program; other tools or processes are acceptable as long as the tasks of identification, planning, education, evaluation, and improvement are addressed.

B.12.1.1 Program Development Steps and Activities.

B.12.1.1.1 Review the Hazard Vulnerability Analysis (HVA) and Determine Priorities for Developing Standard Operating Procedures (SOPs). Using the input submitted by operating unit managers, the Emergency Management Committee (EMC) must prioritize threats/events and develop a list of SOPs that must be developed to address those hazards. Figure B.12.1.1.1 displays a sample standard operating procedure format.

B.12.1.1.2 Implement Mitigation and Preparedness Strategies. Using the SOPs developed for prioritized threats/events, develop and implement actions that will eliminate or reduce the impact of adverse events to the facility and build capabilities to manage them. The committee should review the SOPs to identify resources needed for mitigation and preparedness, develop cost estimates or resources required, and submit the resource request to the director for funding. The committee is responsible for tracking mitigation and preparedness planning activities until completed.

B.12.1.1.3 Report Results of Mitigation and Preparedness Activities to the Emergency Management Committee. Operating unit managers and the emergency program coordinator should regularly report results of mitigation and preparedness activities to the committee. Reports should include mitigation activities taken that effectively reduced or eliminated adverse impacts to the facility; mitigation activities that did not reduce or eliminate adverse impacts to the facility operation; and recommendations for mitigation and preparedness activities, budget, and timelines.

B.12.1.1.4 Develop, Publish, and Distribute the Emergency Operations Plan (EOP). *NFPA 1600, Standard on Disaster/Emergency Management and Business Continuity Programs,* Section 3.6, describes four types of planning: strategic administrative (preparedness) planning, mitigation planning, recovery planning, and emergency operations planning.

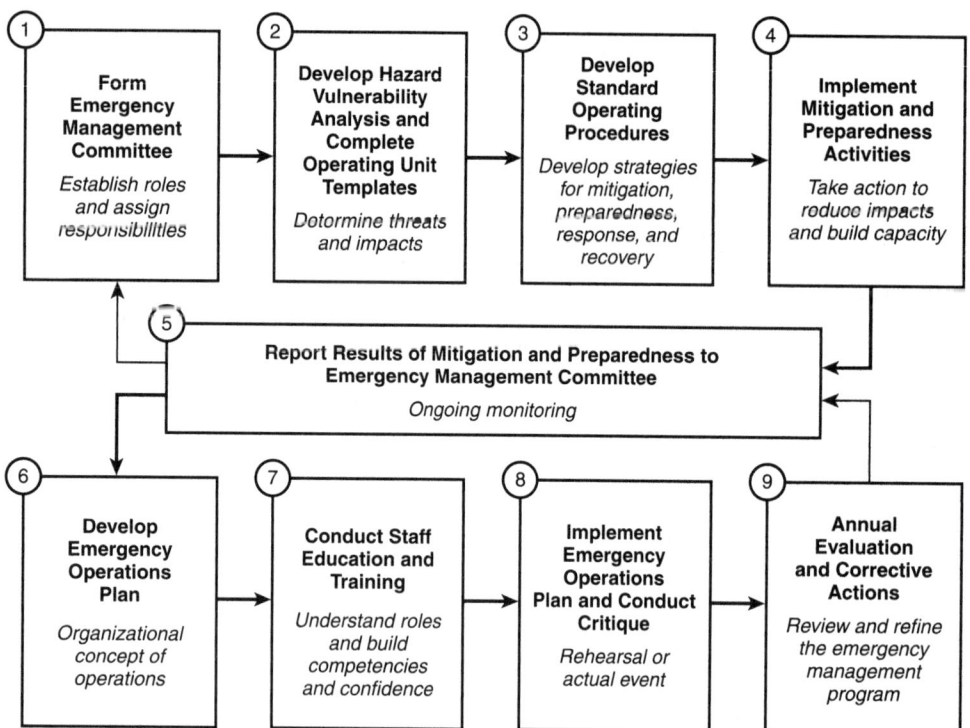

FIGURE B.12.1 Emergency Management Program Development Process.

The Federal Emergency Management Agency, now part of the Department of Homeland Security, issues guidance on the development of emergency operations plans, or EOPs. The EOP is designed to address all hazards, and it accomplishes this through its organization by functions, not departments, hazards, or individuals. Flexibility is a key feature of this type of format, as only the functions needed to address the problems are activated, not the entire plan. This type of EOP format (a basic plan and functional annexes) is that used by communities, states, and the Federal Response Plan. *(See Annex D.)*

Hard copies of the EOP need not be widely distributed. Staff members need access to incident-specific plans, but not the entire document. Several copies of the full EOP should be available in the Hospital Command Center, the administrative offices, and with the chair of the Emergency Management Committee. Posting the EOP on the hospital intranet with linkages to enhance movement through the plan can also be very effective; however, a few hard copies should still be available in the event of computer failure.

B.12.1.1.5 Train Staff on the EOP. See 12.5.3.5.

B.12.1.1.6 Delegate the Emergency Management Committee (EMC), Identify Department Roles, and Assign Responsibilities. The EMC is a multidisciplinary committee established to coordinate and oversee the emergency management program and should have a close relationship with the Safety Committee or Environment of Care Committee, or both.

The functions of the EMC include defining the role of the organization in the communitywide emergency management program; conducting/reviewing a hazard vulnerability analysis (HVA), which addresses all hazards that threaten the facility; developing/reviewing standard operating procedures (SOPs) that address hazards identified in the HVA; developing/reviewing the emergency operations plan and coordinating it with other health care organizations in the communitywide emergency management program; assigning roles and responsibilities of operating unit managers and key operators/managers; overseeing the development and maintenance of the EMP; ensuring that all employees have received appropriate training; conducting an annual evaluation of the effectiveness of the program; and ensuring a telephone roster of key personnel responsible for critical operations is kept current.

B.12.1.1.7 Test and Evaluate the EOP in Response to a Drill or Actual Event. See 12.5.3.3.7.

B.12.2 Personnel Notification and Recall. Medical staff, key personnel, and other personnel needed will be notified and recalled as required. In order to relieve switchboard congestion, it is desirable to utilize a pyramidal system to recall individuals who are off duty or otherwise out of the facility. Under the pyramidal system, an individual who has been notified will notify two other individuals, who, in turn, will each notify two other individuals, and so on. A current copy of the notification and recall roster, with current home and on-call telephone numbers, will be maintained at the hospital switchboard at all times. In case the pyramidal system is to be utilized, each individual involved in the system has to maintain a current copy of the roster at all times, in order that each knows who they are to notify and the telephone numbers concerned. It is essential that key personnel rosters be kept current.

B.12.3 Special Considerations and Protocols.

B.12.3.1 Fire and Explosion. In the event that the health care facility does not need to be completely evacuated immediately, the actions staff should take when they are alerted to a fire are detailed in Sections 18.7, 19.7, 20.7, and 21.7 of NFPA *101, Life Safety Code.*

_____
(Name of Facility)

Standard Operating Procedure #: _____   _____
                                                                              (Date)

SUBJECT: _____
                              (Insert Hazard, Threat, or Event Name)

1. Description of hazard, threat, or event: _____
   _____
   _____

2. Impact on mission-critical systems: _____
   _____
   _____

3. Operating units and key personnel with responsibility: _____
   _____
   _____

4. Mitigation and preparedness activities
   a. Hazard reduction strategies and resource issues: _____
      _____

   b. Preparedness strategies and resource issues: _____
      _____

5. Response and recovery activities
   a. Hazard control strategies and resource issues: _____
      _____

   b. Hazard monitoring strategies: _____
      _____

   c. Recovery strategies and resource issues: _____
      _____

6. Notification procedures
   a. Internal: _____
      _____

   b. External: _____
      _____

7. Specialized staff training: _____
   _____
   _____

8. References and further assistance: _____
   _____
   _____

9. Review date: _____

_____   _____
(Name)   (Position/Title)

© 2014 National Fire Protection Association   NFPA 99

FIGURE B.12.1.1.1 Sample Standard Operating Procedure Format.

B.12.3.2 During the past decade, the health care industry has been struck by numerous natural disasters. A study of these disasters has provided a series of "lessons learned." Examples follow.

B.12.3.2.1 Alert. Alerting is often provided through the local Office of Emergency Management and the National Oceanic and Atmospheric Administration. Based on the facility's geographical location, these alerts often come in the form of a "watch," which indicates the strong possibility of a natural disaster, or a "warning," which indicates the immediate threat of a disaster. A hazard vulnerability analysis (HVA) of the area would result in actions that should be taken at different stages of the alert. Such actions could indicate monitoring reports by a liaison with the Office of Emergency Management; the Internal Incident Command of a facility would assess existing staffing and supply issues and make decisions to activate staff call-back plans and augment critical supplies. Some facilities have also indicated they have found it useful to complete a pre-storm patient census reduction to discharge, where possible, or to move preselected high-acuity patients, such as neonatal intensive care patients, bariatric patients, or patients with severe respiratory complications, or a combination thereof. Also, during this pre-storm patient relocation, selected patients could be moved to what are considered to be safer areas within the facility.

Some natural disasters provide absolutely no warning, such as an earthquake. During such a situation, staff would have to assume immediate operations plan procedures.

B.12.3.2.2 Isolation Survival for up to 96 Hours. Recent disasters have shown it could be 96 hours (in some cases, longer) before outside help is able to reach the facility. Those facilities with sufficient resources in-house for critical areas fared the best during disasters. Facilities that had corporate structures or associations with out-of-area organizations also received help in a timely manner.

B.12.3.2.3 Wind. When wind knocked down communications antennas, the result was severe difficulties in facility communication with staff, other health care facilities, and so forth. One solution was construction of antennas with hinges so they could be laid down during a pre-storm warning. In other cases, the history of high winds or the HVA, or both, showed that certain windows should be boarded and certain doors should be braced and sandbagged.

B.12.3.2.4 Flooding. Flooding into a facility's lower levels where utilities are often housed results in a disruption of these services. In a recovery effort, the following procedures were helpful:

(1) Keep the electrical service turned off until the electrical device completely dries.
(2) Secure environmental waste containers and portable medical gas containers.

One contractor dealing with the hospitals of the Gulf states stated that, based on its experience with floods, it would never again use "fail return" elevators to a basement level. Many times elevators returned to the basement level upon loss of power, resulting in the wires that are located in the cab flooring being soaked, rendering the elevator useless.

B.12.3.2.5 Power Loss. Power is lost typically as a result of high winds and flooding. Generators and transfer switches have been lost due to their location below the high water mark. Another problem surfacing with generators is that they often were not powerful enough to provide HVAC services to the facility. In some of the Gulf states, this alone forced evacuation of the hospital or long-term care facility. Another problem encountered was fuel, designed for another type of generator (such as military), contaminating the unit. When ordering backup generators, it is important to know the size needed and the method to be used to connect the generator to the facility.

Services, in priority, which should be up and running for the facility to be operational, are as follows.

(1) Clinical care:
    (a) Life-support systems
    (b) Life safety (exits and fire alarm and fire suppression)
    (c) Lab services
    (d) Pharmaceutical services
    (e) Diagnostic services
(2) Infrastructure:
    (a) Heating, ventilation, and air-conditioning
    (b) Water supply and removal systems
    (c) Information technology
    (d) Food and liquids
    (e) Elevators

B.12.3.2.6 Loss of Communications. Loss of communications is one of the major problems during a disaster. There have been many suggested solutions, including satellite phones and websites to communicate to staff and responsible parties. A successful solution has been communications conducted through Amateur Radio Emergency Services (ARES). Health care facilities have also reported that it was important to have either a liaison to, or a liaison position within, the community emergency operations center.

B.12.3.2.7 Loss of Water. Loss of water can result in problems with sewer service, air-conditioning, generator and refrigeration cooling systems, sterilization, dialysis services, liquid consumption, laundry, dishwashing, staff and patient hygiene, and fire suppression. It is recommended that hospitals perform risk assessments in these areas and prepare themselves for 96 hours of isolation, without outside water delivery. If this is not possible, evacuation might have to be considered.

B.12.3.2.8 Staff Issues. The following staff issues were found to be important to address:

(1) Transportation, including knowledge of which roads are open and actually transporting staff to the facility. The health care organization should work with the state and local agencies to identify qualified volunteer groups or other organizations that could provide transportation services to staff.
(2) Addressing the safety of families of staff.
(3) Provision of food, liquids, and lodging for staff and family members.

B.12.3.2.9 Civil Disturbance. Civil disturbance resulted mostly from members of the disaster-struck community seeking drugs, food, and shelter. During such disturbances, facilities initiate lockdown procedures and request local police assistance to protect the facility. It was noted that police departments were often overburdened during disasters that involved civil disturbance, and facilities had to rely on private security or the National Guard, or both.

B.12.3.2.10 Influx of Patients. Even though the facility is trying to keep its patients alive and safe, it must be recognized that some outside patients might come to the facility due to serious injuries that occur during the disaster or other medical needs. Others might have power at home or home health care providers to assist them with special needs and may not need to go to the health care facility. Community members might come to the facility simply to seek shelter or medications in the event their normal pharmacy is shut down. Some patients might come from an evacuating facility.

Recognizing these situations, facilities might be able to redirect the special needs population to a special needs shelter if the community has set up such provisions. It was found helpful to establish an outside (under cover) pre-emergency/triage room to relieve stress from the critical care emergency room within the hospital.

B.12.3.2.11 Evacuation. Often evacuation was accomplished in two phases. Phase I was a pre-storm evacuation. It was found that, if this was going to be undertaken, it had to begin early enough to avoid traffic congestion to get patients to their ultimate destination in an acceptable length of time; otherwise, Phase II, or full hospital evacuation, could not take place until the disaster had subsided and transportation was once again available to move patients. Additional problems occurred with tracking patients and moving their medical records. Medical records that were on the Electronic Health Records System were at times unattainable if they could not be accessed from a remote site. Hard copy records, in some cases, were lost during transport.

B.12.3.3 Evacuation. Evacuation can be partial or total. It might involve moving from one story to another, from one lateral section or wing to another, or moving out of the structure. Even partial evacuations can involve all categories of patients. Where patients are those who would not routinely be moved, extraordinary measures might be required to support life. It is also necessary to ensure movement of supplies in conjunction with any evacuation. Decisions to evacuate might be made as a result of internal problems or under the menace of engulfing external threats. In all cases, the following considerations should govern:

(1) Move to predesignated areas, whether in the facility, nearby, or in remote zones. Evacuation directives will normally indicate destinations. Note that it is recommended to predesign a mutual aid evacuation plan with other health care facilities in the community. *(See Annex D, U.S. Government Publication 3152, Hospitals and Community Emergency Response — What you Need to Know, on the subject of health care community mutual aid and evacuation planning.)* Receiving facilities can also be designated based on the scope of the event, activation of state or federal resources, or by local emergency medical services (EMS), based on availability.
(2) Ensure movement of equipment, supplies, and medical records to accompany or meet patients and staff in the new location.
(3) Execute predetermined staffing plans. Some staff will accompany patients; others will rendezvous in the new location. Maintenance of shifts is more complex than normal, especially when some hard-to-move patients stay behind in the threatened location, and when staff might be separated from their own relocated families.
(4) Protect patients and staff (during and after movement) against the threatening environment.
(5) When planning, consider transportation arrangements and patient tracking.

B.12.3.3.1 The Emergency Management Committee oversees the HVA process to ensure that all major threats to the facility are accounted for and assessed. Input to the HVA by operating unit managers is very important. Once a list of priority hazards, threats, and events has been compiled, managers should complete an operating unit template for their particular service or department. Some threats to individual operating units are so severe that they might interrupt the continuity of critical operations in the facility. The operating unit template is a unit level contingency plan, useful in staff education, drills, and actual events. Figure B.12.3.3.1(a) illustrates a sample HVA format. Figure B.12.3.3.1(b) shows a sample operating unit template.

B.12.3.4 Activation of Emergency Utility Resources. Loss of utility resources can occur at any time due to a natural disaster, an internal system failure, or even a supply shortage. Redundancy in system design and support is fundamental to avoiding loss of utilities. Critical points of failure in systems and supply chains should be identified, and their malfunction, disruption, or loss mitigated.

The key to a successful response to a loss of utilities is planning. Assessment of the organization's utility systems is an excellent starting point and should consist of identifying essential utility systems, such as electricity, water/sewer, piped medical gas and vacuum systems, HVAC systems, and vertical and horizontal transports. Once the systems are identified, they should be broken down to the component level and evaluated for importance. This allows the organization to establish priorities and the capabilities of the systems and their components.

Organizations should evaluate their self sufficiency for at least 96 hours. This evaluation will assist in establishing gaps in the utility systems. For example, if a facility loses electricity and has backup generators, the organization must establish how long it can operate on those generators. By looking at different components of the generator system, the organization might determine that only enough fuel for 24 hours is maintained. If the same generator is cooled by water and the water supply is disrupted, that is an additional limitation. This concept should be applied to all systems and key components of each utility.

By establishing the limitations of the utility systems and components, an organization can identify the need to mitigate with alternative means, such as extra equipment, generators, bottled medical gas, bottled water, or formal documented agreements and understandings with other organizations. Contingency plans should be established and tested for effectiveness on a regular basis. It is essential that an organization understands its limitations and uses this information to determine if it can sustain itself for 96 hours or if it is appropriate to stop or limit services.

B.12.3.5 Civil Disturbance. Large-scale civil disturbances have shown that health care facilities and their personnel are not immune to the direct effects of human violence during such disturbances. Hospitals in large urban areas have to make special provisions in their disaster plans to ensure the physical safety of their employees in transit from the hospital exit to and from a secure means of transportation to their homes. In extreme cases, it might be necessary to house employees within the health care facility itself during such civil disturbances. Another aspect of civil disturbances not to be overlooked in facility security planning is the possibility that a given health care facility might have to admit and treat large numbers of prisoners during such emergencies; however, security guards for such patients will normally be provided by the local police department.

Type of Event	Severity Classification — Low, Moderate, High				Rank
	Probability	Human Impact	Property Impact	Operational Impact	
	Likelihood this will occur within 1 year	Possibility of death or injury	Physical losses and damage	Interruption of services	Score of 2 or higher in any category requires an SOP
Score	0 = N/A 1 = Low 2 = Moderate 3 = High	0 = N/A 1 = Low 2 = Moderate 3 = High	0 = N/A 1 = Low 2 = Moderate 3 = High	0 = N/A 1 = Low 2 = Moderate 3 = High	SOP required — yes or no?
(Hazard Type)					

© 2014 National Fire Protection Association NFPA 99

FIGURE B.12.3.3.1(a) Sample Hazard Vulnerability Analysis (HVA) Format.

Operating Unit: _____   Operating Unit Manager: _____

Mission-Critical System	Potential Problems	Contact for Assistance in Preparing	Mitigation Actions	If this mission-critical system is interrupted, then:	
				Assess situation for:	Action required:
(Lighting)					
(Electrical Power)					
(Steam Distribution)					
(HVAC)					
(Room or Hood Exhaust)					
(Water Delivery)					
(Waste Stream)					
(Communications)					

© 2014 National Fire Protection Association   NFPA 99

FIGURE B.12.3.3.1(b) Sample Operating Unit Template.

B.12.3.6 Hazardous Materials. There are at least three major sources of concern with regard to nonradioactive hazardous materials unrelated to the intentional use of chemical agents to harm people *(see B.12.3.8)*. The first is the possibility of a large spill or venting of hazardous materials near the facility; this is especially likely near major rail or truck shipping routes, near pipelines, or near heavy manufacturing plants. Second, every facility contains within its boundaries varying amounts of such materials, especially in the laboratory and custodial areas. A spill of a highly volatile chemical can quickly contaminate an entire structure by way of the air ducts. Finally, contaminated patients can pose a risk to staff, though on a more localized basis. Usually, removal of their clothing will reduce the risk materially. In any case, staff has to be prepared to seek advice on unknown hazards. This type of advice is not usually available from poison centers, but rather from a central referral, such as CHEMTREC, and its toll-free emergency information service number (800-424-9300).

See Annex D for publications concerning hazardous materials regulations and reports on various types of chemical protective equipment.

B.12.3.7 Volcanic Eruptions. Although most of the direct effects of a volcanic eruption are covered in other protocols for disasters (e.g., fire, explosion), it is necessary to make special provisions for functioning in areas of heavy to moderate ash fall. This hazard can exist hundreds of miles downwind from the eruption.

Volcanic ash is actually finely pulverized rock blown out of the volcano. Outside the area of direct damage, the ash varies from a fine powder to a coarse sand. General housekeeping measures can exclude much ash. It should be noted, however, that people move about freely during and after ash fall.

Ash fall presents the following four problems for health care facilities:

(1) People require cleanup (brushing, vacuuming) before entering the building.
(2) Electromechanical and automotive equipment and air-filtering systems require special care because of the highly abrasive and fine-penetration nature of the ash.
(3) Increased flow of patients with respiratory complaints can be expected.
(4) Eye protection is required for people who have to be out in the dust. (No contact lenses should be worn; goggles are suggested.) Dust masks are available that are approved by the National Institute for Occupational Safety and Health (NIOSH) and are marked TC-21 plus other digits.

B.12.3.8 Weapons of Mass Destruction. Weapons of mass destruction, or WMD, are defined as any weapon or device that is intended to cause, or has the capability to cause, death or serious bodily injury to a significant number of people through the release, dissemination, or impact of toxic or poisonous chemicals or their precursors; a disease organism; or radiation or radioactivity. A complete index of chemical, biological, and radiological agents and treatment recommendations can be found at the following web site: http://www.bt.cdc.gov/agent/index.asp. Many federal departments and agencies are involved in supporting WMD preparedness and response activities at the state and local level. The Department of Health and Human Services manages two cooperative grant programs administered by the Centers for Disease Control and Prevention (CDC) and the Assistant Secretary for Preparedness and Response (ASPR). These programs are aimed at enhancing the readiness of the public health and hospital system. The following web link describes these initiatives: http://www.bt.cdc.gov/planning/continuationguidance/pdf/activities_attachments.pdf.

The Department of Justice maintains a help line (1-800-368-6498) offering technical assistance in nonemergency cases and provides information on the following subjects: detection equipment; personal protective equipment; decontamination systems and methods; physical properties of WMD materials; signs and symptoms of WMD exposure; treatment of exposure to WMD materials; toxicology information; federal response assets; and applicable laws and regulations. For reporting actual or potential acts of terrorism, health care facilities should contact their local or state health departments. The National Response Center (1-800-424-8802) can link callers to technical experts.

See Annex D for publications relating to WMD preparedness for health systems.

B.12.4 Continuing Operations Plan and Recovery. It has been well documented that a community cannot recover without the health care facilities recovering simultaneously or beforehand. If at all possible, a health care facility needs to remain operational throughout a disaster to care for its patients, as well as for those who are injured during the disaster.

It is important to advertise that the facility is open for business as soon as it is able to operate.

As part of disaster recovery, health care facilities should consider the following:

(1) Know the sources of funding, such as insurance and FEMA, and request loans from those sources to initiate cash flow as soon as possible.
(2) Activate lines of credit with vendors and banks.
(3) Use stock and other investments as collateral for loans and lines of credit.
(4) Have the finance section chief of the Incident Command System work with other section chiefs to keep precise records of disaster-related expenses for reimbursement following the disaster.
(5) Ensure that doctors' offices open simultaneously with the opening of the hospital to provide services to patients. Provide office support as soon as possible for physicians, with the priority being emergency department physicians, general practitioners, orthopedic surgeons, and cardiovascular surgeons.
(6) Activate billing and payroll as soon as possible.
(7) Consider setting up a bank within the hospital for staff.

Credits:

Educational Fact Finding and On-site Research in Houston, Texas, following Tropical Storm Allison, 2001.

National Fire Protection Association.

Russell Phillips & Associates, LLC.

Also see "Atmospheric Pressure," *Fire Journal*, July–August, 2002.

Educational Fact Finding and On-site Research for Hurricane Katrina, Louisiana, Mississippi, 2005.

American Society of Healthcare Engineers: Dale Woodin.

Russell Phillips & Associates, LLC: Russell Phillips & Scott Aronson.

B.13 Reserved.

B.14 Additional Information on Chapter 14.

B.14.1 Nature of Hazards.

B.14.1.1 Fire and Explosion.

B.14.1.1.1 The occurrence of a fire requires the presence of combustible or flammable materials, an atmosphere containing oxygen or other oxidizing agent(s), and heat or energy source of ignition.

Note that certain substances such as acetylenic hydrocarbons can propagate flame in the absence of oxygen.

B.14.1.1.2 Under hyperbaric conditions utilizing compressed air, the partial pressure of oxygen is increased. Leakage of oxygen into the atmosphere of the chamber (for example, from improper application of respiratory therapy apparatus) can further increase markedly the oxygen partial pressure.

B.14.1.1.2.1 The flammability or combustibility of materials generally increases as the partial pressure of oxygen increases, even when the percentage of oxygen in the gas mixture remains constant. Materials that are nonflammable or noncombustible under normal atmospheric conditions can become flammable or combustible under such circumstances.

B.14.1.1.3 Sources of Fuel.

B.14.1.1.3.1 Materials that might not ignite in air at atmospheric pressure or require relatively high temperatures for their ignition but that burn vigorously in 100 percent oxygen include, but are not necessarily limited to, the following: tri-cresyl phosphate (lubricant); certain types of flame-resistant fabrics; silicone rubber; polyvinyl chloride; asbestos-containing paint; glass fiber-sheathed silicone rubber-insulated wire; polyvinyl chloride-insulated asbestos-covered wire and sheet; polyamides; epoxy compounds; and certain asbestos blankets.

Note that flammable lubricants are used widely in equipment designed for conventional use, including shafts, gear boxes, pulleys and casters, and threaded joints, which are coupled and uncoupled.

B.14.1.1.3.2 The flammability of certain volatile liquids and gases containing carbon and hydrogen is well known. Hazards and safeguards for their use in oxygen-enriched atmospheres at ambient pressure are well-documented in 13.4.1. See also NFPA 325, *Guide to Fire Hazard Properties of Flammable Liquids, Gases, and Volatile Solids*, now part of the NFPA *Fire Protection Guide to Hazardous Materials*.

Note that repeated reference to subsection 13.5.1 is made throughout Chapter 14. These references do not imply, and should not be construed to mean, that flammable anesthetics can or should be employed in or around hyperbaric facilities.

B.14.1.1.3.3 Human tissues will burn in an atmosphere of 100 percent oxygen. Body oils and fats, as well as hair, will burn readily under such circumstances.

B.14.1.1.3.4 When a conventional loose cotton outergarment, such as scrub suits, dresses, and gowns employed in hospital operating suites, is ignited in an atmosphere of pure oxygen, the garment will become engulfed in flame rapidly and will be totally destroyed within 20 seconds or less.

If such a garment is ignited in a compressed air atmosphere, the flame spread is increased. When oxygen concentration exceeds 23.5 percent at elevated total pressure, flame spread is much more rapid, and at 6 ATA, is comparable to 95 ± 5 percent at 1 ATA. Flame spread in air (21 percent oxygen) is somewhat increased at 6 ATA, but not to the level of 95 ± 5 percent at 1 ATA.

Combustible fabrics have tiny air spaces that become filled with oxygen when exposed to oxygen-enriched environments. Once removed to atmospheric air (e.g., room air outside the chamber), the fabric will burn, if ignited, almost as rapidly as if it were still in the oxygen environment. This hazard will remain until the oxygen trapped in the air spaces in the fabric has had time to diffuse out and be replaced by air.

B.14.1.1.3.5 Oil-based or volatile cosmetics (facial creams, body oils, hair sprays, and the like) constitute a source of fuel that is highly flammable in an oxygen-enriched atmosphere.

B.14.1.1.4 Sources of Ignition.

B.14.1.1.4.1 Sources of ignition that might be encountered in a hyperbaric chamber include, but are not necessarily limited to, the following: defective electrical equipment, including failure of high-voltage components of radiological or monitoring equipment; heated surfaces in broken vacuum tubes or broken lamps used for general illumination, spot illumination, or illumination of diagnostic instruments; the hot-wire cautery or high-frequency electrocautery; open or arcing switches, including motor switches; bare defibrillator paddles; overheated motors; and electrical thermostats.

B.14.1.1.4.2 Sources of ignition that should not be encountered in a hyperbaric facility, but that might be introduced by inept practice, include the following: lighted matches or tobacco, static sparks from improper use of personal attire, electrical wiring not complying with 14.2.8, cigarette lighters, and any oil-contaminated materials that present a spontaneous heating hazard.

B.14.1.1.4.3 In oxygen-enriched atmospheres, the minimum energy necessary to ignite flammable or combustible materials is reduced in most instances below the energy required in atmospheres of ambient air.

B.14.1.2 Mechanical Hazards.

B.14.1.2.1 General.

B.14.1.2.1.1 A large amount of potential energy is stored in even a small volume of compressed gas. In hyperbaric chambers of moderate or large size, the potential energy of the chamber's compressed atmosphere, if released suddenly, can produce devastating destruction to adjacent structures and personnel, as well as to structures and personnel remote from the site of the chamber. Such sudden release could result from failure of the vessel structure, its parts, or its piping.

B.14.1.2.1.2 A particular hazard can be created if individuals attempt to drill, cut, or weld the vessel in a manner contrary to ASME *Boiler and Pressure Vessel Code*.

B.14.1.2.2 The restriction on escape and the impedance to rescue and fire-fighting efforts posed by the chamber create a significant hazard to life in case of fire or other emergency.

B.14.1.2.2.1 A particular hazard exists to chamber personnel in the event of a fire within the structure housing the chamber. Inability to escape from the chamber and loss of services of the chamber operator would pose serious threats to the lives of all occupants of the chamber.

B.14.1.2.2.2 All personnel involved in hyperbaric chamber operation and therapy, including patients and family, have to be made aware of the risks and hazards involved. Fire prevention is essential. Extinguishment of a fire within a Class B chamber is impossible. Extinguishment of a fire within a Class A chamber is only possible utilizing equipment already installed in such a chamber, and then often only by the efforts of the occupants of such a chamber or the chamber operator.

B.14.1.2.3 The necessity for restricting viewing ports to small size limits the vision of chamber operators and other observers, reducing their effectiveness as safety monitors.

B.14.1.2.4 Containers and enclosures can be subjected to collapse or rupture as a consequence of the changing pressures of the hyperbaric chamber. Items containing entrained gas include, but are not necessarily limited to, the following: ampuls, partially filled syringes, stoppered or capped bottles, cuffed endotracheal tubes, and pneumatic cushions employed for breathing masks or aids in positioning patients. The rupture of such containers having combustible or flammable liquids would also constitute a severe fire or explosion hazard.

B.14.1.2.4.1 The sudden collapse of containers from high external pressures will result in adiabatic heating of the contents. Therefore the collapse of a container of flammable liquid would constitute a severe fire or explosion hazard both from heating and from a spill of the liquid. *(See 14.3.1.5.2 and B.14.1.1.3.2.)*

B.14.1.2.5 Other mechanical hazards relate to the malfunction, disruption, or inoperativeness of many standard items when placed in service under pressurized atmospheres. Hazards that might be encountered in this regard are implosion of illuminating lamps and vacuum tubes; overloading of fans driving gas at higher density; and inaccurate operation of standard flowmeters, pressure gauges, and pressure-reducing regulators.

Note that illuminating lamps or vacuum tubes, which implode, or overloaded fans, are sources of ignition.

B.14.1.3 Pathophysiological, Medical, and Other Related Hazards.

B.14.1.3.1 Exposure of pregnant chamber occupants to hyperbaric atmospheres might result in fetal risk.

B.14.1.3.2 Medical hazards that can be encountered routinely include compression problems, nitrogen narcosis, oxygen toxicity, and the direct effects of sudden pressure changes.

B.14.1.3.2.1 Inability to equalize pressure differentials between nasopharynx (nose) and nasal sinuses or the middle ear can result in excruciating pain and might cause rupture of the eardrum or hemorrhage into the ear cavity or nasal sinus.

B.14.1.3.2.2 The breathing of air (78 percent nitrogen) under significant pressures (as by chamber personnel breathing chamber atmosphere) can result in nitrogen narcosis, which resembles alcoholic inebriation. The degree of narcosis is directly related to the amount of pressurization. Nitrogen narcosis results in impairment of mental functions, loss of manual dexterity, and interference with alertness and ability to think clearly and act quickly and intelligently in an emergency.

B.14.1.3.2.3 Oxygen toxicity can develop from breathing oxygen at partial pressures above 0.50 atmospheres absolute for a significant length of time. Oxygen toxicity can affect the lungs (pain in the chest, rapid shallow breathing, coughing), nervous system (impaired consciousness and convulsions), or other tissues and organs, or combinations thereof.

B.14.1.3.2.4 Direct effects of reduction in pressure can include inability to equalize pressures between the nasopharynx and sinuses or middle ear, expansion of gas pockets in the gastrointestinal tract, and expansion of trapped gas in the lungs.

B.14.1.3.2.5 The presence of personnel within the cramped confines of the hyperbaric chamber in close proximity to grounded metallic structures on all sides creates a definite shock hazard if accidental contact is made with a live electrical conductor or a defective piece of electrical equipment. Such accidental contact also could be a source of ignition of flammable or combustible materials. *(See B.14.1.1.4.)*

B.14.1.3.3 Medical hazards that are not ordinarily encountered during hyperbaric oxygen therapy, but that might arise during malfunction, fire, or other emergency conditions, include electric shock and fouling of the atmosphere of the chamber with oxygen, nitrous oxide, carbon dioxide, carbon monoxide, pyrolysis products from overheated materials, or the toxic products of combustion from any fire.

B.14.1.3.3.1 Increased concentrations of carbon dioxide within the chamber, as might result from malfunction of the systems responsible for monitoring or removal thereof, can be toxic under increased pressures.

B.14.1.3.3.2 The development of combustion products or gases evolved from heated nonmetallics within the closed space of the hyperbaric chamber can be extremely toxic to life because of the confining nature of the chamber and the increased hazards of breathing such products under elevated pressure.

Note that extreme pressure rises have accompanied catastrophic fires in confined atmospheres. These pressures have driven hot, toxic gases into the lungs of victims as well as exceeding the structural limits of the vessel in at least one case.

B.14.1.3.4 Physiological hazards include exposure to high noise levels and decompression sickness. Rapid release of pressurized gases can produce shock waves and loss of visibility.

B.14.1.3.4.1 During hyperbaric therapy, and especially during compression, the noise level within the chamber becomes quite high. Such a level can be hazardous because it is distractive, interferes with communication, and can produce permanent sensory-neural deafness.

B.14.1.3.4.2 Decompression sickness (bends, caisson worker's disease) results from the elution into the bloodstream or extravascular tissues of bubbles of inert gas (mainly nitrogen) that becomes dissolved in the blood and tissue fluids while breathing air at elevated pressures for a significant period of time.

Note that rapid decompression of the chamber can occur if the pressure relief valve is damaged from exposure to a fire external to the chamber or from the venting of hot products of combustion from within the chamber.

B.14.1.3.4.3 The use of decompression procedures will prevent immediate escape from the Class A chamber by occupants during emergency situations.

Note that these procedures are not followed if chamber occupants are exposed to a "no-decompression exposure" [compression to less than 2 atmospheres absolute (ATA) air], or when compressed to 2 ATA or higher pressures and breathing 100 percent oxygen.

B.14.1.3.4.4 The sudden release of gas, whether by rupture of a container or operation of a device such as used in fire fighting, will produce noise, possible shock waves, reduced or obscured visibility, and temperature changes. The initial effect might be to cool the air, but resulting pressure rises will cause adiabatic heating.

B.14.1.3.5 In summary, the hazards of fire and related problems in hyperbaric systems are real. By the very nature of the hyperbaric atmosphere, increased partial pressures of oxygen are present routinely. Flammability and combustibility of materials are increased. Ignition energy is lowered. Both immediate escape and ready entry for rescue are impeded. Finally, attendants within the chamber, through effects of the elevated noise level and nitrogen pressure, might be unable to respond to emergencies quickly and accurately.

B.14.2 Suggested Procedures to Follow in Event of Fire in Class A Chambers.

B.14.2.1 Fire Inside Chamber. For fire inside the chamber the following procedures should be performed:

(1) *Inside Observer:*
   (a) Activate fire suppression system and/or hand-held hoses.
   (b) Advise outside.
   (c) Don breathing air mask.
(2) *Chamber Operator:*
   (a) Activate the fire suppression system, if needed.
   (b) Switch breathing gas to air.
   (c) Decompress the chamber as rapidly as possible.
(3) *Medical Personnel (Outside):*
   (a) Direct operations and assist crew members wherever necessary.
   (b) Provide medical support as required.
(4) *Other Personnel (Outside):*
   (a) Notify the fire department by activating fire signaling device.
   (b) Stand by with a fire extinguisher.
   (c) Assist in unloading chamber occupants.

B.14.2.2 Fire Outside Chamber. For fire outside the chamber the following procedures should be performed:

(1) *Chamber Operator:*
   (a) Notify the inside observer to stand by for emergency return to normal atmospheric pressure.
   (b) Notify fire department by activating fire signaling device.
   (c) Switch breathing gas to air.
   (d) Don the operator's source of breathable gas.
(2) *Medical Personnel (Outside):*
   (a) Determine whether procedure should be terminated.
   (b) Provide medical support as required.
(3) *Other Personnel (Outside):*
   (a) Stand by with a fire extinguisher.
   (b) Assist in unloading chamber occupants.

B.14.3 Suggested Procedures for Hyperbaric Chamber Operator to Follow in Event of Fire in Class B Chambers.

B.14.3.1 For fires within the facility not involving the chamber, the following procedure should be performed:

(1) If there is smoke in the area, don the operator's source of breathable gas.
(2) Decompress the chamber. The urgency of decompression should be determined by the location of the fire.
(3) Remove the patient and evacuate to safe area.
(4) Turn off the oxygen zone valve to the chamber room and close any smoke/fire barrier doors. These steps are consistent with the Rescue and Confine elements of the Rescue, Alarm, Confine, Extinguish (R.A.C.E.) procedure. It is assumed that other personnel will evacuate other patients and visitors from the area and activate a fire alarm signaling device (if not already activated).

B.14.3.2 For fire within the chamber, the following procedure should be performed:

(1) Stop oxygen from flowing into the chamber by switching off the chamber (if the chamber is compressed with oxygen) or switching the supply gas of a breathing device from oxygen to air (if the chamber is compressed with air).
(2) Decompress the chamber as rapidly as possible.
(3) Stand by with a hand-held fire extinguisher and spray into the chamber (if necessary) when the chamber door is opened.
(4) Remove the patient and evacuate to a safe area.
(5) Turn off the oxygen zone valve to the chamber room and close any smoke/fire barrier doors.

These steps are consistent with the Rescue and Confine elements of the Rescue, Alarm, Confine, Extinguish (R.A.C.E.) procedure. It is assumed that other personnel will evacuate other patients and visitors from the area and activate a fire alarm signaling device (if not already activated). The injured patient should have appropriate medical attention immediately after evacuation to a safe area. Many Class B chambers require oxygen supply pressure to operate a rapid decompression feature. If this is the case, do not turn off the oxygen zone valve or any inline oxygen supply shutoff valve until all patients have been removed from the chamber(s).

B.14.4 See Table B.14.4.

Table B.14.4 Pressure Table

Atmosphere Absolute (ATA)	mm Hg	psia	psig	Equivalent Depth in Seawater		mm Hg Oxygen Pressure of Compressed Air	mm Hg Oxygen Pressure of Oxygen-Enriched Air (23.5%)
				ft	m		
1	760	14.7	0	0	0	160	179
1.5	1140	22	7.35	16.5	5.07	240	268
2.0	1520	29.4	14.7	33.1	10.13	320	357
2.5	1900	36.7	22.0	49.7	15.20	400	447
3.0	2280	44.1	29.4	66.2	20.26	480	536
3.5	2660	51.4	36.7	82.7	25.33	560	625
4.0	3040	58.8	44.1	99.2	30.40	640	714
5.0	3800	73.5	58.8	132.3	40.53	800	893

Notes:
1. The oxygen percentage in the chamber environment, not the oxygen partial pressure, is of principal concern, as concentrations above 23.5 percent oxygen increase the rate of flame spread. Thirty percent oxygen in nitrogen at 1 ATA (228 mm Hg pO_2) increases burning rate. However, 6 percent oxygen in nitrogen will not support combustion, regardless of oxygen partial pressure (at 5 ATA, 6 percent oxygen gives 228 mm Hg pO_2).
2. The Subcommittee on Hyperbaric and Hypobaric Facilities recommends that one unit of pressure measurement be employed. Since a variety of different units are now in use, and since chamber operators have not settled upon one single unit, the above table includes the five units most commonly employed in chamber practice.

Annex C   Sample Ordinance Adopting NFPA 99

This annex is not a part of the requirements of this NFPA document but is included for informational purposes only.

C.1 The following sample ordinance is provided to assist a jurisdiction in the adoption of this code and is not part of this code.

ORDINANCE NO. _____

An ordinance of the *[jurisdiction]* adopting the *[year]* edition of NFPA *[document number]*, *[complete document title]*, and documents listed in Chapter 2 of that *[code, standard]*; prescribing regulations governing conditions hazardous to life and property from fire or explosion; providing for the issuance of permits and collection of fees; repealing Ordinance No. _____ of the *[jurisdiction]* and all other ordinances and parts of ordinances in conflict therewith; providing a penalty; providing a severability clause; and providing for publication; and providing an effective date.

BE IT ORDAINED BY THE *[governing body]* OF THE *[jurisdiction]*:

SECTION 1 That the *[complete document title]* and documents adopted by Chapter 2, three (3) copies of which are on file and are open to inspection by the public in the office of the *[jurisdiction's keeper of records]* of the *[jurisdiction]*, are hereby adopted and incorporated into this ordinance as fully as if set out at length herein, and from the date on which this ordinance shall take effect, the provisions thereof shall be controlling within the limits of the *[jurisdiction]*. The same are hereby adopted as the *[code, standard]* of the *[jurisdiction]* for the purpose of prescribing regulations governing conditions hazardous to life and property from fire or explosion and providing for issuance of permits and collection of fees.

SECTION 2 Any person who shall violate any provision of this code or standard hereby adopted or fail to comply therewith; or who shall violate or fail to comply with any order made thereunder; or who shall build in violation of any detailed statement of specifications or plans submitted and approved thereunder; or fail to operate in accordance with any certificate or permit issued thereunder; and from which no appeal has been taken; or who shall fail to comply with such an order as affirmed or modified by a court of competent jurisdiction, within the time fixed herein, shall severally for each and every such violation and noncompliance, respectively, be guilty of a misdemeanor, punishable by a fine of not less than $ _____ nor more than $_____ or by imprisonment for not less than _____ days nor more than _____ days or by both such fine and imprisonment. The imposition of one penalty for any violation shall not excuse the violation or permit it to continue; and all such persons shall be required to correct or remedy such violations or defects within a reasonable time; and when not otherwise specified the application of the above penalty shall not be held to prevent the enforced removal of prohibited conditions. Each day that prohibited conditions are maintained shall constitute a separate offense.

SECTION 3 Additions, insertions, and changes — that the *[year]* edition of NFPA *[document number]*, *[complete document title]* is amended and changed in the following respects:

List Amendments

SECTION 4 That ordinance No. _____ of *[jurisdiction]* entitled *[fill in the title of the ordinance or ordinances in effect at the present time]* and all other ordinances or parts of ordinances in conflict herewith are hereby repealed.

SECTION 5 That if any section, subsection, sentence, clause, or phrase of this ordinance is, for any reason, held to be invalid or unconstitutional, such decision shall not affect the validity or constitutionality of the remaining portions of this ordinance. The *[governing body]* hereby declares that it would have passed this ordinance, and each section, subsection, clause, or phrase hereof, irrespective of the fact that any one or more sections, subsections, sentences, clauses, and phrases be declared unconstitutional.

SECTION 6 That the *[jurisdiction's keeper of records]* is hereby ordered and directed to cause this ordinance to be published. [NOTE: An additional provision may be required to direct the number of times the ordinance is to be published and to specify that it is to be in a newspaper in general circulation. Posting may also be required.]

SECTION 7 That this ordinance and the rules, regulations, provisions, requirements, orders, and matters established and adopted hereby shall take effect and be in full force and effect *[time period]* from and after the date of its final passage and adoption.

Annex D Informational References

D.1 Referenced Publications. The documents or portions thereof listed in this annex are referenced within the informational sections of this code and are not part of the requirements of this document unless also listed in Chapter 2

D.1.1 NFPA Publications. National Fire Protection Association, 1 Batterymarch Park, Quincy, MA 02169-7471.

NFPA 10, *Standard for Portable Fire Extinguishers*, 2013 edition.

NFPA 13, *Standard for the Installation of Sprinkler Systems*, 2013 edition.

NFPA 30, *Flammable and Combustible Liquids Code*, 2015 edition.

NFPA 49, *Hazardous Chemicals Data*, 1994 edition. (No longer in print; appears in NFPA *Fire Protection Guide to Hazardous Materials*, 13th edition, 2002.)

NFPA 53, *Recommended Practice on Materials, Equipment, and Systems Used in Oxygen-Enriched Atmospheres*, 2011 edition.

NFPA 55, *Compressed Gases and Cryogenic Fluids Code*, 2013 edition.

NFPA 58, *Liquefied Petroleum Gas Code*, 2014 edition.

NFPA 70®, *National Electrical Code®*, 2014 edition.

NFPA 72®, *National Fire Alarm and Signaling Code*, 2013 edition.

NFPA 90B, *Standard for the Installation of Warm Air Heating and Air-Conditioning Systems*, 2015 edition.

NFPA 99B, *Standard for Hypobaric Facilities*, 2015 edition.

NFPA 101®, *Life Safety Code®*, 2015 edition.

NFPA 110, *Standard for Emergency and Standby Power Systems*, 2013 edition.

NFPA 220, *Standard on Types of Building Construction*, 2015 edition.

NFPA 252, *Standard Methods of Fire Tests of Door Assemblies*, 2012 edition.

NFPA 259, *Standard Test Method for Potential Heat of Building Materials*, 2013 edition.

NFPA 325, *Guide to Fire Hazard Properties of Flammable Liquids, Gases, and Volatile Solids*, 1994 edition. (No longer in print; appears in NFPA *Fire Protection Guide to Hazardous Materials*, 13th edition, 2002.)

NFPA 491, *Guide to Hazardous Chemical Reactions*, 1997 edition.

NFPA 551, *Guide for the Evaluation of Fire Risk Assessments*, 2013 edition.

NFPA 730, *Guide for Premises Security*, 2014 edition.

NFPA 1600®, *Standard on Disaster/Emergency Management and Business Continuity Programs*, 2013 edition.

NFPA *Fire Protection Guide to Hazardous Materials*, 2008 edition.

D.1.2 Other Publications.

D.1.2.1 ACS Publications. American College of Surgeons, 633 N. Saint Clair Street, Chicago, IL 60611-3211.

04–GR-0001, Guidelines for Optional Ambulatory Surgical Care and Office-Based Surgery, 2000.

D.1.2.2 ASHRAE Publications. ASHRAE, 1791 Tullie Circle, NE, Atlanta, GA 30329-2305.

ASHRAE Handbook of Fundamentals, 2001.

ASHRAE Guideline 0, *The Commissioning Process*, 2005.

ASHRAE Guideline 1.1, *HVAC&R Technical Requirements for the Commissioning Process*, 2007.

D.1.2.3 ASME Publications. American Society of Mechanical Engineers, Two Park Avenue, New York, NY 10016-5990.

ASME B16.22, *Wrought Copper and Copper Alloy Solder Joint Pressure Fitting*, 2001.

ANSI/ASME B16.50, *Wrought Copper and Copper Alloy Braze Joint Pressure Fitting*, 2001.

ASME Boiler and Pressure Vessel Code, 2001.

D.1.2.4 ASSE Publications. American Society of Sanitary Engineering, 901 Canterbury Road, Suite A, Westlake, OH 44145-1480.

ASSE 6040, *Professional Qualification Standard for Medical Gas Maintenance Personnel*, 2001.

D.1.2.5 ASTM Publications. ASTM International, 100 Barr Harbor Drive, P.O. Box C700, West Conshohocken, PA 19428-2959.

ASTM E 119, *Standard Test Method for Fire Tests of Building Construction and Materials*, 2012.

ASTM G 63, *Standard Guide for Evaluating Nonmetallic Materials for Oxygen Service*, 2007.

ASTM G 88, *Standard Guide for Designing Systems for Oxygen Service*, 2005.

ASTM G 93, *Standard Practice for Cleaning Methods and Cleanliness Levels for Material and Equipment Used in Oxygen-Enriched Environments*, 1999 (2007).

ASTM G 94, *Standard Guide for Evaluating Metals for Oxygen Service*, 2005.

D.1.2.6 CGA Publications. Compressed Gas Association, 4221 Walney Road, 5th Floor, Chantilly, VA 20151-2923.

CGA G-8.1, *Standard for Nitrous Oxide Systems at Consumer Sites*, 1990.

CGA P-2.7, *Guide for the Safe Storage, Handling, and Use of Small Portable Liquid Oxygen Systems in Health Care Facilities*, 2011.

CGA V-1, *Standard for Compressed Gas Cylinder Valve Outlet and Inlet Connections* (ANSI B57.1), 1994.

D.1.2.7 FGI Publications. Facility Guidelines Institute, 1919 McKinney Avenue, Dallas, TX 75201.

Guidelines for Design and Construction of Hospitals and Outpatient Facilities, 2014.

D.1.2.8 IEC Publications. International Electrotechnical Commission, 3, rue de Varembé, P.O. Box 131, CH-1211 Geneva 20, Switzerland.

ANSI/IEC/ISO 80001-1-1, *Risk Management of Medical IT-Networks*, 2010.

ANSI/IEC/ISO 80001-2-5, *Guidance for the Risk Management of Distributed Alarm Systems Utilizing Medical IT-Networks*, 2010.

IEC, 60601-1-1, *Medical Electrical Equipment — Part 1: General Requirements for Basic Safety and Essential Performance*, 2007.

IEC 60601-1-2 *Medical Electrical Equipment– Part 1-2: General Requirements for Safety – Collateral Standard: Electromagnetic Compatibility – Requirements and Tests,* 2007.

D.1.2.9 IEEE Publications. IEEE, Three Park Avenue, 17th Floor, New York, NY 10016-5997.

ANSI/IEEE 493-2007, *Recommended Practice for the Design of Reliable Industrial and Commercial Power System,* 2007.

IEEE 602, *Recommended Practice for Electric Systems in Health Care Facilities,* 2007.

D.1.2.10 Ocean Systems, Inc. Publications. Ocean Systems, Inc., Research and Development Laboratory, Tarrytown, NY 10591. Work carried out under U.S. Office of Contract No. N00014-67-A-0214-0013.

Ocean Systems, Inc., "Technical Memorandum UCRI-721, Chamber Fire Safety." (Figure A.3.3.11.2 is adapted from Figure 4, "Technical Memorandum UCRI-721, Chamber Fire Safety," T. C. Schmidt, V. A. Dorr, and R. W. Hamilton, Jr., Ocean Systems, Inc., Research and Development Laboratory, Tarrytown, NY 10591. Work carried out under US Office of Naval Research, Washington, DC, Contract No. N00014-67-A-0214-0013.) (G. A. Cook, R. E. Meierer, and B. M. Shields, "Screening of Flame-Resistant Materials and Comparison of Helium with Nitrogen for Use in Dividing Atmospheres." First summary report under ONR Contract No. 0014-66-C-0149. Tonawanda, NY: Union Carbide, 31 March 1967. DDC No. Ad-651583.)

D.1.2.11 SAE Publications Society of Automotive Engineers, 400 Commonwealth Drive, Warrendale, PA 15096.

AMS QQ-N290, *Nickel Plating (Electrodeposited),* 2009.

D.1.2.12 TIA Publications. Telecommunications Industry Association, 2500 Wilson Boulevard, Suite 300, Arlington, VA 22201.

TIA/EIA 569-B, *Commercial Building Standard for Telecommunications Pathways and Spaces,* 2004.

D.1.2.13 UL Publications. Underwriters Laboratories Inc., 333 Pfingsten Road, Northbrook, IL 60062-2096.

UL 263, *Fire Resistance Ratings,* 2011.

ANSI/UL 1069, *Safety Standard for Hospital Signaling and Nurse Call Equipment,* 2012.

D.1.2.14 U.S. Government Publications. U.S. Government Printing Office, Washington, DC 20402.

"Crisis Standards of Care: A Systems Framework for Catastrophic Disaster Response," Institute of Medicine (IOM) Report, 2012.

Medical Surge Capacity and Capability Handbook, Department of Health and Human Services, 2007.

D.1.2.15 Other Publications.

D.2 Informational References. The following documents or portions thereof are listed here as informational resources only. They are not a part of the requirements of this document.

D.2.1 Published Articles on Fire Involving Respiratory Therapy Equipment and Related Incidents.

Benson, D. M., and Wecht, C. H. Conflagration in an ambulance oxygen system. *Journal of Trauma,* vol. 15, no. 6:536-649, 1975.

Dillon, J. J. Cry fire! *Respiratory Care,* vol. 21, no. 11:1139-1140, 1976.

Gjerde, G. E., and Kraemer, R. An oxygen therapy fire. *Respiratory Care,* vol. 25, no. 3 3:362-363, 1980.

Walter, C. W. Fire in an oxygen-powered respirator. *JAMA* 197:44-46, 1960.

Webre, D. E., Leon, R., and Larson, N W. Case History; Fire in a nebulizer. *Anesthesia and Analgesia* 52:843-848, 1973.

D.2.2 References for A.10.2.13.4.3.

Dalziel, C. F., and Lee, W. R., Reevaluation of lethal, electric currents effects of electricity on man. *Transactions on Industry and General Applications,* vol. IGA-4, no. 5, September/October 1968.

Roy, O. A., Park, G. R., and Scott, J. R., Intracardiac catheter fibrillation thresholds as a function of duration of 60 Hz current and electrode area. *IEEE Trans. Biomed. Eng.* BME 24:430-435, 1977.

Roy, O. A., and Scott, J. R., 60 Hz ventricular fibrillation and pump failure thresholds versus electrode area. *IEEE Trans. Biomed. Eng.* BME 23:45-48, 1976.

Watson, A. B., Wright, J. S., and Loughman, J., Electrical thresholds for ventricular fibrillation in man. *Med. J. Australia* 1:1179-1181, 1973.

Weinberg, D. I., et al., Electric shock hazards in cardiac catheterization. *Elec. Eng.* 82:30-35, 1963.

D.2.3 References for A.14.3.1.5.4.3.

NASA BMS Document GRC-M8300.001. (2005). Ch. 5, Para. 5.6.3.

Raleigh, G., et al. (2005). *Air-Activated Chemical Warming Devices: Effects of Oxygen and Pressure.* Undersea Hyper Med, 32(6).

Workman, W.T. (1999). *Hyperbaric Facility Safety: A Practical Guide* (p. 531). Flagstaff (AZ): Best Publishing.

Kindwall, E.P., & Whelan, H.T. (2004). *Hyperbaric Medicine Practice* (p. 86). Flagstaff (AZ): Best Publishing.

Burman, F. (2006). *Risk Assessment Guide for the Installation and Operation of Clinical Hyperbaric Facilities;* 4th edition. San Antonio (TX): International ATMO.

D.2.4 Addresses of Other Organizations that Publish Standards or Guidelines.

American Conference of Governmental Industrial Hygienists, 1330 Kemper Meadow Drive, Cincinnati, OH 45240-1634.

American Industrial Hygiene Assoc., 475 Wolf Ledges Parkway, Akron, OH 44311.

Department of Health and Human Services, ASPR, National Disaster Medical System (NDMS), http://www.phe.gov/preparedness/responders/ndms/pages/default.aspx.

George Washington University, School of Engineering and Applied Sciences, Institute for Crisis, Disaster and Risk Management. *Medical and Health Incident Management (maHim) System: A Comprehensive Functional System Description for Mass Casualty Medical and Health Incident Management,* http://www.seas.gwu.edu/~icdm/MaHIM%20V2%20final%20report%20sec%202.pdf.

National Emergency Management Association, Council of State Governments, Lexington, KY, Emergency Management Assistance Compact, http://www.emacweb.org/emac/index.cfm?CFID=5327&CFTOKEN=28115803.

Scientific Apparatus Makers Assoc., 1101 16th Street, NW, Washington, DC 20036.

University of Colorado, Natural Hazards and Information Applications Center, Disaster Research Clearinghouse, www.colorado.edu/hazards.

University of Delaware, Disaster Research Center, http://www.udel.edu/DRC/.

American Society for Healthcare Engineering (www.ashe.org), 155 North Wacker Drive, Chicago, IL 60606.

D.2.5 Addresses of Organizations and Agencies That Provide Health Care Emergency Preparedness Educational Materials.

D.2.5.1 Publications.

National Fire Protection Association, 1 Batterymarch Park, Quincy, MA 02169-7471.

American Health Care Association, 1201 L Street, Washington, DC 20005.

American Hospital Association, 155 N. Wacker Drive, Suite 400, Chicago, IL 60606.

American Medical Association, 515 N. State Street, Chicago, IL 60610.

American Nurses' Association, 8515 Georgia Avenue, Suite 400, Silver Spring, MD 20910.

American Red Cross, National Headquarters, 2025 E Street, NW, Washington, DC 20006.

Family Disaster Planning http://www.redcross.org/services/disaster/beprepared/familyplan.html

Disaster Preparedness for People with Disabilities, http://www.redcross.org/services/disaster/beprepared/disability.html

Association of American Railroads, 50 F Street, Washington, DC 20001-1564.

Charles C. Thomas Publisher, 2600 South First Street, Springfield, IL 62704.

Dun-Donnelley Publishing Corp., 666 Fifth Avenue, New York, NY 10019.

Federal Emergency Management Agency, 500 C Street, SW, Washington, DC 20472.

Florida Health Care Association, 307 W. Park Avenue, P.O. Box 1459, Tallahassee, FL 32301.

Helicopter Association International, 1635 Prince Street, Alexandria, VA 22314-2818.

Hospital Emergency Incident Command System, State of California Emergency Medical Services Authority, 1930 9th Street, Sacramento, CA 95814. http://www.emsa.ca.gov/dms2/heics3.htm

International Association of Fire Chiefs, 4025 Fair Ridge Drive, Suite 300, Fairfax, VA 22033-2868.

Joint Commission on Accreditation of Healthcare Organizations (JCAHO), One Renaissance Blvd., Oakbrook Terrace, IL 60181.

National Interagency Incident Management System, Incident Command System, National Interagency Fire Coordination Center, Boise, ID. http://www.nwcg.gov/pms/forms/ics_cours/ics_courses.htm

Pan American Health Organization, 525 23rd Street, NW, Washington, DC 20037 (Attn.: Editor, Disaster Preparedness in the Americas).

Standardized Emergency Management System, State of California Governor's Office of Emergency Services, 3650 Schreiber Avenue, Mather, CA 95655. http://www.oes.ca.gov/Operational/OESHome.nsf/Content/B49435352108954488256C2A0071E038?OpenDocument

University of Delaware, Disaster Research Center (Publications), Newark, DE 19716.

U.S. Department of Transportation (available from U.S. Government Printing Office, Washington, DC 20402).

D.2.5.2 Audiovisual Materials.

National Fire Protection Association, 1 Batterymarch Park, Quincy, MA 02169-7471.

Abbott Laboratories, Audiovisual Services, 565 Fifth Avenue, New York, NY 10017.

Brose Productions, Inc., 10850 Riverside Drive, N. Hollywood, CA 91602.

Federal Emergency Management Agency, Office of Public Affairs, Washington, DC 20472.

Fire Prevention Through Films, Inc., P.O. Box 11, Newton Highlands, MA 02161.

General Services Administration, National Audiovisual Center, Reference Section, Washington, DC 20409.

Helicopter Association International, 1635 Prince Street, Alexandria, VA 22314-2818.

Pyramid, P.O. Box 1048, Santa Monica, CA 90406.

University of Illinois Medical Center, Circle Campus, Chicago, IL 60612.

D.2.6 Additional U.S. Government Informational Sources.

Kidney Community Emergency Response Coalition, www.kcercoalition.com

Health Professional Predisaster Identification (ESAR-VHP), www.phe.gov/esarvhp

Hospital Available Beds for Emergencies and Disasters HAvBED, havbedhhs.gov

National Response Framework, www.fema.gov/national-response-framework

National Recovery Framework, www.fema.gov/recovery-framework

Department of Health and Human Services, ASPR National Health Security Strategy, http://www.phe.gov/Preparedness/planning/authority/nhss/Pages/default.aspx

Department of Health and Human Services, ASPR Hospital Preparedness Program, http://www.phe.gov/preparedness/planning/hpp/pages/default.aspx

U.S. Government Printing Office, Washington, DC 20402.

Biological Threat Interrogatories, http://www.va.gov/emshg/page.cfm?ID=BioThreatInterr.

Title 29, Code of Federal Regulations, Part 1910, Subpart 1030, *Bloodborne Pathogens.*

Title 29, Code of Federal Regulations, Part 910, Subpart 1910, *Occupational Exposures to Chemical Laboratories.*

Title 49, Code of Federal Regulations, Parts 171 through 190 (U.S. Dept. of Transportation, Specifications for Transportation of Explosives and Dangerous Articles). (In Canada, the regulations of the Board of Transport Commissioners, Union Station, Ottawa, Canada, apply.)

Title 49, Code of Federal Regulations, Part 173, *Shippers — General Requirements for Shipments and Packagings.*

Commercial Standard 223-59, *Casters, Wheels, and Glides for Hospital Equipment.*

Environmental Protection Agency, Chemical Emergency Preparedness and Prevention, http://yosemite.epa.gov/oswer/ceppoweb.nsf/content/homelandSecurity.htm?OpenDocument.

National Research Council Publication 1132, *Diesel Engines for Use with Generators to Supply Emergency and Short Term Electric Power.* (Also available as Order No. O.P.52870 from University Microfilms, P.O. Box 1366, Ann Arbor, MI 48106.)

U.S. Department of Defense:

U.S. Army Medical Research Institute of Chemical Defense (USAMRICD), http://chemdef.apgea.army.mil/.

U.S. Army Medical Research Institute of Infectious Diseases (USAMRIID), http://www.usamriid.army.mil/general/index.html.

U.S. Army Soldier and Biological Chemical Command (SBCCOM), http://hld.sbccom.army.mil/ip/detectors.

U.S. Department of Health and Human Services:

Centers for Disease Control and Prevention: HHS Publication No. 93–8395, *Biosafety in Microbiological and Biomedical Laboratories*.

Centers for Disease Control and Prevention, Public Health Preparedness and Response for Bioterrorism Program, http://www.bt.cdc.gov/planning/continuationguidance/index.asp.

National Institute for Occupational Health and Safety, Personal Protection Equipment, http://www.cdc.gov/niosh/topics/emres/ppe.html.

Protecting Building Environments from Airborne Chemical, Biologic and Radiologic Agents (page 9). http://www.cdc.gov/mmwr/PDF/wk/mm5135.pdf.

U.S. Department of Homeland Security:

Capability Assessment for Readiness, http://www.fema.gov/pdf/rrr/car.pdf.

Exercise Design Course, http://training.fema.gov/emiweb/IS/is120.asp.

Guide for All-Hazard Emergency Operations Planning, http://www.fema.gov/pdf/rrr/slg101.pdf.

Metropolitan Medical Response System, Resources, http://mmrs.hhs.gov/main/Resources.aspx.

National Disaster Medical System, Conference Library, http://ndms.dhhs.gov/NDMS%20Conference/conf2k3/previous_confe_03/previous_confe_03.html.

Strategic National Stockpile, http://www.bt.cdc.gov/stockpile/index.asp.

U.S. Department of Justice, Office of Domestic Preparedness, Publications Library, http://www.ojp.usdoj.gov/odp/library/bulletins.htm.

U.S. Department of Labor, Occupational Health and Safety Administration, Washington, DC:

Title 29, Code of Federal Regulations, Part 1910: *Employee Protection Plans*, 1910.38; Subpart H, *Hazardous Materials* (1910.101-126), specifically 1910.120, *Hazardous Waste Operations and Emergency Response* (HAZWOPER) and Appendices A-E; Subpart I, *Personal Protective Equipment* (1910.132-139 and Appendix B), specifically: 1910.132, *General Provisions*; 1910.133, *Eye and Face Protection*; 1910.134, *Respiratory Protection* (and Appendices A-D); 1910.136, *Occupational Foot Protection*; 1910.138, *Hand Protection*; Subpart Z - *Toxic and Hazardous Substances* (1910.1000-1450 and Appendix B), specifically 1910.1200–*Hazard Communication* (and Appendices A-E).

Publication 3114, *Hazardous Waste Operations and Emergency Response*, http://www.osha.gov/Publications/OSHA3114/osha3114.html.

Publication 3152, *Hospitals and Community Emergency Response – What You Need to Know*, http://www.osha.gov/Publications/OSHA3152/osha3152.html.

D.2.7 Additional Resources for Emergency Management.

Emergency Management Principles and Practices for Health Care Systems, Institute of Crisis, Disaster and Risk Management, The George Washington University, Washington, DC, 2010, for the Veterans Health Administration, Principal Investigator Joseph Barbera, MD; http://www.gwu.edu/~icdrm/publications/index.html#books.

D.2.8 Other Publications.

DuPont Safety News, June 14, 1965.

Dasler and Bauer, Ind. Eng. Chem. Anal., Ed. 18, 52 (1964).

Hoeltge, G. A., Miller, A., Klein, B. R., Hamlin, W. B., *Accidental fires in clinical laboratories*.

ISO/IEC 31010, *Risk Management — Risk Assessment Techniques*, 2009.

SEMI S10-0307E, *Safety Guideline for Risk Assessment and Risk Evaluation Process*.

D.3 References for Extracts in Informational Sections.

NFPA 70®, *National Electrical Code*®, 2014 edition.

NFPA 101®, *Life Safety Code*®, 2015 edition.

NFPA 110, *Standard for Emergency and Standby Power Systems*, 2013 edition.

Index

Copyright © 2014 National Fire Protection Association. All Rights Reserved.

The copyright in this index is separate and distinct from the copyright in the document that it indexes. The licensing provisions set forth for the document are not applicable to this index. This index may not be reproduced in whole or in part by any means without the express written permission of NFPA.

-A-

Absolute atmosphere (ATA) .... B.14.1.3.4.3, Table B.14.4
   Definition 3.3.12.1
Absolute pressure (definition) 3.3.135.1
Access 13.6, A.13.6.3.1
Adapters 10.2.4, 10.5.2.3
Adiabatic heating B.14.1.2.4.1, B.14.1.3.4.4
   Definition 3.3.1
Aerosols 11.5.1.2.1, 15.13.3.1
   Definition 3.3.2
Aftercoolers .... 5.1.3.6.3.5, 5.1.3.6.3.9(C), 5.1.3.6.3.9(H), 5.1.13.3.5.10(1), 5.2.3.5(1)
Air
   Compressors .... *see also* Dental air compressors; Medical air compressors
      Hyperbaric chambers ...... 14.2.4.2, A.14.2.4.2.1 to A.14.2.4.5.3
      Instrument air 5.1.13.3.5.13
      Starting devices 6.4.1.1.15
   Dental .... *see* Dental air
   Medical .... *see* Medical air
   Monitoring, hyperbaric chamber 14.2.9.6, A.14.2.9.6
   Nonmedical compressed air 8.3.5
Air conditioning .... *see* Heating, ventilating, and air conditioning (HVAC) systems
Air dryers .... 5.1.3.6.3.3, 5.1.3.6.3.7, 5.1.3.6.3.9(E), 5.1.3.6.3.9(F), 5.1.3.6.3.9(I), 5.1.13.3.5.10(3), 5.2.3.5(1), 5.3.3.6.1.1(12), A.5.1.3.6.3.9(F)
Alarm systems .... 15.7, A.15.7.2.4 to A.15.7.4.3.2; *see also* Nurse call systems
   Annunciator, remote 6.4.1.1.18, A.6.4.1.1.18.1
   Area 5.1.8.1.6, 5.1.9.1, 5.1.9.4, 5.1.12.3.5.3, 5.1.13.8.3, A.5.1.9.4
      Definition 3.3.3.1
   Category 1 .... 5.1.8.1.6, 5.1.9, A.5.1.9, B.5.2.9 to B.5.2.13, B.5.2.15, B.5.2.16; *see also subheads:* Area, Local, Master
      Instrument air 5.1.13.3.5.12
      Medical air compressors; .... *see* Medical air compressors
      Support gases 5.1.13.8
      Tests 5.1.12.3.5, 5.1.14.4.8
      Waste anesthetic gas disposal (WAGD); .... *see* Waste anesthetic gas disposal (WAGD) systems
   Category 2 .... 5.2.9
   Category 3 .... 5.3.9
   Definition 3.3.3.2
   Changeover signals .... 5.1.9.2.4, B.5.2.10 to B.5.2.13
   Essential electrical systems .... 6.4.2.2.3.2(7), 6.5.2.2.2.1(3), 7.3.1.2.2.7(B)
   Generator sets 6.4.1.1.17.1
   Hyperbaric chambers ... 14.2.1.4.4.5, 14.2.5.1.4, 14.2.5.4, 14.2.7.2, 14.2.8.3.16, A.14.2.5.1.4, A.14.2.5.4.2, A.14.2.8.3.16.1
   Labeling 5.1.11.4, A.5.1.11.4.2
   Local .... 5.1.3.5.9, 5.1.3.6.3.12, 5.1.3.7.7, 5.1.9.1, 5.1.9.5, 5.1.13.8.4, A.5.1.9.5; *see also* Local signal
      Definition 3.3.3.3
   Master .... 5.1.3.5.14.4, 5.1.3.6.3.14(C)(9), 5.1.8.1.6, 5.1.9.1, 5.1.9.2, 5.1.12.3.5.2, 5.1.13.8.2, A.5.1.3.5.14.4, A.5.1.3.6.3.14(C)(9), A.5.1.9.2, B.5.2.10 to B.5.2.13
      Computer systems as substitutes for 5.1.9.3
      Definition 3.3.3.4
   Occupant notification .... 15.7.4, A.15.7.4.2.1 to A.15.7.4.3.2
   Remote control 6.4.1.1.18, A.6.4.1.1.18.1
   Signal initiation 15.7.2, A.15.7.2.4
   Silencing means 6.4.1.1.18.4, 6.4.1.1.18.5

Alternate power sources .... 6.4.1.1.4 to 6.4.1.1.6, 6.4.2.2.2; *see also* Generator sets
   Connection to 6.4.2.2.5.2
   Definition 3.3.4
   Failure of 6.4.3.2.6, 6.4.3.2.7, 6.5.3.2.5, 6.5.3.2.6
   Maintenance and testing of .... 6.4.3.2.2 to 6.4.3.2.8, 6.4.4.1.1, 6.5.3.2.3 to 6.5.3.2.7, 6.5.4.1.1, A.6.4.4.1.1.4(A)
   Transfer to; .... *see* Transfer switches
Ambient atmosphere .... *see* Atmosphere
Ambient pressure .... *see* Pressure
Ambient temperature .... *see* Temperature
Ambulatory health care centers 15.7.4.3.2, A.15.7.4.3.2, A.15.13
   Definition 3.3.5
Ampacity (definition) 3.3.6
Anesthetic apparatus .... 11.4.1, A.11.4.1.1 to A.11.4.1.4
Anesthetics .... *see also specific gases*
   Definition 3.3.7
   Flammable 5.1.14.1.1, A.5.1.14.1.1
   Nonflammable anesthetic agent (definition) 3.3.114
   Waste .... *see* Waste anesthetic gas disposal (WAGD) systems
Anesthetizing locations .... 1.3.4.2, 15.13.3; *see also* Waste anesthetic gas disposal (WAGD) systems
Animal chambers .... 14.1.2.2(3), 14.2.2.3, A.14.1.2.2
Antiseptics 15.13.3
Apparatus .... *see* Anesthetic apparatus; Equipment
Appliances .... *see also* Electrical equipment; Equipment
   Battery-powered 14.2.8.3.17.5
   Definition 3.3.8
   Double-insulated 10.2.2.1.2, 10.4.2.3
      Definition 3.3.38, A.3.3.38
Applicable code
   Definition 3.3.9
Application of standard 1.3
Applicator (definition) 3.3.10, A.3.3.10
Approved (definition) 3.2.1, A.3.2.1
Area of administration 11.5.1.1.2, A.11.5.1.1.2
   Definition 3.3.11
Atmosphere
   Absolute (ATA) .... B.14.1.3.4.3, Table B.14.4
      Definition 3.3.12.1
   Ambient .... 5.1.10.4.5.5, 14.2.4.3.1, B.14.1.1.4.3; *see also* Temperature, Ambient
   Chamber .... 14.2.4.1.3, 14.2.4.2, 14.2.8.2.3, 14.2.9.3.1, 14.2.9.4.1, 14.2.9.5, 14.3.1.5.2.2, 14.3.2.6.1, A.14.2.4.2.1 to A.14.2.4.5.3, A.14.3.1.5.2.2(2), B.14.1.3.2.2
      Definition 3.3.12.3
   Definition 3.3.12, A.3.3.12
   Of increased burning rate (definition) 3.3.12.2, A.3.3.12.2
   Oxygen-enriched .... *see* Oxygen-enriched atmosphere
Attachment plugs .... *see* Plugs (attachment plugs, caps)
Authority having jurisdiction
   Definition 3.2.2, A.3.2.2
   Emergency planning responsibilities 12.2.1, A.12.2.1
Automatic (definition) 3.3.13
Automatic sprinklers .... *see* Sprinklers, automatic

-B-

Basic care rooms .... *see* Patient care spaces
Bathrooms
   Definition 3.3.14

Electrical receptacles .... 6.3.2.2.6.2(D)
Nurse and staff emergency assistance calls 7.3.3.1.2.4, 7.3.3.1.8.2
Wet procedure locations 6.3.2.2.8.3
Batteries 6.4.1.2
   Appliances, battery-powered 14.2.8.3.17.5
   Automatic detection system, for 14.2.5.4.3
   Cranking 6.4.1.1.14
   Maintenance 6.4.4.1.3, 6.5.4.1.3
Battery-powered lighting units 6.3.2.2.11
   Definition 3.3.15
Bends B.14.1.3.4
   Definition 3.3.16
Bilge, hyperbaric facilities 14.2.2.4.2
Branch circuits
   Definition 3.3.17
   Patient care spaces 6.3.2.2.1, A.6.3.2.2.1
   Special-purpose outlets, serving 6.3.2.2.1.4
Branch (lateral) lines, piping .... *see* Pipe and piping, gas
Brazing .... 5.1.6.8, 5.1.10.3.1(1), 5.1.10.4, 5.1.10.11.11, A.5.1.10.4.5
Breathing apparatus, hyperbaric chambers 14.2.4.1.3, 14.2.4.5.3, 14.2.6, A.14.2.4.5.3
Building system categories
   Application 4.3
   Plumbing system 8.2, A.8.2.1
Bulk systems .... 5.1.3.6.1(1)
   Cryogenic liquids .... *see* Cryogenic liquids, Bulk systems
   Definition 3.3.19, A.3.3.19.3
   Inert gas 5.1.3.3.1.10, A.5.1.3.3.1.10
      Definition 3.3.19.1
   Micro-bulk or small bulk systems 5.1.3.5.13, A.5.1.3.5.13.3
      Definition 3.3.19.4
   Nitrous oxide (definition) 3.3.19.2
   Oxygen (definition) .... 3.3.19.3, A.3.3.19.3; *see also* Cryogenic liquids, Bulk systems
Bypass-isolation switch 6.4.2.1.8, A.6.4.2.1.8.3

-C-

Cannula, nasal (definition) 3.3.106
Carbon dioxide
   Designation colors/operating pressures .... Table 5.1.11
   Monitoring 14.2.9.5
   Supply systems .... 5.1.1.3, 5.1.3.3.1.8, 5.1.3.3.1.9, 5.1.3.5.2, A.5.1.1.3; *see also* Medical gas systems
Carbon monoxide monitors .... 5.1.14.4.7(2)
Cardiac defibrillators .... *see* Defibrillators
Carts, for cylinders 11.4.3.1
Casters 11.4.2.1, 14.3.6.2.1.2, 14.3.6.2.1.3, A.14.3.6.2.1.2
Category 3 drive gas systems 5.3.12.2.13, 5.3.12.2.14
   Definition 3.3.20
Category 1 emergency management requirements .... 12.3, 12.5, A.12.5.3.1.2 to A.12.5.3.5
Category 2 emergency management requirements .... 12.3, 12.5, A.12.5.3.1.2 to A.12.5.3.5
Category 1 information technology and communications systems .... 7.3, A.7.3.1.2 to A.7.3.3.1.4.1
   Infrastructure 7.3.1, A.7.3.1.2
      Cabling pathways and raceway requirements 7.3.1.2.4
      Entrance facility (EF) .... 7.3.1.2.1, A.7.3.1.2.1.3(B) to A.7.3.1.2.1.8(B)
      Premises distribution system (fiber and copper) 7.3.1.1
      Remote primary data center .... 7.3.1.2.1.3, A.7.3.1.2.1.3(B)
      Telecommunications equipment room (TER) .... 7.3.1.2.1.1, 7.3.1.2.2, A.7.3.1.2.2.2, A.7.3.1.2.2.4(E)
      Telecommunications room (TR) ..... 7.3.1.2.3, A.7.3.1.2.3.4(D)
      Telecommunications systems spaces and pathways 7.3.1.2, A.7.3.1.2
   Nurse call systems .... 7.3.3.1, A.7.3.3.1.1 to A.7.3.3.1.4.1
Category 2 information technology and communications systems 7.4
   Infrastructure 7.4.1
   Nurse call systems 7.4.3.1

Category 3 information technology and communications systems 7.5
Category 1 patient care spaces .... *see* Patient care spaces, Category 1
Category 2 patient care spaces ...... *see* Patient care spaces, Category 2
Category 3 patient care spaces 6.3.2.2.10.4
   Definition 3.3.127.3, A.3.3.127.3
Category 4 patient care spaces 6.3.2.2.10.4
   Definition 3.3.127.4, A.3.3.127.4
Category 1 piped gas and vacuum systems .... 5.1, A.5.1.1.1 to A.5.1.14.2.3, B.5.1, B.5.2; *see also* Central supply systems; Medical-surgical vacuum systems; Waste anesthetic gas disposal (WAGD) systems
   Alarms .... *see* Alarm systems
   Applicability of requirements 5.1.1, A.5.1.1
   Cross-connections 5.1.12.2.4, 5.1.12.3.3, B.5.1
   Designation colors/operating pressures .... Table 5.1.11
   Distribution .... 5.1.10, 5.1.13.9, A.5.1.10.1.4 to A.5.1.10.8(3)
   Hazards, nature of 5.1.2, 5.1.13.2
   Labeling and identification 5.1.11, 5.1.12.3.13, 5.1.13.10, 5.1.14.3, A.5.1.11
   Maintenance ..... 5.1.14.2, 5.1.14.4, A.5.1.14.2.1 to 5.1.14.2.3, B.5.2
   Manufactured assemblies .... *see* Manufactured assemblies
   Operation and management 5.1.14, A.5.1.14
   Performance criteria and tests 5.1.6.1, 5.1.6.2, 5.1.12, 5.1.13.11, A.5.1.12, B.5.1
   Pressure and vacuum indicators 5.1.8, 5.1.13.7, A.5.1.8.1.3, B.5.2.5, B.5.2.14
   Reserve supply .... 5.1.9.2.4, B.5.2.10 to B.5.2.13
   Sources .... 5.1.3, A.5.1.3, B.5.2.1 to B.5.2.4, B.5.2.6 to B.5.2.8; *see also* Central supply systems
      Auxiliary source connection 5.1.3.5.7
      Instrument air supply systems .... *see* Instrument air, supply systems
      Medical air supply systems .... 5.1.3.6, A.5.1.3.6, B.5.2.6 to B.5.2.8
      Medical-surgical vacuum supply systems ...... 5.1.3.5(5), 5.1.3.7, 5.1.12.3.14.5, A.5.1.3.7
      Reserve supply .... *see* Reserve supply
      Support gases 5.1.13.3, A.5.1.13.3.5
      Verification of equipment 5.1.12.3.14, 5.1.14.4.6
      Waste anesthetic gas disposal (WAGD) 5.1.3.8, A.5.1.3.8
   Station inlets .... *see* Station inlets
   Station outlets .... *see* Station outlets
   Support gases .... 5.1.13, 5.1.14.2, A.5.1.13.1, A.5.1.14.2.1 to 5.1.14.2.3
      Breaching or penetrating of medical gas piping 5.1.10.11.12.1
      Central supply system for 5.1.3.5.3
      Joints 5.1.10.3.1, A.5.1.10.3.1
   Surface-mounted medical gas rails 5.1.7, A.5.1.7
   Tests
      Maintenance program .... 5.1.14.2.3, 5.1.14.4.1, 5.1.14.4.5 to 5.1.14.4.7, A.5.1.14.2.3, B.5.2
      Performance .... *see subhead:* Performance criteria and tests
      Retests B.5.2
   Valves .... *see* Valves
Category 2 piped gas and vacuum systems .... 5.2, A.5.2.1, A.5.2.14; *see also* Central supply systems
   Alarms 5.2.9
   Applicability of requirements 5.2.1, A.5.2.1
   Distribution 5.2.10
   Hazards, nature of 5.2.2
   Labeling and identification 5.2.3.1, 5.2.11
   Manufactured assemblies 5.2.6
   Operation and management 5.2.3.2, 5.2.14, A.5.2.14
   Performance criteria and tests 5.2.12
   Pressure and vacuum indicators 5.2.8
   Sources 5.2.3
   Station outlets/inlets 5.2.5
   Support gases 5.2.13
   Surface-mounted medical gas rails 5.2.7
   Valves 5.2.4

Category 3 piped gas and vacuum systems .... 5.3, A.5.3; *see also* Category 3 drive gas systems; Category 3 vacuum systems; Central supply systems; Scavenging
  Alarms 5.3.9
    Definition 3.3.3.2
  Applicability of requirements 5.3.1, A.5.3.1
  Dental air supply systems 5.3.3.6, A.5.3.3.6
  Dental vacuum supply systems .... 5.3.3.10, A.5.3.3.10.1.3 to A.5.3.3.10.1.4(8)
  Distribution .... 5.3.10, 5.3.12.1.3, 5.3.12.1.4, 5.3.12.2, A.5.3.12.2.4(6)
  Gas-powered devices .... *see* Category 3 drive gas systems
  Hazards, nature of 5.3.2
  Labeling and identification .... 5.3.3.1, 5.3.4.1(3), 5.3.11
  Manufactured assemblies 5.3.6
  Medical-surgical vacuum systems 5.3.3.9
  Operations and management 5.3.3.2
  Performance criteria and tests .... 5.3.12, A.5.3.12.2.4(6)
  Pressure and vacuum indicators .... 5.3.8, 5.3.9(3)
  Scavenging systems .... *see* Scavenging
  Sources .... 5.3.3, A.5.3.3.6 to A.5.3.3.10.1.4(8)
    Drive gas systems .... *see* Category 3 drive gas systems
    Medical air supply systems 5.3.3.5
    Tests 5.3.12.2.16
  Station inlets 5.3.5, A.5.3.5
  Station outlets 5.3.5, A.5.3.5
  Support gases .... 5.3.12, A.5.3.12.2.4(6)
  Surface-mounted medical gas rail systems 5.3.7
  Tests, performance .... 5.3.12, A.5.3.12.2.4(6)
  Valves .... *see* Valves
  Waste anesthetic gas disposal (WAGD) systems .... 5.3.3.11, 5.3.12, A.5.3.12.2.4(6)

Category 3 vacuum systems .... 5.3.1.5; *see also* Category 3 piped gas and vacuum systems
  Definition 3.3.21
  Drains 5.3.3.10.1.3, A.5.3.3.10.1.3
  Exhaust systems .... 5.3.3.10.1.4, A.5.3.3.10.1.4(8)
  Performance criteria and tests .... 5.3.12, A.5.3.12.2.4(6)
  Source equipment 5.3.3.3, 5.3.3.3.1, 5.3.3.4, 5.3.3.9, 5.3.3.10

Central supply systems
  Category 1
    Components .... 5.1.3.5, A.5.1.3.5, B.5.2.1 to B.5.2.4
    Control equipment 5.1.3.4
    Headers .... *see* Headers
    Labeling and identification 5.1.3.1, A.5.1.3.1.1, A.5.1.3.1.2
    Locations 5.1.3.3, A.5.1.3.3
    Operations 5.1.3.2
    Reserve supply .... *see* Reserve supply
  Category 2 .... 5.2.3.1 to 5.2.3.4
    Labeling and identification 5.2.3.1
    Locations 5.2.3.3
    Operations 5.2.3.2
  Category 3 .... 5.3.3.4
    Labeling and identification 5.3.3.1
    Locations 5.3.3.3
    Operations 5.3.3.2
  Hyperbaric oxygen systems 14.2.1.4.4.2, 14.2.1.4.4.4, 14.2.1.4.4.7

Chamber atmosphere .... *see* Atmosphere
Check valves .... 5.1.3.5.15.2(4), 5.1.4.9, 5.3.3.6.1.1(11), 5.3.3.6.2.3(4), 5.3.3.8.1(4)
Chutes, waste/linen 15.6
Circuit breakers 6.4.4.1.2.1, A.6.4.4.1.2.1
Circuits
  Branch .... *see* Branch circuits
  Maintenance and testing of 6.4.4.1.2, 6.5.4.1.2, 14.3.5.1.1, A.6.4.4.1.2.1
  Patient care spaces .... 6.3.2.2, A.6.3.2.2.1 to A.6.3.2.2.8.7
  Restoration of .... 6.4.1.1.1.2(2)
  Telecommunications facility 7.3.1.2.1.7, 7.3.1.2.2.7, 7.3.1.2.3.7

Cleaning, hyperbaric facilities 14.3.6.4, A.14.3.6.4
Clinical support activities, emergency plan for 12.5.3.3.6.4
Closets, sprinklers in 15.8.1.4, A.15.8.1.4
Clothing in hyperbaric chambers 14.3.1.5.5.1, B.14.1.1.3.4
Code
  Adoption
    Requirements 1.6
    Sample ordinance .... Annex C
  Definition 3.2.3, A.3.2.3
  Enforcement 1.6.2
Code calls 7.3.3.1.4, A.7.3.3.1.4
Combustible (definition) 3.3.22
Combustible liquids
  Definition 3.3.23, A.3.3.23
  In hyperbaric chambers .... 14.3.1.5.2, A.14.3.1.5.2.2(2)
  Use of 11.5.1.2.3
Combustion (definition) 3.3.24, A.3.3.24
Communications .... *see* Information technology and communications systems
Compact storage
  Definition 3.3.25
  Fire protection 15.10, 15.11, A.15.10, A.15.11.3
  Mobile 15.11, A.15.11.3
Compartmentation of facility 15.2
Compressed air starting devices 6.4.1.1.15
Conductive accessories 14.3.6.2.2
Conductive surfaces, exposed
  Definition 3.3.47
  Tests 6.3.3.1.1.2, 6.3.3.1.1.3
Conductors
  Isolated ungrounded systems 6.3.2.6.4
  Isolation of neutral 6.4.2.1.5.14, A.6.4.2.1.5.14
Connections
  Exterior 12.5.3.3.6.6
  Future, valves for 5.1.4.8
Connectors, flexible .... *see* Flexible connectors
Construction
  Central supply systems and storage 5.1.3.3.2, A.5.1.3.3.2
  Health care facility 15.2
  Hyperbaric facilities .... 14.2, A.14.2.1.1.1 to A.14.2.10.2
Containers .... *see also* Cylinders
  Connections 11.6.3
  Definition 3.3.26
  Filling 11.7.3.5.3, 11.7.3.6, A.11.7.3.6.1.1
  Handling .... 5.1.3.2.2, 5.1.3.2.12, 5.1.3.2.13, 5.1.3.3.1.8, 5.1.3.3.1.9, 11.4.2.3, 11.5.2, A.11.5.2.1.1 to A.11.5.2.4
  Labeling 5.1.3.1, A.5.1.3.1.1, A.5.1.3.1.2
  Liquid oxygen ambulatory containers (definition) 3.3.26.1
  Liquid oxygen base reservoir containers .... 11.7, A.11.7.3.1 to A.11.7.3.6.1.1
    Definition 3.3.26.2
  Liquid oxygen portable containers .... 11.7, A.11.7.3.1 to A.11.7.3.6.1.1, A.11.7.3.5
    Definition 3.3.26.3
  Safety mechanisms 11.6.4
  Source 11.2, A.11.2.8
  Storage 5.1.3.2.3, 5.1.3.2.7, 5.1.3.2.11, 11.3, 11.6.5, A.11.3.1, A.11.3.2
Controls
  Central supply system 5.1.3.4
  Electrical equipment .... 14.2.9.1.3(1)
Conveyors 15.5.3
Cooking operations 15.5.2.3
Cooling, essential electrical system generator sets 6.4.1.1.13, A.6.4.1.1.13
Cords .... *see* Power cords
Cover plates .... 6.4.2.2.6.2(C), 6.5.2.2.4.2, A.6.4.2.2.6.2(C), A.6.5.2.2.4.2
Cranking batteries 6.4.1.1.14
Critical branch ...... 6.4.1.1.3, 6.4.2.2.1.1(2), 6.4.2.2.4, 6.4.2.2.5.3(B), 7.3.1.2.3.7(B), 14.2.5.4.3, 14.2.8.2.2, A.6.4.2.2.4
  Definition 3.3.27
  Mechanical protection of 6.4.2.2.6.4

2015 Edition

Critical care area (definition) 3.3.28;
  see also Patient care spaces, Category 1
Critical equipment 6.4.1.1.7.5
  Definition 3.3.29
Cross-connections, tests 5.1.12.2.4, 5.1.12.3.3, 5.3.12.2.6,
  5.3.12.2.7, 5.3.12.2.13, B.5.1
Crowd control 13.8, A.13.8
Cryogenic liquids ...., 5.1.3.1.4 to 5.1.3.1.6, 5.1.3.2.7, 5.1.3.2.8,
  5.1.3.3.1.1, 5.1.3.3.1.2(2), 5.1.3.5(2), 5.1.3.5(3)
  Bulk systems ..... 5.1.3.3.1.10, 5.1.3.5.9.1(4), 5.1.3.5.13, 5.1.3.5.14,
    5.1.9.2.4, 5.1.14.4.2, 5.1.14.4.3, A.5.1.3.3.1.10,
    A.5.1.3.5.13.3, A.5.1.3.5.14, B.5.2.4, B.5.2.10 to B.5.2.13
  Manifolds, cylinder .... 5.1.3.5.9.1(3), 5.1.3.5.12, A.5.1.3.5.12,
    B.5.2.2, B.5.2.3
  Micro-bulk or small bulk systems 5.1.3.5.13, A.5.1.3.5.13.3
  Definition 3.3.19.4
  Storage or transfilling, ventilation for 9.3.6, A.9.3.6.3,
    A.9.3.6.5.1
Cylinders
  Connections .... 11.2.2 to 11.2.4, 11.2.9, 11.2.10, 11.6.3
  Definition 3.3.30
  Handling .... 5.1.3.2.2, 5.1.3.2.6 to 5.1.3.2.10, 5.1.3.2.12,
    5.1.3.2.13, 11.4.2.3, 11.5.2, 11.6, A.11.5.2.1.1
    to A.11.5.2.4
  Headers .... see Headers
  Labeling 5.1.3.1, A.5.1.3.1.1, A.5.1.3.1.2
  Management 11.6
  Manifolds for .... see Manifolds
  Regulators .... 11.2.5 to 11.2.7
  Reserves .... 5.1.9.2.4, B.5.2.10 to B.5.2.13
  Safety mechanisms 11.6.4
  Source 11.2, A.11.2.8
  Storage .... 5.1.3.2.3 to 5.1.3.2.5, 5.1.3.2.7, 5.1.3.2.9, 5.1.3.2.11,
    5.1.3.2.12, 5.1.3.3.4, 11.3, 11.6.5, A.11.3.1, A.11.3.2,
    B.11.4
  Supply systems .... 5.1.3.6.1(1), 5.3.3.6.2, 14.2.1.6.4.7(2)
  Transfilling .... 11.5.2.2 to 11.5.2.4, 11.7.3.6, A.11.5.2.2.2
    to A.11.5.2.4, A.11.7.3.6.1.1

-D-

Decompression sickness B.14.1.3.4
  Definition 3.3.31
Defend in place 15.7.4.3, 15.8.1.3, A.15.7.4.3.1, A.15.7.4.3.2,
  A.15.8.1.3
  Definition 3.3.32, A.3.3.32
Defibrillators 10.5.2.4, 10.5.4.6, A.10.5.4.6
Definitions .... Chap. 3
Delayed-automatic connection, equipment for 6.4.2.2.5.3,
  6.4.2.2.5.4, 6.5.2.2.3.3, 6.5.2.2.3.4, A.6.4.2.2.5.3,
  A.6.4.2.2.5.4, A.6.5.2.2.3.4
Deluge systems 14.2.5.2, 14.2.5.4.4, 14.2.5.5, A.14.2.5.2.4,
  A.14.2.5.2.6, A.14.2.5.5
Demand check 5.1.8.2.3, 5.1.8.2.4, B.5.2.5
  Definition 3.3.33
Dental air
  Distribution 5.3.10
  Performance criteria and tests .... 5.3.12, A.5.3.12.2.4(6)
Dental air compressors 5.3.3.3.3, 5.3.3.3.4, 5.3.3.6.1,
  5.3.3.6.1.3, A.5.3.3.6.1.5
Dental air supply systems 5.3.3.6, A.5.3.3.6
  Central supply systems .... 5.3.3.3.2 to 5.3.3.3.6, 5.3.3.4
  Compressor supply systems ... 5.3.3.6.1, A.5.3.3.6.1.3, A.5.3.3.6.1.5
  Cylinder supply systems 5.3.3.6.2
  Receivers 5.3.3.6.1.2
  Tests .... 5.3.12.2.8 to 5.3.12.2.11
Dental office (definition) 3.3.98, A.3.3.98
Dental vacuum systems 5.3.1.5
  Distribution 5.3.10
  Supply systems 5.3.3.3.1, 5.3.3.3.3, 5.3.3.4, 5.3.3.10,
    A.5.3.3.10.1.3 to A.5.3.3.10.1.4(8)
  Tests .... 5.3.12, A.5.3.12.2.4(6)

Detonation (definition) 3.3.34
Direct electrical pathway to heart
  Definition 3.3.35, A.3.3.35
  Protection of patients 10.5.2.2
Disasters (definition) 3.3.36, A.3.3.36;
  see also Emergency management
D.I.S.S. connectors .... 5.1.14.2.3.1(D)
  Definition 3.3.37
Distribution piping 5.1.13.9
Documentation
  Brazing procedure 5.1.10.11.10.5, 5.1.10.11.11.4
  Electrical equipment, records of .... 10.2.3.6(4), 10.5.2.6(2),
    10.5.3.1, 10.5.3.2, 10.5.6, 14.2.8.3.2, 14.2.8.3.17.2,
    A.10.2.3.6(4), A.10.5.6.2
  Emergency management .... 12.4.1, 12.5.3.3.2, 12.5.3.3.6.4(9),
    14.3.1.4.5, A.14.3.1.4.5
  HVAC 9.3.3.2, A.9.3.3.2
  Installers 5.1.10.1.6
  Liquid oxygen 11.7.2
  Maintenance 5.1.14.2.1, 5.1.14.4.2, 10.5.3.1, 10.5.3.2,
    14.3.4.2, A.5.1.14.2.1
  Manufacturers .... 5.1.6.3, 10.5.2.6(2), 10.5.3.1, 10.5.3.2
  Test procedures .... 5.1.12.1.1, 5.1.14.2.3.1(F), 6.3.4.1.2
  Test records and reports .... see Tests
Double-insulated appliances 10.2.2.1.2, 10.4.2.3
  Definition 3.3.38, A.3.3.38
Drains
  Category 3 vacuum equipment 5.3.3.10.1.3, A.5.3.3.10.1.3
  Hyperbaric chamber wiring enclosures 14.2.8.3.8
Drills
  Emergency plans and operations 12.5.3.3.8, A.12.5.3.3.8,
    B.12.1.1.7
  Fire 14.2.4.5.4, 14.3.1.4.5, 15.3.3.10.3, A.14.3.1.4.5
  Security management 13.12.1, 13.12.2, A.13.12.1

-E-

Egress 13.6, A.13.6.3.1
Electrical equipment .... 5.1.3.3.2(7), 5.1.3.3.2(10), Chap. 10,
  A.5.1.3.3.2(7), A.5.1.3.3.2(10); see also
  Patient-care-related electrical equipment
  Administration .... 10.5, A.10.5.2.5 to A.10.5.8.3
  Cord- and plug-connected (portable) 10.2.2, 10.4.2,
    14.2.8.3.17.6
  Fire protection 15.5.1.2
  Hazards 6.2, A.6.2.1
  Hyperbaric facilities 14.2.9.1.2, 14.3.5
  Instructions for 10.5.3.1, 10.5.3.2, 10.5.6.1
  Laboratory 10.5.5, A.10.5.5.1
  Life support 6.3.2.5.1
  Definition 3.3.39
  Maintenance 10.5.3
  Nonpatient 10.4
  Not supplied by facility 10.5.2.7
  Performance criteria .... 10.2, A.10.2.3.2.4 to A.10.2.6
  Permanently connected (fixed) 10.2.1
  Portable 10.2.2, 10.2.6, 10.3, 10.4.2, 14.2.8.3.17,
    A.10.2.6, A.10.3.1 to A.10.3.6, A.14.2.8.3.17
  Qualification and training of personnel 10.5.8, A.10.5.8.1,
    A.10.5.8.3
  Recordkeeping 10.5.6, A.10.5.6.2
  Respiratory therapy apparatus .... B.11.3.4.1.1 to B.11.3.4.2
  System demonstration 10.5.2.5, A.10.5.2.5
  Testing .... 10.3, 10.5.2.1.2, A.10.3.1 to A.10.3.6
Electrical equipment branch .... 6.4.2.2.1.1(3), 6.4.2.2.5,
  A.6.4.2.2.5.3 to A.6.4.2.2.5.4(9)
Electrical hazards 6.2, A.6.2.1
Electrical receptacles .... see Receptacles
Electrical systems .... 1.1.4, Chap. 6;
  see also Essential electrical systems
  Abnormal currents .... 6.4.1.1.1.2(1), 6.4.1.1.2
  Administration of 6.3.4, A.6.3.4

Design considerations 6.4.1.1.1, A.6.4.1.1.1
Distribution .... 6.3.2, A.6.3.2.1.1 to A.6.3.2.6.3.4
    Ground-fault protection 6.3.2.5
    Installation 6.3.2.1, A.6.3.2.1.1
    Isolated power systems .... see Isolated power systems
    Laboratories 6.3.2.3
    Patient care spaces .... 6.3.2.2, A.6.3.2.2.1 to A.6.3.2.2.8.7
Emergency plan .... 12.5.3.3.6.5(1), A.12.5.3.3.6.5
Equipment branch .... 6.4.2.2.1.1(3), 6.4.2.2.5, A.6.4.2.2.5.3
    to A.6.4.2.2.5.4(9)
Fire protection .... 15.5.1.2 to 15.5.1.4
Hyperbaric facilities 14.2.9.1.2
Information technology and communications
    systems 7.3.1.2.1.7, 7.3.1.2.2.7, 7.3.1.2.3.7, 7.5.1.1.2
Instrument air supply systems 5.1.13.3.5.13
Isolated power systems .... see Isolated power systems
Laboratories 6.3.2.3
Maintenance 6.3.4.1
Medical air systems 5.1.3.6.3.10
Medical-surgical vacuum supply systems 5.1.3.7.5
Patient care spaces .... 6.3.2.2, A.6.3.2.2.1 to A.6.3.2.2.8.7
    Grounding systems .... 6.3.3.1, A.6.3.3.1.1 to A.6.3.3.1.4
    Instructions for appliances 10.5.3.1, 10.5.3.2, 10.5.6.1
Performance criteria and tests ...... 6.3.3, A.6.3.3.1.1 to A.6.3.3.1.4
    Grounding systems in patient care rooms 6.3.3.1, A.6.3.3.1.1
        to A.6.3.3.1.4
    Isolated power systems 6.3.3.3
    Receptacle testing in patient care rooms 6.3.3.2
Sources of power 6.3.1, 6.5.1
    Alternate .... see Alternate power sources
    Failure of .... 6.4.3.2.2 to 6.4.3.2.8, 6.5.3.2.2 to 6.5.3.2.7
    Performance criteria and tests 6.4.3.1, 6.5.3.1
    Power loss, emergency management of B.12.3.2.5
    Transfer to .... see Transfer switches
    Type 1 essential electrical systems 6.4.1, A.6.4.1.1.1
        to A.6.4.1.1.18.1
Tests .... 6.3.3, 6.3.4, A.6.3.3.1.1 to A.6.3.3.1.4, A.6.3.4
Waste anesthetic gas disposal (WAGD) systems 5.1.3.8.4
Electrical wiring B.6.1
    Essential electrical systems 6.4.2.2.6, 6.5.2.2.4, A.6.4.2.2.6.1,
        A.6.4.2.2.6.2(C), A.6.5.2.2.4.1, A.6.5.2.2.4.2
    Fire protection 15.5.1.2
    Hyperbaric facilities .... 14.2.8.3 to 14.2.8.6, A.14.2.8.3
        to A.14.2.8.6.1.1
    Isolated power systems 6.3.2.6.2, A.6.3.2.6.2.1
    Low-voltage 6.3.2.2.5
    Patient care spaces 6.3.2.2.5
Electrocautery equipment .... 14.3.2.1.2(2), 15.13.3
Electrode 5.1.14.1.2, 6.3.2.6.3.7, 14.2.8.4.1
    Definition 3.3.40
Electrostatic safeguards, hyperbaric chambers 14.3.6, A.14.3.6
Electrosurgery 15.13.3
Elevators .... 15.5.3, A.15.5.3.3, B.12.3.2.5(2)
    Alarm systems 15.7.4.2.1, A.15.7.4.2.1
    Essential electrical systems .... 6.4.2.2.3.2(5), 6.4.2.2.5.4(3),
        6.5.2.2.2.1(7), 7.3.1.2.2.7(B)
    Temporary operation of .... 6.5.2.2.3.4(B), A.6.5.2.2.3.4(B)
Emergency care, policies for 1.3.3
Emergency generators 15.5.1.3;
    see also Generator sets
Emergency management .... 1.1.10, Chap. 12, A.1.1.10;
    see also Emergency procedures
    Administration 12.5.3.6
    Category 1 and Category 2 requirements ... 12.3, 12.5, A.12.5.3.1.2
        to A.12.5.3.5, A.12.5.3.4.12.1
    Committee for 12.2.3, A.12.2.3, B.12.1.1.3, B.12.1.1.6,
        B.12.3.3.1
    Continuing operations plan and recovery B.12.4
    Definition 3.3.41
    General requirements 12.4
    Medical air, loss of .... 5.2.3.5(2)
    Mitigation 12.5.3.2, B.12.1.1.2

    Preparedness .... 12.5.3.3, A.12.5.3.3.6.1(5) to A.12.5.3.3.9.8,
        B.12.1.1.2
    Program development B.12.1
    Responsibilities .... 12.2, A.12.2.1 to A.12.2.3.3
        Authority having jurisdiction 12.2.1, A.12.2.1
        Emergency management committee 12.2.3, A.12.2.3,
            B.12.1.1.3, B.12.1.1.6, B.12.3.3.1
        Senior management 12.2.2
    Safety and security 12.5.3.3.6.3, A.12.5.3.3.6.3
    Special considerations and protocols B.12.3
    Staff education 12.5.3.3.7
    Surge capacity of victims 12.5.3.4.12, A.12.5.3.4.12.1
    Waste anesthetic gas disposal (WAGD), loss of .... 5.2.3.7(2)
Emergency operations plan 12.2.3.3, 12.4.1, 12.5.3.3.5, 12.5.3.3.8,
    A.12.2.3.3, A.12.5.3.3.8, B.12.1.1.4 to B.12.1.1.7
Emergency oxygen supply connection (EOSC) 5.1.3.5.15,
    A.5.1.3.5.15
    Definition 3.3.42
Emergency power system (EPS) room HVAC systems 9.3.9
Emergency procedures .... see also Emergency management
    Hyperbaric facilities .... 14.3.1.4.3 to 14.3.1.4.6, A.14.3.1.4.4.1,
        A.14.3.1.4.5, B.14.2, B.14.3
    Operating room 15.13.3.9
Emergency recovery 12.5.3.5, A.12.5.3.5
Emergency response .... 12.5.3.4, A.12.5.3.4.1 to A.12.5.3.4.12.1,
    B.12.3
Emergency systems .... see Critical branch;
    Equipment branch; Life safety branch
Employees
    Emergency roles 12.5.3.3.6.7, 12.5.3.3.7, B.12.2, B.12.3.2.8
    Gas equipment cylinders and containers, handling of 11.5.2.1,
        A.11.5.2.1.1
    Gas equipment servicing 11.5.3.3.5, 11.6.1.2
    Hyperbaric chambers 14.3.1.3, 14.3.1.5.3, 14.3.1.5.5,
        A.14.3.1.3.2 to A.14.3.1.3.5, A.14.3.1.5.3, B.14.1.1.3.3 to
        B.14.1.1.3.5, B.14.1.2.2, B.14.1.3, B.14.3
    Medical gas and vacuum systems 5.1.10.11.10, 5.1.10.11.11,
        5.1.11.2, 5.1.14.2.2.5, 5.1.14.4.4, 5.3.12.1.2,
        A.5.1.12.2.3.5 to A.5.1.12.2.7.5, B.5.1
    Operating rooms 15.13.3.10
    Respiratory therapy B.11.1.1.1, B.11.1.1.2, B.11.3.5, B.11.3.6
    Security management .... 13.4, 13.10, 13.11, A.13.4.2(3)(c),
        A.13.10, A.13.11
    Training .... see Training, employee
Enclosures, gas system 5.3.3.3.2, 11.3.2.1, 11.4.2.2
Endotracheal tube B.14.1.2.4
    Definition 3.3.165.1, A.3.3.165.1
Equipment .... see also Electrical equipment
    Critical 6.4.1.1.7.5
        Definition 3.3.29
    Hyperbaric .... 14.2, 14.3.2, A.14.2.1.1.1 to A.14.2.10.2,
        A.14.3.2.1.6 to A.14.3.2.6
    Portable
        Electrical .... see Electrical equipment
        Patient care gas equipment .... 11.4.1, A.11.4.1.1 to A.11.4.1.4
    Rack or cart 11.4.3.1
Equipment branch .... 6.5.2.2.1.2(1), 6.5.2.2.3, 7.3.1.2.2.8(B),
    A.6.5.2.2.3.4
    Definition 3.3.43
Equipment grounding bus 10.2.1
    Definition 3.3.44
Equivalency to standard 1.4
Escalators 15.5.3
Essential electrical systems .... 5.1.3.3.2(10), A.5.1.3.3.2(10),
    B.6.2, B.6.3
    Definition 3.3.45, A.3.3.45
    Fuel cell systems 6.4.1.1.7, A.6.4.1.1.7.3
    Information technology and communications
        systems .... see Information technology and
        communications systems
    Patient care spaces 6.3.2.2.10
    Sources 6.5.1

Type 1 .... 6.4, A.6.4.1.1.1 to A.6.4.4.1.2.1
   Administration .... 6.4.4, A.6.4.4.1.1.4(A), A.6.4.4.1.2.1
   Branches .... 6.4.2.2, A.6.4.2.2.1 to A.6.4.2.2.6.2(C)
   Distribution 6.4.2, A.6.4.2
   Fuel cell systems 6.4.1.1.7, A.6.4.1.1.7.3
   Generator sets .... 6.4.1.1, 6.4.4.1.1, A.6.4.1.1.1 to A.6.4.1.1.18.1, A.6.4.4.1.1.4(A)
   Maintenance and testing .... 6.4.4.1, A.6.4.4.1.1.4(A), A.6.4.4.1.2.1
   Performance criteria and tests 6.4.3
   Sources .... 6.4.1, A.6.4.1.1.1 to A.6.4.1.1.18.1
   Uses 6.4.1.1.8, A.6.4.1.1.8.3
Type 2 .... 6.4.1.1.6.1, 6.5, A.6.5.2.1.1 to A.6.5.2.2.4.2
   Administration 6.5.4
   Distribution .... 6.5.2, A.6.5.2.1.1 to A.6.5.2.2.4.2
   Generator sets 6.5.4.1.1
   Maintenance and testing 6.5.4.1
   Performance criteria and tests 6.5.3
   Sources 6.5.1, 6.5.3.1
Type 3 .... 6.4.1.1.6.2
Evacuation, waste gas ....see Waste anesthetic gas disposal (WAGD) systems
Evacuation of occupants
   Emergency plan B.12.3.2.11, B.12.3.3
   Hyperbaric facilities 14.2.4.5.4, B.14.3.1
   Natural disasters, due to B.12.3.2.11
   Oxygen-enriched atmosphere (OEA) B.11.3.2.1
Exhaust systems .... 6.4.2.2.5.3(A)(5), 6.4.2.2.5.3(A)(6), 6.4.2.2.5.4(4); see also Ventilation
   Emergency systems .... 6.4.2.2.5.4(4)
   Hyperbaric facilities 14.2.10.2, A.14.2.10.2
   Medical gases storage or transfilling areas 9.3.6.5.3
   Vacuum or scavenging source .... 5.1.3.7.6, 5.3.3.10.1.4, A.5.3.3.10.1.4(8)
   Waste anesthetic gas disposal (WAGD) systems 5.1.3.8.5
Exposed conductive surfaces
   Definition 3.3.47
   Tests 6.3.3.1.1.2, 6.3.3.1.1.3
Extension cords .... 10.2.4, 10.5.2.3; see also Power cords
Exterior connections 12.5.3.3.6.6
Extinguishers, portable fire 14.2.5.1.5, 15.9.1, A.14.2.5.1.5, A.15.9.1, B.11.3.4.2, B.14.2.2, B.14.3.2
Extinguishing systems .... see Fire extinguishing systems

-F-

Facility fire plan .... B.11.3.2.1(2), B.12.3.1; see also Emergency operations plan
   Definition 3.3.48, A.3.3.48
Failure
   Definition 3.3.49, A.3.3.49
   Power sources .... 6.4.3.2.2 to 6.4.3.2.8, 6.5.3.2.2 to 6.5.3.2.7
Fault current 6.3.2.2.8.2, 6.3.2.6.3.1, 6.4.2.1.3, 14.2.8.3.7.4, A.6.3.2.6.3.1
   Definition 3.3.50
Fault hazard current .... 6.3.2.6.3.2 to 6.3.2.6.3.6, A.6.3.2.6.3.3, A.6.3.2.6.3.4
   Definition 3.3.66.1
Feeders
   From alternative source 6.4.2.2.2
   Circuit breakers, inspection of 6.4.4.1.2.1, A.6.4.4.1.2.1
   Definition 3.3.51
   Disconnecting means 6.3.2.5.2, 6.3.2.5.3
   Emergency power system .... 6.4.2.1.5.9(B)
   Insulation resistance 6.4.4.1.2.2
Filters, air
   Instrument air 5.1.13.3.5.9
   Medical air .... 5.1.3.6.3.8, 5.1.3.6.3.9(E), 5.1.3.6.3.9(F), 5.1.3.6.3.9(I), 5.2.3.5(1), A.5.1.3.6.3.9(F)
Final tie-in tests 5.1.12.3.9
Fire alarm systems .... see Alarm systems

Fire blankets 14.2.5.1.5, A.14.2.5.1.5
Fire dampers or shutters ..... 6.4.1.1.13.2(C), 6.4.1.1.13.5, 15.7.4.2.2, A.15.7.4.2.2
Fire detection systems .... 15.7, A.15.7.2.4 to A.15.7.4.3.2; see also Alarm systems; Smoke alarms/detectors
   Hyperbaric facilities 14.2.5.4, 14.2.7.2, 14.3.6.3, A.14.2.5.4.2
Fire extinguishing systems and equipment ..... 15.8 to 15.10, 15.11.3, 15.12, A.15.8.1.3 to A.15.10, A.15.11.3, B.12.3.2.7; see also Fire protection
   Hyperbaric chambers ..... 14.2.5, 14.2.6, A.14.2.5.1.4 to A.14.2.5.5, B.14.2.1
   Respiratory therapy apparatus B.11.3.4
   Telecommunications facilities .... 7.3.1.2.1.8(D), 7.3.1.2.3.8(B)
Fire fighters' emergency operations (elevator) 15.5.3.2
Fire hazards
   Electrical systems 6.2.1, A.6.2.1
   Gas systems 5.1.2, 5.3.2
   Hyperbaric facilities B.14.1.1
   Oxygen delivery equipment 11.5.1.1.4, 11.5.1.2.2
   Oxygen-enriched atmospheres (OEAs) 11.5.1.2.1
   Respiratory therapy 11.5.1.1, 11.5.1.2, A.11.5.1.1.2, A.11.5.1.1.3, B.11.3.1
Fire prevention
   Hyperbaric facilities .... 14.3.1.5, A.14.3.1.5.1.1 to A.14.3.1.5.4.3
   Operating rooms 15.13, A.15.13
Fire protection .... 12.5.3.3.6.5(5), Chap. 15, A.12.5.3.3.6.5
   Construction and compartmentation 15.2
   Hyperbaric facilities .... 14.2.5 to 14.2.7, 14.3.1.4.7, 14.3.6.3, A.14.2.5.1.4 to A.14.2.5.5, B.14
   Respiratory therapy B.11.3
Flammable (definition) 3.3.52, A.3.3.52
Flammable gases .... 14.3.1.5.2, A.14.3.1.5.2.2(2); see also Cylinders; Gases
   Definition .... 3.3.53; see also Cylinders
   Hyperbaric facilities 14.3.2.6.1, 14.3.3.3, 14.3.4.1.4.1
   Special hazard protection for 15.3, A.15.3.2
   Storage 14.3.3.3, 14.3.4.1.4.1
Flammable liquids 15.13.3
   Definition 3.3.54
   In hyperbaric chambers .... 14.3.1.5.2, A.14.3.1.5.2.2(2)
   Special hazard protection for 15.3, A.15.3.2
   Use of 11.5.1.2.3
Flammable substances and agents, misuse of 11.5.1.2
Flash point (definition) 3.3.55, A.3.3.55
Flexible connectors
   Category 1 .... 5.1.6.4, 5.1.6.7, 5.1.10.11.6, 5.1.14.2.3.1
   Category 3 .... 5.3.3.6.2.6, 5.3.3.6.2.7, 5.3.3.8.2
Flexible cords .... see Power cords
Flow-control valves 11.5.3.3.1
   Definition 3.3.56
Flowmeters
   Definition 3.3.57, A.3.3.57
   Labels on 11.5.3.1.3
Footwear, nails in 14.3.1.5.3.3
Frequency (definition) 3.3.58, A.3.3.58
Fuel supply or sources 6.4.1.1.16, 12.5.3.3.2, 12.5.3.3.6.5, A.12.5.3.3.6.5, B.14.1.1.3
Fume hoods .... 6.4.2.2.5.3(A)(6)
   Definition 3.3.59, A.3.3.59
Fundamentals .... Chap. 4
Furniture, hyperbaric facilities 14.3.6.2.1, A.14.3.6.2.1.2

-G-

Gas cylinders .... see Cylinders
Gas equipment .... Chap. 11
   Administration .... 11.5, A.11.5.1.1.2 to A.11.5.3.2
   Cylinders and containers 11.2, 11.6, A.11.2.8
   Fire protection 15.5.1.1
   Performance criteria and tests .... 11.4, A.11.4.1.1 to A.11.4.3.2
   Qualification and training of personnel 11.5.3.3.5
   Source 11.2, A.11.2.8
   Transport, storage, and use 11.5.3.3
   Use 11.5.3, A.11.5.3.2

Gases .... *see also* Waste anesthetic gas disposal
        (WAGD) systems; *specific gases*
   Flammable .... *see* Flammable gases
   Handling 14.3.3, A.14.3.3.4
   Medical .... *see* Medical gases
   Monitoring, hyperbaric chambers .... 14.2.9.3 to 14.2.9.8,
        A.14.2.9.6 to A.14.2.9.8
   Nonflammable 11.3.1, 11.3.2, A.11.3.1, A.11.3.2;
        see also Medical gases
   Oxidizing .... *see* Oxidizing gas
Gasket material 14.2.3.1.2
Gas-powered systems (definition) .... 3.3.60; *see also*
        Category 3 drive gas systems
Gas systems .... *see* Medical gas systems
Gas tungsten arc welding (GTAW) 5.1.10.5.1, A.5.1.10.5.1.5
General anesthesia and levels of sedation/analgesia
   Deep sedation/analgesia .... 5.1.1.2(1), 5.1.3.3.1.5(2),
        5.1.4.6.8, 5.1.9.4(1), 5.1.9.4.4(2), 6.3.2.2.11.1, 6.3.4.1.1,
        6.3.4.1.3, A.5.1.9.4.4(2)
      Definition 3.3.61.2
   Definition A.3.3.61
   General anesthesia .... 5.1.1.2(1), 5.1.3.3.1.5(2), 5.1.4.6.8,
        5.1.9.4(1), 5.1.9.4.4(2), 6.3.2.2.11.1, 6.3.4.1.1, 6.3.4.1.3,
        A.5.1.9.4.4(2)
      Definition 3.3.61.1
   Minimal sedation (anxiolysis) (definition) 3.3.61.4
   Moderate sedation/analgesia
      (conscious sedation) .... 5.1.3.3.1.5(2),
        5.1.4.6.8, 5.1.9.4(1), 5.1.9.4.4(2), A.5.1.9.4.4(2)
      Definition 3.3.61.3
Generator sets
   Alarm annunciator 6.4.1.1.18, A.6.4.1.1.18.1
   Design considerations 6.4.1.1.1, A.6.4.1.1.1
   Exclusive use for essential electrical systems 6.4.1.1.8.1
   Fire protection 15.5.1.3
   Fuel cell systems 6.4.1.1.7.5
   Fuel supply 6.4.1.1.3, 6.4.1.1.16
   Generator load shed circuits 6.4.1.1.3
   Hyperbaric chambers 14.2.8.2.1.2
   Maintenance and testing .... 6.4.4.1.1, 6.5.4.1.1, A.6.4.4.1.1.4(A)
   Power loss, emergency management of B.12.3.2.5
   Safety devices 6.4.1.1.17
   Type 1 essential electrical systems 6.4.1.1, 6.4.2, A.6.4.1.1.1
        to A.6.4.1.1.18.1, A.6.4.2
Germicides 15.13.3
Governing body ..... 6.3.2.2.8.4, 6.4.2.2.4.2(8), 8.2, 9.2, A.6.3.2.2.8.4,
        A.8.2.1, A.9.2.1
   Definition 3.3.62
Grease interceptors 8.3.7
Ground-fault circuit interrupters (GFCIs) 6.3.2.2.4.2, 6.3.2.2.8.7,
        6.3.2.2.8.8, A.6.3.2.2.8.7
   Definition 3.3.63, A.3.3.63
Ground-fault protection 6.3.2.2.4, 6.3.2.5, 6.3.3.4, 14.2.8.4,
        A.6.3.2.2.4.1
Grounding .... *see also* Grounding systems
   Appliances 10.2.2.1, 10.4.2.3
   Hyperbaric chambers 14.2.2.4.3, 14.2.8.4, 14.2.10.1
   Patient care spaces .... 6.3.2.2.2, 6.3.2.2.5(2), 6.3.2.2.7, 6.3.3.1,
        A.6.3.2.2.7.1, A.6.3.2.2.7.3, A.6.3.3.1.1 to A.6.3.3.1.4
   Piping system 5.1.14.1.2
   Special 6.3.2.2.7, A.6.3.2.2.7.1, A.6.3.2.2.7.3
Grounding circuit integrity 6.3.2.2.2.1
Grounding conductors 10.2.3.2, A.10.2.3.2.4
Grounding point
   Patient equipment 6.3.2.2.7.2
   Definition 3.3.129
   Reference .... *see* Reference grounding point
Grounding systems
   Definition 3.3.65, A.3.3.65
   Tests 6.3.3.1.1, A.6.3.3.1.1
Guide (definition) 3.2.4

-H-

Handlines 14.2.5.3, 14.2.5.5, A.14.2.5.5
Hand trucks 11.4.3.1
Hazard current
   Definition 3.3.66
   Fault hazard current .... *see* Fault hazard current
   Monitor 6.3.2.6.3.3, 6.3.2.6.3.4, A.6.3.2.6.3.3, A.6.3.2.6.3.4
      Definition 3.3.66.2
   Total 6.3.2.6.3.2, 6.3.2.6.3.4, 6.3.2.6.3.6, A.6.3.2.6.3.4
      Definition 3.3.66.3
Hazards
   Fire .... *see* Fire hazards
   Nature of
      Hyperbaric facilities 14.3.1.2, 14.3.1.5, A.14.3.1.2,
        A.14.3.1.5.1.1 to A.14.3.1.5.4.3, B.14.1
      Piped gas and vacuum systems 5.1.2, 5.2.2, 5.3.2
Hazard vulnerability analysis (HVA) 12.5.2, 12.5.3.1, A.12.5.3.1.2,
        A.12.5.3.1.3(1), B.12.1.1.1, B.12.3.2.1, B.12.3.3.1
Headers
   Central supply systems .... 5.1.3.5.13.2 to 5.1.3.5.13.5,
        A.5.1.3.5.13.3
   Instrument air .... 5.1.3.5.10, 5.1.3.5.11.4 to 5.1.3.5.11.6,
        5.1.3.5.12.2 to 5.1.3.5.12.9, 5.1.13.3.5.7, 5.1.13.3.5.11,
        5.1.13.3.5.12(B), A.5.1.3.5.10, B.5.2.1
   Local signals .... 5.1.3.5.9.1(6)
   Reserve 5.1.3.3.1.4, 5.1.3.3.4.2, 5.1.3.5.12.3, 5.1.3.5.12.4,
        5.1.3.5.12.9, 5.1.3.5.16.3(1), 5.1.9.2.4(4)
   Standby .... 5.1.3.3.1.2(4)
   Medical air .... 5.1.3.6.3.4(C), 14.2.1.6.4.7
Health care facilities (definition) 3.3.67, A.3.3.67
Heart, direct electrical pathway to
   Definition 3.3.35
   Protection of patients 10.5.2.2
Heating, adiabatic .... *see* Adiabatic heating
Heating, ventilating, and air conditioning (HVAC)
      systems .... Chap. 9
   Commissioning 9.3.3, A.9.3.3.2
   Ductwork 9.3.5
   Emergency plan ...... 12.5.3.3.6.5(4), A.12.5.3.3.6.5, B.12.3.2.5(2)
   Emergency power system room 9.3.9
   Energy conservation 9.3.2
   Fire protection 15.5.2, A.15.5.2.1
   Information technology and communications
      systems .... 7.3.1.2.1.8(B), 7.3.1.2.2.7(E), 7.3.1.2.2.8(B),
        7.4.1.1.1, 7.5.1.1.3, A.7.3.1.2.1.8(B)
   Medical plume evacuation 9.3.7
   Piping 9.3.4
   Smoke detectors 15.7.4.2.2, A.15.7.4.2.2
   System category criteria 9.2.1, A.9.2.1
   Waste gas 9.3.7
Heating equipment
   Essential electrical system generator sets 6.4.1.1.13,
        6.5.2.2.3.4(A), A.6.4.1.1.13, A.6.5.2.2.3.4(A)(1)
   Water heating 8.3.3, A.8.3.3
Helium 5.3.1.3;
      See also Medical gas systems
High-energy devices .... 14.3.2.1.2(3)
High pressure .... *see* Pressure
Home care 1.3.1
   Definition 3.3.68
Hoods
   Fume .... 6.4.2.2.5.3(A)(6)
      Definition 3.3.59, A.3.3.59
   Oxygen A.11.5.1.1.3
      Definition 3.3.122, A.3.3.122
Hose
   Gas and vacuum systems 5.1.6.4, 5.1.6.6, 5.1.6.7, 5.1.10.11.6
   Standpipe and hose systems 15.9.2, A.15.9.2.2
Hospitals
   Definition 3.3.69
   Governing body .... *see* Governing body

Humidification 11.4.2.4, 14.2.4.3.5
Humidifiers 11.4.2.5, 11.4.2.6
   Definition 3.3.70
Humidity control .... 6.4.1.1.13.8(4), 7.3.1.2.1.8(A), 7.3.1.2.2.8(A),
   7.3.1.2.3.8(A), 14.2.4.3, A.14.2.4.3.2
Hydrostatic tests 14.2.1.1.9.1
Hyperbaric facilities .... 1.1.12, Chap. 14, A.1.1.12
   Administration and maintenance 14.3, A.14.3.1.2
       to A.14.3.1.5.1.1
   Air supply, monitoring of 14.2.9.6, A.14.2.9.6
   Chambers, classification 14.1.2, A.14.1.2.2
   Communications and monitoring equipment 14.2.9,
       A.14.2.9.2 to A.14.2.9.8
   Construction and equipment 14.2, 14.3.2, A.14.2.1.1.1
       to A.14.2.10.2, A.14.3.2.1.6 to A.14.3.2.6
   Definition 3.3.71
   Electrical equipment 10.5.4, 14.3.5, A.10.5.4.5, A.10.5.4.6
   Electrical systems .... 6.4.2.2.5.4(5), 14.2.8, A.14.2.8.1.4.1
       to A.14.2.8.6.1.1
   Emergency depressurization capability 14.2.4.5, A.14.2.4.5.3
   Emergency procedures ..... 14.3.1.4.3 to 14.3.1.4.6, A.14.3.1.4.4.1,
       A.14.3.1.4.5, B.14.2, B.14.3
   Fabrication of chamber .... 14.2.2, A.14.2.2.1 to A.14.2.2.6
   Fire prevention .... 14.3.1.5, A.14.3.1.5.1.1 to A.14.3.1.5.4.3
   Fire protection .... 14.2.5 to 14.2.7, 14.3.6.3, A.14.2.5.1.4
       to A.14.2.5.5, B.14
   Gases
       Handling of .... 14.3.2.2 to 14.3.2.4, 14.3.3, A.14.3.2.2,
           A.14.3.3.4
       Monitoring of .... 14.2.9.3 to 14.2.9.8, A.14.2.9.6 to A.14.2.9.8
   Hazards, nature of 14.3.1.2, 14.3.1.5, A.14.3.1.2, A.14.3.1.5.1.1
       to A.14.3.1.5.4.3, B.14.1
   Housing for .... 14.2.1, A.14.2.1.1.6 to A.14.2.1.3.1
   Ignition sources 14.3.1.5.1, A.14.3.1.5.1.1
   Illumination 14.2.3
   Intercommunications equipment 14.2.9.2, A.14.2.9.2
   Other fixtures and equipment 14.2.10, 14.3.2, A.14.2.10.2,
       A.14.3.2.1.6 to A.14.3.2.6
   Personnel 14.3.1.3, 14.3.1.5.3, 14.3.1.5.5, A.14.3.1.3.2
       to A.14.3.1.3.5, A.14.3.1.5.3
   Rules and regulations 14.3.1.3.4, A.14.3.1.3.4
   Ventilation of chambers 14.2.4, 14.2.10.2, A.14.2.4.1.2
       to A.14.2.4.5.3, A.14.2.10.2
   Viewports 14.2.2.6, A.14.2.2.6, B.14.1.2.3
   Wiring and service equipment ...... 14.2.8.3 to 14.2.8.6, A.14.2.8.3
       to A.14.2.8.6.1.1
Hyperbaric stand-alone oxygen system 14.2.1.4.4.3, 14.2.1.4.4.6,
   14.2.1.6.4.3, 14.2.1.6.4.6
   Definition 3.3.72
Hypobaric facilities
   Definition 3.3.73
   Electrical systems .... 6.4.2.2.5.4(6)
Hypoxia (definition) 3.3.74, B.11.2.13

-I-

Identification .... *see* Labeling and identification
Ignition sources .... *see also* Fire hazards
   Hyperbaric facilities 14.3.1.5.1, A.14.3.1.5.1.1, B.14.1.1.4
   Respiratory therapy locations 11.5.1.1, A.11.5.1.1.2,
       A.11.5.1.1.3
Immediate restoration of service (definition) 3.3.75
Impedance
   Definition 3.3.76, A.3.3.76
   Of isolated wiring 6.3.2.6.2, A.6.3.2.6.2.1
   Limit 6.3.3.1.6.2
   Measuring 6.3.3.1.4, A.6.3.3.1.4
Incident command system (ICS) 12.2.3.3, 12.5.3.3.7.3,
   12.5.3.3.7.5, A.12.2.3.3, B.12.3.2.1
   Definition 3.3.77
Incinerators 15.6

Inert gas bulk systems 5.1.3.3.1.10, A.5.1.3.3.1.10
   Definition 3.3.19.1
Inert gases, distribution of A.5.1.1.3
Information technology and communications systems .... Chap. 7,
   15.7, A.15.7.2.4 to A.15.7.4.3.2; *see also* Alarm systems
   Category 1 .... *see* Category 1 information
       technology and communications systems
   Category 2 .... 7.4
       Infrastructure 7.4.1
       Nurse call systems 7.4.3.1
   Category 3 .... 7.5
   Circuits, mechanical protection of 6.4.2.2.6.6
   Emergency plan ...... 12.5.3.3.6.1, A.12.5.3.3.6.1(5), B.12.3.2.5(2),
       B.12.3.2.6
   Essential electrical systems .... 6.5.2.2.2.1(4), 7.3.1.2.2.7(B),
       7.3.1.2.2.7(C), 7.3.1.2.2.8(B), 7.4.1.1.1, A.6.5.2.2.2.1(4)
   Hyperbaric facilities .... 14.2.9, A.14.2.9.2 to A.14.2.9.8
   Life safety branch .... 6.4.2.2.3(2), A.6.4.2.2.3.2(3)
Inhalation anesthetics .... *see* Anesthetic apparatus;
       Anesthetics; Anesthetizing locations
Inlets; .... *see* Station inlets
In-line valves 5.1.4.7, 5.1.4.9
Inspections
   Category 1 piped gas and vacuum systems 5.1.14.2.2.2,
       5.1.14.2.2.3, 5.1.14.2.3, 5.1.14.4.4, A.5.1.14.2.2.2,
       A.5.1.14.2.3
   Circuit breakers 6.4.4.1.2.1, A.6.4.4.1.2.1
Instructions
   Category 1 piped gas and vacuum systems 5.1.3.5.1, 5.1.10.3.2,
       5.1.10.6.2, 5.1.10.7.2, 5.1.10.11.8, 5.1.12.3.14.3(A),
       5.1.12.3.14.4(B)
   Electrical equipment 10.5.3.1, 10.5.3.2, 10.5.6.1
   Liquid oxygen equipment 11.7.2
Instrument air .... *see also* Support gases
   Definition 3.3.78, A.3.3.78
   Supply systems 5.1.1.3, A.5.1.1.3;
       see also Medical gas systems
       Category 1 piped gas and vacuum systems .... 5.1.3.5(7),
           Table 5.1.11, 5.1.13.3.5, A.5.1.13.3.5
           Alarms 5.1.9.5, 5.1.13.3.5.12, A.5.1.9.5
           Headers .... 5.1.13.3.5.7, 5.1.13.3.5.11, 5.1.13.3.5.12(B)
       Category 2 piped gas and vacuum systems 5.2.3.8
       Category 3 piped gas and vacuum systems 5.3.3.7
Insulation resistance, circuitry 6.4.4.1.2.2
Integrity tests 10.3.1, A.10.3.1
Intentional expulsion, site of .... *see* Site of intentional expulsion
Intermittent positive-pressure breathing (IPPB) A.5.3.3.6
   Definition 3.3.79, B.11.2.15
Internal combustion engines 6.4.1.1.17.1
Intrinsically safe
   Circuits, hyperbaric chamber 14.2.5.1.3.1, 14.2.8.3.7.5
   Definition 3.3.80, A.3.3.80
Invasive procedure (definition) 3.3.81
Isolated ground receptacles 6.3.2.2.7.1, A.6.3.2.2.7.1
Isolated patient leads 10.5.2.2
   Definition 3.3.82
Isolated power systems 6.3.2.2.8.6, 6.3.2.2.8.7, 6.3.2.2.9, 6.3.2.6,
   A.6.3.2.2.8.7, A.6.3.2.6
   Definition 3.3.83, A.3.3.83
   Tests 6.3.3.3, 6.3.4.2.2
Isolation switch, bypass 6.4.2.1.8, A.6.4.2.1.8.3
Isolation transformers 6.3.2.6.1
   Definition 3.3.84

-J-

Joints 5.1.10.3, A.5.1.10.3.1
   Axially swaged .... 5.1.10.3.1(4), 5.1.10.7
   Brazed ..... 5.1.10.3.1(1), 5.1.10.4, 5.1.10.11.11, A.5.1.10.4.5
   Memory metal .... 5.1.10.3.1(3), 5.1.10.6
   Prohibited 5.1.10.10
   Special fittings 5.1.10.9
   Threaded .... 5.1.10.3.1(5), 5.1.10.8, A.5.1.10.8(3)
   Welded .... 5.1.10.3.1(2), 5.1.10.5, A.5.1.10.5.1.5

INDEX

-L-

Labeled (definition) 3.2.5
Labeling and identification
   Cylinders and containers 5.1.3.1, 11.3.4, A.5.1.3.1.1, A.5.1.3.1.2
   Gas equipment 11.5.3.1, 11.5.3.2, A.11.5.3.2
   Gas/vacuum systems 5.1.11, 5.1.14.3, 5.2.3.1, 5.2.11,
      5.3.3.1, 5.3.11, A.5.1.11
      Alarm panels 5.1.11.4, A.5.1.11.4.2
      Outlets and inlets 5.1.11.3
      Pipes 5.1.11.1
      Shutoff valves .... 5.1.11.2, 5.3.4.1(3), A.5.1.11.2.7
      Support gas 5.1.13.10
      Verification of 5.1.12.3.13
   Hyperbaric facilities 14.2.5.1.7, 14.2.7.1, 14.3.1.4.7
   Oxygen 11.5.3.1, 11.5.3.2, A.11.5.3.2
   Respiratory therapy 11.5.3.1, 11.5.3.2, A.11.5.3.2
   Valves 5.1.4.1.3
Laboratories 6.3.2.3
   Definition 3.3.85, A.3.3.85
   Electrical equipment and appliances 10.5.5, A.10.5.5.1
   Fire protection 15.4, 15.5.2.4
   Heating, ventilating, and air conditioning (HVAC)
      systems 9.3.1.2
   Ventilation 15.5.2.4
Lasers 14.3.2.1.4, 15.13.3
Leakage current .... see also Touch current
   Fixed equipment 10.2.5, 10.3.4
   Limits 10.3.3.4, 10.3.6, A.10.3.6
   Portable equipment 10.3.6, A.10.3.6
   Tests 10.3.3, 10.3.4, 10.3.6, A.10.3.3, A.10.3.6
Leak detectant 5.1.12.2.3.5, 5.1.12.3.9.2, A.5.1.12.2.3.5
   Definition 3.3.86
Leak tests, plastic vacuum piping systems 5.3.12.2.5
Life safety branch .... 6.4.1.1.3, 6.4.2.2.1.1(1), 6.4.2.2.3,
     6.5.2.2.1.2(1), 6.5.2.2.2, 7.3.1.2.2.7(B), A.6.4.2.2.3.2(3),
     A.6.5.2.2.2.1(4)
   Definition 3.3.87
   Mechanical protection of 6.4.2.2.6.4
Life support equipment, electrical 6.3.2.5.1, 6.4.1.1.7.5,
   B.12.3.2.5(1)
   Definition 3.3.39
Lighting B.6.3
   Hyperbaric chambers 14.2.8.3.15, A.14.2.8.3.15
   Photographic lighting equipment 14.3.2.1.3
Limited-combustible material 4.4.2, A.4.4.2
   Definition 3.3.88, A.3.3.88
Line isolation monitor .... 6.3.2.6.3, A.6.3.2.6.3.1 to A.6.3.2.6.3.4
   Definition 3.3.89
   Tests 6.3.3.3.2, 6.3.4.1.4
Linen chutes 15.6
Liquids (definition) .... 3.3.90, A.3.3.90; see also
   Combustible liquids; Flammable liquids; Oxygen
Listed (definition) 3.2.6, A.3.2.6
Local signal
   Bulk system operating status 5.1.3.5.14.4, A.5.1.3.5.14.4
   Definition 3.3.91, A.3.3.91
   Proportioning system for medical air .... 5.1.3.6.3.14(C)(9),
     A.5.1.3.6.3.14(C)(9)
Lubricants .... 14.3.2.3(4), 14.3.2.4, 14.3.6.2.1.4

-M-

mA (definition) 3.3.92
Main lines, piping
   Definition 3.3.133.2
   Valves 5.1.4.3, 5.1.11.2.4, A.5.1.4.3
Maintenance
   Essential electrical systems B.6.2
   Fire protection systems 15.12
   Piped gas and vacuum systems, Category 1 .... see Category 1
     piped gas and vacuum systems

Manifolds .... 5.1.3.3.1.1, 5.1.3.3.1.2, 5.1.3.5(1), 5.1.3.5(2),
    5.1.3.5.9.1(2), 5.1.3.5.11, 5.1.3.5.13.5, 5.1.9.2.4(1),
    5.3.3.6.2.3(2), 5.3.3.6.2.5, 5.3.3.8.1(2), 5.3.4.1(4),
    11.6.2, A.5.1.3.5.11
   Cryogenic liquid containers ...... 5.1.3.5.9.1(3), 5.1.3.5.12, 9.3.6.5,
     A.5.1.3.5.12, A.9.3.6.5.1, B.5.2.2, B.5.2.3
   Cylinders without reserve supply, for .... 5.1.3.5.9.1(1)
   Definition 3.3.93
   Ventilation of manifold areas 9.3.6.5, A.9.3.6.5.1
Manual extinguishing equipment 15.9, A.15.9.1, A.15.9.2.2
Manufactured assemblies
   Category 1 piped gas and vacuum systems 5.1.6, 5.1.13.6,
     5.1.14.2.3.1, A.5.1.6
   Category 2 piped gas and vacuum systems 5.2.6
   Category 3 piped gas and vacuum systems 5.3.6
   Definition 3.3.94, A.3.3.94
Manufacturers
   Documentation .... 5.1.6.3, 10.5.2.6(2), 10.5.3.1, 10.5.3.2
   Instructions .... see Instructions
Masks
   Definition 3.3.95
   Oxygen mask microphones 14.2.9.2.2
Measurement, units of 1.5, A.1.5
Media control 13.7, A.13.7
Medical air 5.1.1.3, A.5.1.1.3;
   see also Dental air; Medical gas systems
   Alarms 5.1.3.6.3.12
   Category 1 supply systems 5.1.3.5.2, 5.1.3.6, A.5.1.3.6, B.5.2.6
     to B.5.2.8
   Category 2 supply systems 5.2.3.5
   Definition 3.3.96, A.3.3.96
   Designation colors/operating pressures .... Table 5.1.11
   Dryers .... 5.1.3.6.3.3, 5.1.3.6.3.7, 5.1.3.6.3.9(E), 5.1.3.6.3.9(F),
     5.1.3.6.3.9(I), 5.2.3.5(1), A.5.1.3.6.3.9(F)
   Filters ..... 5.1.3.6.3.8, 5.1.3.6.3.9(E), 5.1.3.6.3.9(F), 5.1.3.6.3.9(I),
     A.5.1.3.6.3.9(F)
   Hyperbaric facilities 14.2.1.6
   Proportioning system for medical air (USP) .... 5.1.3.5(8),
     5.1.3.6.3.14, A.5.1.3.6.3.14(A)(9), A.5.1.3.6.3.14(C)(9),
     B.5.2.6, B.5.2.7
     Components .... 5.1.3.6.3.14(C), A.5.1.3.6.3.14(C)(9)
     Definition 3.3.96.1
     Location .... 5.1.3.6.3.14(B)
     Tests 5.1.12.3.14.4
   Purity test 5.1.12.3.12
   Quality monitoring 5.1.3.6.3.13
   Receivers .... 5.1.3.6.3.6, 5.1.3.6.3.9(D), A.5.1.3.6.3.9(D)
   Regulators .... 5.1.3.6.3.9(E), 5.1.3.6.3.9(F), 5.1.3.6.3.9(I),
     A.5.1.3.6.3.9(F)
Medical air compressors .... see also Dental air compressors
   Category 1 .... 5.1.3.5(4), 5.1.3.6.1(1), 5.1.3.6.3, 5.1.3.6.3.9(B),
     5.1.14.4.7(1), A.5.1.3.6.3, B.5.2.6, B.5.2.7
     Alarms 5.1.3.6.3.12, 5.1.9.5, A.5.1.9.5
     Electrical power and control 5.1.3.6.3.10
     Intake 5.1.3.6.3.11
     Tests 5.1.12.3.14.3
   Category 2 .... 5.2.3.5(1)
   Definition 3.3.97
   Electrical power and control .... 6.4.1.1.3, 6.4.2.2.5.3(A)(3)
   Intake 14.2.4.2.2
Medical/dental office (definition) 3.3.98, A.3.3.98
Medical devices not for patient care 11.4.3.2, A.11.4.3.2
Medical gases .... 5.1.1.3, A.5.1.1.3; see also specific gases
   Concentration test .... 5.1.12.3.11, A.5.1.12.3.11(3)
   Definition 3.3.99
   Patient medical gas; .... see Patient medical gas
   Permitted locations 5.1.3.5.2
Medical gas rails (MGRs) .... see Surface-mounted
   medical gas rail systems

Medical gas systems .... Chap. 5; *see also* Cylinders; Gas equipment; Medical gases; Medical-surgical vacuum systems; Supply systems; Waste anesthetic gas disposal (WAGD) systems
   Category 1 ....*see* Category 1 piped gas and vacuum systems
   Category 2 ....*see* Category 2 piped gas and vacuum systems
   Category 3 ....*see* Category 3 piped gas and vacuum systems
   Definition 3.3.100
   Distribution, piped ....*see* Piped distribution systems
   Emergency plan .... 12.5.3.3.6.5(8), A.12.5.3.3.6.5
   Manufactured assemblies to use with ....*see* Manufactured assemblies
   Storage or transfilling of gases, ventilation for 9.3.6, A.9.3.6.3, A.9.3.6.5.1
Medical plume evacuation 9.3.8
Medical-surgical vacuum (definition) 3.3.102
Medical-surgical vacuum systems .... Chap. 5
   Alarms 5.1.3.7.7
   Category 1 .... 5.1.3.5(5), 5.1.3.7, 5.1.10.3.1, 5.1.10.3.2, 5.1.12.3.14.5, 5.1.14.3, A.5.1.3.7, A.5.1.10.3.1
      Area alarms 5.1.9.4, A.5.1.9.4
      Conversion to gas system 5.1.10.11.9.2
      Designation colors .... Table 5.1.11
      Electrical power and control 5.1.3.7.5
      Maintenance .... 5.1.14.2, 5.1.14.4.7(3), A.5.1.14.2.1 to 5.1.14.2.3
      Operating pressures .... Table 5.1.11
      Piping .... 5.1.3.7.4, 5.1.10.2, 5.1.14.4.7(3)
      Shutoff valves 5.1.4.1
      Special precautions 5.1.14.1.3, 5.1.14.1.4, A.5.1.14.1.3, A.5.1.14.1.4
      Standard designation colors and operating pressures .... Table 5.1.11
      Station outlets/inlets 5.1.5, A.5.1.5, B.5.2.18
   Category 2 .... 5.2.3.6
   Category 3 .... 5.3.1.5, 5.3.3.9, 5.3.10, 5.3.12, A.5.3.12.2.4(6)
   Definition 3.3.103
   Emergency plan .... 12.5.3.3.6.5(8), A.12.5.3.3.6.5
   Exhaust 5.1.3.7.6
   Installers, qualification of 5.1.10.11.10, 5.1.10.11.11
   Labeling 5.1.14.3
   Manufactured assemblies to use with ....*see* Manufactured assemblies
   Tests .... 5.3.12, A.5.3.12.2.4(6); *see also* Cross-connections, tests
      Initial pressure tests 5.3.12.2.2
      Leak tests 5.3.12.2.5
      Source system 5.1.12.3.14.5
      Standing vacuum tests 5.1.12.2.7, 5.3.12.2.12, A.5.1.12.2.7.5
Micro-bulk cryogenic systems 5.1.3.5.13, A.5.1.3.5.13.3
   Definition 3.3.19.4
Microphones, oxygen mask 14.2.9.2.2
Moisture indicators .... 5.3.3.6.1.1(16), 5.3.3.6.1.3, A.5.3.3.6.1.3
Monitor hazard current ....*see* Hazard current
Motor load transfer 6.4.2.1.5.13
Multiple treatment facility (definition) 3.3.104
mV (definition) 3.3.105

-N-

Nasal cannula (definition) 3.3.106
Natural disasters 12.3.4, B.12.3.2, B.12.3.7
Nebulizers 11.4.2.5, 11.4.2.6
   Definition 3.3.107
Negative pressure ....*see* Pressure
Nitrogen .... 5.1.1.3, A.5.1.1.3; *see also* Medical gases; Medical gas systems; Support gases
   Category 1 medical air proportioning system .... 5.1.3.6.3.14, A.5.1.3.6.3.14(A)(9), A.5.1.3.6.3.14(C)(9), B.5.2.6, B.5.2.7
   Category 1 medical air supply systems .... 5.1.3.6.1(1)

   Category 3 piped gas and vacuum systems .... 5.3.3.8, 5.3.12.2.8 to 5.3.12.2.11, 5.3.12.2.15, 5.3.12.2.16, A.5.3.3.8
   Definition 3.3.109
   Joints, purging of 5.1.10.4.5, A.5.1.10.4.5.12
   NF (oil-free, dry) .... 5.3.12.2.8(2)
      Category 1 medical air supply systems .... 5.1.3.6.1(1)
      Category 3 piped gas and vacuum systems .... 5.3.3.8.1(1)
      Definition 3.3.109.1
      For purge 5.1.10.4.5.1
      For tests .... 5.1.12.2.1.2, 5.1.12.2.2, 5.1.12.3.1.2, 5.1.12.3.5.1(F), 5.1.12.3.7.5, 5.3.12.1.3, 5.3.12.2.13(7)
Nitrogen narcosis B.14.1.3.2
   Definition 3.3.110
Nitrous oxide
   Definition 3.3.111
   Handling 11.6.1.3
   Storage 11.3.2.2, 11.6.1.3
Nitrous oxide systems .... 5.1.1.3, A.5.1.1.3; *see also* Medical gas systems
   Bulk (definition) 3.3.19.2
   Designation colors/operating pressures .... Table 5.1.11
   Indoor locations for 5.1.3.3.1.5
   Supply systems .... 5.1.3.3.1.7, 5.1.3.5.2, 5.1.3.5.4(2)
Noncombustible material 4.4.1, A.4.4.1
   Definition 3.3.112
Nonflammable (definition) 3.3.113
Nonflammable anesthetic agent (definition) .... 3.3.114; *see also* Anesthetics
Nonflammable anesthetizing location
   ....*see* Anesthetizing locations
Nonflammable gases .... 11.3.1, 11.3.2, A.11.3.1, A.11.3.2; *see also* Medical gases
Nonflammable medical gas systems
   ....*see* Medical gas systems
Nonmedical compressed air 8.3.5
   Definition 3.3.116
Nurse call systems .... 6.4.2.2.4.2(5), 7.3.3.1, 7.4.3.1, 10.5.2.6, A.7.3.3.1.1 to A.7.3.3.1.4.1
Nursing homes (definition) 3.3.117

-O-

Occupant notification .... 15.7.4, A.15.7.4.2.1 to A.15.7.4.3.2
Odor, positive pressure outlets 5.1.12.3.6.4, A.5.1.12.3.6.4
Operating rooms
   Electrical receptacles for .... 6.3.2.2.6.2(C)
   Emergency procedures 15.13.3.9
   Fire loss prevention 15.13, A.15.13
   Hyperbaric chambers 14.2.4.3.3
   Personnel orientation and training 15.13.3.10
   Wet procedure locations 6.3.2.2.8.4, 6.3.2.2.8.7, 6.3.2.2.8.8, A.6.3.2.2.8.4, A.6.3.2.2.8.7
Operating supply (definition) 3.3.159.1
Outlets, electrical ....*see* Receptacles
Outlets, gas ....*see* Station outlets
Overcurrent protection 6.3.2.2.1.3, 6.4.2.1.2.1
Oxidizing gas .... 11.3.2.2, 11.3.2.3; *see also* Nitrous oxide; Oxygen
   Definition 3.3.118, A.3.3.118
Oxygen ....*see also* Medical air; Medical gases
   Ambulatory patients receiving 11.5.2.5
   Compatibility 14.3.2.3, 14.3.2.4
   Definition 3.3.119, A.3.3.119
   Designation colors/operating pressures .... Table 5.1.11
   Gaseous (definition) 3.3.119.1
   Handling .... 11.6.1.3, 11.6.2, 14.3.2.2 to 14.3.2.4, A.14.3.2.2
   Liquid 14.3.3.2
      Definition 3.3.119.2, A.3.3.119.2
      Equipment .... 11.7, A.11.7.3.1 to A.11.7.3.6.1.1
      Transferring 11.5.2.3, 11.7.3.5.3, 11.7.3.6, A.11.5.2.3.2, A.11.7.3.6.1.1
   Monitoring 14.2.9.4

Storage 5.1.3.1.9, 11.3.2.2, 11.6.1.3, 14.3.2.2, A.14.3.2.2
Therapy .... see Hyperbaric facilities
Toxicity (hyperbaric) B.14.1.3.2
  Definition 3.3.123, A.3.3.123
Transferring .... 11.5.2.2 to 11.5.2.4, 11.7.3.5.3, 11.7.3.6,
  A.11.5.2.2 to A.11.5.2.4, A.11.7.3.6.1.1
Oxygen cylinders .... see Cylinders
Oxygen delivery equipment .... 10.5.4.5, A.10.5.4.5; see also
  Cylinders; Pipe and piping, gas
  Definition 3.3.120, A.3.3.120
  Hazards 11.5.1.1.4, 11.5.1.2.2
  Maintenance of 11.5.1.3.2
Oxygen enclosures 11.4.2.2
Oxygen-enriched atmosphere (OEA) .... see also
  Hyperbaric facilities
  Definition 3.3.121
  Electrical appliances in 10.5.4, A.10.5.4.5, A.10.5.4.6
  Equipment transport, storage, and use 11.5.3.3
  Fire protection B.11.3
  Hazards of 11.5.1.2.1
  Labeling of equipment for use in 11.5.3.1.1
Oxygen hoods A.11.5.1.1.3
  Definition 3.3.122, A.3.3.122
Oxygen systems .... 5.1.1.3, A.5.1.1.3, B.5.3; see also Cylinders;
  Medical gas systems; Oxygen delivery equipment; Pipe
  and piping, gas
  Bulk (definition) 3.3.19.3, A.3.3.19.3;
    see also Cryogenic liquids, Bulk systems
  Hyperbaric facilities 14.2.1.4
  Indoor locations for 5.1.3.3.1.5
  Supply systems 5.1.3.3.1.6, 5.1.3.5.2, 5.1.3.5.4, A.5.1.3.5.4
    Closing off supply .... B.11.3.2.1(3), B.11.3.3, B.14.3.2(1)
    Emergency oxygen supply connection (EOSC) .... see
      Emergency oxygen supply connection (EOSC)
Oxygen USP
  Category 1 medical air proportioning system 5.1.3.6.3.14,
    A.5.1.3.6.3.14(A)(9), A.5.1.3.6.3.14(C)(9), B.5.2.6,
    B.5.2.7
  Category 1 medical air supply systems .... 5.1.3.6.1(1)
  Definition 3.3.124

-P-

Paper 14.3.2.1.6, 14.3.2.1.7, A.14.3.2.1.6
Partial pressure .... see Pressure
Particulate test, piping 5.1.12.3.7
Patient area call stations .... 7.3.3.1.2, A.7.3.3.1.2.1 to A.7.3.3.1.2.3
Patient bed locations
  Call station 7.3.3.1.2.1, 7.3.3.1.2.3, A.7.3.3.1.2.1, A.7.3.3.1.2.3
  Definition 3.3.125
  Electrical receptacles for .... 6.3.2.2.6.2(A), 6.3.2.2.6.2(B)
  Liquid oxygen storage 11.7.4
Patient-care-related electrical equipment 10.2, 10.3.2, 10.5.6,
  14.2.8.3.17, A.10.2.3.2.4 to A.10.2.6, A.10.3.2, A.10.5.6.2,
  A.14.2.8.3.17
  Definition 3.3.126
Patient care spaces .... 1.3.4, 6.5.2.2.3.4(A), A.6.5.2.2.3.4(A)(1)
  Category 1 .... 5.1.1.2(3), 6.3.2.2.1.2, 6.3.2.2.1.3
    Definition 3.3.127.1, A.3.3.127.1
    Electrical receptacles .... 6.3.2.2.6.2(B)
    Essential electrical systems 6.3.2.2.10.1, 6.3.2.2.10.3
    Generator load shed circuits 6.4.1.1.3
    Ground-fault protection 6.3.2.5.1
  Category 2 .... 6.3.2.2.1.3(B)
    Definition 3.3.127.2, A.3.3.127.2
    Electrical receptacles .... 6.3.2.2.6.2(A)
    Essential electrical systems 6.3.2.2.10.2, 6.3.2.2.10.3
  Category 3 .... 6.3.2.2.10.4
    Definition 3.3.127.3, A.3.3.127.3
  Category 4 .... 6.3.2.2.10.4
    Definition 3.3.127.4, A.3.3.127.4
  Definition 3.3.127, A.3.3.127

Electrical systems 6.3.2.2, 6.3.3.1, 6.3.3.2, 6.3.3.3.1, A.6.3.2.2.1
  to A.6.3.2.2.8.7, A.6.3.3.1.1 to A.6.3.3.1.4
Liquid oxygen storage 11.7.4
Sprinklers 15.8.1.4, A.15.8.1.4
Patient care vicinity
  Definition 3.3.128
  Electrical equipment, testing of 10.3.4.1
Patient equipment grounding point 6.3.2.2.7.2
  Definition 3.3.129
Patient leads 10.3.6, A.10.3.6;
    see also Electrical wiring
  Definition 3.3.130, A.3.3.130
  Isolated 10.5.2.2
  Definition 3.3.82
Patient medical gas .... see also Medical air;
  Nitrous oxide; Oxygen
  Definition 3.3.131
  Identification 5.1.11.1.1
  Permitted locations for 5.1.3.5.2
  Positive pressure piping 5.1.10.11.12
Performance tests
  Category 1 piped gas and vacuum systems 5.1.6.1, 5.1.6.2,
    5.1.12, 5.1.13.11, A.5.1.12, B.5.1
  Category 2 piped gas and vacuum systems 5.2.12
  Category 3 piped gas and vacuum systems 5.3.12,
    A.5.3.12.2.4(6)
Photographic lighting equipment 14.3.2.1.3
Physicians, discretionary use of nonconforming
  materials 14.3.1.5.4.4
Pin-index safety system 11.4.1.2, A.11.4.1.2
Pipe and piping, gas
  Branch (lateral) line .... 5.1.3.6.3.9(G), 5.1.4.5, 5.1.10.3.2,
    5.1.10.11.1.2, 5.1.10.11.1.3
    Definition 3.3.133.1
  Breaching or penetrating 5.1.10.11.12, 5.1.14.4.6
  Category 1 .... 5.1.3.6.3.9, A.5.1.3.6.3.9(D), A.5.1.3.6.3.9(F)
    Distribution .... 5.1.10, 5.1.13.9, A.5.1.10.1.4 to A.5.1.10.8(3)
    Supply .... 5.1.3.6.3.9, 5.1.3.7.4, A.5.1.3.6.3.9(D),
      A.5.1.3.6.3.9(F)
  Category 2 distribution piping 5.2.10
  Category 3 distribution piping 5.3.10
  Definitions 3.3.133
  Fire protection 15.5.1.1
  Ground, use as 5.1.14.1.2
  Hyperbaric facilities 14.2.1.3, 14.2.10.3, A.14.2.1.3.1
  Installation 5.1.10.11, 5.3.10.1
  Interconnections, prohibited 5.1.10.11.7
  Labeling 5.1.11.1
  Main line 5.1.4.3, 5.1.11.2.4, A.5.1.4.3
    Definition 3.3.133.2
  Manufacturer's instructions 5.1.10.11.8
  Special precautions .... 5.1.14.1, A.5.1.14.1.1 to A.5.1.14.1.4
  Support 5.1.10.11.4, 5.3.10.1.4
  Tests 5.3.12.1.3, 5.3.12.2.2
    Gas-powered devices .... 5.3.12.2.1, 5.3.12.2.2 to 5.3.12.2.4,
      A.5.3.12.2.4(6)
    Installer performed tests 5.1.12.2, 5.3.12.1.2, A.5.1.12.2.3.5
      to A.5.1.12.2.7.5, B.5.1
    System verification 5.1.12.3, 5.1.14.4.6, A.5.1.12.3.2
      to A.5.1.12.3.11(3)
  Underground piping outside of buildings 5.1.10.11.5,
    5.3.10.1.5
  Underground piping within buildings/within
    floor slabs 5.3.10.1.6, 5.3.10.1.7
Piped distribution systems .... 5.1.10, 5.2.10, 5.3.10, 5.3.12.2.1(1),
  A.5.1.10.1.4 to A.5.1.10.8(3); see also Central
  supply systems
  Definition 3.3.132
Piped gas systems .... see Medical gas systems
Piping, heating, cooling, ventilating, and process systems 9.3.4
Plugs (attachment plugs, caps) .... 10.2.2.2, 14.2.8.3.17.6(3)
  Definition 3.3.134

Plumbing .... Chap. 8
  Building system categories, criteria for 8.2, A.8.2.1
  Fixtures 8.3.8
  General requirements 8.3, A.8.3.3
  Grease interceptors 8.3.7
  Nonmedical compressed air 8.3.5
  Potable and nonpotable water 8.3.1, 8.3.2
  Special use water systems 8.3.6
  Wastewater .... 8.3.9 to 8.3.11
  Water conditioning 8.3.4
  Water heating 8.3.3, A.8.3.3
Portable equipment
  Electrical .... see Electrical equipment
  Patient care gas equipment 11.4.1, 11.4.2, A.11.4.1.1
      to A.11.4.1.4
Positive pressure
  Breaching or penetrating of medical gas piping 5.1.10.11.12
  Change in system 5.1.10.11.9.1
  Definition 3.3.135.7
  Indicators 5.1.8.1.3, A.5.1.8.1.3
  Joints for positive pressure gas systems 5.1.10.3.1, A.5.1.10.3.1
  Labeling of systems .... 5.1.11.1.1(3), 5.1.11.2.2
  Odor from outlets 5.1.12.3.6.4, A.5.1.12.3.6.4
  Storage of positive-pressure gases 5.1.3.3.2, 5.1.3.3.2
  Telecommunications facilities, for .... 7.3.1.2.1.8(C),
      7.3.1.2.2.8(C)
Power cords .... 10.2.3, 10.5.2.3, A.10.2.3.2.4 to A.10.2.3.6(4)
  Appliances 6.4.2.2.6.5
  Hyperbaric chambers 14.2.8.3.9, 14.2.8.3.17.6
  Patient care spaces 6.3.2.2.8.5
Power systems .... see Electrical systems
Pressure
  Absolute (definition) 3.3.135.1
  Ambient 14.2.4.5.1, 14.2.4.5.2, B.14.1.1.3.2
    Definition 3.3.135.2
  Gauge (definition) 3.3.135.3
  High
    Definition 3.3.135.4
    Intermixing of gases 11.2.8, A.11.2.8
    Pressure reducing regulators and gauges 11.2.5, 11.2.6,
        11.5.3.3.4
    Valve 11.6.2.2
  Multiple, central supply system 5.1.3.5.8
  Negative 9.3.6.5.3.1, 9.3.7.1.1
    Definition 3.3.108
  Operating .... 5.1.12.2.3.4, 11.2.6, 11.6.3.1(6), 14.2.5.3.4
    Definition 3.3.135.5, A.3.3.135.5
  Partial .... B.14.1.1.2, B.14.1.3.2.3, B.14.1.3.5, Table B.14.4
    Definition 3.3.135.6, A.3.3.135.6
  Positive .... see Positive pressure
  Standard operating pressures, gas and vacuum
    systems .... Table 5.1.11
  Working (rated) .... 5.1.3.5.6.1(3), 14.2.8.3.15.1
    Definition 3.3.135.8
Pressure indicators .... 5.1.8, 5.1.13.7, 5.2.8, 5.3.3.6.1.1(16),
    5.3.8, 5.3.9(3), A.5.1.8.1.3, B.5.2.5, B.5.2.14
Pressure-reducing regulators .... 11.2.5 to 11.2.7, 11.5.3.1.2,
    11.5.3.1.3, 11.5.3.3.4
  Definition 3.3.136, A.3.3.136
Pressure regulators .... 5.1.12.3.3.2(F); see also
    Pressure-reducing regulators
  Category 3 dental air compressor .... 5.3.3.6.1.1(14)
  Category 3 dental gas source equipment .... 5.3.3.6.2.3(3)
  Category 3 nitrogen source equipment .... 5.3.3.8.1(3)
  Final line .... 5.1.3.5.5, 5.1.3.5.11.4(1), 5.1.3.5.12.4, 5.1.3.6.3.9(K),
      5.1.3.6.3.14(C)(10)
  Headers .... 5.1.3.5.10(8), 5.1.3.5.11.4, 5.1.3.5.12.4
  Instrument air .... 5.1.3.5.5(4)
  Nitrogen purge 5.1.10.4.5.3
Pressure relief valves 14.3.4.1.1.1
  Air compressors 5.3.3.6.1.4
  Central supply systems 5.1.3.3.3.2, 5.1.3.5.6, 5.1.3.5.13.7

Dental air supply systems .... 5.3.3.6.1.1(15), 5.3.3.6.1.4,
    5.3.3.6.2.3(5), 5.3.3.6.2.8
  Emergency oxygen supply connection (EOSC) .... 5.1.3.5.15.2(5)
  Medical air supply systems .... 5.1.3.6.3.9(J), 5.1.3.6.3.14(C)(8)
  Nitrogen supply systems 5.3.3.8.1
Pressure tests
  Individual pressurization 5.1.12.3.3.1
  Initial pressure .... 5.1.12.2.3, 5.3.12.2.2 to 5.3.12.2.4,
      A.5.1.12.2.3.5, A.5.3.12.2.4(6)
  Operational pressure 5.1.12.3.10, A.5.1.12.3.10.6
  Pressure differential 5.1.12.3.3.2
  Standing pressure .... 5.1.6.1(4), 5.1.6.2, 5.1.12.2.6,
      5.1.12.3.2, 5.3.12.2.11, A.5.1.12.2.6.5, A.5.1.12.3.2
Pressure vessels .... see Hyperbaric facilities
Prime movers .... 6.4.1.1.1.2(4)
Procedure room 15.7.4.3.7
  Definition 3.3.137
Psia (pounds per square inch absolute) (definition) 3.3.138
Psig (pounds per square inch gauge) (definition) 3.3.139,
    A.3.3.139
Pumps, vacuum .... see Vacuum pumps
Purge test, piping 5.1.12.2.5, 5.1.12.3.6, 5.3.12.2.8, 5.3.12.2.14,
    A.5.1.12.3.6.4
Purity test
  Medical air 5.1.12.3.12
  Piping 5.1.12.3.8, A.5.1.12.3.8
Purpose of standard 1,2

-Q-

Qualified person (definition) 3.3.140

-R-

Radiation equipment, hyperbaric chambers 14.3.2.6, A.14.3.2.6
Reactance (definition) 3.3.141
Receptacles B.6.3
  Anesthetizing locations 6.3.4.1.1, 6.3.4.1.3
  Definition 3.3.142
  Emergency systems .... 6.4.2.2.6.2, A.6.4.2.2.6.2(C)
  Grounding of 6.3.2.2.2.4, 6.3.2.2.7.1, A.6.3.2.2.7.1
  Hyperbaric chambers 14.2.8.3.10, A.14.2.8.3.10
  Laboratories 6.3.2.3
  Life safety and equipment branches 6.5.2.2.4.2, A.6.5.2.2.4.2
  Multiple outlet connection, for .... 10.2.3.6, A.10.2.3.6(2),
      A.10.2.3.6(4)
  Number of 6.3.2.2.6.2
  Other services, for 6.3.2.2.6.4
  Patient care spaces 6.3.2.2.6, A.6.3.2.2.6.1
  Polarity of 6.3.2.2.6.3
  Selected (definition) 3.3.148
  Special-purpose outlets .... 6.3.2.2.1.4, 6.3.2.2.6.1(B)
  Telecommunications facilities .... 7.3.1.2.1.7(C), 7.3.1.2.2.7(D),
      7.3.1.2.3.7(C)
  Tests .... 6.3.3.2, 6.3.4.1.1 to 6.3.4.1.3
  Types of 6.3.2.2.6.1, A.6.3.2.2.6.1
  Wet procedure locations 6.3.2.2.8.5, 6.3.2.2.8.8
Reference grounding point .... 6.3.2.6.2.1, 6.3.2.6.3.6, 6.3.3.1.2(1),
    A.6.3.2.6.2.1
  Definition 3.3.143
Reference point 6.3.3.1.2, 6.3.3.1.4.1, 6.3.3.1.4.2
References .... Chap. 2, Annex D
Refrigerating equipment B.12.3.2.7
Regulators
  Cylinders .... 11.2.5 to 11.2.7
  Instrument air .... 5.1.13.5.10(4)
  Medical air .... 5.1.3.6.3.9(E), 5.1.3.6.3.9(F), 5.1.3.6.3.9(I),
      5.2.3.5(1), A.5.1.3.6.3.9(F)
  Pressure-reducing .... see Pressure-reducing regulators
Remote (Level 3 supply source) .... 5.3.3.4(3)
  Definition 3.3.144, A.3.3.144

Reserve supply
  Category 1
    Bulk cryogenic liquid systems 5.1.3.5.14.3, 5.1.3.5.14.4,
      A.5.1.3.5.14.4
    In-building emergency reserves .... 5.1.3.5.9.1(5), 5.1.3.5.16
    Warning signals .... 5.1.9.2.4, B.5.2.10 to B.5.2.13
    Definition 3.3.145
  Micro-bulk or small bulk cryogenic systems .... 5.1.3.5.13.5(5)
Reservoir jars 11.4.2.4
Resistance tests 10.3.2, 10.3.3.2, A.10.3.2
Respiratory therapy .... B.11; *see also*
  Oxygen-enriched atmosphere (OEA)
  Administration and maintenance 11.5, A.11.5.1.1.2
    to A.11.5.3.2
  Apparatus for 11.4.2
  Equipment labeling 11.5.3.1
  Equipment servicing and maintenance 11.5.1.3, A.11.5.1.3.3
  Equipment transport, storage, and use 11.5.3.3
  Fire hazards 11.5.1.1, 11.5.1.2, A.11.5.1.1.2,
    A.11.5.1.1.3, B.11.3.1
Risers, piping
  Definition 3.3.133.3
  Valves 5.1.4.4, 5.1.11.2.5
Risk assessment 4.2, A.4.2
Risk categories 4.1, A.4.1
  Category 1 .... 4.1.1, A.4.1, A.4.1.1
    Definition 3.3.146.1
  Category 2 .... 4.1.2, A.4.1, A.4.1.2
    Definition 3.3.146.2
  Category 3 .... 4.1.3, A.4.1
    Definition 3.3.146.3
  Category 4 .... 4.1.4, A.4.1
    Definition 3.3.146.4
  Risk assessment procedure, determination using 4.2, A.4.2

-S-

Safety mechanisms 11.6.4
Scavenging .... *see also* Waste anesthetic gas disposal
  (WAGD) systems
  Category 3 systems
    Exhaust systems .... 5.1.3.7.6, 5.3.3.10.1.4, A.5.3.3.10.1.4(8)
    Performance criteria and tests 5.3.12.2.13
  Definition 3.3.147
  Ventilation system 9.3.7
    Active system 9.3.7.1.1
    Passive systems 9.3.7.1.2
Scope of standard 1.1, A.1.1.10, A.1.1.12
Security management .... Chap. 13
  Emergency 12.5.3.3.6.3, A.12.5.3.3.6.3
  Information technology and communications
    systems 7.3.1.2.2.6, 7.3.1.2.3.6
  Operations 13.11, A.13.11
  Plan .... 13.2, 13.12.3 to 13.12.5, A.13.2
  Program evaluation 13.12, A.13.12.1
  Responsible person .... 13.4, A.13.4.2(3)(c)
  Security equipment 13.9, A.13.9.3
  Security-sensitive areas .... 13.5, A.13.5.2(1) to A.13.5.7(1)
  Security vulnerability assessment 13.3, A.13.3.1
Selected receptacles (definition) 3.3.148
Self-extinguishing (definition) 3.3.149
Semipermanent connection (definition) 3.3.150
Sensors, inside hyperbaric chambers 14.2.9.1.1
Service inlets (definition) 3.3.151
Service outlets (definition) 3.3.152
Service valves 5.1.4.5, 5.1.11.2.6
Shall (definition) 3.2.7
Should (definition) 3.2.8
Shutoff valves
  Air compressor .... 5.3.3.6.1.1(10)
  Emergency oxygen supply connection (EOSC) .... 5.1.3.5.15.2(3)

Gas 5.1.4.1.1, 5.1.4.2.1, 5.1.4.3.1, 5.1.4.4, 5.1.4.7,
  5.3.4.1, 5.3.12.2.1(2), 5.3.12.2.5(2), A.5.3.4.1(1)
  Hyperbaric facilities 14.2.1.3.2
  Labeling and identification .... 5.1.11.2, 5.3.4.1(3), A.5.1.11.2.7
  Vacuum .... 5.1.4.1.1, 5.1.4.1.6(1)
Signs .... *see* Labeling and identification
Single treatment facilities (definition) 3.3.153, A.3.3.153
Site of intentional expulsion 10.5.4.1, 10.5.4.2, 11.5.1.1.4,
  11.5.1.2.3
  Definition 3.3.154, A.3.3.154
Smoke 14.2.4.5.3, A.14.2.4.5.3
Smoke alarms/detectors 15.7.3, 15.7.4.2.1, 15.7.4.2.2, 15.11.2,
  A.15.7.4.2.1, A.15.7.4.2.2
Smoke compartments 15.7.4.3.1, 15.8.1.3, A.15.7.4.3.1,
  A.15.8.1.3
Smoke control .... 6.4.2.2.5.3(A)(4), 6.5.2.2.3.3(4)
Smoke zones 15.7.4.3.3
Smoking ...... 11.5.1.1.1, 11.5.2.5, 11.5.3.2, 14.3.1.5.1.1(1), A.11.5.3.2
Sound-deadening materials 14.2.2.5.4, A.14.2.2.5.4
Source valves 5.1.4.2, 5.1.11.2.3
Spaces (definition) .... 3.3.155; *see also*
  Patient care spaces
Sprinklers, automatic ...... 7.3.1.2.1.8(D), 7.3.1.2.3.8(B), 15.8, 15.10,
  15.11.3, A.15.8.1.3, A.15.8.1.4, A.15.10, A.15.11.3
Staff .... *see* Employees
Staff emergency calls 7.3.3.1.3, A.7.3.3.1.3.1
Standard cubic feet per minute (SCFM) (definition) 3.3.156
Standard (definition) 3.2.9
Standby power systems 15.5.1.3
Standpipe and hose systems 15.9.2, A.15.9.2.2
Station inlets
  Category 1 ..... 5.1.4.6.1, 5.1.5, 5.1.6.6 to 5.1.6.8, 5.1.7.5 to 5.1.7.7,
    5.1.10.10, 5.1.10.11.4, 5.1.11.3, 5.1.12.1.2, 5.1.12.1.10,
    5.1.12.2.1.3(3), 5.1.12.2.2, 5.1.12.2.4.6, 5.1.12.3.3.1(D),
    5.1.12.3.3.2(E), 5.1.12.3.10, 5.1.12.3.13, 5.1.14.4.9,
    A.5.1.5, A.5.1.12.3.10.6
  Category 2 .... 5.2.5
  Category 3 .... 5.3.5, A.5.3.5
  Definition 3.3.157
Station outlets
  Category 1 .... 5.1.4.6.1, 5.1.5, 5.1.6.6 to 5.1.6.8, 5.1.7.5
    to 5.1.7.7, 5.1.10.1.1, 5.1.10.10, 5.1.10.11.4, 5.1.11.3,
    5.1.12.1.2, 5.1.12.1.10, 5.1.12.2.1.3(3), 5.1.12.2.2,
    5.1.12.2.4.6, 5.1.12.2.6.1, 5.1.12.3.3.1(D),
    5.1.12.3.3.2(E), 5.1.12.3.3.2(G), 5.1.12.3.6.1,
    5.1.12.3.6.3, 5.1.12.3.6.4, 5.1.12.3.7.2 to 5.1.12.3.7.4,
    5.1.12.3.8.2, 5.1.12.3.8.6, 5.1.12.3.10, 5.1.12.3.13,
    A.5.1.5, A.5.1.12.3.6.4, A.5.1.12.3.10.6, B.5.2.18
  Category 2 .... 5.2.5
  Category 3 .... 5.3.5, A.5.3.5
  Definition 3.3.158
Storage
  Compact .... *see* Compact storage
  Containers .... *see* Containers
  Cryogenic liquids, ventilation for storage
    of 9.3.6, A.9.3.6.3, A.9.3.6.5.1
  Cylinders .... *see* Cylinders
  Flammable and combustible liquids 15.3, A.15.3.2
  Flammable gases 14.3.3.3, 14.3.4.1.4.1, 15.3, A.15.3.2
  Gas equipment 11.5.3.3
  Hyperbaric facilities 14.2.1.5, 14.3.2.2, 14.3.3.3, 14.3.4.1.4,
    A.14.3.2.2
  Nonflammable gases 11.3.1, 11.3.2, A.11.3.1, A.11.3.2
  Oxygen 5.1.3.1.9, 14.3.2.2, A.14.3.2.2
  Ventilation for medical gas storage 9.3.6, A.9.3.6.3, A.9.3.6.5.1
Stored electrical energy systems 15.5.1.4
Supply source .... *see also* Category 1 piped gas
  and vacuum systems, sources; Category 3 piped gas and
  vacuum systems, sources
  Category 2 piped gas and vacuum systems 5.2.3
  Operating supply (definition) 3.3.159.1
  Primary supply (definition) 3.3.159.2
  Reserve supply (definition) 3.3.159.3;
    see also Reserve supply
  Secondary supply (definition) 3.3.159.4

Supply systems .... *see also* Category 1 piped gas and vacuum systems; Category 2 piped gas and vacuum systems; Category 3 piped gas and vacuum systems; Central supply systems; Dental air supply systems; Pipe and piping, gas; Piped distribution systems
   Dental vacuum systems .... 5.3.3.10, A.5.3.3.10.1.3 to A.5.3.3.10.1.4(8)
   Hyperbaric facilities 14.2.4.1.3.2
   Portable .... 11.4.1, A.11.4.1.1 to A.11.4.1.4
Support gases .... *see also* Instrument air; Nitrogen
   Category 1 piped gas and vacuum systems .... *see* Category 1 piped gas and vacuum systems
   Category 2 piped gas and vacuum systems 5.2.13
   Category 3 piped gas and vacuum systems .... 5.3.12, A.5.3.12.2.4(6)
   Medical .... 5.1.1.3, 5.3.12, A.5.1.1.3, A.5.3.12.2.4(6)
   Definition 3.3.101
Surface-mounted medical gas rail systems 5.1.7, 5.2.7, 5.3.7, A.5.1.7
   Definition 3.3.160, A.3.3.160
Switches .... *see also* Transfer switches
   Essential electrical systems .... 6.4.2.1, 6.4.2.2.1.2, 6.4.2.2.1.4, 6.4.2.2.6.3, 6.5.3.2, A.6.4.2.1.2 to A.6.4.2.1.8.3
   Hyperbaric facilities 14.2.8.3.11, A.14.2.8.3.11.1
   Isolation switch, bypass 6.4.2.1.8, A.6.4.2.1.8.3
   Test 6.4.2.1.5.11
Synthetic materials .... *see* Textiles

-T-

Task illumination .... 6.4.2.2.3.2(4), 6.4.2.2.4.2, 6.5.2.2.2.1(6), 6.5.2.2.3.3(1), A.6.4.2.2.4.2(7)
   Definition 3.3.161
Telecommunications entrance facility (EF) .... 7.3.1.2.1, 7.4.1.1.1, 7.5.1.1.1 to 7.5.1.1.3, A.7.3.1.2.1.3(B) to A.7.3.1.2.1.8(B)
   Definition 3.4.1
Telecommunications equipment room (TER) ... 7.3.1.2.1.1, 7.3.1.2.2, 7.4.1.1.1, 7.5.1.1.2, A.7.3.1.2.2.2, A.7.3.1.2.2.4(E)
   Definition 3.4.2
Telecommunications room (TR) .... 7.3.1.2.3, 7.4.1.1.1, 7.5.1.1.2, 7.5.1.1.3, A.7.3.1.2.3.4(D)
   Definition 3.4.3
Temperature
   Ambient .... 5.1.3.6.3.1(3), 5.1.3.7.1.1(3), 5.1.3.8.1.3(4), 5.1.3.8.1.4(2), 5.1.10.11.3.3, 5.1.12.2.6.5, 5.1.12.2.7, 6.4.1.1.13, A.5.1.12.2.6.5, A.5.1.12.2.7.5, A.6.4.1.1.13
   Generator sets and 6.4.1.1.12
   Hyperbaric chambers 14.2.4.3, 14.2.8.3.12, 14.3.2.1.8, A.14.2.4.3.2, A.14.2.8.3.12
Temperature control .... 7.3.1.2.1.8(A), 7.3.1.2.2.8(A), 14.2.4.3, A.14.2.4.3.2
Terminals
   Definition 3.3.162
   Inlets .... *see* Station inlets
Test equipment 6.3.3.1.5
Tests .... *see also* Performance tests
   Elevators 15.5.3.4
   Emergency plans and operations 12.5.3.3.8, A.12.5.3.3.8
   Fire protection systems 14.2.5.5, 14.3.1.4.6, 15.12, A.14.2.5.5
   Hyperbaric chambers, electrical equipment for 14.2.8.3.2, 14.3.5.1.1
   Maintenance program
     Electrical systems 6.3.4.1
     Gas and vacuum systems .... 5.1.14.2.3, 5.1.14.4.1, 5.1.14.4.5 to 5.1.14.4.7, A.5.1.14.2.3, B.5.2
     Records and reports .... 5.1.12.1.6 to 5.1.12.1.9, 5.1.12.1.11, 5.1.12.2.1.1, 5.1.14.2.3.1(F), 5.1.14.4.1, 6.3.4.1.2, 6.3.4.2, 6.4.4.2, 6.5.4.2, 10.5.6.2, 10.5.6.3, A.6.3.4.2.1, A.10.5.6.2
     Retests B.5.2
Test switches 6.4.2.1.5.11
Textiles, hyperbaric chamber use 14.3.1.5.4, A.14.3.1.5.4

Total hazard current .... *see* Hazard current
Touch current
   Definition 3.3.163
   Limits 10.3.3.4
   Portable equipment 10.2.6, 10.3.5, A.10.2.6
   Test procedure 10.3.5.3
Training, employee
   Electrical equipment 10.5.8, A.10.5.8.1, A.10.5.8.3
   Emergency management .... 12.5.3.3.6.7(B), 12.5.3.3.7
   Emergency operations plan B.12.1.1.5
   Gas equipment and systems 11.5.2.1, 11.6.1.2, A.11.5.2.1.1
   Hyperbaric facilities .... 14.3.1.4.3 to 14.3.1.4.5, 14.3.6.4.1, A.14.3.1.4.4.1, A.14.3.1.4.5
   Operating rooms 15.13.3.10
   Security personnel 13.11.2
Transfer switches .... 6.3.2.2.1.2, 6.4.2.1.1, 6.4.2.2.1.2, 6.4.2.2.1.4, B.6.1, B.12.3.2.5; *see also* Switches
   Automatic .... 6.4.2.1.4, 6.4.2.1.5, 6.4.3.2.5, 6.5.3.2.4, A.6.4.2.1.5.1(A) to A.6.4.2.1.5.15
   Isolating 6.4.2.1.8, A.6.4.2.1.8.3
   Nonautomatic 6.4.2.1.5.15, 6.4.2.1.6, 6.4.2.1.7, 6.4.3.2.8, 6.5.3.2.7, A.6.4.2.1.5.15
   Performance criteria and tests .... 6.4.3.2, 6.4.4.1.1, 6.5.3.2, 6.5.4.1.1, A.6.4.4.1.1.4(A)
   Rating 6.4.2.1.3
Transfilling .... 9.3.6, 11.5.2.2 to 11.5.2.4, 11.7.3.5.3, 11.7.3.6, A.9.3.6.3, A.9.3.6.5.1, A.11.5.2.2.2 to A.11.5.2.4, A.11.7.3.6.1.1
   Definition 3.3.164
Transformers, isolation 6.3.2.6.1
   Definition 3.3.84
Tube, endotracheal B.14.1.2.4
   Definition 3.3.165.1, A.3.3.165.1, B.14.1.2.4
Tube (pipe) .... *see* Pipe and piping, gas

-U-

Units of measurement 1.5, A.1.5
Use points (definition) 3.3.166
Utilities
   Emergency plan for .... 12.3.2.5 to 12.3.2.7, 12.5.3.3.6.5, 12.5.3.3.6.6, A.12.5.3.3.6.5, B.12.3.4
   Fire protection for 15.5, A.15.5.2.1, A.15.5.3.3

-V-

Vacuum indicators 5.1.8, 5.2.8, 5.3.8, A.5.1.8.1.3, B.5.2.5, B.5.2.14
Vacuum pumps
   Alarms 5.1.9.5, A.5.1.9.5
   Category 1 systems .... 5.1.3.7.2, 5.1.3.7.5.1 to 5.1.3.7.5.5
   Category 3 systems .... 5.3.3.10.1.1(1)
   Electrical load reduction 6.4.1.1.3
Vacuum receivers 5.1.3.7.3
Vacuum systems .... *see* Medical-surgical vacuum systems
Vacuum terminals .... *see* Station inlets
Valves .... *see also* Check valves; Pressure relief valves; Shutoff valves; Zone valves
   Accessibility 5.1.4.1.4
   Category 1 .... 5.1.3.2.10, 5.1.3.3.2, 5.1.3.5.6, 5.1.3.5.15.2, 5.1.3.6.3.9(G), 5.1.3.6.3.9(J), 5.1.3.6.3.14(C)(8), 5.1.4, 5.1.8.2.2(3), 5.1.8.2.3.1, 5.1.11.2, 5.1.12.1.2, 5.1.12.2.5.2, 5.1.12.3.2(1), 5.1.12.3.4, 5.1.13.4, A.5.1.4, A.5.1.11.2.7, B.5.2.17
   Category 2 .... 5.2.4
   Category 3 .... 5.3.3.6.1.4, 5.3.3.6.2.3(4), 5.3.3.6.2.3(5), 5.3.3.6.2.8, 5.3.3.8.1, 5.3.4, 5.3.12.2.9, A.5.3.4.1(1)
   Flow-control 11.5.3.3.1
     Definition 3.3.56
   Future connections, for 5.1.4.8
   Gas equipment, cylinder and container sources for 11.2.2, 11.2.9, 11.2.10, 11.6.2.2
   High pressure 11.6.2.2

INDEX

Hyperbaric facilities .... 14.2.5.5.1, 14.3.2.2(2), 14.3.2.3(1), 14.3.4.1.1.1
In-line 5.1.4.7, 5.1.4.9
Labeling 5.1.4.1.3
Protection caps 5.1.3.2.10
Security 5.1.4.1.2
Service 5.1.4.5, 5.1.11.2.6
Source 5.1.4.2, 5.1.11.2.3
Types 5.1.4.1.6
Vaporizers 5.1.10.4.1.8
Definition 3.3.167
Ventilation .... 6.4.2.2.5.4(4); *see also* Exhaust systems; Heating, ventilating, and air conditioning (HVAC) systems
Alternate power sources .... 6.4.1.1.13, 6.4.2.2.5.3(A)(5), 6.4.2.2.5.3(A)(6), 6.4.2.2.5.3(B), A.6.4.1.1.13
Central supply systems and storage, gas 5.1.3.3.3, 5.3.3.3.1
Construction, during 9.3.10
Cryogenic liquids, storage or transfilling of 9.3.6, A.9.3.6.3, A.9.3.6.5.1
Definition 3.3.168
Emergency systems .... 6.4.2.2.5.4(4)
Hyperbaric chambers .... 14.2.4, 14.2.10.2, A.14.2.4.1.2 to A.14.2.4.5.3, A.14.2.10.2
Laboratories 15.5.2.4
Medical gas, storage or transfilling of 9.3.6, A.9.3.6.3, A.9.3.6.5.1
Waste gas 9.3.7
Venting .... *see* Ventilation
Voltage limit 6.3.3.1.6.1
Voltages
Abnormal .... 6.4.1.1.1.2(1), 6.4.1.1.2
Measuring .... 6.3.3.1.2 to 6.3.3.1.5, A.6.3.3.1.3, A.6.3.3.1.4
Voltage sensing .... 6.4.2.1.5.1(D)

-W-

Warning signs .... *see* Labeling and identification
Warning systems .... *see* Alarm systems
Waste anesthetic gas disposal (WAGD) systems
Alarms 5.1.3.8.3
Area 5.1.9.4, A.5.1.9.4
Local 5.1.9.5, A.5.1.9.5
Master .... 5.1.9.2.4(11)
Category 1 piped gas and vacuum systems .... 5.1.3.3.1.3(3), 5.1.3.5(6), 5.1.3.8, 5.1.10.2.3, 5.1.10.3, 5.1.14.1.3, 5.1.14.1.4, 5.1.14.2, 5.1.14.4.7(4), A.5.1.3.8, A.5.1.10.3.1, A.5.1.14.1.3, A.5.1.14.1.4, A.5.1.14.2.1 to 5.1.14.2.3

Category 2 piped gas and vacuum systems 5.2.3.7
Category 3 piped gas and vacuum systems .... 5.3.3.11, 5.3.12, A.5.3.12.2.4(6)
Definition 3.3.169
Designation colors .... Table 5.1.11
Electrical power and control 5.1.3.8.4
Exhaust 5.1.3.8.5
Maintenance .... 5.1.14.2, 5.1.14.4.7(4), A.5.1.14.2.1 to 5.1.14.2.3
Operating pressures .... Table 5.1.11
Producers 5.1.3.8.1, 5.1.3.8.2, A.5.1.3.8.1
Sources 5.1.3.8.1, A.5.1.3.8.1
Special precautions 5.1.14.1.3, 5.1.14.1.4, A.5.1.14.1.3, A.5.1.14.1.4
Ventilation system 9.3.7
Waste chutes 15.6
Waste water .... 8.3.9 to 8.3.11
Black 8.3.9
Definition 3.3.170.1
Clear 8.3.11
Definition 3.3.170.2
Gray 8.3.10
Definition 3.3.170.3
Water, potable and nonpotable .... 8.3.1, 8.3.2, 12.5.3.3.6.5, A.12.5.3.3.6.5, B.12.3.2.5(2), B.12.3.2.7
Water conditioning 8.3.4
Water heating 8.3.3, A.8.3.3
Welded joints .... 5.1.10.3.1(2), 5.1.10.5, A.5.1.10.5.1.5
Wet procedure locations .... 1.3.4.3, 6.3.2.2.8, A.6.3.2.2.8.1 to A.6.3.2.2.8.7
Definition 3.3.171, A.3.3.171
Wiring .... *see* Electrical wiring
Working pressure .... *see* Pressure
Work space or room
Generators 6.4.1.1.9
Telecommunications facility 7.3.1.2.2.5, 7.3.1.2.3.5

-X-

X-ray equipment .... 14.3.2.1.2(1)

-Z-

Zone valves
Category 1 .... 5.1.4.1.4(A), 5.1.4.1.5, 5.1.4.6, 5.1.8.2.2(3), 5.1.8.2.3.1, 5.1.12.1.2, 5.1.12.2.5.2, 5.1.12.3.2(1)
Gas equipment B.11.3.3.1, B.11.3.6.1
Hyperbaric facilities B.14.3.1, B.14.3.2

Sequence of Events for the Standards Development Process

As soon as the current edition is published, a Standard is open for Public Input

Step 1: Input Stage

- Input accepted from the public or other committees for consideration to develop the First Draft
- Committee holds First Draft Meeting to revise Standard (23 weeks)
  Committee(s) with Correlating Committee (10 weeks)
- Committee ballots on First Draft (12 weeks)
  Committee(s) with Correlating Committee (11 weeks)
- Correlating Committee First Draft Meeting (9 weeks)
- Correlating Committee ballots on First Draft (5 weeks)
- First Draft Report posted

Step 2: Comment Stage

- Public Comments accepted on First Draft (10 weeks)
- If Standard does not receive Public Comments and the Committee does not wish to further revise the Standard, the Standard becomes a Consent Standard and is sent directly to the Standards Council for issuance
- Committee holds Second Draft Meeting (21 weeks)
  Committee(s) with Correlating Committee (7 weeks)
- Committee ballots on Second Draft (11 weeks)
  Committee(s) with Correlating Committee (10 weeks)
- Correlating Committee First Draft Meeting (9 weeks)
- Correlating Committee ballots on First Draft (8 weeks)
- Second Draft Report posted

Step 3: Association Technical Meeting

- Notice of Intent to Make a Motion (NITMAM) accepted (5 weeks)
- NITMAMs are reviewed and valid motions are certified for presentation at the Association Technical Meeting
- Consent Standard bypasses Association Technical Meeting and proceeds directly to the Standards Council for issuance
- NFPA membership meets each June at the Association Technical Meeting and acts on Standards with "Certified Amending Motions" (certified NITMAMs)
- Committee(s) and Panel(s) vote on any successful amendments to the Technical Committee Reports made by the NFPA membership at the Association Technical Meeting

Step 4: Council Appeals and Issuance of Standard

- Notification of intent to file an appeal to the Standards Council on Association action must be filed within 20 days of the Association Technical Meeting
- Standards Council decides, based on all evidence, whether or not to issue the Standards or to take other action

Committee Membership Classifications[1,2,3,4]

The following classifications apply to Committee members and represent their principal interest in the activity of the Committee.

1. M   *Manufacturer:* A representative of a maker or marketer of a product, assembly, or system, or portion thereof, that is affected by the standard.
2. U   *User:* A representative of an entity that is subject to the provisions of the standard or that voluntarily uses the standard.
3. IM   *Installer/Maintainer:* A representative of an entity that is in the business of installing or maintaining a product, assembly, or system affected by the standard.
4. L   *Labor:* A labor representative or employee concerned with safety in the workplace.
5. RT   *Applied Research/Testing Laboratory:* A representative of an independent testing laboratory or independent applied research organization that promulgates and/or enforces standards.
6. E   *Enforcing Authority:* A representative of an agency or an organization that promulgates and/or enforces standards.
7. I   *Insurance:* A representative of an insurance company, broker, agent, bureau, or inspection agency.
8. C   *Consumer:* A person who is or represents the ultimate purchaser of a product, system, or service affected by the standard, but who is not included in (2).
9. SE   *Special Expert:* A person not representing (1) through (8) and who has special expertise in the scope of the standard or portion thereof.

NOTE 1: "Standard" connotes code, standard, recommended practice, or guide.

NOTE 2: A representative includes an employee.

NOTE 3: While these classifications will be used by the Standards Council to achieve a balance for Technical Committees, the Standards Council may determine that new classifications of member or unique interests need representation in order to foster the best possible Committee deliberations on any project. In this connection, the Standards Council may make such appointments as it deems appropriate in the public interest, such as the classification of "Utilities" in the National Electrical Code Committee.

NOTE 4: Representatives of subsidiaries of any group are generally considered to have the same classification as the parent organization.

Submitting Public Input / Public Comment through the Electronic Submission System (e-Submission):

As soon as the current edition is published, a Standard is open for Public Input.

Before accessing the e-Submission System, you must first sign-in at www.NFPA.org. *Note: You will be asked to sign-in or create a free online account with NFPA before using this system*:

a. Click in the gray Sign In box on the upper left side of the page. Once signed-in, you will see a red "Welcome" message in the top right corner.
b. Under the Codes and Standards heading, Click on the Document Information pages (List of Codes & Standards), and then select your document from the list or use one of the search features in the upper right gray box.

OR

a. Go directly to your specific document page by typing the convenient short link of www.nfpa.org/document#, (Example: NFPA 921 would be www.nfpa.org/921) Click in the gray Sign In box on the upper left side of the page. Once signed in, you will see a red "Welcome" message in the top right corner.

To begin your Public Input, select the link The next edition of this standard is now open for Public Input (formally "proposals") located on the Document Information tab, the Next Edition tab, or the right-hand Navigation bar. Alternatively, the Next Edition tab includes a link to Submit Public Input online

At this point, the NFPA Standards Development Site will open showing details for the document you have selected. This "Document Home" page site includes an explanatory introduction, information on the current document phase and closing date, a left-hand navigation panel that includes useful links, a document Table of Contents, and icons at the top you can click for Help when using the site. The Help icons and navigation panel will be visible except when you are actually in the process of creating a Public Input.

Once the First Draft Report becomes available there is a Public comment period during which anyone may submit a Public Comment on the First Draft. Any objections or further related changes to the content of the First Draft must be submitted at the Comment stage.

To submit a Public Comment you may access the e-Submission System utilizing the same steps as previous explained for the submission of Public Input.

For further information on submitting public input and public comments, go to: http://www.nfpa.org/publicinput

Other Resources available on the Doc Info Pages

Document information tab: Research current and previous edition information on a Standard

Next edition tab: Follow the committee's progress in the processing of a Standard in its next revision cycle.

Technical committee tab: View current committee member rosters or apply to a committee

Technical questions tab: For members and Public Sector Officials/AHJs to submit questions about codes and standards to NFPA staff. Our Technical Questions Service provides a convenient way to receive timely and consistent technical assistance when you need to know more about NFPA codes and standards relevant to your work. Responses are provided by NFPA staff on an informal basis.

Products/training tab: List of NFPA's publications and training available for purchase.

Community tab: Information and discussions about a Standard

Information on the NFPA Standards Development Process

I. Applicable Regulations. The primary rules governing the processing of NFPA standards (codes, standards, recommended practices, and guides) are the NFPA *Regulations Governing the Development of NFPA Standards (Regs)*. Other applicable rules include NFPA *Bylaws*, NFPA *Technical Meeting Convention Rules*, NFPA *Guide for the Conduct of Participants in the NFPA Standards Development Process*, and the NFPA *Regulations Governing Petitions to the Board of Directors from Decisions of the Standards Council*. Most of these rules and regulations are contained in the *NFPA Standards Directory*. For copies of the *Directory*, contact Codes and Standards Administration at NFPA Headquarters; all these documents are also available on the NFPA website at "www.nfpa.org."

The following is general information on the NFPA process. All participants, however, should refer to the actual rules and regulations for a full understanding of this process and for the criteria that govern participation.

II. Technical Committee Report. The Technical Committee Report is defined as "the Report of the responsible Committee(s), in accordance with the Regulations, in preparation of a new or revised NFPA Standard." The Technical Committee Report is in two parts and consists of the First Draft Report and the Second Draft Report. (See *Regs* at 1.4)

III. Step 1: First Draft Report. The First Draft Report is defined as "Part one of the Technical Committee Report, which documents the Input Stage." The First Draft Report consists of the First Draft, Public Input, Committee Input, Committee and Correlating Committee Statements, Correlating Input, Correlating Notes, and Ballot Statements. (See *Regs* at 4.2.5.2 and Section 4.3) Any objection to an action in the First Draft Report must be raised through the filing of an appropriate Comment for consideration in the Second Draft Report or the objection will be considered resolved. [See *Regs* at 4.3.1(b)]

IV. Step 2: Second Draft Report. The Second Draft Report is defined as "Part two of the Technical Committee Report, which documents the Comment Stage." The Second Draft Report consists of the Second Draft, Public Comments with corresponding Committee Actions and Committee Statements, Correlating Notes and their respective Committee Statements, Committee Comments, Correlating Revisions, and Ballot Statements. (See *Regs* at Section 4.2.5.2 and 4.4) The First Draft Report and the Second Draft Report together constitute the Technical Committee Report. Any outstanding objection following the Second Draft Report must be raised through an appropriate Amending Motion at the Association Technical Meeting or the objection will be considered resolved. [See *Regs* at 4.4.1(b)]

V. Step 3a: Action at Association Technical Meeting. Following the publication of the Second Draft Report, there is a period during which those wishing to make proper Amending Motions on the Technical Committee Reports must signal their intention by submitting a Notice of Intent to Make a Motion. (See *Regs* at 4.5.2) Standards that receive notice of proper Amending Motions (Certified Amending Motions) will be presented for action at the annual June Association Technical Meeting. At the meeting, the NFPA membership can consider and act on these Certified Amending Motions as well as Follow-up Amending Motions, that is, motions that become necessary as a result of a previous successful Amending Motion. (See 4.5.3.2 through 4.5.3.6 and Table1, Columns 1-3 of *Regs* for a summary of the available Amending Motions and who may make them.) Any outstanding objection following action at an Association Technical Meeting (and any further Technical Committee consideration following successful Amending Motions, see *Regs* at 4.5.3.7 through 4.6.5.3) must be raised through an appeal to the Standards Council or it will be considered to be resolved.

VI. Step 3b: Documents Forwarded Directly to the Council. Where no Notice of Intent to Make a Motion (NITMAM) is received and certified in accordance with the Technical Meeting Convention Rules, the standard is forwarded directly to the Standards Council for action on issuance. Objections are deemed to be resolved for these documents. (See *Regs* at 4.5.2.5)

VII. Step 4a: Council Appeals. Anyone can appeal to the Standards Council concerning procedural or substantive matters related to the development, content, or issuance of any document of the Association or on matters within the purview of the authority of the Council, as established by the *Bylaws* and as determined by the Board of Directors. Such appeals must be in written form and filed with the Secretary of the Standards Council (See *Regs* at 1.6). Time constraints for filing an appeal must be in accordance with 1.6.2 of the *Regs*. Objections are deemed to be resolved if not pursued at this level.

VIII. Step 4b: Document Issuance. The Standards Council is the issuer of all documents (see Article 8 of *Bylaws*). The Council acts on the issuance of a document presented for action at an Association Technical Meeting within 75 days from the date of the recommendation from the Association Technical Meeting, unless this period is extended by the Council (See *Regs at* 4.7.2). For documents forwarded directly to the Standards Council, the Council acts on the issuance of the document at its next scheduled meeting, or at such other meeting as the Council may determine (See *Regs* at 4.5.2.5 and 4.7.4).

IX. Petitions to the Board of Directors. The Standards Council has been delegated the responsibility for the administration of the codes and standards development process and the issuance of documents. However, where extraordinary circumstances requiring the intervention of the Board of Directors exist, the Board of Directors may take any action necessary to fulfill its obligations to preserve the integrity of the codes and standards development process and to protect the interests of the Association. The rules for petitioning the Board of Directors can be found in the *Regulations Governing Petitions to the Board of Directors from Decisions of the Standards Council* and in 1.7 of the *Regs*.

X. For More Information. The program for the Association Technical Meeting (as well as the NFPA website as information becomes available) should be consulted for the date on which each report scheduled for consideration at the meeting will be presented. For copies of the First Draft Report and Second Draft Report as well as more information on NFPA rules and for up-to-date information on schedules and deadlines for processing NFPA documents, check the NFPA website (www.nfpa.org/aboutthecodes) or contact NFPA Codes & Standards Administration at (617) 984-7246.

CLEARANCE - 3 FT
PIPE - ROLLED SMOOTH, DEBURRED
INTAKE - 25 FT FROM EXHAUST, 10 FT DR, 20 FT UP
REGULATORS - SIZED 100%
OXYGEN - 350 PSI